Pearson
BTEC National
Engineering

Student Book

Andrew Buckenham

Gareth Thomson

Natalie Grifiths

Steve Singleton

Alan Serplus

Mike Ryan

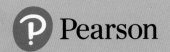 Pearson

Published by Pearson Education Limited, 80 Strand, London, WC2R 0RL.

www.pearsonschoolsandfecolleges.co.uk

Copies of official specifications for all Edexcel qualifications may be found on the website: www.edexcel.com

Text © Andrew Buckenham, Natalie Griffiths, Michael Ryan, Alan Serplus, Stephen Singleton, Gareth Thomson, 2016
Edited by Eleanor Barber
Designed by Andy Magee
Typeset by Tech-Set Ltd, Gateshead
Original illustrations © Pearson Education Ltd, 2016
Illustrated by Tech-Set Ltd, Gateshead
Cover design by Vince Haig
Picture research by Alison Prior

The rights of Andrew Buckenham, Natalie Griffiths, Michael Ryan, Alan Serplus, Stephen Singleton, Gareth Thomson to be identified as authors of this work have been asserted by them in accordance with the Copyright, Designs and Patents Act 1988.

First published 2016

19 18 17
10 9 8 7 6 5 4 3 2

British Library Cataloguing in Publication Data
A catalogue record for this book is available from the British Library

ISBN 978 1 292 14100 8

Acknowledgements
The publisher would like to thank the following for their kind permission to reproduce their photographs:
(Key: b-bottom; c-centre; l-left; r-right; t-top)
123RF.com: 358; **Alamy Images:** Antonino Miroballo 109, Antony Nettle 206, Cultura Creative RF 1, 189, Crisitan M Vela 79tl, David J Green 79br, Derek Meijer 67, Image Source Salsa 188, imageBROKER 107, 323, incamerastock 211, Ingram Publishing 11, Jason Langley 329, Jeremy Sutton-Hibbert 99, Jim West 84, John Baud 388b, laboratory 389, Sciencephotos 445l, THEERAVAT BOONNUANG 462, Tony Watson 29, Yuriy Zahachevskyy 82cr; **Anthony Shearer:** 423; **Baileigh Industrial Ltd:** 82tl; **Billy May:** 270; **David Roberts:** 459; **Department of Materials Science and Metallurgy:** 442c, 442b, 443; **Fotolia.com:** AndG 25, Artem Merzlenko 233, dietwalther 219, photomic 325, sorapolujjin 82bl, wattanaphob 78tl, womg yu liang 187; **Getty Images:** Arch Men 359, Bloomberg 169, 227, brozova 83cl, Chris Hunter 199, Fox Photos 455, Future Publishing 196, makaule 388t, Monty Rakusen 175, 382, Neil Farrin 192, PABLO PORCIUNCULA 73, PangeaPics 207; **James Davies:** 419; **Martyn F. Chillmaid:** 6t, 6b, 83tr, 86, 398, 445r; **Mike Ryan:** 383; **One Eighty Materials Testing (Pty) Ltd:** 450; **Pearson Education Ltd:** Gareth Boden 466, Trevor Clifford 78br; **Shaun Spilsbury:** 357; **Shutterstock.com:** a_v_d 79bl, Billion Photos 218, Djomas 324, Fotosenmee 465, goodcat 225, India Picture 217, Lucas Photo 287, Monkey Business Images 108, 381, 418, mr Hansen 78bl, Zeljko Radojko 34; **Simon Bentley:** 89, 417; **Thomas Howes:** 162

Cover images: *Front:* **Shutterstock.com:** F.Schmidt
All other images © Pearson Education

The authors and publisher would like to thank the following individuals and organisations for their approval and permission to reproduce their materials:
Figures
p. 380: Figure 10.19 from https://knowledge.autodesk.com/support/autocad/learn-explore/caas/CloudHelp/cloudhelp/2015/ENU/AutoCAD-Core/files/GUID-0B829612-4961-41C2-8602-93021D81E12D-htm.html, Reproduced with permission from Autodesk Inc. under Creative Commons Attribution-NoDerivatives 4.0 International License; **p. 387**: Figure 19.1 from an educational licensed National Instruments Multisim 12.0 , Photo Courtesy of National Instruments.
Text
p. 162: 'Think Future' Box from Thomas Howes, Reproduced with permission; **p. 211**: Extract from Definition of ISO 9001, Permission to reproduce extracts from British Standards is granted by BSI Standards Limited (BSI). No other use of this material is permitted. British Standards can be obtained in PDF or hard copy formats from the BSI online shop: www.bsigroup.com/Shop; **p. 270**: 'Think Future' Box from Stephen Serplus, Reproduced with permission; **p. 323**: 'Think Future' Box from Gary Capewell, Reproduced with permission; **p. 357**: 'Think Future' Box Courtesy Shaun Spilsbury BEng (Hons) CEng MIMechE, Reproduced with permission ; **p. 417**: 'Think Future' Box from Simon Bentley, Senior Electrical Engineer , Reproduced with permission.

Websites
Pearson Education Limited is not responsible for the content of any external internet sites. It is essential for tutors to preview each website before using it in class so as to ensure that the URL is still accurate, relevant and appropriate. We suggest that tutors bookmark useful websites and consider enabling students to access them through the school/college intranet.

Every effort has been made to trace copyright holders and we apologise for any omissions. We would be pleased to insert the appropriate acknowledgement in any subsequent editions.

A note from the publisher
In order to ensure that this resource offers high-quality support for the associated Pearson qualification, it has been through a review process by the awarding body. This process confirms that this resource fully covers the teaching and learning content of the specification or part of a specification at which it is aimed. It also confirms that it demonstrates an appropriate balance between the development of subject skills, knowledge and understanding, in addition to preparation for assessment.

Endorsement does not cover any guidance on assessment activities or processes (e.g. practice questions or advice on how to answer assessment questions), included in the resource nor does it prescribe any particular approach to the teaching or delivery of a related course.

While the publishers have made every attempt to ensure that advice on the qualification and its assessment is accurate, the official specification and associated assessment guidance materials are the only authoritative source of information and should always be referred to for definitive guidance.

Pearson examiners have not contributed to any sections in this resource relevant to examination papers for which they have responsibility.

Examiners will not use endorsed resources as a source of material for any assessment set by Pearson.

Endorsement of a resource does not mean that the resource is required to achieve this Pearson qualification, nor does it mean that it is the only suitable material available to support the qualification, and any resource lists produced by the awarding body shall include this and other appropriate resources.

Contents

How to use this book

Congratulations!

By choosing to study Engineering you have chosen to follow in the footsteps of the pioneering men and women who developed the technologies that have shaped the modern world. By choosing to study a BTEC National you have chosen a suite of qualifications designed to provide the knowledge, skills and practical experience to enable you to take these technologies forward into the future.

Hundreds of thousands of BTEC qualified engineers have successful careers across the dizzying spectrum of engineering sectors and specialisms, more diverse and challenging than in any other discipline. For those wanting to join them a BTEC National in Engineering might be their ticket into a rewarding hands on technician level job role or apprenticeship in industry. For others it will be a first step on the ladder towards a HND or degree in engineering and perhaps a career in management, design or research and development.

Whatever the future might hold, BTEC Nationals are designed to ensure that you're well prepared to overcome the challenges you will face when studying and working in this demanding but highly rewarding discipline.

The future belongs to engineers and the technological solutions they will develop to overcome problems that are as yet unknown. You will be part of that future.

How your BTEC is structured

Your BTEC National is divided into **mandatory units** (the ones you must do) and **optional units** (the ones you can choose to do).

The number of mandatory and optional units will vary depending on the type of BTEC National you are doing.

This book supports 7 of the mandatory units from across the Engineering qualification. Units 1, 3 and 6 are externally assessed by examination or controlled task. Units 2, 4, 5 and 7 are internally assessed using a series of assignments. You will be required to study some or all of these in the following qualifications:

▶ Extended Certificate
▶ Foundation Diploma
▶ Diploma
▶ Extended Diploma

In addition, 4 popular optional units are also covered. These are Units 8, 10, 19 and 25. Their inclusion enables valid courses to be structured using just the units covered in this book for learners studying:

▶ Extended Certificate
▶ Foundation Diploma
▶ Diploma* (Engineering pathway)

* In the Diploma, Unit 7 can be used as the fifth optional unit to complete the qualification.

Your learning experience

You may not realise it but you are always learning. Your educational and life experiences are constantly shaping your ideas and thinking, and how you view and engage with the world around you.

You are the person most responsible for your own learning experience so you must understand what you are learning, why you are learning it and why it is important both to your course and to your personal development. .

Your learning can be seen as a journey with four phases.

Phase 1	Phase 2	Phase 3	Phase 4
You are introduced to a topic or concept and you start to develop an awareness of what learning is required.	You explore the topic or concept through different methods (e.g. research, questioning, analysis, deep thinking, critical evaluation) and form your own understanding.	You apply your knowledge and skills to a task designed to test your understanding.	You reflect on your learning, evaluate your efforts, identify gaps in your knowledge and look for ways to improve.

During each phase, you will use different learning strategies to secure the core knowledge and skills you need. This student book has been written using similar learning principles, strategies and tools. It has been designed to support your learning journey, to give you control over your own learning, and to equip you with the knowledge, understanding and tools you need to be successful in your future studies or career.

Features of this book

This student book contains many different features. They are there to help you learn about key topics in different ways and understand them from multiple perspectives. Together, these features:

▶ explain what your learning is about
▶ help you to build your knowledge
▶ help you to understand how to succeed in your assessment
▶ help you to reflect on and evaluate your learning
▶ help you to link your learning to the workplace.

Each individual feature has a specific purpose, designed to support important learning strategies. For example, some features will:

▶ encourage you to question assumptions about what you are learning
▶ help you to think beyond what you are reading about
▶ help you to make connections between different areas of your learning and across units
▶ draw comparisons between your own learning and real-world workplace environments
▶ help you to develop some of the important skills you will need for the workplace, including team work, effective communication and problem solving.

Features that explain what your learning is about

Getting to know your unit

This section introduces the unit and explains how you will be assessed. It gives an overview of what will be covered and will help you to understand why you are doing the things you are asked to do in this unit.

Getting started

This is designed to get you thinking about the unit and what it involves. This feature will also help you to identify what you may already know about some of the topics in the unit and act as a starting point for understanding the skills and knowledge you will need to develop to complete the unit.

Features that help you to build your knowledge

Key terms

Concise and simple definitions are provided for key words, phrases and concepts, giving you, at a glance, a clear understanding of the key ideas in each unit.

Link

Link features show any links between content in different units or within the same unit, helping you to identify knowledge you have learned elsewhere that will help you to achieve the requirements of the unit. Remember, although your BTEC National is made up of several units, there are common themes that are explored from different perspectives across the whole of your course.

Research

This asks you to research a topic in greater depth. These features will help to expand your understanding of a topic and develop your research and investigation skills. All of this will be invaluable for your future progression, both professionally and academically.

Theory into practice

In this feature, you will be asked to consider the workplace or industry implications of a topic or concept from the unit. This will help you to understand the relevance of your current learning and the ways in which it may affect a future career in your chosen sector.

Discussion

Discussion features encourage you to talk to other learners about a topic, working together to increase your understanding of the topic and to understand other people's perspectives on an issue. These features will also help to build your teamworking skills, which will be invaluable in your future professional and academic career.

Safety tip

These tips give advice about health and safety when working on the unit. They will help to build your knowledge about best practice in the workplace, as well as making sure that you stay safe.

Worked Example

Worked examples show the process you need to follow to solve a problem, such as a maths or science equation, or the process for writing a letter or memo. They will help you to develop your understanding and your numeracy and literacy skills.

Further reading and resources

This feature lists other resources – such as books, journals, articles or websites – you can use to expand your knowledge of the unit content. This is a good opportunity for you to take responsibility for your own learning and prepare for research tasks you may need to complete academically or professionally.

Features connected to your assessment

Your course is made up of mandatory and optional units. There are two different types of mandatory unit:

▶ externally assessed
▶ internally assessed.

The features that support you in preparing for assessment are below. But first, what is the difference between these two different types of unit?

Externally assessed units

These units will give you the opportunity to demonstrate your knowledge and understanding, or your skills, in a direct way. For these units you will complete a task, set directly by Pearson, in controlled conditions. This could take the form of an exam or it could be another type of task. You may have the opportunity to prepare in advance, to research and make notes about a topic which can be used when completing the assessment.

Internally assessed units

Most of your units will be internally assessed and will involve you completing a series of assignments, set and marked by your tutor. The assignments you complete will allow you to demonstrate your learning in a number of different ways, from a written report to a presentation to a video recording and observation statements of you completing a practical task. Whatever the method, you will need to make sure you have clear evidence of what you have achieved and how you did it.

Assessment practice

These features give you the opportunity to practise some of the skills you will need during the unit assessment. They do not fully reflect the actual assessment tasks but will help you to prepare for them.

Plan – Do – Review

You will also find handy advice on how to plan, complete and evaluate your work. This is designed to get you thinking about the best way to complete your work and to build your skills and experience before doing the actual assessment. These questions will prompt you to think about the way you work and why particular tasks are relevant.

Getting Ready for Assessment

For internally assessed units, this is a case study of a BTEC National student, talking about how they planned and carried out their assignment work and what they would do differently if they were to do it again. It will give you advice on preparing for your internal assessments, including Think about it points for you to consider for your own development.

Getting ready for assessment

This section will help you to prepare for external assessment. It gives practical advice on preparing for and sitting exams or a set task. It provides a series of sample answers for the types of question you will need to answer in your external assessment, including guidance on the good points of these answers and ways in which they could be improved.

Features to help you reflect on and evaluate your learning

PAUSE POINT
Pause Points appear regularly throughout the book and provide opportunities to review and reflect on your learning. The ability to reflect on your own performance is a key skill you will need to develop and use throughout your life, and will be essential whatever your future plans are.

Hint
Extend
These sections give you suggestions to help cement your knowledge and indicate other areas you can look at to expand it.

Features which link your learning with the workplace

Case study

Case studies throughout the book will allow you to apply the learning and knowledge from each unit to a scenario from the workplace or industry. Case studies include questions to help you consider the wider context of a topic. They show how the course content is reflected in the real world and help you to build familiarity with issues you may find in a real-world workplace.

THINK ▶▶FUTURE

This is a case study in which someone working in the industry talks about their job role and the skills they need. The *Focusing your skills* section suggests ways for you to develop the employability skills and experiences you will need to be successful in a career in your chosen sector. This will help you to identify what you could do, inside and outside your BTEC National studies, to build up your employability skills.

Engineering Principles 1

Getting to know your unit

Assessment

This unit is externally assessed using an unseen paper-based examination that is marked by Pearson.

To make an effective contribution to the design and development of engineered products and systems, you must be able to draw on the principles laid down by the pioneers of engineering science. The theories developed by the likes of Newton and Ohm are at the heart of the work carried out by today's multi-skilled engineering workforce. This unit covers a range of both mechanical and electrical principles and some of the necessary mathematics that underpins their application to solve a range of engineering problems.

How you will be assessed

This unit is externally assessed by an unseen paper-based examination. The examination is set and marked by Pearson. Throughout this unit you will find practice activities that will help you to prepare for the examination. At the end of the unit you will also find help and advice on how to prepare for and approach the examination. The examination must be taken under examination conditions, so it is important that you are fully prepared and familiar with the application of the principles covered in the unit. You will also need to learn key formulae and be confident in carrying out calculations accurately. A scientific calculator and knowledge of how to use it effectively will be essential.

The examination will contain a number of short- and long-answer questions. Assessment will focus on applying appropriate principles and techniques to solving problems. Questions may be focused on a particular area of study or require the combined use of principles from across the unit. An *Information Booklet of Formulae and Constants* will be available during the examination. As the guidelines for assessment can change, you should refer to the official assessment guidance on the Pearson Qualifications website for the latest definitive guidance.

This table contains the skills that the examination will be designed to assess.

Assessment objectives	
AO1	Recall basic engineering principles and mathematical methods and formulae
AO2	Perform mathematical procedures to solve engineering problems
AO3	Demonstrate an understanding of electrical, electronic and mechanical principles to solve engineering problems
AO4	Analyse information and systems to solve engineering problems
AO5	Integrate and apply electrical, electronic and mechanical principles to develop an engineering solution

This table contains the areas of essential content that learners must be familiar with prior to assessment.

Essential content	
A1	Algebraic methods
A2	Trigonometric methods
B1	Static engineering systems
B2	Loaded components
C1	Dynamic engineering systems
D1	Fluid systems
D2	Thermodynamic systems
E1	Static and direct current electricity
E2	Direct current circuit theory
E3	Direct current networks
F1	Magnetism
G1	Single-phase alternating current theory

Getting started

To get started, have a quick look through each topic in this unit and then in small groups discuss why you think these areas are considered important enough to be studied by everyone taking a BTEC National Engineering course. Pick one or more topics and discuss how they might be relevant to a product, activity or industry you are familiar with.

 # A Algebraic and trigonometric mathematical methods

Engineers have to be confident that the solutions they devise to address practical problems are based on sound scientific principles. In order to do this, they often have to solve complex mathematical problems, so they must be comfortable and competent when working with algebra and trigonometry.

A1 Algebraic methods

Algebra allows relationships between variables to be expressed in mathematical shorthand notation that can be manipulated to solve problems. In this part of the unit, we will consider algebraic **expressions** involving indices and logarithms.

Indices

Even if you do not yet recognise the term, you will already be familiar with the use of indices in common mathematical expressions. For example:

▶ 3×3 is otherwise known as 'three squared' or 3^2 in mathematical notation using indices.
▶ $5 \times 5 \times 5$ is otherwise known as 'five cubed' or 5^3 in mathematical notation using indices.

The two parts of the notation used to describe indices are called the base and the index. For example:

▶ In the expression 3^2 the **base** is 3 and the **index** is 2.

Often in engineering mathematics, we have to consider situations where we do not know values for the numbers involved. We use algebra to represent unknown numbers with letters or symbols. When applied to indices:

▶ $a \times a = a^2$, where a is the base and 2 is the index.
▶ $b \times b \times b = b^3$, where b is the base and 3 is the index.

Where the index is also unknown, it can be represented by a letter as well, for example:

▶ a^n, where a is the base and n is the index.
▶ b^m, where b is the base and m is the index.

The laws of indices

▶ When dealing with **equations** that contain terms involving indices, there is a set of basic rules that you can apply to simplify and help solve them. These are the laws of indices. They are summarised in **Table 1.1**.

▶ **Table 1.1** The laws of indices

Operation	Rule
Multiplication	$a^m \times a^n = a^{m+n}$
Division	$\dfrac{a^m}{a^n} = a^{m-n}$
Powers	$(a^m)^n = a^{m \times n}$
Reciprocals	$\dfrac{1}{a^n} = a^{-n}$
Index = 0	$a^0 = 1$
Index = $\frac{1}{2}$ or 0.5	$a^{\frac{1}{2}} = \sqrt{a}$
Index = $\dfrac{1}{n}$	$a^{\frac{1}{n}} = \sqrt[n]{a}$
Index = 1	$a^1 = a$

Key terms

Expression – a mathematical statement such as $a^2 + 3$ or $3a - t$. Expressions can easily be recognised because they do not contain an equals sign.

Base – the term that is raised to an index, power or exponent. For example, in the expression 4^3 the base is 4.

Index – the term to which the base is raised. For example, in the expression 4^3 the index is 3. The index may also be called the power or exponent. The plural of 'index' is 'indices'.

Equation – used to equate two expressions that have equal value, such as $a^2 + 3 = 19$ or $t - 1 = 3a + 12$. Equations can easily be recognised because they always contain an equals sign.

Worked Example

You can use the laws of indices to simplify expressions containing indices.

- $a^2 a^4 a^0 a^{-3.2} = a^{2+4+0-3.2} = a^{2.8}$
- $(\sqrt{a})^3 a^{-1} = (a^{0.5})^3 a^{-1} = a^{1.5} a^{-1} = a^{1.5-1} = a^{0.5}$

  or $\sqrt{a}$

- $\dfrac{a^{\frac{3}{2}} b^2 a^{-1}}{b} = a^{1.5} b^2 a^{-1} b^{-1} = a^{1.5-1} b^{2-1} = a^{0.5} b^1$

  or $b\sqrt{a}$

- $\dfrac{a^{-1} b^{-1} a^{\frac{1}{2}}}{b^{-2}} = a^{-1} b^{-1} a^{0.5} b^2 = a^{-1+0.5} b^{-1+2} = a^{-0.5} b^1$

  or $\dfrac{b}{\sqrt{a}}$

- $a^{-3}(b^{-2})^2 a^{3.5} = a^{-3+3.5} b^{-4} = a^{0.5} b^{-4}$ or $\dfrac{\sqrt{a}}{b^4}$

Logarithms

Logarithms are very closely related to indices.

The **logarithm** of a number (N) is the power (x) to which a given base (a) must be raised to give that number.

▶ In general terms, where $N = a^x$, then $\log_a N = x$.

In engineering, we encounter mainly **common logarithms**, which use base 10, and **natural logarithms**, which use base e.

The Euler number (e) is a mathematical constant that approximates to 2.718. You will come across this again later in the section dealing with natural exponential functions.

Common logarithms

Common logarithms are logarithms with base 10.

▶ Where $N = 10^x$, then $\log_{10} N = x$.

▶ When using common logarithms, there is no need to include the base 10 in the notation, so where $N = 10^x$, you can write simply $\log N = x$.

This corresponds to the log function on your calculator.

Natural logarithms

In a similar way, when dealing with natural logarithms with base e:

▶ Where $N = e^x$, then $\log_e N = x$.

▶ Natural logarithms with base e are so important in mathematics that they have their own special notation, where $\log_e N$ is written as $\ln N$. So where $N = e^x$, then $\ln N = x$.

This corresponds to the ln function on your calculator.

The laws of logarithms

There are a number of standard rules that can be used to simplify and solve equations involving logarithms. These are the laws of logarithms. They are summarised in **Table 1.2**.

▶ **Table 1.2** The laws of logarithms

Operation	Common logarithms	Natural logarithms
Multiplication	$\log AB = \log A + \log B$	$\ln AB = \ln A + \ln B$
Division	$\log \dfrac{A}{B} = \log A - \log B$	$\ln \dfrac{A}{B} = \ln A - \ln B$
Powers	$\log A^n = n \log A$	$\ln A^n = n \ln A$
Logarithm of 0	$\log 0 = $ not defined	$\ln 0 = $ not defined
Logarithm of 1	$\log 1 = 0$	$\ln 1 = 0$

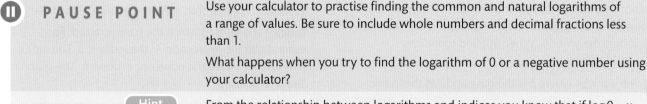

Ⅱ PAUSE POINT Use your calculator to practise finding the common and natural logarithms of a range of values. Be sure to include whole numbers and decimal fractions less than 1.

What happens when you try to find the logarithm of 0 or a negative number using your calculator?

Hint From the relationship between logarithms and indices you know that if $\log 0 = x$, then $0 = 10^x$. What value must x take if $10^x = 0$?

Extend Take a few moments to think about why $\log 0$ and $\log -1$ are not defined and will produce an error on your calculator.

Worked Examples

1 Use the laws of logarithms to solve the equation $\log 14^{(1-x)} = \log 9^{(x+3)}$.

Solution

$$\log 14^{(1-x)} = \log 9^{(x+3)}$$

$$(1-x)\log 14 = (x+3)\log 9$$

$$\log 14 - x\log 14 = x\log 9 + 3\log 9$$

$$\log 14 - 3\log 9 = x\log 9 + x\log 14$$

$$\log 14 - 3\log 9 = x(\log 9 + \log 14)$$

$$x = \frac{\log 14 - 3\log 9}{\log 9 + \log 14} = -0.817 \text{ (rounded to 3 significant figures (s.f.))}$$

Always check your solution by substituting the unrounded value of the solution back into the original equation:

$$\log 14^{(1-x)} = \log 14^{[1-(-0.817...)]} = 2.08 \text{ (to 3 s.f.)}$$

$$\log 9^{(x+3)} = \log 9^{(-0.817...+3)} = 2.08 \text{ (to 3 s.f.)}$$

2 Use the laws of logarithms to solve the equation $\ln 3t = 2\ln\frac{12}{t} + 2$.

Solution

$$\ln 3t = 2\ln\frac{12}{t} + 2$$

$$\ln 3 + \ln t = 2(\ln 12 - \ln t) + 2$$

$$\ln 3 + \ln t = 2\ln 12 - 2\ln t + 2$$

$$3\ln t = 2\ln 12 - \ln 3 + 2$$

$$\ln t = \frac{2\ln 12 - \ln 3 + 2}{3} = 1.957...$$

You can now use the general relationship that where $\ln N = x$, then $N = e^x$.

In this case, $\ln t = 1.957...$, so $t = e^{1.957...}$, which you can use your calculator to evaluate.

$$t = 7.08 \text{ (to 3 s.f.)}$$

Always check your solution by substituting the unrounded value of the solution back into the original equation:

$$\ln 3t = \ln 21.235... = 3.056 \text{ (to 4 s.f.)}$$

$$2\ln\frac{12}{t} + 2 = 2\ln 1.695... + 2 = 2 \times 0.5278... + 2$$

$$= 3.056 \text{ (to 4 s.f.)}$$

Exponential growth and decay

Exponential functions where x and y are variables and N is a constant take the general form $y = N^x$.

Some common systems found in engineering and in nature – such as charging capacitors, radioactive decay and light penetration in oceans – can be defined by exponential functions.

Several important growth and decay processes in engineering are defined by a special type of exponential function that uses the Euler number (e) as its base (N).

This is known as the natural exponential function and takes the basic form $y = e^x$.

If you plot the function $y = e^x$ as a graph (see **Figure 1.1**), then the curve it describes has two special characteristics that other exponential functions do not have.

▸ At any point on the curve the slope or gradient of the graph is equal to e^x.

▸ At $x = 0$ (where the curve intersects the y-axis) the graph has a slope or gradient of exactly 1.

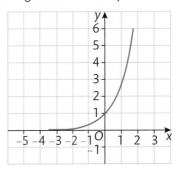

▸ **Figure 1.1** Graph of the natural exponential function $y = e^x$

Examples of engineering formulae containing natural exponential functions are shown in **Table 1.3** for charging/discharging a capacitor.

▸ **Table 1.3** Example engineering formulae containing natural exponential functions

Description	Formula
Capacitor charge voltage	$V = V_0(1 - e^{-t/RC})$
Capacitor discharge voltage	$V = V_0 e^{-t/RC}$

Key term

Tangent – a straight line with a slope equal to that of a curve at the point where they touch.

 **PAUSE POINT**   Examine the characteristics of exponential functions by plotting a graph of $y = 2^x$ for values of x between 1 and 10.

Hint   Tabulate values of x and y. Ensure that your axes are of appropriate size and scale.

Extend   The gradient of the **tangent** at any point on the curve reflects the rate of change in y with respect to x at that point. What happens to this rate of change as x increases?

PAUSE POINT Examine the characteristics of the natural exponential function by plotting a graph of $y = e^x$ for values of x between 1 and 10.

Extend The gradient of the tangent at any point on the curve reflects the rate of change in y with respect to x at that point. Use your graph to demonstrate that for $y = e^x$ the tangent at any point has gradient e^x.

Problems involving exponential growth and decay

Worked Example – Exponential growth

When a capacitor with capacitance C is charged through a resistance R towards a final potential V_0, the equation giving the voltage V across the capacitor at any time t is $V = V_0(1 - e^{-t/RC})$.

Given the values $C = 100\,\mu F$, $R = 10\,k\Omega$ and $V_0 = 3\,V$, calculate V when $t = 3\,s$.

▶ A 100 μF capacitor

Solution

Substitute the given values into the formula, remembering to convert all values into appropriate units.

$$C = 100\,\mu F = 100 \times 10^{-6}\,F$$
$$R = 10\,k\Omega = 10 \times 10^3\,\Omega$$
$$V_0 = 3\,V$$
$$t = 3\,s$$

Substituting these values into $V = V_0(1 - e^{-t/RC})$ gives $V = 3(1 - e^{-3})$.

So $V = 3(1 - 0.049...) = 2.85\,V$ (to 3 s.f.)

Worked Example – Exponential decay

When a capacitor with capacitance C is discharged through a resistance R from an initial potential V_0, the voltage V across the capacitor at any time t is given by the equation $V = V_0e^{-t/RC}$.

Given the values $C = 47\,\mu F$, $R = 30\,k\Omega$ and $V_0 = 4\,V$, calculate the time t at which $V = 3.5\,V$.

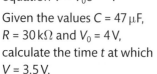
▶ A 47 μF capacitor

Solution

Substitute the given values into the formula, remembering to convert all values into appropriate units.

$$C = 47\,\mu F = 47 \times 10^{-6}\,F$$
$$R = 30\,k\Omega = 30 \times 10^3\,\Omega$$
$$V_0 = 4\,V$$
$$V = 3.5\,V$$

Rearranging the formula $V = V_0e^{-t/RC}$ to make t the subject:

$$e^{-t/RC} = \frac{V}{V_0}$$
$$-\frac{t}{RC} = \ln\frac{V}{V_0}$$
$$t = -RC\ln\frac{V}{V_0}$$

Substituting in the values gives:

$$t = -[30 \times 10^3 \times 47 \times 10^{-6}]\ln\frac{3.5}{4} = -1.41 \times -0.133...$$
$$= 0.188\,s \text{ (to 3 s.f.)}$$

Linear equations and straight-line graphs

In engineering, many simple systems behave in a linear fashion. **Table 1.4** shows some examples. When plotted graphically, linear relationships are characterised by a straight line with a constant gradient.

▶ **Table 1.4** Examples of engineering formulae describing linear relationships

Description	Formula
Standard general form	$y = mx + c$
Ohm's law	$V = IR$
Gravitational potential energy	$PE = mgh$
Electrical power	$P = IV$
Resistivity	$\rho = \dfrac{RA}{l}$
Linear motion	$v = u + at$

Linear equations

Linear equations contain unknowns that are raised to their first power only, such as x raised to its first power or x^1 (which is usually written simply as x). Linear equations take the general form $y = mx + c$.

Linear equations containing a single unknown quantity can be solved by rearranging to make the unknown quantity the **subject** of the equation.

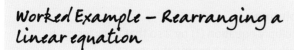

Worked Example – Rearranging a linear equation

Rearrange $4x + 3 = 12$ to make x the subject of the equation.

Solution

$4x = 12 - 3 = 9$ Subtract 3 from both sides of the equation.

$x = \dfrac{9}{4}$ Divide both sides of the equation by 4.

If the same unknown quantity occurs more than once in a linear equation, then all the terms containing the unknown quantity need to be isolated on one side of the equation with all the other terms on the other side. This is called 'gathering like terms'.

Worked Example – Gathering like terms in a linear equation

Rearrange $9b + 6 = 14 - 2b$ to make b the subject of the equation.

Solution

$9b + 2b = 14 - 6$ Add $2b$ and subtract 6 on both sides of the equation.

$11b = 8$ Combine like terms.

$b = \dfrac{8}{11}$ Divide both sides of the equation by 11.

Straight-line graphs

Graphs in mathematics usually use a horizontal x-axis and a vertical y-axis (known as Cartesian axes).

A linear equation with two unknowns can be represented graphically by a straight line as shown in **Figure 1.2**.

In the general formula used to describe a linear equation, $y = mx + c$, m is the gradient of the line and c is the value of y where the line intercepts the y-axis (when $x = 0$).

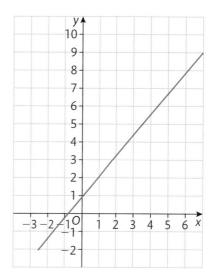

▶ **Figure 1.2** The graphical representation of a linear equation is a straight line

The **gradient** of a linear equation can be:
- ▶ positive – the y values increase linearly as the x values increase (the line slopes up from left to right)
- ▶ negative – the y values decrease linearly as the x values increase (the line slopes down from left to right)
- ▶ zero – the y values stay the same as the x values increase (the line is parallel to the x-axis).

Key terms

Subject of an equation – a single term becomes the subject of an equation when it is isolated on one side of the equation with all the other terms on the other side. For example, in the equation $y = 4x + 3$, the y term is the subject.

Gradient – also called 'slope', measures how steep a line is. It is calculated by picking two points on the line and dividing the change in height by the change in horizontal distance, or $\dfrac{\text{change in } y \text{ value}}{\text{change in } x \text{ value}}$.

Solving pairs of simultaneous linear equations

Sometimes in engineering it is necessary to solve systems that involve pairs of independent equations that share two unknown quantities. There are several examples of this in the assessment activity practice questions at the end of this section.

When you solve simultaneous equations, you are determining values for the unknowns that satisfy both equations. This is easiest to see by considering two independent linear equations in x and y that are plotted graphically on the same axes (see **Figure 1.3**). The only position where the same values of x and y satisfy both equations is where the lines intersect. You can read these values from the x-axis and the y-axis. In this example, the solution is $x = 2$ and $y = 4$.

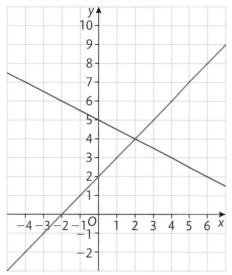

▶ **Figure 1.3** The solution of the simultaneous linear equations $y = x + 2$ and $2y = -x + 10$ is given by the intersection point of the two lines

Of course, you could always plot graphs of the linear equations when solving simultaneous equations, but this is a time-consuming approach and less accurate than the alternative algebraic methods.

The two main algebraic methods – **substitution** and **elimination** – are best explained by working through examples.

Worked Example – Substitution method

Solve this pair of simultaneous equations:

$y = x + 2$ (1)

$2y = -x + 10$ (2)

Solution

Substitute equation (1) into equation (2) to get $2(x + 2) = -x + 10$.

So $2x + 4 = -x + 10$ Multiply out the brackets.

    $2x + x = 10 - 4$ Collect like terms.

      $3x = 6$

       $x = 2$ Divide both sides by 3.

Now substitute $x = 2$ into equation (1) in order to find y: $y = 2 + 2 = 4$.

To check your solution, substitute $x = 2$ into equation (2) and see if you get the same value for y:

    $2y = -2 + 10 = 8$

      $y = 4$

Worked Example – Elimination method

Solve the same pair of simultaneous equations as before:

$y = x + 2$ (1)

$2y = -x + 10$ (2)

Solution

Multiply equation (1) by 2 so that the y terms in both equations become the same:

$2y = 2x + 4$ (3)

Subtract equation (2) from equation (3) to eliminate the identical terms in y:

$(2x + 4) - (-x + 10) = 0$

      $3x - 6 = 0$

        $3x = 6$

         $x = 2$

As before, substitute $x = 2$ into equation (1) to find $y = 4$, and then check your solution in equation (2).

Coordinates on Cartesian grid

Any point on a plotted linear equation can be expressed by its **coordinates** (x, y).

In the example shown in **Figure 1.3**, the two lines cross at the point where $x = 2$ and $y = 4$. Its coordinates are (2, 4).

Any point on the x-axis has a y coordinate of zero. For example, the point (4, 0) lies on the x-axis.

Similarly, any point on the y-axis has an x coordinate of zero, so the point (0, 4) lies on the y-axis.

The point (0, 0), where both x- and y coordinates are zero, is called the **origin** and is represented by 'O'.

Quadratic equations

Quadratic equations contain unknowns that are raised to their second power, such as x^2. They take the general form $y = ax^2 + bx + c$. In this equation x and y are unknown variables, and coefficients a, b and c are constants that will be discussed later.

A quadratic equation with two unknowns can be represented graphically by a curve. An example is shown in **Figure 1.4**.

🔘 **PAUSE POINT**   Show the efficiency of using an analytical approach by solving the simultaneous equations in the above worked examples with the graphical method.

> Hint   Tabulate values of x and y for each equation. Ensure that the scales of your axes are appropriate.

> Extend   Compare and evaluate the use of graphical and analytical methods in solving simultaneous equations. Are there any situations where a graphical approach might be preferred?

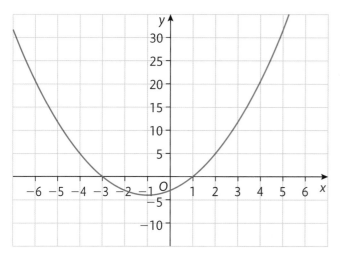

▸ **Figure 1.4** Graph of the quadratic equation $y = x^2 + 2x - 3$

You will often have to solve quadratic equations in engineering problems. This can be done by arranging the equation in the general form for quadratics and finding the two values of x for which $y = 0$. These are called the **roots** of the equation and are shown on the graph as the points where the line intersects the x-axis.

Two algebraic methods of solving quadratic equations are **factorisation** and using the **quadratic formula**. Before you consider factorising, it is a good idea to remind yourself how to multiply an expression by a number or by a symbol or more complicated expression.

Worked Example – Multiplying an expression by a number

Multiply each term of the expression by the number:

- $4(3x - 2) = 12x - 8$

- $5\left(\dfrac{x}{3} - 2\right) = \dfrac{5x}{3} - 10$.

Worked Example – Multiplying an expression by a symbol or another expression

Multiply out the brackets term by term, then collect like terms:

- $2x(3x - 2) = 6x^2 - 4x$

- $(x + 1)(x - 2) = x^2 - 2x + x - 2 = x^2 - x - 2$

- $(x + 2)(3x - 2) = 3x^2 - 2x + 6x - 4 = 3x^2 + 4x - 4$.

Factorisation

You can think of factorisation as the opposite of multiplying out, although it is generally somewhat more difficult. Factorising can be done in several ways, using some or all of the following techniques.

▸ **Extracting common factors**
 Common factors are terms by which some or all parts of an expression can be divided. These can be extracted as follows:

- $ax + ay = a(x + y)$ where a is the common factor
- $2x + 4y = 2(x + 2y)$ where 2 is the common factor
- $a(x + 2) + b(x + 2) = (x + 2)(a + b)$ where $(x + 2)$ is the common factor.

▸ **Grouping**
 When two variables x and y are present in the same expression, they must be grouped together before factorising. For example:

$$32x^2 + 18y^3 + 8x + 9y^2 = 18y^3 + 9y^2 + 32x^2 + 8x$$
$$= 9y^2(2y + 1) + 8x(4x + 1)$$

Sometimes this might mean that you have to multiply out the expression first. For example:

$$4(2y^2 - x^2) + y(y^2 + 7) = 8y^2 - 4x^2 + y^3 + 7y$$
$$= y^3 + 8y^2 + 7y - 4x^2$$
$$= y(y^2 + 8y + 7) - 4x^2$$

The second example can be factorised further by using the methods for factorising quadratics described in the next section.

Solving quadratic equations using factorisation

Let us take an example of a quadratic equation that you might need to solve as part of an engineering problem:

$$3x + 8 = 5x + x^2$$

▸ First, rearrange the equation (if necessary) to make one side of the equation zero.

| $0 = 5x + x^2 - 3x - 8$ | Move all the terms to one side of the equation. |
| $0 = x^2 + 2x - 8$ | Arrange into the standard form for quadratic equations. |

▸ The next step is to factorise the right-hand side of this equation.

> **Key term**
>
> **Coefficient** – a number or symbol that multiplies a variable. For example, in the expression $3x$, 3 is the coefficient of the variable x.

To do this, you need to find two expressions that when multiplied together give $x^2 + 2x - 8$. This looks difficult, but there are a few general guidelines that will help:

▸ If the **coefficient** of the x^2 term (a) is 1, then the coefficient of each of the x terms in the factors will also be 1.

▸ When the number terms in the two factors are multiplied together, the product must equal the number term (c) in the quadratic expression.

▸ If the coefficient of the x^2 term (a) is 1, then the coefficient of the x term (b) in the quadratic expression is equal to the sum of the number terms in the two factors.

Applying these guidelines to $x^2 + 2x - 8$, you know that:

▸ the coefficients of the x terms in the factors will be 1

▸ the product of the number terms in the factors will be −8

▸ the sum of the number terms in the factors will be 2.

Finding the terms to put in the brackets is then often a case of trying different values until you identify those that meet the required criteria:

$$x^2 + 2x - 8 = (x - 2)(x + 4)$$

▸ Equate each of the factors to zero to obtain the roots of the equation.

$$(x - 2) = 0, \text{ so } x = 2$$
$$(x + 4) = 0, \text{ so } x = -4$$

▸ Check that each root satisfies the original equation by substitution.

$$3x + 8 = 5x + x^2$$

When $x = -4$: $-12 + 8 = -20 + 16$
or $-4 = -4$

When $x = 2$: $6 + 8 = 10 + 4$
or $14 = 14$

▸ Check to ensure that each solution is reasonable in the context of the question and clearly state the solution.

There are two values of x for which the quadratic equation $3x + 8 = 5x + x^2$ is true. These are $x = 2$ and $x = -4$.

See the worked example on page 11.

Some examples of engineering formulae describing quadratic relationships are given in **Table 1.5**.

▸ **Table 1.5** Examples of engineering formulae describing quadratic relationships

Description	Formula
Standard general form	$ax^2 + bx + c = 0$
Displacement	$s = ut + \frac{1}{2}at^2$
Total surface area of a cylinder	$A = 2\pi r^2 + 2\pi rh$

Solving quadratic equations using the quadratic formula

Sometimes it is difficult to use factorisation to find the roots of a quadratic equation. In such cases, you can use an alternative method.

The formula for the roots of a quadratic equation arranged in the standard form is:

$$x = \frac{-b \pm \sqrt{b^2 - 4ac}}{2a}$$

Worked Example

A train runs along a level track with a velocity of $5\,\text{m}\,\text{s}^{-1}$. The driver presses the accelerator, causing the train to increase speed by $2\,\text{m}\,\text{s}^{-2}$. The motion of the train is defined by the equation $s = ut + \frac{1}{2}at^2$

where: s is displacement (the distance travelled) (m)

       u is initial velocity ($\text{m}\,\text{s}^{-1}$)

       t is time (s)

       a is acceleration ($\text{m}\,\text{s}^{-2}$).

Calculate the time the train takes to travel a distance of 6 m.

▶   Train running along a track

Solution

Rearrange the equation into the general form for a quadratic, making one side equal to zero:

$\frac{1}{2}at^2 + ut - s = 0$

Substitute the values $u = 5$, $s = 6$ and $a = 2$ into the equation:

$\frac{1}{2}(2)t^2 + 5t - 6 = 0$

$t^2 + 5t - 6 = 0$

The left-hand side can be factorised by inspection to give $(t + 6)(t - 1) = 0$.

This equation is true when either of the factors is equal to zero, so $t = -6$ or $t = 1$.

It is important to check these solutions by substituting back into the original quadratic equation given in the question: $s = ut + \frac{1}{2}at^2 = 5t + t^2$.

You will find that when $t = 1$, $s = 6$ and when $t = -6$, $s = 6$.

This confirms that the solutions determined by finding the roots of the quadratic equation are mathematically correct, because they both give the distance required in the question.

However, you must now consider whether both solutions actually fit the practical situation in the real world. In this case $t = -6$ doesn't make sense as a solution, because it would mean that the train reached the required distance 6 seconds before it started accelerating.

The roots can be seen more clearly when illustrated graphically, as in **Figure 1.5**.

Finally, you should clearly state the solution to the problem:

The time taken for the train to travel a distance of 6 m is 1 s.

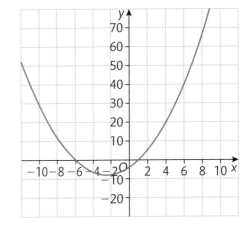

▶   **Figure 1.5** Graph showing roots of the quadratic expression $t^2 + 5t - 6$ at $t = -6$ and $t = 1$

Worked Example

A ball is thrown vertically upwards with an initial velocity of $16\,\mathrm{m\,s^{-1}}$. The height reached by the ball is given by the equation $h = ut - \frac{1}{2}gt^2$

where: h is the vertical height (m)

u is the initial vertical velocity ($\mathrm{m\,s^{-1}}$)

t is time (s)

g is the gravitational field strength ($9.81\,\mathrm{m\,s^{-2}}$)

Calculate the values of t at which the height of the ball is 6 m above the point where it was released.

Solution

Rearrange the equation into the general form for a quadratic equation, making one side equal to zero:

$\frac{1}{2}gt^2 - ut + h = 0$

Substitute the values $u = 16$, $h = 6$ and $g = 9.81$ into the equation:

$\frac{1}{2} \times 9.81t^2 - 16t + 6 = 0$

$4.905t^2 - 16t + 6 = 0$

It would be really difficult to find the roots of this equation by factorisation, so use the

general formula for solving quadratics: $x = \dfrac{-b \pm \sqrt{b^2 - 4ac}}{2a}$.

By comparing the formula with the standard form $ax^2 + bx + c$, you can see that

$a = 4.905$, $b = -16$, $c = 6$

So $t = \dfrac{-(-16) \pm \sqrt{(-16)^2 - (4 \times 4.905 \times 6)}}{2 \times 4.905}$

$= \dfrac{16 \pm \sqrt{256 - 117.72}}{9.81}$

$= \dfrac{16 \pm 11.76}{9.81}$

$= 0.432$ or 2.83

It is important to check these solutions by substituting back into the original quadratic equation given in the question: $h = ut - \frac{1}{2}gt^2 = 16t - 4.905t^2$.

You will find that when $t = 0.432...$, $h = 6$ and when $t = 2.83...$, $h = 6$. Be careful to use the unrounded values when you do the checks.

This confirms that the solutions determined by finding the roots of the quadratic equation are correct because they both give the required height stated in the question.

The next thing to do is consider whether these solutions actually fit the practical situation for which they were generated. In this case, it makes sense that as the ball moves up it will pass the height of 6 m. Gravity will slow this ascent until the ball actually stops momentarily before it falls back towards the ground. On its way down it will pass the height of 6 m once again.

This can be seen more clearly when illustrated graphically, as in **Figure 1.6**.

Finally, you should clearly state the solution to the problem:

The values of t at which the height of the ball is 6 m above the point where it was released are 0.432 s and 2.83 s.

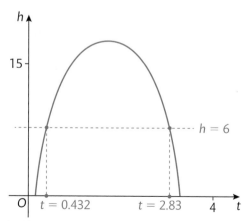

▶ **Figure 1.6** Graph of $h = 16t - 4.905t^2$, showing when it reaches the height of 6 m

▶ **PAUSE POINT** Discuss with a colleague or as a group why methods of finding the roots of quadratic equations are important and useful mathematical tools for engineers.

Hint Consider how else you might solve problems that involve quadratic equations.

Extend Suppose that you want to solve the quadratic equation $y = x^2 + 4x + 6$ to find values of x when $y = 1$. This would mean finding the roots of the quadratic expression $x^2 + 4x + 5$. Can you solve this? Draw a graph of the function $y = x^2 + 4x + 5$ to help explain why not.

A2 Trigonometric methods

Angular measurement

You are already familiar with angular measurements made in **degrees**. In practical terms, this is the most common way to define an angle on an engineering drawing that a technician might use when manufacturing a component in the workshop.

However, there is another unit of angular measurement, called the **radian**, which is used extensively in engineering calculations.

Key terms

Degree (symbol: °) – one degree is $\frac{1}{360}$th of a complete circle. A complete circle contains 360°.

Radian (symbol: rad or ᶜ) – one radian is the angle **subtended** at the centre of a circle by two radii of length r that describe an arc of the same length r on the circumference. A complete circle contains 2π rad.

Subtend – to form an angle between two lines at the point where they meet.

Circular measurement

One revolution of a full circle contains 360° or 2π radians.

It is reasonably straightforward to convert angles stated in degrees to radians and vice versa.

Given that 2π radians = 360°

$$1 \text{ radian} = \frac{360\,°}{2\pi} \approx 57.3° \text{ (to 3 s.f.)}$$

$$\text{and } 1° = \frac{2\pi}{360} \approx 0.0175 \text{ rad (to 3 s.f.)}$$

The use of radians makes it straightforward to calculate some basic elements of circles with the general formulae shown in **Table 1.6**, where the angle θ is measured in radians.

▶ **Table 1.6** General formulae for circular measurements (see **Figure 1.7**)

Arc length	$= r\theta$
Circumference of a circle	$= r(2\pi) = 2\pi r$
Area of a sector	$= \frac{1}{2}r^2\theta$
Area of a full circle	$= \frac{1}{2}r^2(2\pi) = \pi r^2$

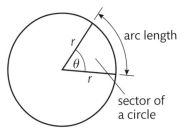

▶ **Figure 1.7** Arc length and sector of a circle

Triangular measurement

In right-angled triangles we name the three sides in relation to the right angle and one of the other two angles, θ (see **Figure 1.8**).

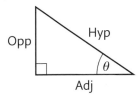

▶ **Figure 1.8** Trigonometric naming conventions for a right-angled triangle

▶ The side opposite the right angle is the **hypotenuse** (hyp).

▶ The side next to the angle θ is the **adjacent** (adj) side.

▶ The side opposite the angle θ is the **opposite** (opp) side.

The ratios of the lengths of these sides are given specific names and are widely used in engineering (see **Figures 1.9–1.11**):

▶ **sine** (sin), where

$$\sin\theta = \frac{\text{opp}}{\text{hyp}}$$

- **cosine** (cos), where
$$\cos \theta = \frac{adj}{hyp}$$

- **tangent** (tan), where
$$\tan \theta = \frac{opp}{adj}$$

From these definitions it can also be deduced that
$$\tan \theta = \frac{\sin \theta}{\cos \theta}$$

Graphs of the trigonometric functions

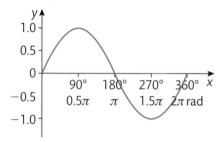

▶ **Figure 1.9** Graph of $y = \sin \theta$

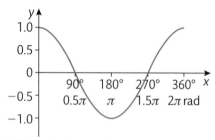

▶ **Figure 1.10** Graph of $y = \cos \theta$

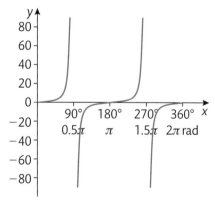

▶ **Figure 1.11** Graph of $y = \tan \theta$

When plotted graphically, both the sine and the cosine functions generate periodic waveforms. Both functions vary between a maximum of 1 and a minimum of –1 and so are said to have an **amplitude** of 1. Both functions have a **period** of 360° or 2π radians, after which the cycle repeats.

The tangent function does not generate a smooth waveform, although the graph is still periodic with a period of 180° or π radians.

Some values for the trigonometric ratios are given in **Table 1.7**.

▶ **Table 1.7** Values of the trigonometric ratios for angles between 0° and 360°

$\theta°$	θrad	$\sin \theta$	$\cos \theta$	$\tan \theta$
0	0	0	1	0
30	0.52	0.50	0.87	0.58
60	1.05	0.87	0.50	1.73
90	1.57	1.00	0	±∞
120	2.09	0.87	–0.50	–1.73
150	2.62	0.50	–0.87	–0.58
180	π	0	–1.00	0
210	3.67	–0.50	–0.87	0.58
240	4.19	–0.87	–0.50	1.73
270	4.71	–1.00	0	±∞
300	5.24	–0.87	0.50	–1.73
330	5.76	–0.50	0.87	–0.58
360	2π	0	1.00	0

Sine and cosine rules

The basic definitions of the trigonometric functions sine, cosine and tangent only apply to right-angled triangles. However, the sine and cosine rules can be applied to any triangle of the form shown in **Figure 1.12**.

▶ The sine rule:
$$\frac{a}{\sin A} = \frac{b}{\sin B} = \frac{c}{\sin C}$$

PAUSE POINT

Check that you can set up and use a scientific calculator in both degree and radian modes.

Hint

Look for the set-up screen on your calculator and select Rad. You will find that the D (for degrees) usually displayed at the top of the screen changes to R (for radians).

Extend

Before carrying out any work involving trigonometry, you should check that you are using the appropriate setting on your calculator. Use some simple examples to show what might happen if you get it wrong.

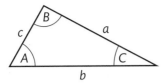

▶ **Figure 1.12** Naming conventions when applying the sine rule and the cosine rule

▶ The cosine rule can take three different forms depending on the missing value to be determined:

$$a^2 = b^2 + c^2 - 2bc\cos A$$
$$b^2 = a^2 + c^2 - 2ac\cos B$$
$$c^2 = a^2 + b^2 - 2ab\cos C$$

Vectors and their applications

Many quantities encountered in engineering, such as force and velocity, are only fully described when magnitude, direction and sense are known (see **Figure 1.13**). Such quantities are called **vectors**. When adding or subtracting vectors you must always take into account the direction in which they act.

Diagrammatic representation of vectors

▶ The length of the arrow represents the magnitude of the vector.
▶ The angle θ specifies the direction of the vector.
▶ The head of the arrow specifies the positive sense of the vector.

Vector addition

To find the sum (or resultant) of two vectors v_1 and v_2, you can represent the situation graphically by drawing a vector diagram. In **Figure 1.14** the two vectors are drawn to scale, forming a triangle or parallelogram from which the characteristics of the resultant vector (v_T) can be measured.

Phasors

Phasors are rotating vectors that are useful in analysing sinusoidal (sine-shaped) waveforms. **Figure 1.15** shows the relationship between a phasor and the sine wave it represents.

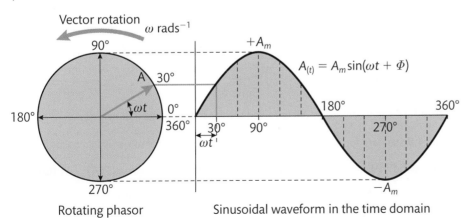

Rotating phasor   Sinusoidal waveform in the time domain

▶ **Figure 1.15** How the rotating phasor relates to the sinusoidal waveform

When using phasors in the analysis of alternating current, the length of the phasor represents the peak voltage (V) or amplitude of the sinusoidal waveform, and the phasor rotates about a point of origin with an **angular velocity** of ω.

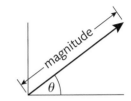

▶ **Figure 1.13** A vector has magnitude, direction and sense

Link

Graphical and analytical techniques for vector addition are discussed further in the section **B1 Static engineering systems**.

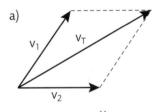

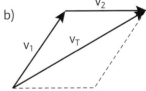

▶ **Figure 1.14** Vector addition can be done using a) the parallelogram rule or b) the triangle rule

At any fixed instant in time (t), the phasor will have a phase angle (θ) and its vertical component will be equal to the instantaneous voltage (v) on the corresponding sine wave.

So the instantaneous voltage (v) at any point on the waveform is

$$v = V\sin\theta$$

In terms of angular velocity, this gives

$$v = V\sin(\omega t)$$

This relationship is true when the waveform begins its cycle when $t = 0$. However, it is common to have a waveform that is said to lead or lag the standard waveform. This **phase difference** (see **Figure 1.16**) is expressed as an angle (Φ), so that

$$v = V\sin(\omega t + \Phi)$$

The angular velocity of the phasor (ω) is related to the **frequency** of the waveform (f) by

$$\omega = 2\pi f$$

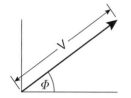

▶ **Figure 1.16** Phasor representing the sinusoidal waveform $v = V\sin(\omega t + \Phi)$, where Φ is the phase shift from the standard waveform

PAUSE POINT   Explain the difference between a vector and a phasor.

Hint   Consider the different types of systems they are used to describe.

Extend   Use a diagram to explain the relationship between a phasor and the sinusoidal waveform it describes.

Mensuration

It is very common for an engineer to need to calculate the surface area or volume of three-dimensional shapes (see **Figure 1.17**); for example, to determine the number of tiles required to line a swimming pool or to find the capacity of a cylindrical storage tank. There are several important formulae that can help you do this (see **Table 1.8**).

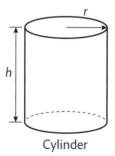
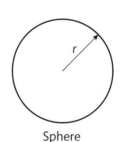
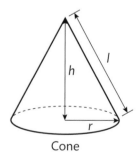

Cylinder   Sphere   Cone

▶ **Figure 1.17** Some commonly encountered regular solids and their dimensions

▶ **Table 1.8** Standard formulae for the surface area and volume of some regular solids

Regular solid	Curved surface area (CSA)	Total surface area (TSA)	Volume
Cylinder	$2\pi rh$	$2\pi rh + 2\pi r^2 = 2\pi r(h + r)$	$\pi r^2 h$
Cone	πrl	$\pi r^2 + \pi rl = \pi r(r + l)$	$\frac{1}{3}\pi r^2 h$
Sphere	$4\pi r^2$		$\frac{4}{3}\pi r^3$

Often, an apparently complex object can be broken down into a series of regular solids.

Assessment practice 1.1

1 A small boat is capable of a maximum velocity in still water of $v\,km\,h^{-1}$. On a journey upriver against a current with velocity $v_c\,km\,h^{-1}$, the boat travels at its full speed for 2 hours and covers 16 km. On a journey back downriver with the same current flow, the boat travels at full speed for 1 hour and 20 minutes and covers 18 km. Establish and then solve a pair of simultaneous linear equations to determine the maximum velocity of the boat v and the current v_c. (4 marks)

2 Scrap transformers have been collected for recycling. In total, 203 transformers weigh 403.4 kg. It is found that the transformers come in two different types, which weigh 1.8 kg and 2.3 kg, respectively. Establish and then solve a pair of simultaneous linear equations to determine the number of each type of transformer collected. (4 marks)

3 The motion of two vehicles is described by the linear equations

$$27 = v - 3t$$
$$13 = v - 4t$$

By solving this pair of simultaneous equations determine:

a) the time t at which both vehicles will be travelling with the same velocity

b) the corresponding velocity v at that point. (2 marks)

4 The time t (years) taken for a radioactive isotope contained in stored nuclear waste to decay to 5% of its original quantity is given by the equation

$$5 = 100 \times 2^{\frac{-t}{1622}}$$

Solve the equation to find the time t.

Show evidence of the use of the laws of logarithms in your answer. (2 marks)

5 An engineer has been given a drawing (see **Figure 1.18**) of a triangular plate with minimal dimensional detail.

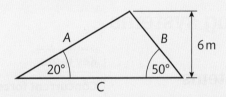

▶ **Figure 1.18** Engineering drawing of a triangular plate

Calculate the lengths of the sides A, B and C. (4 marks)

6 A pair of dividers used for marking out has legs that are 150 mm long, as shown in **Figure 1.19**. The angle α between the legs can be set to a maximum of 60° by rotating the adjustment screw.

a) Use the sine rule to calculate the maximum distance that can be set between the points of the dividers when $\alpha = 60°$.

b) Use the cosine rule to calculate the value of angle α when the points of the dividers are set to a distance of 38 mm.

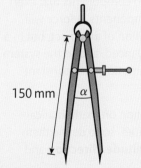

▶ **Figure 1.19** A pair of dividers (4 marks)

7 The displacement s of an accelerating body with respect to time t is described by the equation

$$s = t^2 - 7t + 12$$

Using factorisation, solve this equation to determine the values of t for which $s = 2$. (3 marks)

8 Use the formula for solving quadratic equations to determine the radius r of an enclosed cylinder which has a total surface area (*TSA*) of 12.25 m² and a height h of 1.2 m, where

$$TSA = 2\pi rh + 2\pi r^2$$ (4 marks)

9 A technician has been given an engineering drawing of a circular component (see **Figure 1.20**) where the arc length *AB* is 23 cm and the radius *r* is 16 cm.

Calculate the angle θ in degrees to allow the technician to use a Vernier protractor to mark out the component accurately. **(2 marks)**

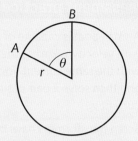

▶ **Figure 1.20** Engineering drawing of a circular component

For each of the problems in this assessment practice, use the following stages to guide your progress through the task.

Plan
- What important information is provided in the question? Can this information be summarised or collated to make it easier to digest?
- Will a sketch help me to visualise what is being asked?

Do
- I am confident that I have interpreted the question correctly, and I have a plan to approach finding a solution.
- I recognise when a solution is clearly incorrect and will challenge my previous approach and try an alternative.

Review
- I can explain the engineering principles that underpin my approach to a task.
- I can clearly follow the method I used to complete a task and explain my reasoning at each stage.

B Static engineering systems

B1 Static engineering systems

Concurrent and non-concurrent coplanar forces

Mechanical systems often contain components that exert forces that push or pull other components. A force will tend to cause a change in the motion of an object, but in a static system all the forces are balanced, and so the system will either remain at rest or be in motion with constant velocity.

As mentioned in the previous section on mathematical methods, forces are vector quantities, so to define them fully we need to know their **magnitude**, **direction** and **sense**.

Where systems contain multiple forces, it is possible to represent the forces visually in a space diagram, free body diagram or vector diagram.

Consider a mass suspended by two wires under tension.

A **space diagram** is a sketch of the physical arrangement of the system being considered, like that shown in **Figure 1.21**.

> **Key terms**
>
> **Concurrent forces** – forces that all pass through a common point.
>
> **Non-concurrent forces** – forces that do not all pass through the same point.
>
> **Coplanar forces** – forces acting in the same two-dimensional plane.
>
> **Magnitude** – the size of a force.
>
> **Direction** – the orientation of the line in which the force is acting (the line of action).
>
> **Sense** – the direction along the line of action in which the force acts.

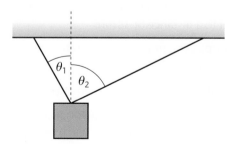

▶ **Figure 1.21** Space diagram showing an object suspended by two wires

A **free body diagram** is a sketch containing just the forces acting on the system or the part of the system that you are interested in, as shown in **Figure 1.22**.

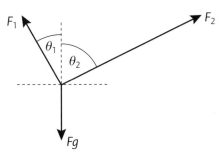

▶ **Figure 1.22** Free body diagram showing the forces acting on the suspended mass (which is represented by a point)

A **vector diagram** is a sketch in which the lengths of the lines representing the force vectors correspond to the magnitudes of the respective forces. The vectors may also have been rearranged to form a triangle of forces as shown in **Figure 1.23**. (Similarly, larger numbers of force vectors can be arranged into a polygon of forces.)

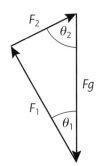

▶ **Figure 1.23** Vector diagram with the three forces rearranged into a triangle

Resolution of forces in perpendicular directions

When analysing a system in which multiple forces act on a body, it may not be obvious what the overall effect of those forces will be. What we need is a method of adding the forces together to find their combined effect, known as the **resultant** force acting on the body. We can combine forces graphically by drawing a vector diagram, but here we will consider an alternative approach using the resolution of forces.

Key term

Resultant – the force that represents the combined effect of all the forces in a system.

One way to add or subtract vector quantities is to split each vector into components acting in specific, perpendicular directions. A force acting in any direction can be resolved into a vertical and a horizontal component. The horizontal components of multiple forces can be simply added to calculate a single horizontal force. Similarly, the vertical components of multiple forces can be simply added to calculate a single vertical force. These can then be recombined into a single resultant force that represents the combined effect of all the original individual forces.

Splitting any vector into its horizontal and vertical components will involve the use of trigonometry and/or Pythagoras' theorem.

Figure 1.24 shows how a single force F can be resolved into a horizontal component F_h and a vertical component F_v.

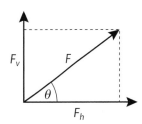

▶ **Figure 1.24** Force vector resolved into vertical and horizontal components

In the arrangement shown in **Figure 1.24**:

$$F_v = F\sin\theta$$
$$F_h = F\cos\theta$$
$$\frac{F_v}{F_h} = \tan\theta$$
$$F^2 = F_v^2 + F_h^2$$

When several forces act on a body, apply the same principles to each of the forces.

Ⅱ   PAUSE POINT   Explain the difference between a space diagram, a free body diagram and a vector diagram.

    Hint   Draw an example of each for the same system.

    Extend   How will the diagrams differ between concurrent and non-concurrent systems of forces?

Worked Example

Find the resultant and **equilibrant** for the system of concurrent forces shown in **Figure 1.25**.

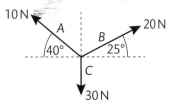

▶ **Figure 1.25** System of three concurrent coplanar forces

Solution

First, establish the convention that the positive vertical direction is up and the positive horizontal direction is to the right.

Resolve each vector into its vertical and horizontal components using the formulae on the previous page.

Find the sum of the vertical components (the Greek letter Σ is shorthand notation for 'sum'):

$$\Sigma F_v = A_v + B_v + C_v = 10\sin 40 + 20\sin 25 - 30$$
$$= -15.12\,\text{N}$$

Find the sum of the horizontal components:

$$\Sigma F_h = A_h + B_h + C_h = -10\cos 40 + 20\cos 25 + 0$$
$$= +10.47\,\text{N}$$

Figure 1.26 shows that the resultant force F_r is acting in the positive sense in a direction θ below the horizontal, where:

$$\tan\theta = \frac{15.12}{10.47} = 1.44$$

$$\theta = 55.3°\ \text{(using unrounded values)}$$

The magnitude of the resultant can be calculated using Pythagoras' theorem:

$$F_r = \sqrt{15.12^2 + 10.47^2} = 18.39\,\text{N}$$

So the resultant of this system of forces has magnitude 18.39 N and acts in a direction 55.3° below the horizontal with a positive sense.

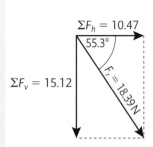

The equilibrant is the force required to bring this system into equilibrium, so it also has magnitude 18.39 N and direction 55.3° below the horizontal, but it will have a negative sense.

▶ **Figure 1.26** Vector diagram showing how the sums of the vertical and horizontal components are combined to give the resultant

Conditions for static equilibrium

A system of forces in static equilibrium will have no tendency to move because all the forces acting in the system are perfectly balanced. The resultant force for a system in static equilibrium is zero. This also means that the horizontal and vertical components of all the forces are such that:

$$\Sigma F_v = 0$$
$$\Sigma F_h = 0$$

For a system of concurrent forces, these equations are enough to define static equilibrium. However, in a non-concurrent system of forces, where the lines of action do not intersect at the same point, there could be a tendency for the system to rotate.

The **moment** of a force describes the tendency of a force to produce rotation about a pivot point or centre of rotation.

The moment of a force (M) about a pivot is calculated by multiplying the magnitude of the force (F) by the perpendicular distance (s) from the pivot point to the line of action of the force.

$$M = Fs$$

So, for a system of non-concurrent forces, there are three conditions that must be met for static equilibrium:

$$\Sigma F_v = 0$$
$$\Sigma F_h = 0$$
$$\Sigma M = 0$$

Worked Example

Find the equilibrant for the system of non-concurrent forces shown in **Figure 1.27** .

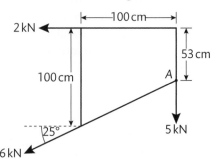

▶ **Figure 1.27** System of three non-concurrent coplanar forces

(Note that to define the equilibrant fully for a non-concurrent system of forces, you will need to find its magnitude, direction, sense and the perpendicular distance from its line of action to the centre of rotation of the body.)

Solution

Find the sum of the vertical components:
$$\Sigma F_v = 0 - 6\sin 25 - 5 = -7.54\,\text{kN}$$

Find the sum of the horizontal components:
$$\Sigma F_h = -2 - 6\cos 25 + 0 = -7.44\,\text{kN}$$

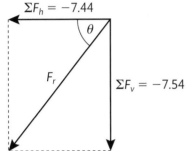

▶ **Figure 1.28** Vector diagram showing how the sums of the vertical and horizontal components are combined to give the resultant

Figure 1.28 shows that the resultant F_r is acting in the negative sense in a direction θ below the horizontal, where:

$$\tan\theta = \frac{7.54}{7.44} = 1.01$$
$$\theta = 45.4°$$

The magnitude of the resultant can be calculated using Pythagoras' theorem:
$$F_r = \sqrt{7.54^2 + 7.44^2} = 10.59\,\text{kN}$$

You can now say that the equilibrant has magnitude 10.59 kN, direction 45.4° above the horizontal and positive sense.

To complete the description of the equilibrant you must now take moments about some point in the system. This could be anywhere, but selecting point A in **Figure 1.27** will simplify the calculations because the moments of the 6 kN and 5 kN forces are zero about this point.

Before starting the calculations, you should establish the convention that clockwise moments are positive and anti-clockwise moments are negative.

Let the perpendicular distance from the equilibrant's line of action to point A be x m.

Taking moments about A:
$$\Sigma M_A = (10.59 \times x) - (2 \times 0.53) + (6 \times 0) + (5 \times 0)$$

For equilibrium, $\Sigma M_A = 0$, so
$$10.59x = 2 \times 0.53$$
$$x = 0.1\,\text{m}$$

You can now fully define the equilibrant, which when applied to this system of forces would bring it into static equilibrium (see **Figure 1.29**).

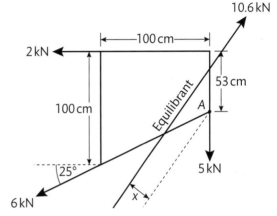

▶ **Figure 1.29** The system of forces, showing the equilibrant that would bring it into static equilibrium

The equilibrant has magnitude 10.59 kN, direction 45.4° above the horizontal and positive sense, and the perpendicular distance from its line of action to point A is 0.1 m.

Simply supported beams

Beams have been used extensively for thousands of years to support structures and can be seen in everything from bridges and aircraft carriers to the playground see-saw.

When dealing with beams, we often need to consider two different ways in which they are loaded:

▶ **Concentrated loads** – a narrowly focused force that can be assumed to act at a specific point along the length of the beam; for example, the weight of a car parked on a bridge.

▶ **Uniformly distributed loads** (**UDL**) – a force distributed along the full length (or a defined section) of the beam, for example, the weight of the bridge itself.

Reactions

A simply supported beam is supported from below at two points, *A* and *B*. At each of these points, **support reaction** forces act on the beam to maintain the static equilibrium of the system. These can be calculated if we know the magnitude and position of the forces acting on the beam.

The type of support used dictates the direction in which the support reaction can act.

▶ **Rollers** provide a support **reaction normal** (perpendicular) to their point of contact with the beam. This is vertically upwards in the case of a horizontal beam (see **Figure 1.30**).

▶ **Pins** can provide a support reaction in any direction, so there can be a vertical and a horizontal component to the support reaction provided by a pinned joint (see **Figure 1.31**).

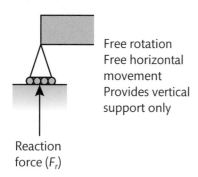

Free rotation
Free horizontal movement
Provides vertical support only

Reaction force (F_r)

▶ **Figure 1.30** Roller support

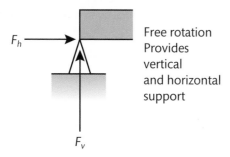

F_h

Free rotation
Provides vertical and horizontal support

F_v

▶ **Figure 1.31** Pin support

Key terms

Normal reaction – the force that acts perpendicular to a surface upon an object that is in contact with the surface.

Support reactions – the forces that are maintaining the equilibrium of a beam or structure.

Worked Example

Determine the support reaction for the simply supported beam shown in **Figure 1.32**.

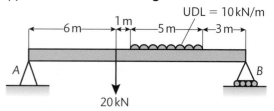

▶ **Figure 1.32** Space diagram of a simply supported beam

Solution

Draw a free body diagram of the beam, as in **Figure 1.33**.

(Note that the UDL can be treated as a point load acting at the centre of the distribution.)

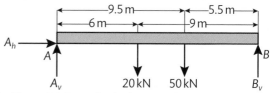

▶ **Figure 1.33** Free body diagram of the simply supported beam

To find the support reaction B_v take moments about point *A*. Because the beam is in static equilibrium, you know that these moments must sum to zero.

$$\Sigma M_A = (6 \times 20) + (9.5 \times 50) - (15 \times B_v) = 0$$

$$B_v = \frac{120 + 475}{15} = 39.67\,\text{kN}$$

To find the support reaction A_v take moments about point *B*.

$$\Sigma M_B = -(5.5 \times 50) - (9 \times 20) + (15 \times A_v) = 0$$

$$A_v = \frac{275 + 180}{15} = 30.33\,\text{kN}$$

Now check that the other requirements for static equilibrium are met by considering the horizontal and vertical components of the forces acting on the beam. For equilibrium these must sum to zero.

Find the sum of the vertical components:

$$\Sigma F_v = A_v + B_v - 20 - 50 = 0$$

Find the sum of the horizontal components:

$$\Sigma F_h = A_h = 0$$

Once calculated and checked, state the solution clearly:

The support reaction at point *A* is 30.33 kN vertically upwards.

The support reaction at point *B* is 39.67 kN vertically upwards.

PAUSE POINT Describe what is meant by static equilibrium in a non-concurrent system of forces.

> Hint Consider all the conditions that must be fulfilled for a body to be in static equilibrium.

> Extend Where a non-concurrent system is not in equilibrium, explain all the characteristics required to fully define the resultant and describe its relationship to the equilibrant.

B2 Loaded components

Table 1.9 summarises the formulae needed to calculate direct and shear loading.

▶ **Table 1.9** Summary of direct and shear loading

Operation	Direct loading	Shear loading
Stress	$\sigma = \dfrac{F}{A_\sigma}$	$\tau = \dfrac{F}{A_\tau}$
Strain	$\varepsilon = \dfrac{\Delta L}{L}$	$\gamma = \dfrac{a}{b}$
Elastic constants	Modulus of elasticity, $E = \dfrac{\sigma}{\varepsilon}$	Modulus of rigidity, $G = \dfrac{\tau}{\gamma}$

Forces acting on a body may be separated into two different types:

▶ **direct forces** – include tensile forces which tend to stretch and pull apart a material, and compressive forces (acting in the opposite direction) which tend to squash or compress a material.

▶ **shear forces** – forces cutting across a material which tend to shear or cut it apart.

Stress

When either direct or shear forces act on a body, opposing reaction forces are distributed within the material and it is said to be stressed or under stress.

Stress is therefore a measure of the load distribution within a material and is expressed as the load carried per unit area of material. The unit of stress is newton per square metre ($N\,m^{-2}$) or pascal (Pa), where $1\,N\,m^{-2} = 1\,Pa$.

We can consider two types of stress – direct and shear.

▶ direct stress (σ) = $\dfrac{\text{normal force } (F)}{\text{area } (A_\sigma)}$ (see **Figure 1.34**)

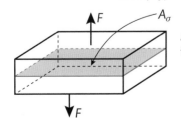

▶ **Figure 1.34** The area $A\sigma$ considered in direct stress

▶ shear stress (τ) = $\dfrac{\text{shear force } (F)}{\text{shear area } (A_\tau)}$ (see **Figure 1.35**)

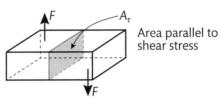
Area parallel to shear stress

▶ **Figure 1.35** The area A_τ considered in shear stress

Worked Example

A rivet has diameter (d) 20 mm and secures two rods together in the arrangement shown in **Figure 1.36**. Find the shear stress in the rivet.

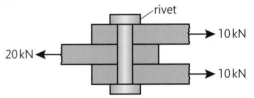

▶ **Figure 1.36** Rivet in double shear

The rivet in this example is said to be in double shear. To solve this type of problem, consider either half the load acting on a single shear area (in this case the cross-sectional area of the rivet) or the whole load acting over the two shear areas (in this case twice the cross-sectional area of the rivet).

First, calculate the cross-sectional area of the rivet in m^2.

$$A = \pi \times \frac{d^2}{4} = \pi \times \frac{0.02^2}{4} = 3.14 \times 10^{-4}\,m^2$$

In the first method, the shear area (A_τ) is the cross-section of the rivet and the shear force (F) is only half the applied load.

$$\tau = \frac{F}{A_\tau} = \frac{10 \times 10^3}{3.14 \times 10^{-4}}$$
$$= 31.8 \times 10^6\,Pa \text{ or } 31.8\,MPa$$

In the second method, the shear area (A_τ) is equal to twice the cross-section of the rivet and the shear force (F) is the whole of the applied load.

$$\tau = \frac{F}{A_\tau} = \frac{20 \times 10^3}{2 \times 3.14 \times 10^{-4}}$$
$$= 31.8 \times 10^6\,Pa \text{ or } 31.8\,MPa$$

Both methods give the same solution, which is that the shear stress in the rivet is 31.8 MPa.

Tensile and shear strength

Tensile and shear strength are material-specific properties that specify the maximum tensile and shear stresses that can be applied to the material. If either the tensile or the shear strength of a material is exceeded, the material will rupture.

Strain

When a material is subjected to an externally applied stress, it has a tendency to change shape. Strain quantifies the deformation of a body as a proportion of its original length. As strain is the ratio of two lengths, it has no units.

Again, we can consider two types of strain – direct and shear.

▸ direct strain $(\varepsilon) = \dfrac{\text{change in length } (\Delta L)}{\text{original length } (L)}$

(see **Figure 1.37**).

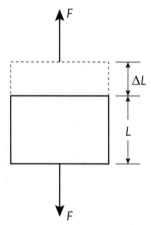

▸ **Figure 1.37** Direct strain

▸ shear strain $(\gamma) = \dfrac{\text{change in length } (\Delta L)}{\text{original length } (L)}$

(see **Figure 1.38**).

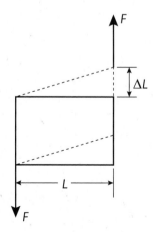

▸ **Figure 1.38** Shear strain

Worked Example

A press tool is required to punch out 40 mm diameter discs from a brass plate that is 1.5 mm thick and has a shear strength of 200 MPa.

a) Calculate the force required to punch out the disc.

b) Calculate the compressive stress in the punch.

Solution

a) You are given the following values, which have been converted into standard units where necessary:

- diameter d = 40 mm = 0.04 m
- thickness t = 1.5 mm = 0.0015 m
- maximum shear strength τ_{max} = 200 MPa = 200×10^6 Pa

For the punch to cut through the brass, the maximum shear strength of 200 MPa must be applied over the shear area.

In this case, the shear area (A_τ) will be equal to the circumference of the brass disc multiplied by its thickness:

$$A_\tau = \pi d \times t = \pi \times 0.04 \times 0.0015$$
$$= 1.88 \times 10^{-4} \, m^2$$

You know that $\tau = \dfrac{F}{A_\tau}$,

so $F = \tau_{max} A_\tau = (200 \times 10^6) \times (1.88 \times 10^{-4})$

$$= 37\,600 \, N = 37.6 \, kN$$

The force required to punch out the brass disc is 37.6 kN.

b) Compressive stress is a direct stress.

The area (A_σ) in this case will be the cross-sectional area of the punch:

$$A_\sigma = \frac{\pi d^2}{4} = \frac{\pi \times 0.04^2}{4} = 1.26 \times 10^{-3} \, m^2$$

Using the formula for direct stress:

$$\sigma = \frac{F}{A_\sigma} = \frac{37\,600}{1.26 \times 10^{-3}}$$

$$= 29.8 \times 10^6 \, Pa = 29.8 \, MPa$$

The compressive stress in the punch is 29.8 MPa.

Elastic constants

A material is considered to perform in its **elastic region** while any change in strain brought about as a consequence of an applied stress will reduce back to zero once that stress is removed. **Figure 1.39** shows a typical stress–strain curve for a ferrous metal. The elastic region is that part of the curve below the yield point.

A Elastic region   *D* Ultimate tensile strength (UTS)
B Limit of proportionality   *E* Fracture point
C Yield point

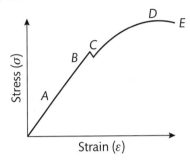

▶ **Figure 1.39** Typical stress–strain curve for a ferrous metal

The **modulus of elasticity** (*E*), also known as **Young's modulus**, expresses the linear relationship between direct stress and direct strain exhibited by a material in the part of the stress–strain curve below the elastic limit. Its unit is newton per square metre ($N\,m^{-2}$).

$$\text{modulus of elasticity } (E) = \frac{\text{direct stress } (\sigma)}{\text{direct strain } (\varepsilon)}$$

The **modulus of rigidity** (*G*) is a similar ratio, which expresses the linear relationship between shear stress and shear strain observed in many materials. Its unit is newton per square metre ($N\,m^{-2}$).

$$\text{modulus of rigidity } (G) = \frac{\text{shear stress } (\tau)}{\text{shear strain } (\gamma)}$$

PAUSE POINT   Explain the difference between shear stress and direct stress.

     Hint   Use a sketch to illustrate each case.

     Extend   Sketch a system in which a component is in both direct and shear stress.

Case study

The UK's longest suspension bridge

The Humber Bridge in Humberside is a suspension bridge crossing the Humber estuary. The bridge's structure relies on the strength of the two main cables used to support it.

Each of the main cables supporting the 1410 m central span is made up of bundles of 14 948 individual 5 mm diameter steel wires.

The steel used in the cables has an ultimate tensile strength (UTS) of 1540 MPa and a modulus of elasticity of 200 GPa.

When fully loaded, each of the main cables carries a load equivalent to 19 400 tonnes in tension.

The length of the unloaded cables was 1890 m (at 15°C) when they were manufactured.

▶ **Humber Bridge**

Check your knowledge

1 Calculate the direct stress in each cable when it is fully loaded.

2 Calculate the corresponding direct strain for the cable.

3 Calculate the corresponding extension of the cable and its total length when in service.

1 **Figure 1.40** shows a system of four non-concurrent coplanar forces acting on the corners of a rectangular plate measuring 3 m × 8 m.

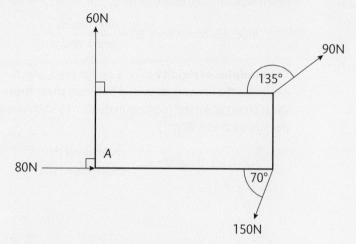

▶ **Figure 1.40** A system of four non-concurrent coplanar forces

Calculate the magnitude, direction and sense of the resultant force for this system, and find the position of its line of action relative to point *A*. (7 marks)

2 **Figure 1.41** shows a simply supported beam in static equilibrium.

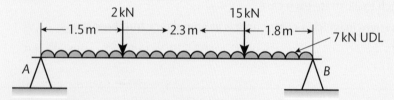

▶ **Figure 1.41** A beam in static equilibrium

Calculate the support reaction forces at points *A* and *B*. (4 marks)

3 On a suspension bridge the vertical support wires attaching the roadway to the main suspension cables are each designed to support 2×10^6 N when the bridge is fully loaded. The support wires are connected to the roadway using pinned joints as shown in **Figure 1.42**. Each pin is 10^5 mm in diameter and has a shear modulus of 160×10^9 Pa.

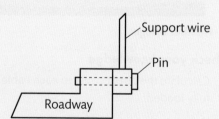

▶ **Figure 1.42** Pinned joint connecting a support wire of the bridge to the roadway

a) Calculate the shear stress in the pin when the bridge is fully loaded.

b) Calculate the corresponding shear strain for the pin. (4 marks)

For each of the problems in this assessment practice, use the following stages to guide your progress through the task.

Plan

- What engineering principles will be involved in answering the question?
- Do I understand exactly what will be required to form a complete solution to the problem? Am I comfortable in how to approach each part?

Do

- I can organise my work logically and systematically, annotating my solutions with notes that explain my approach.
- I can use sketches effectively to help me visualise a problem and explain my approach to solving it.

Review

- I can give possible real-world applications of the principles used to solve the problem and explain why these are important in an engineering context.
- I can explain to others how best to approach similar problems.

C Dynamic engineering systems

C1 Dynamic engineering systems

As well as working with static systems, engineers are also interested in moving objects. This section introduces some of the basic principles and techniques that are used to understand dynamic systems.

Dynamic systems, by definition, are systems that involve the relative movement of several component parts or movement of the system as a whole. In this section, we will see how displacement, time, velocity and acceleration are related and how you can calculate the work done and the power required to overcome a resistance.

Kinetic parameters and principles

Kinetic parameters describe the uniform linear motion of an object over time. They include **displacement**, **velocity** and **acceleration**.

> **Key terms**
>
> **Displacement** – the straight-line distance between the start and finish positions of a moving object. Unit: metre (m).
>
> **Velocity** – the rate at which the displacement of an object changes over time. Unit: metre per second ($m\,s^{-1}$).
>
> **Acceleration** – the rate at which the velocity of an object changes over time. Unit: metre per second squared ($m\,s^{-2}$).

The relationships between the kinetic parameters that describe the motion of an object are defined using a set of equations that are based on the definitions of displacement, velocity and acceleration.

These are often referred to as the SUVAT equations from the letters used to represent the variables involved:

▶ Displacement (s) – the distance travelled by an object in time t.

▶ Initial velocity (u) – the starting velocity of an object when $t = 0$.

▶ Final velocity (v) – the final velocity of the object at time t.

▶ Acceleration (a) – the uniform acceleration of an object over time t.

▶ Time (t) – the period of time over which you will consider the motion of an object.

The SUVAT equations of motion are:

▶ $s = \frac{1}{2}(u + v)t$

▶ $s = ut + \frac{1}{2}at^2$

▶ $v = u + at$

▶ $v^2 = u^2 + 2as.$

Applying the SUVAT equations

A huge range of problems involving the linear motion of objects can be solved using the SUVAT equations.

Worked Example

A cyclist accelerates from $10\,km\,h^{-1}$ to $35\,km\,h^{-1}$ over 2000 m of track. Calculate the acceleration of the cyclist.

Solution

When approaching this problem, you should analyse the question carefully and identify the SUVAT variables that are mentioned:

$s = 2000\,m$, $u = 10\,km\,h^{-1}$, $v = 35\,km\,h^{-1}$, $a = ?$, $t = $ not given

If you can identify three out of the five SUVAT variables you will be able to calculate the missing values.

Next, examine the units for each variable given in the question. Where these are not standard units, they **must** be converted before you begin your calculations. In this example both u and v are in $km\,h^{-1}$, so you need to convert them to $m\,s^{-1}$.

$$u = 10\,km\,h^{-1} = \frac{10 \times 1000\ m}{3600\ s} = 2.78\,m\,s^{-1}$$

$$v = 35\,km\,h^{-1} = \frac{35 \times 1000\ m}{3600\ s} = 9.72\,m\,s^{-1}$$

To choose the appropriate equation to solve this particular problem, find the one that includes only the variables involved. Here these are a, s, u and v, which corresponds to the SUVAT equation $v^2 = u^2 + 2as$.

Rearrange the equation to make a the subject:

$$a = \frac{v^2 - u^2}{2s}$$

To calculate a, substitute the numerical values from the question into the formula:

$$a = \frac{9.72^2 - 2.78^2}{2 \times 2000} = 0.022\,m\,s^{-2} \text{ (to 2 s.f.)}$$

The cyclist's acceleration is $0.022\,m\,s^{-2}$.

Dynamic parameters and principles

Dynamic parameters link the motion of an object to the forces involved in causing, influencing or stopping that motion. There are a number of important dynamic parameters. Their definitions and associated formulae are listed below.

Parameters of linear motion

Force (F) – a push or pull acting on an object. In physics a force is defined as any influence that tends to cause or alter the motion of an object. Unit: newton (N).

Static frictional force (F_s) – a force that must be overcome to set a sliding body in motion. The force (F_s) required to overcome static friction depends on the normal force (N) between the surfaces and their coefficient of static friction (μ_s). Unit: newton (N).

$$F_s = \mu_s N$$

Sliding (kinematic) frictional force (F_k) – a sliding body once in motion must overcome kinematic frictional resistance to its motion. The force (F_k) required to overcome kinematic friction depends on the normal force (N) between the surfaces and their coefficient of kinematic friction (μ_k). Unit: newton (N).

$$F_k = \mu_k N$$

Inertia force (F_i) – the resistance that an object of mass m has to any acceleration (a) that changes its state of motion. Unit: newton (N).

$$F_i = ma$$

Momentum (p) – the product of the mass (m) of a moving body and its velocity (v). Unit: kilogram metre per second ($kg\,m\,s^{-1}$).

$$p = mv$$

Work done (W) – the energy used when a force moves an object. It is the product of the applied force (F) and the associated displacement (s). Unit: newton metre (N m) or joule (J).

$$W = Fs$$

Power (P) – the average rate of doing work (W) over time (t). Unit: joules per second ($J\,s^{-1}$) or watt (W).

$$P = \frac{W}{t} = \frac{Fs}{t}$$

Instantaneous power (P_i) – the product of force (F) and velocity (v). Unit: joules per second ($J\,s^{-1}$) or watt (W).

$$P_i = Fv$$

Weight (F_g) – the force exerted by a gravitational field (g) on a body with mass (m). Unit: newton (N).

$$F_g = mg$$

Gravitational potential energy (E_p) – the potential energy possessed by a body of mass m in a gravitational field (g) when raised to a vertical height (h). Unit: joule (J).

$$E_p = mgh$$

Kinetic energy (E_k) – the energy possessed by a body of mass (m) travelling with velocity (v). Unit: joule (J).

$$E_k = \tfrac{1}{2}mv^2$$

Newton's laws of motion

Newton's laws of motion are fundamental to our understanding of dynamics.

▶ Newton's first law of motion states that 'a body continues in its present state of rest or uniform motion in a straight line unless it is acted upon by an external force'.

▶ Newton's second law of motion states that 'the rate of change of momentum of a body is directly proportional to the resultant force that is producing the change'.

▶ Newton's third law states that 'to every acting force there is an equal and opposite reacting force'.

It is Newton's second law that underpins the derivation of an important equation that establishes the relationship between force (F), mass (m) and acceleration (a):

$$F = ma$$

Principle of conservation of momentum

The principle of conservation of momentum states that 'the total amount of momentum in a system remains constant unless the system is acted upon by an external force'.

Consider two bodies of known masses m_1 and m_2 moving in the same direction with velocities v_1 and v_2, respectively, as shown in **Figure 1.43**. The two bodies then collide and begin to move together with velocity v_3.

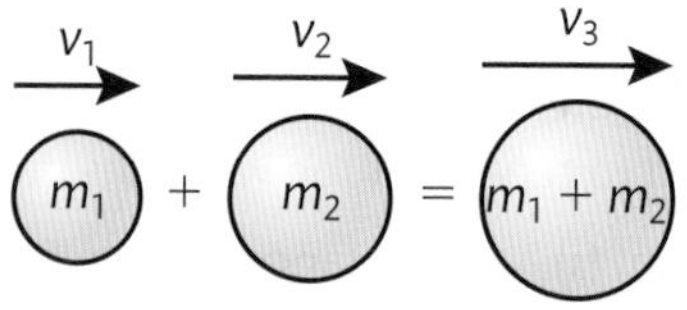

▶ **Figure 1.43** Conservation of momentum of two moving bodies that collide and then move together

Given that no external forces have been applied, the momentum prior to impact equals the momentum afterwards:

$$p_1 + p_2 = p_3$$

$$m_1 v_1 + m_2 v_2 = (m_1 + m_2)v_3$$

(II) PAUSE POINT   Draw a diagram of a body in motion and annotate it to show how each of the parameters of linear motion is calculated.

> Hint   As a starting point, you should give your body a mass, an acceleration, an instantaneous velocity and a displacement from a starting point.

> Extend   Mark any external factors acting on the body that are required in order to determine the parameters of linear motion.

Case study

Flood defences

In 2016, many parts of the UK suffered serious flooding. The damage to property and infrastructure such as roads and bridges was severe, and the lives of many people were badly affected.

▶ **Hydraulic drop** hammer piling rig

A common way of reinforcing river banks damaged by floodwater is by using steel piles driven deep into the ground to prevent further erosion. The piles are installed side by side, forming a protective wall from the surface of the water down and through the river bed.

Piles are installed using piling rigs (which are effectively large hammers) that are able to drive the piles vertically downwards. Each blow of the drop hammer moves the pile a little further into the ground.

The piling rig in the photograph has a 200 kg drop hammer, which is raised 3 m before being dropped onto the pile. It is inserting a pile that weighs 500 kg, which moves 125 mm further into the ground with each strike of the hammer.

Check your knowledge

1   What principles might you apply to the analysis of inserting a pile in this way?

2   Calculate the velocity with which the drop hammer hits the pile.

3   Calculate the initial velocity of the pile and hammer moving together.

4   Determine the force with which the ground is resisting the penetration of the pile.

Principle of conservation of energy

The principle of conservation of energy states that 'energy cannot be created or destroyed, but it can be changed from one form into another'.

In dynamics this means that the total energy within a system always remains the same. Any decrease in one type of energy leads to a corresponding increase in another.

Worked Example

As shown in **Figure 1.44**, A rollercoaster car with mass 200 kg is stationary at point A before rolling down the track, passing point B at the bottom of the curve and then continuing up to point C at the top of the next curve. Throughout its motion along the track, the rollercoaster car has to overcome a frictional force of 200 N.

Using the principle of conservation of energy, calculate the velocities at points B and C.

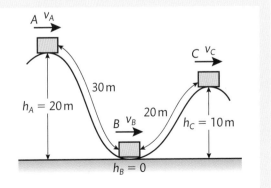

▶ **Figure 1.44** The motion of a rollercoaster car along the rollercoaster track

Solution

By the principle of conservation of energy, the total energy in the system is the same at points A, B and C.

In this case the total energy (E_{total}) at each stage is made up of gravitational potential energy (E_p), kinetic energy (E_k) and work done to overcome friction (W):

$$E_{total} = E_p + E_k + W$$

$$E_{total} = mgh + \tfrac{1}{2}mv^2 + Fs$$

Consider point A where $h_A = 20$ m, $s_A = 0$ m, $v_A = 0$ m s^{-1}.

Then

$$E_{total} = (200 \times 9.81 \times 20) + (\tfrac{1}{2} \times 200 \times 0^2) + (200 \times 0)$$

$$= 39\,240 + 0 + 0 = 39\,240\,J$$

At point A, all the parameters are defined, allowing us to establish that the total energy in the system is 39 240 J. At A this is made up entirely of gravitational potential energy.

Now that you know the total energy in the system, you can calculate the velocity of the rollercoaster car at point B, where

$$E_{total} = 39\,240\,J,\ h_B = 0\,m,\ s_{AB} = 30\,m,\ v_B = ?$$

$$E_{total} = E_p + E_k + W$$

$$39\,240 = (200 \times 9.81 \times 0) + (\tfrac{1}{2} \times 200 \times v_B^2) + (200 \times 30)$$

$$39\,240 = 0 + 100v_B^2 + 6000$$

$$v_B = \sqrt{\frac{39\,240 - 6000}{100}} = \pm 18.23\,m\,s^{-1}$$

The negative value would mean that the rollercoaster car is travelling backwards, so we use only the positive value, $v_B = 18.23$ m s^{-1}.

At point B, there is no longer any gravitational potential energy, 6000 J of work has been done to overcome the friction forces acting on the system, and the remaining energy has been transformed into kinetic energy now present in the moving rollercoaster car.

A similar procedure can be carried out to determine the velocity at point C, where

$$E_{total} = 39\,240\,J,\ h_C = 10\,m,\ s_{AC} = 50\,m,\ v_C = ?$$

$$E_{total} = E_p + E_k + W$$

$$39\,240 = (200 \times 9.81 \times 10) + (\tfrac{1}{2} \times 200 \times v_C^2) + (200 \times 50)$$

$$39\,240 = 19\,620 + 100v_C^2 + 10\,000$$

$$v_C = \sqrt{\frac{39\,240 - 19\,620 - 10\,000}{100}} = \pm 9.81\,m\,s^{-1}$$

Again you want the positive value, so $v_C = 9.81$ m s^{-1}.

At point C, 19 620 J of gravitational potential energy has been retained in the system, a total of 10 000 J of work has been done to overcome friction forces, and the remaining energy has been transformed into kinetic energy present in the moving rollercoaster car.

Rotational motion

The fundamental parameters describing rotational motion are shown in **Figure 1.45**.

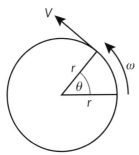

▶ **Figure 1.45** Basic parameters describing rotational motion

Angular displacement (θ) – the angle in radians through which a point or line has been rotated about a specific point (the centre of rotation). Unit: radian (rad).

Angular velocity (ω) – the rate at which angular displacement (θ) changes over time (t). Unit: radian per second (rad s^{-1}).

$$\omega = \frac{\Delta\theta}{\Delta t}$$

Angular acceleration (α) – the rate at which angular velocity (ω) changes over time (t). Unit: radian per second squared (rad s^{-2}).

$$\alpha = \frac{\Delta\omega}{\Delta t}$$

Tangential velocity (v) – the linear velocity of a point moving in a circular path. All points in a rotating body share the same angular velocity (ω), but the tangential velocity of each point will depend on its distance from the centre of rotation (r). Unit: metre per second (m s^{-1}).

$$v = \omega r$$

Centripetal acceleration (a) – the linear acceleration acting on a rotating body towards its centre of rotation. It is defined in terms of the distance of the body from the centre of rotation (r) and either the angular velocity (ω) or the tangential velocity (v). Unit: metre per second squared (m s^{-2})

$$a = \omega^2 r = \frac{v^2}{r}$$

Torque (τ) – a turning force or moment that tends to cause rotational movement. It is defined by the sum force (F)

acting at a distance (r) from the centre of rotation. Unit: newton metre (N m).

$$\tau = Fr$$

Work done (in uniform circular motion) (W) – the work done by a torque (τ) moving through an angular displacement (θ). Unit: joule (J).

$$W = \tau\theta$$

Power (in uniform circular motion) (P) – the rate of work done expressed as the product of torque (τ) and angular velocity (ω). Unit: watt (W).

$$P = \tau\omega$$

Moment of inertia (I) – sometimes called angular mass, is used in calculations relating to rotational motion in a similar way to which mass is used for linear motion. The moment of inertia for a particular rotating body is dependent on the mass distribution around the point of rotation and can be complicated to determine. Its calculation is beyond the scope of this unit.

Kinetic energy (of uniform circular motion) (E_k) – the energy possessed by a rotating body is calculated using its moment of inertia (I) and angular velocity (ω). Unit: joule (J).

$$E_k = \tfrac{1}{2}I\omega^2$$

Lifting machines

Lifting machines are used to allow relatively small forces to lift heavy objects. They include inclined planes, pulleys and scissor jacks. For instance, a person is unlikely to be able to lift and support a vehicle to change a wheel without using a lifting machine. Anyone can exert enough force to operate a scissor jack, which by virtue of a simple mechanism is able to amplify this force to a level sufficient to lift the car.

For all lifting machines, **mechanical advantage** (MA) is the ratio between the output force or load (F_l) and the input force or effort (F_e):

$$MA = \frac{F_l}{F_e}$$

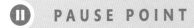

PAUSE POINT   Draw a diagram of a rotating body and annotate it to show how each of the parameters of rotational motion is calculated.

> Hint   As a starting point, you should give your body a moment of inertia, an angular velocity and a distance from the centre of rotation.

> Extend   Mark on the diagram any external factors acting on the body that are required in order to determine the parameters of rotational motion.

The **velocity ratio** (VR) is the ratio between the distance moved by the effort (s_e) and the distance moved by the load (s_l). Equivalently, it is the ratio of the velocity with which the effort moves (v_e) to the velocity with which the load moves (v_l):

$$VR = \frac{s_e}{s_l} = \frac{v_e}{v_l}$$

You already know that work done (W) is given by $W = Fs$. Applying this to the effort and load gives

$$W_e = F_e s_e$$
$$W_l = F_l s_l$$

The **efficiency** of a simple machine is given by the ratio of useful work output to work input.

Useful work output is the work done in actually moving the load. All machines will also have to overcome friction forces. Work done against friction is wasted as heat, noise or other undesirable effects.

The efficiency (η) of a simple machine with work input (W_e) and useful work output (W_l) is given by

$$\eta = \frac{W_l}{W_e} \times 100\%$$

Efficiency can also be stated purely in terms of the mechanical advantage (MA) and the velocity ratio (VR):

$$\eta = \frac{MA}{VR} \times 100\%$$

Inclined plane

It is thought that in the absence of cranes and other modern machinery, the ancient Egyptians used an inclined plane (see **Figure 1.46**) to lift the enormous stones used to construct the pyramids. By pulling the load up an inclined plane sloping at an angle (θ), the effort moves a distance (a) while the load is lifted through a vertical distance (b).

$$VR = \frac{a}{b} = \frac{a}{a \sin\theta} = \frac{1}{\sin\theta}$$

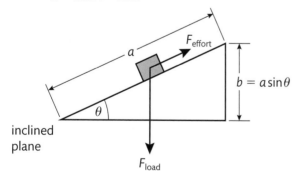

▶ **Figure 1.46** Inclined plane that can be used to lift a load

Pulleys

The velocity ratio of a pulley system is equal to the number of rope sections supporting the load.

Figure 1.47 shows a pulley system with VR = 4.

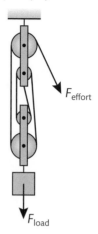

▶ **Figure 1.47** A system of pulleys with four rope sections supporting the load

Screw jack

In a screw jack like the one shown in **Figure 1.48**, one complete rotation of a handle of length (r) causes the load to move a distance equal to the pitch of the screw thread (p). The velocity ratio (VR) is given by

$$VR = \frac{2\pi r}{p}$$

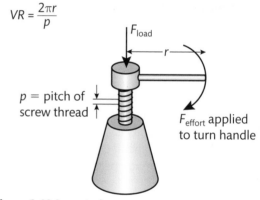

▶ **Figure 1.48** Screw jack

Scissor mechanism

The velocity ratio of a scissor mechanism is not constant – it varies continuously as the angle (θ) changes.

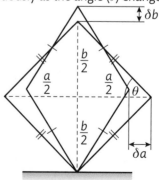

▶ **Figure 1.49** Scissor mechanism

In the example shown in **Figure 1.49**, a small change in distance moved by the effort (δa) leads to a corresponding (but not equal) small change in the distance moved by the load (δb). The velocity ratio depends on the geometry of the mechanism, but in this case

$$VR = \frac{\delta a}{\delta b} = \frac{1}{\tan \theta}$$

This means that:

▶ if $\theta < 45°$ then $VR > 1$, so a movement of effort (a) will produce a larger movement in load (b)

▶ if $\theta = 45°$ then $VR = 1$, so a movement of effort (a) will produce an equal movement in load (b)

▶ if $\theta > 45°$ then $VR < 1$, so a movement of effort (a) will produce a smaller movement in load (b).

Scissor jack

Car drivers often use a jack that incorporates a scissor mechanism when fitting the spare wheel to their vehicle. **Figure 1.50** shows a typical scissor jack, which uses a screw thread turned by a handle (just as in a screw jack, but here the screw does not lift the load directly) to

operate a scissor mechanism. For each rotation of the screw, the dimension a reduces by a distance equivalent to the pitch of the thread (p), and this motion is transferred by the scissor mechanism to increase the dimension b.

By combining the velocity ratios for the scissor and screw mechanisms, we get the velocity ratio for the scissor jack:

$$VR = \frac{1}{\tan \theta} \times \frac{2\pi r}{p} = \frac{2\pi r}{p \tan \theta}$$

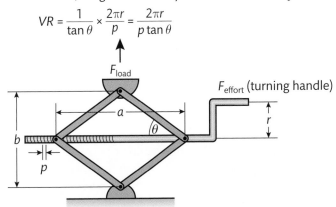

▶ **Figure 1.50** Scissor jack that combines a screw jack with a scissor mechanism

Assessment practice 1.3

1 A car with mass 700 kg accelerates uniformly from rest to a velocity of 20 m s⁻¹ in a time of 10 s. The frictional resistance to the motion of the vehicle is assumed to be constant throughout at 0.5 kN.

    a) Calculate the force accelerating the car.

    b) Calculate the average power developed by the engine. (2 marks)

2 A satellite launch vehicle travelling at 1000 m s⁻¹ uses a controlled explosion to release its payload into orbit. The empty launch vehicle has a mass of 7500 kg and the satellite released has mass 250 kg. Both continue travelling in the same direction, with the launch vehicle now moving at 900 m s⁻¹.

    a) Calculate the initial momentum of the launch vehicle immediately before the satellite was released.

    b) Calculate the final velocity of the satellite after release. (2 marks)

3 A lift cage with mass 500 kg accelerates vertically upwards from rest to a velocity of 6 m s⁻¹ in a distance of 12 m. The frictional resistance to motion is assumed to be constant throughout at 200 N.

    a) Calculate the work done raising the lift.

    b) Calculate the tension in the lifting cable.

    c) Calculate the power developed by the winch. (3 marks)

4 A lorry with mass 3500 kg is parked at the top of a steep hill with a 1-in-8 gradient when its handbrake fails. Assume that the lorry has a constant frictional resistance to motion of 500 N. The lorry rolls 40 m down the hill before crashing into a lamp post.

    Use the principle of conservation of energy to calculate the velocity of the lorry immediately prior to its impact with the lamp post. (3 marks)

For each of the problems in this assessment practice, use the following stages to guide your progress through the task.

Plan
- Which engineering principles need to be applied to this problem and what are the key formulae involved?
- How does the information given in the question dictate or otherwise influence my approach?

Do
- I have identified all the information I need to answer the question, including any formulae required.
- I have laid out my solution logically and explained each step so that my method can be followed easily.

Review
- I can identify the parts of my knowledge and understanding that require further development.
- I can identify the type and style of questions that I find most challenging and devise strategies, such as additional purposeful practice, that will help me to overcome any difficulties.

D Fluid and thermodynamic engineering systems

D1 Fluid systems

There are many situations in which engineers must design ways to contain or transport fluids (liquids or gases), such as oil tankers, dams or pipelines. This topic will develop your knowledge of fluid dynamics as applied in a range of scenarios, including how water exerts a force on retaining walls, up-thrust and its effects on submerged objects, and flow in tapering pipes.

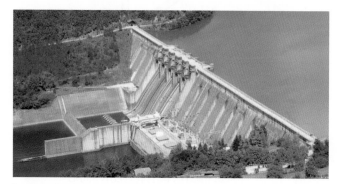

▶ Dam holding back a body of water

Fluid system parameters

The following are definitions of and formulae for the main parameters used to describe fluid systems.

Pressure (p) – a measure of force (F) distributed over an area (A). Unit: newton per square metre ($N\,m^{-2}$) or pascal (Pa), where $1\,N\,m^{-2} = 1$ Pa.

$$p = \frac{F}{A}$$

Gauge pressure (p_{gauge}) – the pressure in a system relative to the ambient (or atmospheric) pressure. A pressure gauge is an instrument that measures the difference between the pressure of a contained fluid and its surroundings. If a pressure gauge on the surface of the Earth reads zero, it does not mean that there is a total vacuum or absence of pressure, only that the pressure being measured is the same as that of its surroundings (in this case atmospheric pressure). Unit: pascal (Pa).

Standard atmospheric pressure (p_{atm}) – the pressure on the surface of the Earth as a result of the weight of the atmosphere above our heads. Unit: pascal (Pa). Standard atmospheric pressure is defined as 101.325 kPa.

Absolute pressure (p_{abs}) – the pressure relative to a perfect vacuum. It is equal to gauge pressure plus standard atmospheric pressure. The equations describing

thermodynamic processes use absolute pressure, which takes atmospheric pressure into account. Unit: pascal (Pa).

$$p_{abs} = p_{gauge} + 101.325 \times 10^3\,Pa$$

Mass (m) – a measure of the amount of matter contained within a body. Unit: kilogram (kg).

Volume (V) – a measure of the amount of space occupied by a body. Unit: cubic metre (m^3).

Density (ρ) – describes the compactness of a material by measuring the amount of matter or mass (m) that is contained within a unit volume (V). Unit: kilogram per cubic metre ($kg\,m^{-3}$).

$$\rho = \frac{m}{V}$$

Weight (F_g) – the force exerted by a gravitational field (g) on a mass (m). Unit: newton (N).

$$F_g = mg$$

On Earth, the gravitational field strength (g) is usually taken as $9.81\,N\,kg^{-1}$.

Hydrostatic pressure

The hydrostatic pressure in an ocean increases with depth due to the weight of the water above. Submarines will be crushed if they descend too deep, and some deep-sea wrecks, such as the Titanic, can only be reached by specialist remotely operated vehicles.

The hydrostatic pressure (p) exerted by a column of fluid is dependent on the height (h) of the fluid column, the density of the fluid (ρ) and the gravitational field strength (g) (see **Figure 1.51**).

$$p = \rho g h$$

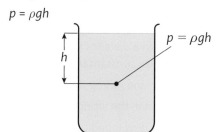

▶ **Figure 1.51** Hydrostatic pressure in a beaker of fluid

Notice that hydrostatic pressure is independent of the cross-sectional area of the fluid column. This means that the pressure felt at 10 cm depth in the Atlantic Ocean is the same as the pressure at 10 cm depth in your bath, which is also the same as the pressure at 10 cm depth in a tall glass (assuming they are all filled with sea water).

Submerged surfaces in fluid systems

Dams are used extensively in civil engineering projects to retain or redirect water in lakes or rivers (see **Figure 1.52**). It is vital that engineers build such structures strong enough to withstand the forces that are exerted by the weight of the water they retain.

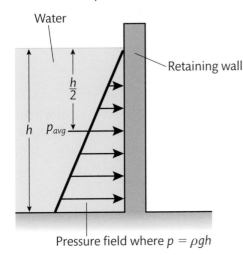

▶ **Figure 1.52** A dam retaining a body of water

The pressure exerted on a submerged rectangular vertical surface, such as a dam, is not uniform over its entire submerged height (h). (Note that h is always measured downwards from the free surface of the fluid.)

Average pressure on a rectangular vertical surface

You know from $p = \rho gh$ that hydrostatic pressure is proportional to submerged height, increasing as h increases. Therefore:

▶ minimum pressure will be at the free surface of the fluid, where $h = 0$ and so $p_{min} = 0$

▶ maximum pressure will exist where $h = h$ and so $p_{max} = \rho gh$.

This means that the average pressure (p_{avg}) exerted on the submerged surface will be:

$$p_{avg} = \frac{p_{min} + p_{max}}{2} = \frac{0 + \rho gh}{2} = \frac{\rho gh}{2}$$

An alternative way of determining p_{avg} is to consider the centroid of the submerged surface. This method is applicable to any shape of vertical submerged surface, not just rectangular.

The **centroid** or geometric centre of a shape is the average position of all the points in the shape. It is important because it determines the height at which the average pressure will act on a submerged plane surface. For a rectangular surface, finding the position of the centroid is straightforward (see **Figure 1.53**). For other geometric shapes it can be more complex.

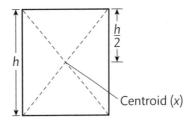

▶ **Figure 1.53** The centroid of a rectangle is at the intersection of the diagonals

The average pressure on a submerged vertical surface will act at its centroid. For a simple rectangle the height of the centroid (x) is $\frac{h}{2}$, so:

$$p_{avg} = \rho gx = \frac{\rho gh}{2}$$

Hydrostatic thrust

If we know the average pressure (p_{avg}), then we can calculate the **average force** or **thrust** (F_T) acting on a submerged plane with surface area (A):

$$F_T = p_{avg} \times A$$

Therefore, the **hydrostatic thrust** (F_T) exerted by a fluid with density (ρ) acting on a submerged plane surface of area (A) whose centroid is at a distance (x) from the free surface of the fluid is given by:

$$F_T = \rho gAx$$

Centre of pressure

Hydrostatic thrust can be treated as a single-point force acting at the **centre of pressure** of the submerged plane.

The centre of pressure for a rectangular submerged plane surface is determined by considering the centroid of the triangular pressure field shown in **Figure 1.54**. For a rectangular plane surface immersed vertically in a fluid, the distance (h_p) of the centre of pressure from the free surface is:

$$h_p = \tfrac{2}{3}h$$

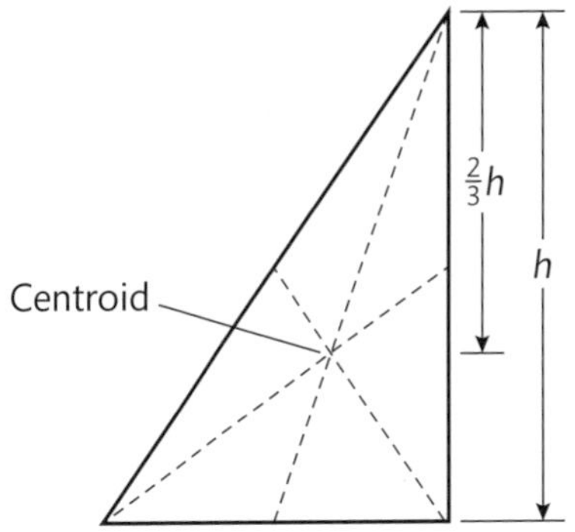

▶ **Figure 1.54** Centroid of triangular pressure field

Explain the difference between the height at which the average pressure on a submerged surface acts and the centre of pressure where the hydrostatic thrust can be considered to act.

Hint   Both values are determined by finding centroids.

Extend   Taking the bottom of a dam as the centre of rotation, derive a formula to calculate the overturning moment acting on the dam as a consequence of the hydrostatic thrust exerted upon it.

Worked Example

A storage tank 2 m wide has a vertical partition across its width. Side 1 of the partition is filled with oil with a density of 900 kg m⁻³ to a depth of 1.8 m, and side 2 is filled with oil with a density of 750 kg m⁻³ to a depth of 0.9 m. Find the resultant hydrostatic thrust on the partition.

Solution
Draw a sketch like the one in **Figure 1.55**.

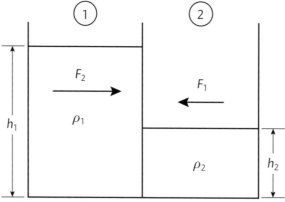

▶ **Figure 1.55** Sketch of the partitioned storage tank

Identify the parameters given in the question:

$\rho_1 = 900$ kg m⁻³, $h_1 = 1.8$ m, $A_1 = 1.8 \times 2 = 3.6$ m², $\rho_2 = 750$ kg m⁻³, $h_2 = 0.9$ m, $A_2 = 0.9 \times 2 = 1.8$ m²

For a rectangular plane the average pressure will occur at a distance of $\frac{h}{2}$ from the free surface:

$$x = \frac{h}{2}$$

Hydrostatic thrust on the partition in side 1:

$$F_1 = \rho_1 g A_1 x_1 = 900 \times 9.81 \times 3.6 \times 0.9 = 28\,606\,\text{N}$$

Hydrostatic thrust on the partition in side 2:

$$F_2 = \rho_2 g A_2 x_2 = 750 \times 9.81 \times 1.8 \times 0.45 = 5960\,\text{N}$$

Let F_R be the resultant thrust acting from left to right. Then

$$F_R = F_1 - F_2 = 28\,606 - 5960 = 22\,646\,\text{N}$$

Immersed bodies

Archimedes' principle

Archimedes' principle states that 'a body totally or partially submerged in a fluid displaces a volume of fluid that weighs the same as the apparent loss in weight of the body'.

Submerged bodies

To help visualise Archimedes' principle, consider the suspended body submerged in a fluid shown in **Figure 1.56**. Given that the suspended body is in static equilibrium, there are three balanced forces acting upon it: the weight of the body (F_g) is balanced by the tension (F_t) in the wire supporting it and the up-thrust (F_{up}) according to Archimedes' principle.

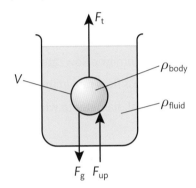

▶ **Figure 1.56** The forces acting on a submerged body in static equilibrium

From the balance of forces we know that

$$F_t = F_g - F_{up}$$

where the weight (F_g) is given by

$$F_g = \rho_{body} V_{body} g$$

and Archimedes' principle tells us that

$$F_{up} = \rho_{fluid} V_{fluid} g$$

Determination of density using floatation methods

If a body is floating on the surface of a fluid (as in **Figure 1.57**), then there is no need to provide support to maintain static equilibrium, and so $F_t = 0$.

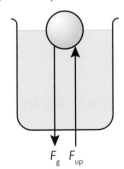

▶ **Figure 1.57** A partially submerged floating body in static equilibrium

In this case the balance of forces gives

$$F_g = F_{up}$$

and hence

$$\rho_{body} V_{body} = \rho_{fluid} V_{fluid}$$

Remember that V_{fluid} is the volume of fluid displaced by the body and so is equal to only the volume of that part of the body which is submerged.

Worked Example

A block of wood 1.8 m long, 300 mm wide and 200 mm thick floats in seawater. The seawater has a density of 1020 kg m⁻³ and 150 mm of the block is below the free surface. Calculate the density and mass of the wood.

Solution

Draw a sketch like the one in **Figure 1.58**.

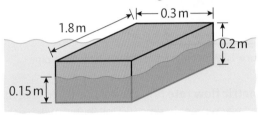

▶ **Figure 1.58** Block of wood partially submerged in seawater

Identify the parameters given in the question:

$$V_{body} = 1.8 \times 0.3 \times 0.2 = 0.108 \, m^3$$

$$\rho_{fluid} = 1020 \, kg \, m^{-3}$$

$$V_{fluid} = 1.8 \times 0.3 \times 0.15 = 0.081 \, m^3$$

For a partially submerged body in static equilibrium:

$$\rho_{body} V_{body} = \rho_{fluid} V_{fluid}$$

$$\text{So } \rho_{body} = \frac{\rho_{fluid} V_{fluid}}{V_{body}} = \frac{1020 \times 0.081}{0.108}$$

$$= 765 \, kg \, m^{-3}$$

We also know that $\rho = \dfrac{m}{V}$, so

$$m_{body} = \rho_{body} V_{body} = 765 \times 0.108 = 82.6 \, kg$$

Relative density

The **relative density** (d) of a substance is defined as the density of the substance compared with the density of pure water. It is given by

$$d_{substance} = \frac{\rho_{substance}}{\rho_{water}}$$

Steel is much denser that water. Explain how it is possible that a steel ship can float.

Hint   Consider the water displaced by a vessel.

Extend   A rowing boat carries a large rock to the centre of a lake. The rower then pushes the rock overboard. It sinks. At this point does the water level in the lake rise, fall or stay the same?

Fluid flow in a gradually tapering pipe

Figure 1.59 shows the flow of an incompressible fluid through a gradually tapering pipe. Analysis of the flow characteristics is simplified considerably if we assume that the density of the fluid, and so its volumetric and mass flow rates (defined below), all remain constant throughout.

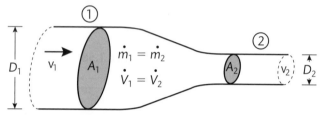

▶ **Figure 1.59** Incompressible flow in a tapering pipe

Volumetric flow rate ($\dot{V}$) – the volume (V) of fluid that passes a given point in time (t). Unit: cubic metre per second ($m^3 s^{-1}$).

$$\dot{V} = \frac{V}{t}$$

For flow through a pipe, $\dot{V}$ can be expressed in terms of the flow velocity (v) and the cross-sectional area of the pipe (A):

$$\dot{V} = Av$$

Mass flow rate ($\dot{m}$) – the mass (m) of fluid that passes a given point in time (t). Unit: kilogram per second ($kg s^{-1}$).

$$\dot{m} = \frac{m}{t}$$

Using the density of the fluid (ρ), the mass flow rate ($\dot{m}$) and volumetric flow rate ($\dot{V}$) are related by

$$\dot{m} = \rho \dot{V}$$

which can also be stated as

$$\dot{m} = \rho Av$$

Equations describing the continuity of flow

In a gradually tapering pipe as in **Figure 1.59**, the volumetric and mass flow rates are the same at points 1 and 2. This means that

$$\dot{V} = A_1 v_1 = A_2 v_2$$

and

$$\dot{m} = \rho A_1 v_1 = \rho A_2 v_2$$

Worked Example

On board a ship, seawater with a density of $1025 \, kg\,m^{-3}$ flows with a velocity of $6 \, m\,s^{-1}$ through a section of horizontal pipe with diameter 150 mm. The water is discharged over the side back into the sea at a velocity of $14 \, m\,s^{-1}$ through a gradually tapering nozzle fitted to the pipe.

a) Calculate the volumetric flow rate in the pipe.

b) Calculate the mass flow rate in the pipe.

c) Calculate the final diameter to which the nozzle must be tapered.

Solution

Identify the parameters given in the question (referring to **Figure 1.59**):

$$\rho = 1025 \, kg\,m^{-3}, v_1 = 6 \, m\,s^{-1}, v_2 = 14 \, m\,s^{-1}$$

$$A_1 = \frac{\pi D_1^2}{4} = \frac{\pi \times 0.15^2}{4} = 0.017\,67 \, m^2$$

a) $\dot{V} = A_1 v_1 = 0.017\,67 \times 6 = 0.106 \, m^3 s^{-1}$

b) $\dot{m} = \rho \dot{V} = 1025 \times 0.106 = 108.68 \, kg\,s^{-1}$ (using unrounded values in the calculation)

c) $\dot{V} = A_1 v_1 = A_2 v_2$

so $A_2 = \dfrac{\dot{V}}{v_2} = \dfrac{0.106}{14} = 0.007\,57 \, m^2$

$$A_2 = \frac{\pi D_2^2}{4}$$

so $D_2 = \sqrt{\dfrac{4 A_2}{\pi}} = \sqrt{\dfrac{4 \times 0.007\,57}{\pi}} = 0.098 \, m$ or 98 mm

D2 Thermodynamic systems

The transfer of heat can significantly affect the operational characteristics of some engineered components. In this section, you will see how the effects can be assessed and how to calculate the amount of heat energy required to complete certain processes.

Heat transfer parameters in thermodynamic systems

Temperature (*T*) – a measure of the kinetic energy of atomic or molecular vibrations within a body. Generally, temperature is measured in degrees Celsius (°C), where 0°C is the freezing point and 100°C is the boiling point of water.

Thermodynamic temperature (*T*) is used in thermodynamic calculations and is measured on the absolute, or kelvin (K), temperature scale. Thermodynamic temperature has the same unit size as the Celsius scale, but the zero point for the kelvin scale is set at absolute zero. Absolute zero is a theoretical minimum possible temperature, at which the kinetic energy of molecular vibrations within a body would be zero. 0°C corresponds to 273 K, 20°C corresponds to 293 K and –20°C corresponds to 253 K.

Pressure (*p*) – a measure of force (*F*) distributed over an area (*A*). Unit: newton per square metre (N m^{-2}) or pascal (Pa), where 1 N m^{-2} = 1 Pa.

Mass (*m*) – a measure of the amount of matter contained within a body. Unit: kilogram (kg).

Thermal conductivity (λ) – a material property that describes a material's ability to conduct heat. Unit: watts per metre per kelvin (W m^{-1} K^{-1}).

Heat transfer processes

Heat transfer within or between bodies can occur by three distinct processes: conduction, convection and radiation.

Conduction

Conduction in solids and liquids involves the transmission of heat energy from one atom to another through physical contact (see **Figure 1.60**). As described previously, heat energy can be thought of as the kinetic energy of atomic vibrations. Atoms with high energy will pass some of this energy on to adjacent low-energy atoms, which establishes a flow of heat energy through the material. A secondary heat transfer process contributes to conduction, where free electrons moving through the material transfer heat energy. This goes some way towards explaining why good electrical conductors tend also to be good thermal conductors.

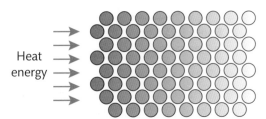

▶ **Figure 1.60** Heat transfer by conduction occurs when heat energy is passed from one atom to another through physical contact.

Conduction is more difficult in gases because the molecules or atoms are not in permanent contact. Heat energy can only be transferred during collisions between high-energy molecules or atoms and those with low heat energy. This helps to explain why gases are generally poor thermal conductors.

When conduction through a material (see **Figure 1.61**) is in steady state (with all variables remaining constant), the heat transfer rate (*Q*) can be expressed in terms of the surface area (*A*), temperature gradient (*T*$_a$ – *T*$_b$), thickness (*x*) and thermal conductivity (λ) of the material:

$$Q = \frac{\lambda A(T_a - T_b)}{x}$$

This formula shows that the rate of heat transfer by conduction (*Q*) is proportional to the surface area of the material (*A*). Often, the surface finish of a component designed for rapid heat transfer is textured so as to maximise its surface area.

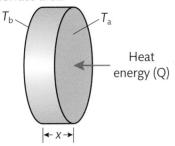

▶ **Figure 1.61** Heat transfer by conduction through a material

Convection

Heat transfer by convection can occur only in liquids or gases where the molecules are free to move (see **Figure 1.62**). Any local heating in one part of such a material (caused by conduction or radiation) will cause localised expansion and a reduction in density. Fluid with lower density than its surroundings tends to rise, carrying its heat energy with it. Low-temperature fluid then flows in to replace the fluid that has risen and it, in turn, is warmed, expands and rises away. A convection current is established in this way.

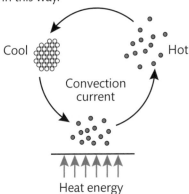

▶ **Figure 1.62** Heat transfer by convection

Forced convection occurs when a fluid is forced to flow over a heat source in order to distribute heat energy to its surroundings. This is the principle employed in a hair dryer – cool ambient air is blown over an electric heating element and emerges at a much higher temperature.

Radiation

Radiation refers to the transfer of heat energy without physical contact. Instead, energy is transmitted in the form of electromagnetic waves (similar to those carrying light or radio signals), which are emitted as a result of energy changes in the orbits of electrons contained within the transmitting material.

This mechanism explains why on a hot sunny day we can feel the warmth of the sun, despite it being millions of miles away, through the vacuum of space.

Linear expansivity

A change in the temperature of a material is associated with a change in its size. This change in size acts in all directions. The amount by which the size changes given the same change in temperature differs from material to material and is defined by the material's **coefficient of linear expansion**.

> **Key term**
>
> **Coefficient of linear expansion** (α) – a material property that describes the amount by which the material expands upon heating with each degree rise in temperature. Unit: inverse kelvin (K^{-1}).

The change in length (ΔL) of a component with coefficient of linear expansion (α) and initial length (L) when subject to a temperature change (ΔT) is given by:

$$\Delta L = \alpha L \Delta T$$

 PAUSE POINT   Explain the three principal mechanisms of heat transfer.

> **Hint**   Use diagrams to illustrate each process.
>
> **Extend**   In terms of heat transfer processes, explain how a vacuum flask keeps hot things hot and cold things cold.

Worked Example

The wheels on a vintage railway carriage have steel tyres that have to be expansion-fitted to cast iron rims. The coefficient of linear thermal expansion for the steel used for the tyres is $13 \times 10^{-6}\ K^{-1}$. If the tyre has an internal diameter of 598.2 mm at 20°C and the diameter of the wheel rim is 600 mm, determine the minimum temperature to which the tyre must be heated for it to be fitted.

Solution

Draw a sketch of one wheel as in **Figure 1.63**.

Identify the parameters given in the question:

$$\alpha = 13 \times 10^{-6}\ K^{-1}$$

$$L = 598.2\,mm = 0.5982\,m \text{ at } 20°C$$

$$\Delta L = 0.6 - 0.5982 = 0.0018\,m$$

$$T_1 = 20°C$$

The important thing to realise here is that the size of the hole will expand in exactly the same way as if it were a solid disk of material.

You know that $\Delta L = \alpha L \Delta T$

$$\text{so } \Delta T = \frac{\Delta L}{\alpha L} = \frac{0.0018}{13 \times 10^{-6} \times 0.5982} = 231.46°C$$

You know that $\Delta T = T_2 - T_1$

$$\text{so } T_2 = \Delta T + T_1 = 231.46 + 20 = 251.46°C$$

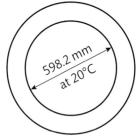

▶ **Figure 1.63** A tyre of the railway carriage

598.2 mm at 20°C

Heat transfer and phase changes

If a solid material is subjected to a continuous input of heat energy, then its temperature will begin to rise until it reaches the melting point of the material (see **Figure 1.64**). The melting point of a solid is the temperature at which a **phase** change from solid to liquid will begin. During the phase change, continued heating will lead to no further increase in temperature until all of the solid has undergone the change into a liquid.

Once the phase change to a liquid is complete, continued heating will once again produce a rise in temperature, but at a different rate, until the boiling point of the liquid is reached. The boiling point of a liquid is the temperature at which a phase change from liquid to gas will begin. Again, during the phase change, continued heating will lead to no further increase in temperature until all the liquid has changed into a gas.

Once the phase change to a gas is complete, continued heating will once again produce a rise in temperature at yet another rate.

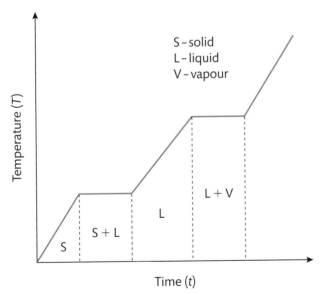

S – solid
L – liquid
V – vapour

▶ **Figure 1.64** Change in temperature observed when heat energy is supplied to a material at a constant rate

Sensible heat

Heat energy that causes a change in the temperature of a material is called **sensible heat**. The amount of sensible heat required to produce a given rise in temperature varies from material to material and even between the solid, liquid and gas phases of the same material.

Specific heat capacity (c) is a material property that describes the amount of heat energy required to raise the temperature of 1 kg of a material by one kelvin. Unit: joules per kilogram per kelvin ($J\,kg^{-1}\,K^{-1}$).

Sensible heat transfer (Q) is the product of the mass of material (m), the specific heat capacity of the material (c) and the change in temperature (ΔT). Unit: joule (J).

$$Q = mc\Delta T$$

Latent heat

Energy that does not cause a temperature rise but instead is absorbed by the material as it undergoes a phase change is called **latent heat**. For a given material, the latent heat of fusion (phase change from solid to liquid) will be different from the latent heat of vaporisation (phase change from liquid to gas).

Latent heat of fusion (L_f) is a material property that describes the amount of heat energy required for 1 kg of the material to undergo a change of phase from solid to liquid.

Latent heat of vaporisation (L_v) is a material property that describes the amount of heat energy required for 1 kg of the material to undergo a change of phase from liquid to gas.

Latent heat transfer (Q) is the product of the mass of material (m) and its latent heat (L). Unit: joule (J).

$$Q = mL$$

Key terms

Phase – the physically distinctive form of a substance: solid, liquid or vapour.

Sensible heat – heat energy that causes a change in the temperature of a substance.

Latent heat – heat energy causing a change of state of a substance without a change in temperature.

Worked Example

Calculate the energy required to convert 1 kg of ice at –5°C into superheated dry steam at 140°C.

For water, the latent heat of fusion is 334 kJ kg^{-1}, the latent heat of vaporisation is 2260 kJ kg^{-1} and the specific heat capacity is 4.2 kJ kg^{-1} K^{-1}.

The specific heat capacity of dry steam is 1.8 kJ kg^{-1} K^{-1} and the specific heat capacity of ice is 2 kJ kg^{-1} K^{-1}.

Solution

Draw a sketch, like the graph in **Figure 1.65**, to show how the temperature of water changes with increasing heat energy.

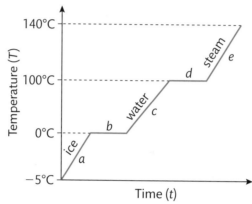

▶ **Figure 1.65** Change in temperature when heat energy is supplied to water

Identify the parameters given in the question:

m = 1 kg, c_{ice} = 2 kJ kg^{-1} K^{-1}, c_{water} = 4.2 kJ kg^{-1} K^{-1}, c_{steam} = 1.8 kJ kg^{-1} K^{-1}, L_f = 334 kJ kg^{-1}, L_v = 2260 kJ kg^{-1}

Considering each of the parts marked a to e on **Figure 1.65**:

$Q_a = mc_{ice}\Delta T = 1 \times 2 \times 5 = 10$ kJ

$Q_b = mL_f = 1 \times 334 = 334$ kJ

$Q_c = mc_{water}\Delta T = 1 \times 4.2 \times 100 = 420$ kJ

$Q_d = mL_v = 1 \times 2260 = 2260$ kJ

$Q_e = mc_{steam}\Delta T = 1 \times 1.8 \times 40 = 72$ kJ

So the total energy required is

Q_{total} = 10 + 334 + 420 + 2260 + 72 = 3096 kJ

Thermal efficiency of heat engines and heat pumps

Heat engines

Heat engines include internal combustion engines, steam engines and any other machine or process that changes heat energy into mechanical work.

In thermodynamic terms, a heat engine takes heat energy supplied at a high temperature and converts some of this energy into useful mechanical work while rejecting the remainder at some lower temperature:

heat received at high temperature = work done + heat rejected to some lower temperature

The **thermal efficiency** of a heat engine is the ratio of useful work done to the equivalent heat energy contained in the amount of fuel used:

$$\text{thermal efficiency} = \frac{\text{mechanical power output (W)}}{\text{equivalent heat energy of fuel (J kg}^{-1}) \times \text{ fuel consumption rate (kg s}^{-1})}$$

Worked Example

An engine under test developed an indicated power of 6.8 kW. Over the 30-minute test the engine used 780 g of fuel with an equivalent heat energy of 41.9 MJ kg^{-1}.

Calculate the thermal efficiency of the engine.

Solution

Identify the parameters given in the question:

mechanical power = 6.8 kW = 6.8×10^3 W

equivalent heat energy of fuel = 41.9 MJ kg^{-1}
$$= 41.9 \times 10^6 \text{ J kg}^{-1}$$

$$\text{rate of fuel consumption} = \frac{780 \times 10^{-3} \text{ kg}}{30 \times 60 \text{ s}}$$
$$= 4.33 \times 10^{-4} \text{ kg s}^{-1}$$

$$\text{So thermal efficiency} = \frac{6.8 \times 10^3}{41.9 \times 10^6 \times 4.33 \times 10^{-4}}$$
$$= 0.375$$

This means that only 37.5% of the potential heat energy in the fuel was converted into useful work.

Heat pumps

Heat pumps include systems such as refrigerators, air conditioning equipment and ground or air source heat pump central heating systems.

In thermodynamic terms, a heat pump performs in the opposite way to a heat engine because it takes in external energy to do work, extracts heat from a low temperature and delivers heat at a higher temperature:

external energy supplied
+ heat extracted from a lower temperature
= heat delivered at a higher temperature

In a heat pump used for refrigeration, the heat extracted from the low-temperature interior of a refrigerator is the useful output of the system. The performance of a refrigerator is not given in terms of an 'efficiency' but as a performance ratio (which can be greater than 1):

refrigeration performance ratio

$$= \frac{\text{heat extracted}}{\text{external energy supplied}}$$

In a heat pump that is used for heating, the heat delivered at a higher temperature is the useful output of the system. The performance of a heat pump is also expressed as a performance ratio:

heat pump performance ratio

$$= \frac{\text{heat delivered}}{\text{external energy supplied}}$$

Worked Example

A domestic ground source heat pump consumes 2 kW of electricity and can deliver 18 MJ of heat energy per hour to a house central heating system.

a) Calculate the heat extracted from the ground per second.

b) Determine the performance ratio of the heat pump when warming the house.

c) Determine the performance ratio of the heat pump as a refrigerator cooling the ground.

Solution

Identify the parameters given in the question:

external energy supplied = 2 kW = 2000 W
$$= 2000 \text{ J s}^{-1}$$

heat delivered at higher temperature = 18 MJ h^{-1}
$$= \frac{18 \times 10^6}{60 \times 60} \text{ J s}^{-1}$$
$$= 5000 \text{ J s}^{-1}$$

So

a) heat extracted from a lower temperature (the ground) per second = 5000 − 2000 = 3000 J

b) heat pump performance ratio = $\frac{5000}{2000}$ = 2.5

c) refrigeration performance ratio = $\frac{3000}{2000}$ = 1.5.

Enthalpy and the first law of thermodynamics

Enthalpy (H) can be used to describe the total energy contained within a closed thermodynamic system. It includes any gravitational potential or kinetic energy.

The idea of enthalpy is reflected in the first law of thermodynamics, which states that 'the net energy supplied by heat to a body is equal to the increase in its internal energy and the energy output due to work done by the body'.

In other words, heat energy added into the system will either increase the internal energy (U) or be accounted for in the work done in expansion quantified by the product of pressure (p) and volume (V):

$$H = U + pV$$

Entropy and the second law of thermodynamics

Entropy (S) can be thought of as a measure of the dispersal of energy present in a thermodynamic system.

The second law of thermodynamics can be expressed in many ways, but it was first put forward in the form 'it is impossible for a self-acting machine unaided by an external agency to convey heat from one body to another at a higher temperature'. You have seen this when you looked at heat pumps. Heat flow would be spontaneous from a high temperature (with low entropy) to a low temperature (with higher entropy). However, in a heat pump, work must be done to reverse this natural process and move heat from a low temperature to a high temperature.

The second law can be expressed in terms of entropy as 'the entropy of any thermodynamic system and its surroundings will always tend to increase'.

This is the fundamental reason why heat flows from hot to cold temperatures and fluids flow from an area of high pressure to an area of low pressure.

The gas laws

Boyle's law states that provided the temperature (T) of a perfect gas remains constant, the volume (V) of a given mass of the gas is inversely proportional to the pressure (p) of the gas. This can be stated as

$$pV = \text{constant}$$

Charles's law states that provided the pressure (p) of a given mass of a perfect gas remains constant, the volume (V) of the gas will be directly proportional to the absolute temperature (T) of the gas. This can be stated as

$$\frac{V}{T} = \text{constant}$$

The **general gas equation** comes from combining Boyle's and Charles's laws and takes the form

$$\frac{pV}{T} = \text{constant}$$

Worked Example

Formula 1 tyres are designed to perform optimally at the full race temperature of 100°C when the gauge pressure in the tyre is 158 kPa. Tyre warmers are used before the race to pre-heat the tyres to 80°C. What pressure needs to be in the tyres when they are in the warmers to ensure optimal performance during the race?

Solution

During the race:

$$T_1 = 100°C = 373\,K$$

$p_1 = 158\,kPa = 158 \times 10^3\,Pa$, which corresponds to an absolute pressure of $259 \times 10^3\,Pa$.

In the tyre warmers:

$$T_2 = 80°C = 353\,K$$

$$p_2 = ?$$

These quantities are related by the general gas equation:

$$\frac{p_1 V_1}{T_1} = \frac{p_2 V_2}{T_2}$$

The tyre itself contains the gas in a constant volume, so $V_1 = V_2$. Eliminating these from the general gas equation gives

$$\frac{p_1}{T_1} = \frac{p_2}{T_2}$$

Make p_2 the subject of the equation: $p_2 = \frac{p_1 T_2}{T_1}$

Substitute in the known quantities:

$$p_2 = \frac{259 \times 10^3 \times 353}{373} = 245\,112\,Pa$$

$$= 245\,kPa$$

The pressure that should be set in the tyres when they are in the warmers is 245 kPa (absolute) or 144 kPa (gauge).

Assessment practice 1.4

1 A dam holding back water in a reservoir consists of a vertical retaining wall 10 m wide and 6 m high. The water has a density of 1000 kg m⁻³ and the water level is 1 m from the top of the wall. When designing and building dams it is essential that engineers understand the forces that will be acting on the dam so it can be made to withstand them.

a) Calculate the resultant force exerted by the water on the dam.

b) Calculate the overturning moment acting around the base of the dam that the structure must be able to withstand.

(8 marks)

2 20 ml of air at atmospheric pressure of 101 kPa and temperature 20°C is contained in a fire piston with a piece of char cloth. The piston is struck and the air is rapidly compressed to 1 ml, causing a temperature increase to 420°C, which is sufficient to ignite the char cloth. Calculate the final pressure in the cylinder.

(2 marks)

3 An aluminium bar with diameter 10 mm measures 1.2 m in length at 120°C. The bar is rigidly clamped at both ends before being allowed to cool to 20°C.

The aluminium has a coefficient of linear expansion of 22.2×10^{-6} °C⁻¹, a tensile strength of 110 MPa and a Young's modulus of 69 GPa.

a) Calculate the force induced in the aluminium bar during cooling.

b) Explain whether or not the bar will break during cooling.

(5 marks)

4 A solid bronze cannon with mass 560 kg is being recovered from a shipwreck and is suspended by a lifting cable. The density of bronze is 8900 kg m⁻³ and the density of the seawater is 1025 kg m⁻³.

a) Calculate the tension in the lifting cable when it is suspending the fully submerged cannon.

b) Calculate the tension in the cable as the cannon leaves the water when it is held with a third of its volume still submerged.

(5 marks)

5 In a water treatment plant, fresh water with density 1000 kg m⁻³ flows at 3 m s⁻¹ through a horizontal section of pipe 60 mm in diameter. The section of pipe gradually tapers to a diameter of 90 mm.

a) Calculate the volumetric flow rate in the pipe.

b) Calculate the mass flow rate in the pipe.

c) Calculate the flow velocity in the section of pipe where the diameter is 90 mm.

(6 marks)

For each of the problems in this assessment practice, use the following stages to guide your progress through the task.

Plan
- How will I approach the task? Which principles and formulae are applicable?
- Do I need clarification around anything? Have I read and fully understood the question?

Do
- I understand my thought process and can explain why I have decided to approach the task in a particular way.
- I can identify where I have gone wrong and adjust my thinking to get myself back on course.

Review
- I can identify which elements I found most difficult and where I need to review my understanding of a topic.
- I can explain the importance of this area of learning in a wider engineering context.

E Static and direct current electricity and circuits

E1 Static and direct current electricity

Static electricity

Static electricity is the build-up of electrical charge on an object. The word 'static' implies that there is no flow of electricity once the object is charged – that is, until the charge finds a suitable pathway through which to move or discharge. When you observe a lightning bolt, you have witnessed the discharge of static electricity from a storm cloud down to the ground or to another cloud. Static electricity is not always so spectacular, though – it is also generated when you rub a balloon on your hair or clothing, allowing you to stick the balloon to a ceiling.

Electrostatic forces are responsible for holding together the atoms and molecules that form the basis of everything that exists, including you.

Matter is made up of atoms. According to the Bohr atomic model, the core or nucleus of an atom contains neutrons and positively charged protons tightly bound together. The positive charge on the protons holds negatively charged electrons in orbit around the nucleus. Usually the positive charge possessed by the protons is balanced by the negative charge of the electrons held in orbit. However, electrons are only weakly held in orbit and have a tendency to move or migrate to adjacent atoms. An object becomes positively charged when electrons move away from it and too few remain to balance the positively charged protons. An object becomes negatively charged when it has an excess of electrons orbiting it, more than are required to balance the positively charged protons.

Electrical charge (q) is measured in coulombs (C). A single electron has an electrostatic charge of 1.602×10^{-19} C.

Electric fields

Wherever a charged particle or object is present, it generates an electric field acting in a particular direction. This can be represented using field lines (see **Figure 1.66**).

Wherever an electric field exists, any object with an electrical charge will experience a force acting upon it.

By convention, the arrows on the field lines indicate the direction of the force that would act on a **positive** test charge. The following are the important parameters used to describe electric fields.

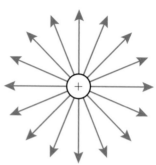

▶ **Figure 1.66** Radial electric field lines from an isolated point charge

Electric field strength (E) – the strength at any point within an electric field, defined by the size of the force (F) that is exerted per unit of charge (q). Unit: newton per coulomb ($N\,C^{-1}$).

$$E = \frac{F}{q}$$

Electric flux (ψ) – measures the amount of flow of an electric field (as represented by electric field lines). The unit of electric flux is defined to be the amount of electric field emanating from a source with a positive charge of 1 C. Unit: coulomb (C), where a charge of 1 C gives rise to an electric flux of 1 C.

Electric flux density (D) – the amount of electric flux (ψ) passing through an area (A) that is perpendicular to the direction of the flux (as represented by the number of electric field lines per unit area). Unit: coulombs per square metre ($C\,m^{-2}$).

$$D = \frac{\psi}{A} = \frac{q}{A}$$

Coulomb's law

Where two objects are electrostatically charged, the electric fields they generate cause an electrostatic force to exist between them. The direction in which this force acts depends on the types of charges involved: like charges repel each other, while opposite charges attract. The magnitude of the force depends on the distance between the objects.

Coulomb's law is used to determine the magnitude of the electrostatic force that exists between two charged particles. It is expressed by the equation

$$F = \frac{q_1 q_2}{4\pi \varepsilon_0 r^2}$$

where F is the force between two particles carrying electrostatic charges q_1 and q_2 that are held in free space with permittivity ε_0 at a distance r apart.

From the formula you can see that force and distance are related by an inverse square law, which means that the force will rapidly increase as the particles move closer together and rapidly decrease as they move apart.

Permittivity

Permittivity is a measure of the resistance of a material to the formation of an electric field. The force exerted between charged particles depends on the medium in which they are contained. The permittivity of free space (see below) is used in the Coulomb's law equation because this approximates to having the particles contained in air.

Absolute permittivity (ε) measures the ability of a material to resist the formation of an electric field within itself. It can also be thought of as a measure of how much electric flux is produced in a material by an electric field. Unit: farads per metre ($F\,m^{-1}$).

Permittivity of free space (ε_0) is the permittivity of a vacuum. For an electric field in a vacuum there is no medium that could affect the formation of electric flux. The permittivity of free space is a constant, $\varepsilon_0 = 8.85 \times 10^{-12}\,F\,m^{-1}$, and is the ratio of electric flux density (D) to electric field strength (E).

$$\varepsilon_0 = \frac{D}{E}$$

Relative permittivity (ε_r) is used to define the permittivity of other media in comparison with that of a vacuum. ε_r is the ratio of a material's absolute permittivity (ε) to the permittivity of free space (ε_0).

$$\varepsilon_r = \frac{\varepsilon}{\varepsilon_0}$$

Uniform electric fields

Coulomb's law deals with the interaction of two point charges that have radial, and so diverging, field lines. In a uniform electric field, such as that produced between two parallel charged plates (see **Figure 1.67**), the field does not vary from place to place and the field lines are parallel and equally spaced.

Capacitance

In a uniform electric field between two electrically charged plates, as in **Figure 1.67**, the field strength (E) can be defined by the potential difference (V) or voltage between the plates and the distance (d) between them:

$$E = \frac{V}{d}$$

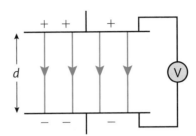

▶ **Figure 1.67** A uniform electric field between two plates forming a parallel plate capacitor

Capacitance (C) is a property of a pair of parallel charged plates. It defines the quantity of electrical charge (q) stored on the plates for each volt of potential difference (V) between them. Unit: farad (F).

$$C = \frac{q}{V}$$

The farad is defined as the capacitance when 1 V of potential difference corresponds to 1 C of electrical charge. It is a very large unit, so, in practice, capacitance is usually expressed in μF ($10^{-6}\,F$), nF ($10^{-9}\,F$) or pF ($10^{-12}\,F$).

Parallel plate capacitor

For a parallel plate **capacitor** like the one in **Figure 1.67**, the capacitance (C) is proportional to the area of the plates (A) and the permittivity (ε) of the material occupying the space between the plates (known as the **dielectric**) and is inversely proportional to the distance between the plates (d) (which also corresponds to the thickness of the dielectric):

$$C = \frac{\varepsilon A}{d}$$

 **PAUSE POINT**
How does the permittivity of a dielectric material affect the capacitance of a parallel flat plate capacitor?

Hint
Consider absolute permittivity, relative permittivity and the permittivity of free space in your explanation.

Extend
Explain permittivity and its use in defining the relationship between electric field strength and electric flux density.

Current electricity

Current electricity deals with the flow of electrons around circuits in order to do work or process information.

Conductors and insulators

For an electric current to flow, a material must allow the free flow of loosely bound electrons between its atoms. Such materials are called electrical **conductors**. Materials called **insulators** have electrons that are more tightly bound to the atomic nucleus and so less free to move.

Electrical parameters

Electromotive force (e.m.f.) drives the movement of electrical charge in a closed circuit and causes current to flow from a point of high electrical potential energy to a point with lower electrical potential energy.

Potential difference (p.d.) is the difference in electrical energy that exists between two points in a circuit. It is usually referred to as voltage and is measured in volts (V).

Voltage (V) is the difference in electrical potential, or the potential difference, between two points in a circuit. Unit: volt (V).

Electrical current (I) is the rate of flow of electrical charge (q) in time (t). Unit: ampere (A), often abbreviated as 'amp'.

$$I = \frac{q}{t}$$

Current cannot flow in an open circuit because the electrons are unable to circulate. Once a circuit is completed, a conducting pathway has been provided and, by convention, current is said to flow from points with high electrical

potential towards points of low potential (that is, from a high voltage to a low voltage). This is shown in the circuit diagram of **Figure 1.68**.

Resistance (R) describes the degree to which a circuit or component resists or opposes the free flow of an electric current. Unit: ohm (Ω), where $1\,\Omega$ is defined as the resistance required so that a voltage of 1 V will result in a current flow of 1 A.

The reciprocal of resistance is **conductance** (G). Unit: siemens (S).

Resistivity (ρ) is a material property that describes the degree to which the material resists or opposes the free flow of electrons. Unit: ohm metre ($\Omega\,$m).

Resistance and resistivity should not be confused. The resistance of a given sample of material is dependent on the size and shape of the sample. Resistivity is a property of the material itself and so is the same for samples of that material of any amount, size or shape. A helpful way to view the relationship between resistivity and resistance is to make a comparison with how density relates to mass.

▶ The resistance (R) of a wire with constant cross-sectional area (A) and length (l) made from a material with resistivity (ρ) is given by:

$$R = \frac{\rho l}{A}$$

The reciprocal of resistivity is **conductivity** (σ). Unit: siemens per metre ($S\,m^{-1}$).

Resistivity (and conductivity) of a material varies with temperature. In general, an increase in temperature will cause an increase in the resistivity of most conductors and a decrease in the resistivity of most insulators.

The **temperature coefficient of resistance** (α) for a material is defined as the increase in the resistance of a $1\,\Omega$ resistor made from that material when its temperature is increased by 1°C.

▶ The change in resistance (ΔR) in a component with standard resistance (R_0) at 0°C and temperature coefficient of resistance (α) caused by a change in temperature (ΔT) is given by

$$\Delta R = \alpha \Delta T R_0$$

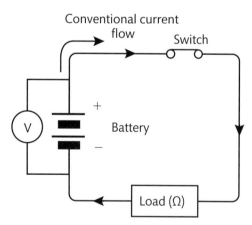

▶ **Figure 1.68** Diagram illustrating measurement of circuit potential difference and conventional current flow

---

❚❚ PAUSE POINT   Explain the difference between resistance and resistivity.

    Hint   Consider the resistance and resistivity of a length of copper wire and how these two characteristics are measured.

    Extend   Explain the effects that temperature has on resistivity.

Resistors

Resistors are components used to limit the flow of current through a circuit. They are available in a wide range of values, from a fraction of an ohm to many mega ohms ($10^6\,\Omega$). There are several types of resistors, including wire wound, metal film and carbon resistors. The choice of which resistor to use in a circuit depends on usage and the power handling requirements – the latter might range from several watts for a wire wound resistor to a few milliwatts (mW) for a tiny surface-mount device used in a mobile phone.

▶ **Fixed resistors** are components designed to have a single specific value and are used extensively in electronic circuits.

▶ **Variable resistors** are resistors designed so that their resistance can be changed (such as a rotary volume control on an amplifier).

The circuit symbols for a fixed resistor and a variable resistor are shown in **Figure 1.69**.

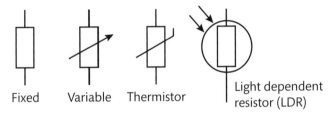

Fixed   Variable   Thermistor   Light dependent resistor (LDR)

▶ **Figure 1.69** Resistor circuit symbols

Capacitors

Capacitors are components designed to have a specific capacitance and so be able to store electrical charge.

Capacitors are used in a variety of forms in electronic circuits and are made from a number of different dielectric materials. **Table 1.10** lists different types of capacitors along with their capacitance range, working voltage and typical uses.

▶ **Table 1.10** Types of capacitors

The circuit symbols for some capacitors are shown in **Figure 1.70**.

Capacitor   Polarised capacitor   Variable capacitor

▶ **Figure 1.70** Capacitor circuit symbols

E2 Direct current circuit theory

The following are some key parameters used when working with capacitors.

▶ **Working voltage** – the voltage that can safely be applied to a capacitor without the dielectric breaking down (failing).

▶ **Charge stored** in a capacitor (q) – given by $q = CV$ for a capacitor with capacitance (C) connected to a voltage (V).

▶ **Energy stored** in a capacitor (W) – given by $W = \frac{1}{2}CV^2$ for a capacitor of capacitance (C) with an applied voltage (V).

Charging a capacitor

Figure 1.71 shows a simple circuit where a direct current (d.c.) voltage is able to charge a capacitor (C) through a series resistor (R). When the switch is closed, current will flow until the capacitor is fully charged and reaches its steady state. Up to the point when current ceases to flow, the variations in circuit current and voltage across the capacitor and resistor are known as **RC transients**.

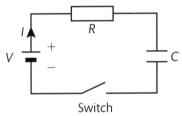

Switch

▶ **Figure 1.71** A capacitor-charging circuit, with a capacitor and a resistor connected in series

Type	Capacitance range	Typical working voltage	Usage
Electrolytic	$1\,\mu F$ to $1\,F$	$3\,V$ to $600\,V$	Large capacitors; power supplies and smoothing
Tantalum	$0.001\,\mu F$ to $1000\,\mu F$	$6\,V$ to $100\,V$	Small capacitors; where space is restricted
Mica	$1\,pF$ to $0.1\,\mu F$	$100\,V$ to $600\,V$	Very stable; high-frequency applications
Ceramic	$10\,pF$ to $1\,\mu F$	$50\,V$ to $1000\,V$	Popular and inexpensive; general usage
Mylar	$0.001\,\mu F$ to $10\,\mu F$	$50\,V$ to $600\,V$	Good performance; general usage
Paper	$500\,pF$ to $50\,\mu F$	$100\,V$	Rarely used now
Polystyrene	$10\,pF$ to $10\,\mu F$	$100\,V$ to $600\,V$	High quality and accuracy; signal filters
Oil	$0.1\,\mu F$ to $20\,\mu F$	$200\,V$ to $10\,kV$	Large, high-voltage filters

If the voltage across the capacitor (V_C) is plotted against time (t), as in **Figure 1.72**, the graph shows that the growth in voltage is exponential, characterised by the curve rising steeply before flattening out as it approaches its maximum value.

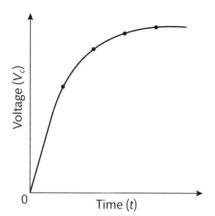

▶ **Figure 1.72** Voltage across a charging capacitor

At the moment when a capacitor starts to charge, the rate of growth of the voltage is high. If the voltage growth were maintained at this initial rate, then the time it would take for the voltage to reach its maximum value is known as the **time constant**.

In fact, more generally, the time constant for an exponential transient can be defined as the time taken for a transient to reach its final value from the time when its rate of change is maintained.

In a series-connected capacitor and resistor circuit, the value of the time constant (τ) is equal to the product of the capacitance (C) and the resistance (R):

$$\tau = RC$$

When a capacitor with capacitance (C) is charged through a resistance (R) towards a final potential (V_0), the equation giving the voltage (V_C) across the capacitor at any time (t) is the **capacitor charge equation**:

$$V_C = V_0(1 - e^{-t/RC})$$

Discharging a capacitor

Once fully charged, if the power supply in the capacitor-charging circuit is replaced by a short circuit, then current will flow out of the capacitor as it discharges. If the voltage (V_C) across the capacitor is plotted against time (t), the graph will show that the reduction or decay in voltage is exponential. Exponential decay is characterised by a curve falling steeply before flattening out as it approaches its minimum value (often zero), as shown in **Figure 1.73**.

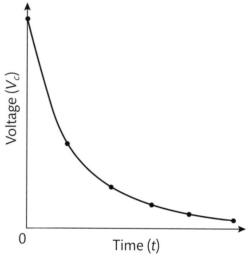

▶ **Figure 1.73** Voltage across a discharging capacitor

When a capacitor with capacitance (C) is discharged through a resistance (R) from an initial potential (V_0), the equation giving the voltage (V_C) across the capacitor at any time (t) is the **capacitor discharge equation**:

$$V_C = V_0 e^{-t/RC}$$

> **Link**
>
> See section in **A1 Algebraic methods** for worked examples of problems involving exponential growth and decay.

Note that when a capacitor is disconnected from a power source, it might retain its charge for a considerable length of time. It is good practice, therefore, to connect a high-value resistor in the circuit across the capacitor terminals. This ensures that the capacitor will be automatically discharged once the supply is switched off.

❚❚ PAUSE POINT   Explain the meaning of the time constant in the exponential equations for the charge and discharge of capacitors.

Hint   What are the two parameters used to calculate the time constant?

Extend   Draw a sketch of an exponential function and explain the relationship between the tangent to the curve at any point and the time constant.

Ohm's law

The relationship between voltage (V) and current (I) at constant temperature obeys **Ohm's law**:

$$\frac{V}{I} = \text{a constant}$$

This constant is called the resistance (R), and the relationship can be expressed as:

$$V = IR$$

Using Ohm's law, electrical power (P) can be expressed in terms of voltage (V), current (I) and resistance (R):

$$P = IV = I^2R = \frac{V^2}{R}$$

Its unit is the watt (W).

The **power efficiency** (E) of a system is the ratio between the power output (P_{out}) and the power input (P_{in}):

$$E = \frac{P_{out}}{P_{in}}$$

Ohm's law can be used to analyse a simple circuit, but in more complex situations involving networks of resistors, Kirchhoff's laws are more useful.

▶ **Kirchhoff's voltage law** – in any closed loop network, the total p.d. across the loop is equal to the sum of the p.d.s around the loop. For example, if three resistors are connected in series, with voltage drops (V_1, V_2 and V_3) across them, then the total supply voltage (V) is given by

$$V = V_1 + V_2 + V_3$$

For n resistors connected in series, using $V = IR$ for each resistor we get:

$$V_1 + V_2 + V_3 + \ldots + V_n = IR_1 + IR_2 + IR_3 + \ldots + IR_n$$

which can also be expressed as:

$$\Sigma \text{p.d.} = \Sigma IR$$

▶ **Kirchhoff's current law** – at any junction of an electric circuit, the total current flowing towards the junction is equal to the total current flowing away from the junction. For example, if three resistors are connected to a common point, with currents (I_1, I_2 and I_3) flowing in towards the junction and current (I) flowing away from the junction, then:

$$I = I_1 + I_2 + I_3$$

Diodes

A diode is a very common electronic component made of a semiconductor material. It allows current to flow in only one direction. The circuit symbols for three types of diodes are shown in **Figure 1.74**. The direction of the arrow indicates the direction in which current can flow through the diode.

Semiconductor diode

Anode ———▶|——— Cathode
(+)   (−)

Zener diode

Anode ———▶|——— Cathode
(+)   (−)

Light-emitting diode

Anode ———▶|——— Cathode
(+)   (−)

▶ **Figure 1.74** Diode symbols

When a voltage is applied to a diode and a current flows in the direction of the arrow, the diode is said to be **forward biased**. With a material such as silicon, a forward bias of approximately 0.7 V is required for a diode to begin to allow current to flow.

Generally, a **reverse bias** of any voltage will result in no current flowing through the diode, with the exception of the Zener diode (see **voltage regulation** on page 52).

Forward bias applications

▶ **Rectification** – one important application of diodes is in the rectification of an alternating current (a.c.) electricity supply into a direct current (d.c.) output. Electronic devices tend to require relatively low d.c. voltages to operate. As the domestic electricity supply in the UK is 230 V a.c., a step-down transformer is used first to reduce the voltage to the required level, usually between 3 V and 15 V depending on the application. The output of the transformer is still a.c., so a rectifier circuit must be used to convert it to d.c. **Figure 1.75** shows examples of simple rectifier circuits. A capacitor is then connected across the output to smooth the varying voltage to approximate a d.c. voltage.

▶ **Component protection** – diodes are also commonly used to protect electronic devices such as integrated circuits (ICs) or transistors from transient voltage spikes induced in relay coils when they are switched off. The current flowing through a relay coil generates a magnetic field that quickly collapses when the current is turned off. The collapsing field induces a back e.m.f. in the coil, which if allowed to build might cause damage to the electronic components. A diode connected across the coil conducts the current away from the circuit components (see **Figure 1.76**).

Half-wave (1 diode)

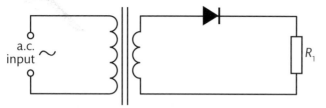

Input

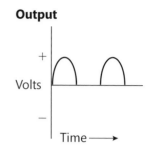

Output

Bi-phase full-wave (2 diodes, tapped transformer)

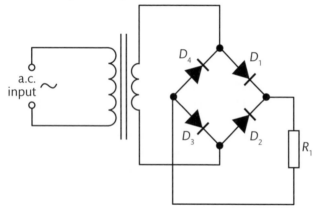

Input

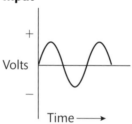

Output

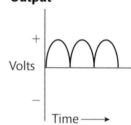

Full-wave (4 diodes)

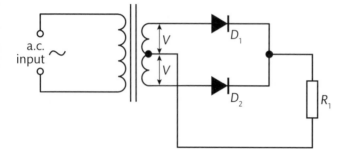

Input

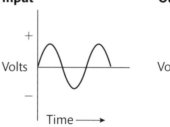

Output

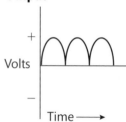

▶ **Figure 1.75** Simple rectifier circuits

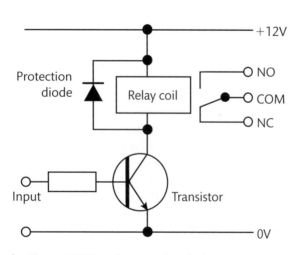

▶ **Figure 1.76** Use of a protection diode

Link

The generation of a back e.m.f. by a collapsing magnetic field is an example of self-inductance, which is dealt with in more detail in the section on 'Self-inductance in a coil' under **F Magnetism and electromagnetic induction**.

Reverse bias applications

▶ **Voltage regulation** – a Zener diode is constructed such that when a reverse bias voltage reaches a defined level, the diode begins to allow the flow of current. The point at which the diode starts to conduct in reverse bias is called the breakdown voltage. This can be set at anywhere from 2.7 V to 150 V. The characteristics of Zener diodes allow them to be used in voltage regulator circuits, which are able to maintain a constant voltage output. A simple voltage regulator circuit using a resistor and a Zener diode is shown in **Figure 1.77**.

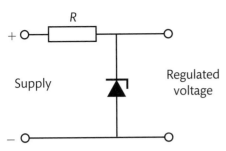

▶ **Figure 1.77** A simple voltage regulator circuit using a Zener diode

Series resistors and diodes

Where a forward bias diode is connected in series with a resistor, it is important to remember that there will be a voltage drop across the diode. For silicon diodes this is typically 0.7 V.

Worked Example

Consider the circuit shown in **Figure 1.78**, where a silicon diode and a resistor are connected in series. Given that the supply voltage is 12 V d.c. and $R_1 = 4.7 \text{ k}\Omega$, find the current flowing in the circuit.

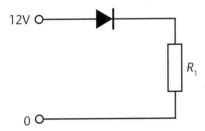

▶ **Figure 1.78** Resistor connected in series with a diode

Solution

The voltage drop across a forward bias silicon diode is 0.7 V. Applying Kirchhoff's voltage law, you can determine the voltage drop across resistor R_1:

$$V_T = V_1 + V_2$$
$$V_1 = 12 - 0.7 = 11.3 \text{ V}$$

You can now determine the current in R_1, which is the current flowing in the whole series circuit, by using Ohm's law:

$$I = \frac{V}{R} = \frac{11.3}{4.7 \times 10^3} = 0.0024 \text{ A or } 2.4 \text{ mA}$$

E3 Direct current networks

It is common for direct current (d.c.) circuits to have a combination of series and parallel elements.

Resistors in series and parallel

Resistors in series

Figure 1.79 shows three resistors connected in series. In this configuration it is important to remember that the same current flows throughout the circuit, and so the current flowing through each resistor will be the same.

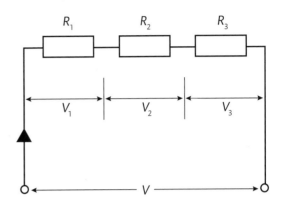

▶ **Figure 1.79** Resistors connected in series

Resistors in parallel

Figure 1.80 shows three resistors connected in parallel. In this configuration, it is important to note that the voltage across each resistor will be the same.

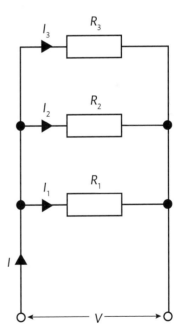

▶ **Figure 1.80** Resistors connected in parallel

By applying Ohm's and Kirchhoff's laws, you can show that the total resistance (R_T) for a number (n) of resistors in parallel is:

$$\frac{1}{R_T} = \frac{1}{R_1} + \frac{1}{R_2} + \dots + \frac{1}{R_n}$$

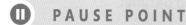

PAUSE POINT Recall the three laws used extensively to analyse d.c. circuits.

> **Hint** Give the name of each law and define it both in words and as a mathematical formula.

> **Extend** Using a combination of these laws, derive the formulae for determining the total resistance for a number of resistors in series.

Series and parallel combinations

If a circuit contains resistors that are connected in a combination of series and parallel configurations, then it is necessary to use a combination of the series and parallel formulae to calculate the various current, voltage and resistor values.

The approach is to split the circuit into smaller elements that can be treated as purely series or purely parallel circuits. Then apply the appropriate formulae and replace each series or parallel element with an equivalent single resistor.

The following worked example demonstrates the principles of simplifying the network to find a single equivalent resistance. The same method can be applied to analyse a network containing five or more resistors.

Worked Example

Consider the circuit shown in **Figure 1.81**, where $R_1 = 47\,\Omega$, $R_2 = 220\,\Omega$, $R_4 = 270\,\Omega$ and the supply voltage is 200 V. The total power dissipated by the circuit is 190 W.

Calculate the value of R_3.

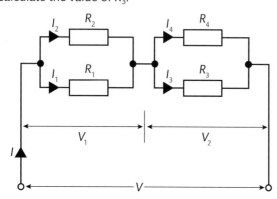

▶ **Figure 1.81** Resistors connected in a series and parallel combination

Solution

Determine the total circuit current using the voltage and power dissipation.

From $P = IV$, $I = \dfrac{P}{V} = \dfrac{190}{200} = 0.95$ A or 950 mA

You can find the total circuit resistance from $V = IR$:

$$R_T = \frac{V}{I} = \frac{200}{0.95} = 210.5\,\Omega$$

Now find the equivalent single resistor R_A that could replace the parallel network involving R_1 and R_2.

$$\frac{1}{R_A} = \frac{1}{R_1} + \frac{1}{R_2} = \frac{1}{47} + \frac{1}{220} = 0.0258$$

$$\text{so } R_A = \frac{1}{0.0258} = 38.8\,\Omega$$

You now know the total resistance R_{TOTAL} and the equivalent resistance of the R_1 and R_2 parallel network (R_A).

The R_3 and R_4 parallel network (R_B) is connected in series with R_A, so $R_T = R_A + R_B$ and therefore $R_B = R_T - R_A = 210.5 - 38.8 = 171.7\,\Omega$.

You know that

$$\frac{1}{R_B} = \frac{1}{R_3} + \frac{1}{R_4}$$

$$\text{so } \frac{1}{R_3} = \frac{1}{R_B} - \frac{1}{R_4} = \frac{1}{171.7} - \frac{1}{270} = 0.00212$$

$$\text{so } R_3 = \frac{1}{0.00212}$$

$$R_3 = 472\,\Omega$$

Capacitors in series and parallel

Capacitors in series

Figure 1.82 shows three capacitors connected in series. In this configuration is it important to remember that the charge (q) on each capacitor will be the same.

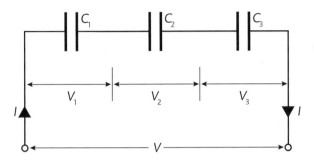

▶ **Figure 1.82** Capacitors connected in series

By applying Kirchhoff's laws, you can show that the total capacitance (C_T) for a number of capacitors (n) in series is:

$$\frac{1}{C_T} = \frac{1}{C_1} + \frac{1}{C_2} \ldots + \frac{1}{C_n}$$

Capacitors in parallel

Figure 1.83 shows three capacitors connected in parallel. By considering the total charge as being the sum of the charges stored in each capacitor, you can show that the total capacitance (C_T) for a number of resistors (n) in series is:

$$C_T = C_1 + C_2 \ldots + C_n$$

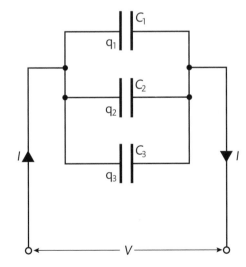

▶ **Figure 1.83** Capacitors connected in parallel

Direct current sources

Cells – these are electrochemical devices able to generate an e.m.f. and so can be used as a source of electrical current. Cells are generally limited by their chemistry to producing relatively low voltages; for example, a typical zinc–carbon cell produces a voltage of only 1.5 V.

Battery – a simple battery can be made up of a single cell or multiple cells. Multiple cells connected in series provide a higher voltage than a single cell; if connected in parallel, they provide a higher current.

Internal resistance – this is the resistance of the internal structure of the battery or cell. It usually manifests itself as heat as the battery discharges.

Stabilised power supply – a device that produces an accurately regulated d.c. voltage output from an a.c. input. It often has a means of limiting the current in case of a circuit malfunction or dead short.

Photovoltaic (PV) cell – a semiconductor device that generates an e.m.f. when exposed to light. Individual cells are able to generate only small voltages and are usually connected together in arrays to enable them to provide more useful voltage or current output characteristics.

Worked Example

A battery with an e.m.f. of 12 V and internal resistance 1 Ω is connected to a resistive load of 9 Ω.

Calculate the voltage present at the battery terminals.

Solution

Calculate the circuit current using Ohm's law:

$$I = \frac{V}{R} = \frac{12}{9 + 1} = 1.2\,A$$

The voltage across the 9 Ω load is therefore:

$$V = IR = 1.2 \times 9 = 10.8\,V$$

By Kirchhoff's law, the voltage being supplied at the terminals of the battery must be 10.8 V.

Assessment practice 1.5

1 The circuit diagram in **Figure 1.84** shows a 12 V d.c. power source connected to a network of resistors.

Note: '4k7' means '4.7 kΩ'. This notation is commonly used in engineering to ensure that decimal points are not missed when printed on small component labels.

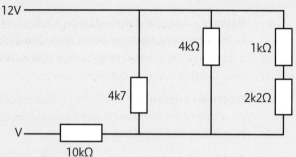

▶ **Figure 1.84** Network of resistors

Calculate the total current flowing in the circuit. (3 marks)

2 The electronic circuit diagram in **Figure 1.85** shows a d.c. power source connected to three capacitors connected in parallel.

  a) Calculate the total equivalent capacitance of the three parallel capacitors. (1 mark)

  b) Calculate the charge stored in each capacitor. (3 marks)

  c) Calculate the total energy stored in the circuit. (1 mark)

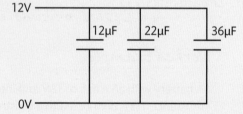

▶ **Figure 1.85** A d.c. power source connected to three capacitors connected in parallel

3 A voltage of 12 V is connected to a fully discharged capacitor through a resistance of 470 kΩ, as shown in **Figure 1.86**. After 2 seconds the voltage across the capacitor is 7.72 V. The voltage (V_C) across a charging capacitor at any time (t) is given by the equation

$$V_C = V_0(1 - e^{-t/RC})$$

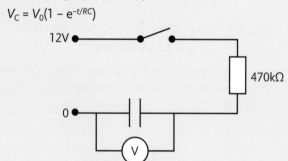

▶ **Figure 1.86** A capacitor-charging circuit

  a) Calculate the size of the capacitor. (6 marks)

  b) Calculate the time constant for the circuit. (1 mark)

For each of the problems in this assessment practice, use the following stages to guide your progress through the task.

Plan

- What are the important engineering concepts that I could apply to this problem?
- Do I feel confident that my understanding of this topic is sufficiently developed to solve the problem?

Do

- I can approach the task logically and methodically, clearly recording each step in my problem-solving process.
- I can identify when I have gone wrong and adjust my approach to get myself back on course.

Review

- I can effectively review and check mathematical calculations to prevent or correct errors.
- I can explain the best way to approach this type of problem.

F Magnetism and electromagnetic induction

F1 Magnetism

Magnetic fields

Magnetic fields are often represented by lines of magnetic flux that seem to flow around sources of magnetism such as a simple bar magnet (see **Figure 1.87**).

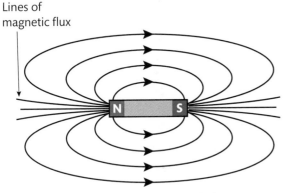

Lines of magnetic flux

▶ **Figure 1.87** Lines of magnetic flux around a bar magnet

The following are important parameters that are used to describe magnetic fields:

▶ **Magnetic flux** (Φ) – a measure of the size of the magnetic field produced by a source of magnetism. Unit: weber (Wb).

▶ **Magnetic flux density** (B) – a measure of how closely packed the lines of magnetic flux (Φ) produced by a source of magnetism with area (A) are. It is used to represent the strength or intensity of a magnetic field. Units: webers per square metre (Wb m^{-2}) or tesla (T).

$$B = \frac{\Phi}{A}$$

▶ **Magnetomotive force** (**m.m.f.**) (F_m) – the force that causes the formation of magnetic flux in a ferromagnetic material placed inside a solenoid or coiled conductor with (N) turns. It is generated by a current flow (I) through the solenoid. Unit: ampere (A).

$$F_m = NI$$

▶ **Magnetic field strength** (H) – a measure of the strength of the magnetising field produced by a solenoid with mean length (L) and number of turns (N) carrying a current (I). Unit: amperes per metre (A m^{-1}).

$$H = \frac{NI}{L}$$

▶ **Permeability** (μ) – a measure of the degree of magnetisation a material undergoes when subject to a magnetic field. In other words, it compares the magnetic field strength (H) with the magnetic flux density (B) generated inside a material. Unit: henrys per metre (H m^{-1}).

$$\mu = \frac{B}{H}$$

▶ **Permeability of free space** (μ_0) – because there is no material to influence a magnetic field in a vacuum, the applied magnetising field strength (H) and flux density (B) will be the same. However, since you measure these two quantities in different units, in practice you must use the value

$$\mu_0 = 4\pi \times 10^{-7} \, \text{H m}^{-1}$$

▶ **Relative permeability** (μ_r) – compares the permeability (μ) calculated for a given material to the permeability of free space (μ_0).

$$\mu_r = \frac{\mu}{\mu_0} \quad \text{so} \quad \mu_0\mu_r = \frac{B}{H}$$

Ferromagnetic materials

Ferromagnetic materials are strongly attracted by a magnetic field. They readily become magnets themselves under the influence of an external magnetic field and are able to retain their magnetic properties even after the external field has been removed.

B/H curves and loops

A **B/H curve** is a graphical representation of the relationship between the magnetic flux density (B) formed in a specific material when the material is exposed to a magnetic field strength (H).

B/H curves are not the straight lines that you might expect. They are curved because the permeability of ferromagnetic materials varies considerably with the strength of the applied magnetic field (H).

Figure 1.88 shows the B/H curves for some common ferromagnetic materials. Each curve starts at a point where B and H are zero and the material is entirely

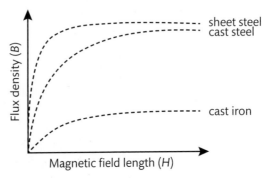

▶ **Figure 1.88** Typical B/H curves for some ferromagnetic materials

demagnetised. In each case, you can see that as the magnetic field strength is increased, there is initially a rapid increase in flux density. The rate of increase in B then slows considerably as the curve flattens out. The material is said to have reached magnetic saturation when any further increase in H leads to a negligible increase in B.

You can gain a greater understanding of the behaviour of ferromagnetic materials by considering the effects of reversing the applied magnetic field once saturation has been reached. Consider the **B/H loop** shown in **Figure 1.89**. The dashed line shows a typical B/H curve, which illustrates how a completely demagnetised material becomes magnetised with the application of a magnetic field.

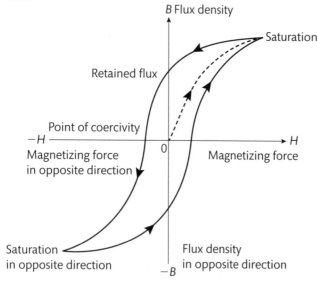

▶ **Figure 1.89** A typical B/H hysteresis loop

As you reduce the magnetic field (H) back to zero, the flux density (B) also decreases, but some retained flux remains in the material. When the applied magnetic field is reversed, this retained flux reduces back to zero. The size of the magnetic field required to eliminate the retained flux is known as the **coercivity** of the material.

As you build the reversed magnetic field further, magnetic saturation in the opposite direction is reached.

Reducing and then reversing the magnetic field once again has a similar effect on the material and so a closed loop is formed.

Hysteresis

Any change in flux density (B) lags behind changes in the applied magnetic field strength (H). This effect is called **magnetic hysteresis** and the loop described in **Figure 1.89** is known as the **hysteresis loop**.

The internal realignments that occur during the magnetisation cycle in a ferromagnetic material are responsible for energy lost as heat. This is called **hysteresis loss**. This energy loss is proportional to the area inside the hysteresis loop and can vary considerably across different materials.

In applications where materials undergo the magnetisation cycle many times a second, as in a.c. transformers, choosing a material with a narrow hysteresis loop is vital to prevent excessive losses and overheating.

Reluctance

Some metals exhibit a resistance to the presence of a magnetic field. The **reluctance** (S) is used as a measure of this magnetic resistance and can be expressed as the ratio of m.m.f. (F_m) to magnetic flux (Φ). Unit: inverse henry (H^{-1}).

$$S = \frac{F_m}{\Phi}$$

Reluctance can also be expressed in terms of the mean length of the flux path (l), the area through which the flux is flowing (A), the permeability of free space (μ_0) and the relative permeability of the material (μ_r) through which the field is passing:

$$S = \frac{l}{\mu_0 \mu_r A}$$

Magnetic screening

Often, it is necessary to screen components or devices from magnetic fields to prevent any unwanted effects. This is a common requirement when dealing with sensitive electronic devices.

Effective screening can be achieved using ferromagnetic materials with low reluctance, which can provide a pathway for lines of magnetic flux around the object or objects being protected.

PAUSE POINT   Explain what is meant by the term magnetic hysteresis.

Hint   Draw a graph to help illustrate your explanation.

Extend   What is the significance of the area inside a hysteresis loop? What possible consequences could this have for the design of components that rely on a continuous magnetisation cycle for their operation, such as a transformer?

F2 Electromagnetic induction

Induced electromotive force

Electromagnetic induction describes the phenomenon by which an electromotive force (e.m.f.) is induced or generated in a conductor when it is subjected to a changing magnetic field according to **Faraday's laws of electromagnetic induction**.

> **Key terms**
>
> **Faraday's laws of induction** – combined, these laws state: 'When a magnetic flux through a coil is made to vary, an e.m.f. is induced. The magnitude of this e.m.f. is proportional to the rate of change of flux.'
>
> This can be readily demonstrated by moving a permanent magnet through a coil of wire, as shown in **Figure 1.90**.
>
> **Lenz's Law** – an induced current always acts in such a direction so as to oppose the change in flux producing the current.

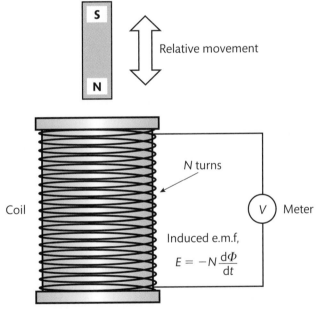

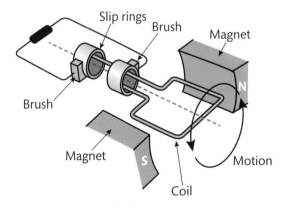

▸ **Figure 1.90** Demonstration of electromagnetic induction using a coil and a bar magnet

An e.m.f. will be generated in the coil whenever the magnet is moved into or out of the coil and perpendicular lines of magnetic flux cut through the conductor. (Alternatively, the magnet could remain still and the coil could be moved.)

The size of the induced e.m.f. (E) depends on the number of turns in the coil (N) and the rate of change of magnetic flux ($\frac{d\Phi}{dt}$):

$$E = -N\frac{d\Phi}{dt}$$

The negative sign is a consequence of **Lenz's law**, which states that the induced current will act to oppose the change in flux.

An alternative approach to determine the e.m.f. (E) considers the rate at which lines of flux are being cut by a conductor of length (L) travelling at velocity (v) through a magnetic field with flux density (B).

$$E = BLv$$

Note that this relationship assumes that the conductor is moving at right angles to the lines of flux.

Eddy currents

As well as inducing an e.m.f., which is able to flow as useful current through the conductor, other localised currents (eddy currents) are also created inside the conductor. These obey Lenz's law by flowing in a direction opposite to the changing magnetic field that created them. As a result, a proportion of useful energy is lost in the form of heat. Devices in which eddy currents occur, such as transformers, have to be carefully designed to prevent excessive energy loss and overheating. For example, transformers are constructed with laminated thin soft iron plates, which are electrically isolated from one another using a thin layer of lacquer to restrict the size of the eddy currents that form.

Electric generators

Electric generators are usually based on the principle of rotating a coiled conductor inside a static magnetic field (see **Figure 1.91**). The ends of the coil are connected to copper slip-rings that maintain contact with stationary carbon brushes through which the induced current flows.

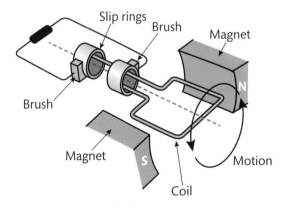

▸ **Figure 1.91** A simple electric generator

The magnitude of the e.m.f. induced in the coil is proportional to the number of turns of the coil. It is also dependent on the speed of rotation and the strength of the magnetic field.

The voltage generated by a rotating coil is sinusoidal because the angle between the coil and the lines of flux varies according to the angle of rotation (see **Figure 1.92**). Peak voltage is achieved where the motion of the coil is perpendicular to the lines of flux, that is, cutting through the flux lines at right angles. Voltage falls to zero when the coil has rotated through 90° because at this stage the motion of the coil becomes parallel to the lines of flux and so does not cut through them at all.

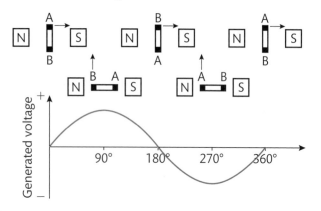

▶ **Figure 1.92** The sinusoidal voltage generated by a rotating coil

Direct current electric motors

An electric motor (see **Figure 1.93**) uses similar principles to an electric generator, but instead produces movement when a current is passed through a coil that is free to rotate inside a magnetic field.

When a current is passed through the coil, its associated magnetic field opposes the field generated by the permanent magnets and causes the coil to rotate until the two fields are aligned. After rotating 180°, the split slip-ring (the commutator), which provides current to the rotating coil via carbon brushes, reverses the current flow in the coil and hence the direction of its magnetic field. Once again the magnetic field generated by the coil opposes the field generated by the permanent magnets, and rotation continues.

In practice, the rotating element (the armature) is mounted on bearings to allow free rotation and to minimise frictional losses. The armature encompasses the main drive shaft of the motor, a laminated iron core, around which several individual coils are wound, and the commutator. The commutator is split into several segments, with each feeding different coils (or windings) in turn as the armature rotates. This arrangement helps to ensure smooth rotation and torque delivery.

Also, large motors do not use permanent magnets in the stationary part of the motor (the stator). Instead, the stator consists of a laminated iron ring with further windings. These field windings are arranged so as to provide the necessary stationary magnetic field when current flows through them.

Self-inductance in a coil

According to Lenz's law, any change in the current flowing in a circuit will generate a changing magnetic field, which in turn induces a back e.m.f. that opposes the current change.

Self-inductance describes a situation where an e.m.f. is induced in the same circuit in which the current is changing.

Self-inductance (*L*) is measured in henrys (H), where 1 H is the inductance present in a circuit where a changing current of $1\,A\,s^{-1}$ induces an e.m.f. of 1 V. The induced e.m.f. (*E*) is therefore related to the rate at which the current changes $(\frac{dI}{dt})$ and the inductance of the circuit (*L*):

$$E = -L\frac{dI}{dt}$$

You already know that the e.m.f. induced in a coil can also be calculated from the rate of change of magnetic flux present:

$$E = -N\frac{d\Phi}{dt}$$

These two relationships allow you to calculate the self-inductance (*L*) using the number of turns in the coil (*N*), the current (*I*) and the generating magnetic flux (*Φ*):

$$L = \frac{N\Phi}{I}$$

Electric current supplied externally through a commutator

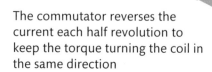

The commutator reverses the current each half revolution to keep the torque turning the coil in the same direction

When electric current (*I*) passes through a coil in a magnetic field, the magnetic force produces a torque which turns the d.c. motor

▶ **Figure 1.93** Basic d.c. electric motor

⏸ **PAUSE POINT** Use notes and an annotated diagram to explain the operation of a d.c. motor.

⬡ Hint A good diagram will often convey much of the necessary information.

⬡ Extend Why do industrial electric motors use field windings instead of permanent magnets in their construction and operation?

Energy storage in an inductor

When you first establish the flow of current in an inductive circuit, a back e.m.f. is induced to oppose the increasing current flow. As a consequence, additional energy is required to establish the current flow. This additional energy is then stored within the magnetic field generated by the current and is recovered when the current stops and the magnetic field collapses.

The energy stored in an inductor (W) is the product of the circuit inductance (L) and the current (I). Unit: joule (J).

$$W = \tfrac{1}{2}LI^2$$

Mutual inductance

Mutual inductance (M) describes a situation where a changing current in one circuit induces an e.m.f. in an adjacent circuit. It is measured in henrys (H), where 1 H is the mutual inductance present where a changing current of $1\,\mathrm{A\,s^{-1}}$ in one circuit induces an e.m.f. of 1 V in the second.

Transformers

The most common and important application of mutual inductance is in transformers (see **Figure 1.94**).

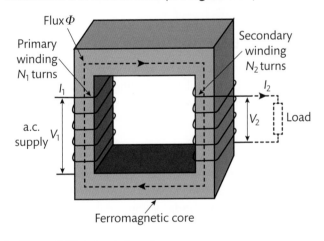

▶ **Figure 1.94** A simple transformer

A transformer consists of two separate coils or windings wound on a common ferromagnetic core (commonly constructed from laminated steel to minimise losses caused by eddy currents). The primary winding is connected to an a.c. electrical supply and the other, secondary, winding is connected to an electrical load.

The varying current in the primary winding generates magnetic flux in the transformer core. The flux flows around the core in a magnetic circuit and induces an e.m.f. in both the primary and the secondary windings. If we assume an ideal transformer (one which is well designed with negligible losses), then the rate of change of flux will be the same for each winding, and so the induced e.m.f. will be dependent on the number of turns in each winding.

Where the primary coil has number of turns (N_1) and voltage (V_1) and the secondary coil has number of turns (N_2) and voltage (V_2), the transformer voltage ratio is given by

$$\frac{V_1}{V_2} = \frac{N_1}{N_2}$$

▶ In a **step-up transformer**, the voltage ratio $\dfrac{V_1}{V_2} < 1$, that is, the secondary voltage is greater than the primary voltage.

▶ In a **step down transformer**, the voltage ratio $\dfrac{V_1}{V_2} > 1$, that is, the secondary voltage is less than the primary voltage.

▶ In an **ideal transformer**, the laws of conservation of energy mean that the power in the primary coil is the same as the power in the secondary coil:

$$V_1 I_1 = V_2 I_2$$

The transformer current ratio is then given by

$$\frac{V_1}{V_2} = \frac{I_2}{I_1}$$

⏸ **PAUSE POINT** Explain two significant sources of losses that will affect the efficiency of a transformer.

⬡ Hint Most losses occur within the transformer core.

⬡ Extend Explain how a transformer core is designed to minimise these losses.

1 The diagram in **Figure 1.95** shows a transformer used in a phone charger that reduces mains voltage at 230 V a.c. to 5 V a.c. There are 800 turns of wire on the primary winding.

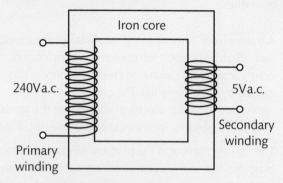

Iron core

240V a.c.

5V a.c.

Secondary winding

Primary winding

▶ **Figure 1.95** A transformer in a phone charger

a) Calculate the number of turns required in the secondary coil. (2 marks)

b) The 5 V a.c. must be converted to 5 V d.c. before it can be used to power the phone. Explain what is meant by full-wave rectification. (2 marks)

c) Draw a circuit diagram for a full-wave rectifier circuit. (3 marks)

2 Calculate the energy stored in an inductor with inductance 0.8 H passing a current of 2.6 A. (1 mark)

For each of the problems in this assessment practice, use the following stages to guide your progress through the task.

Plan

- Have I extracted and summarised all the information available in the question?
- How confident do I feel that I have the knowledge required to approach this task?

Do

- I have spent adequate time planning my approach. I can clearly explain the steps involved and the order in which they should be done.
- I can recognise when my method is leading nowhere and I need to stop and rethink my approach.

Review

- I can appreciate the importance of ensuring that my solution is laid out and explained with sufficient clarity so that it can be followed by someone else.
- I can explain the improvements I would make to my approach the next time I encounter a similar problem.

Single-phase alternating current

Single-phase alternating current theory

Alternating current (a.c.) electricity is usually generated by means of a coil rotating within a magnetic field. For one revolution of the coil, the resulting e.m.f. will alternate between a maximum positive and maximum negative value. This type of electrical generator is also known as an alternator.

Sinusoidal waveforms

When the value of the e.m.f. generated by an alternator is plotted against time, the resulting waveform is sinusoidal. Just like the pure sine functions (see 'Graphs of the trigonometric functions' in **A2 Trigonometric methods**), a sinusoidal waveform is periodic – it is a series of identical repeating cycles.

Two important parameters are used to describe all periodic waveforms.

▶ **Periodic time** (T) – the time taken to complete one cycle, also called simply the **period**. Unit: second (s).

▶ **Frequency** (f) – the number of cycles completed in one second. Unit: hertz (Hz).

The periodic time (T) and the frequency (f) are the reciprocal of each other:

$$T = \frac{1}{f} \text{ and } f = \frac{1}{T}$$

For a sinusoidal a.c. current, there are a variety of measures that can be used to describe the characteristics of the waveform.

▸ **Peak** – the maximum value of voltage or current reached in a positive or negative half-cycle (see **Figure 1.96**).

▸ **Peak-to-peak** – the difference between the positive peak and the negative peak voltage or current in a full cycle (see **Figure 1.96**).

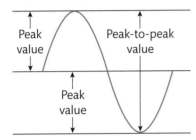

▸ **Figure 1.96** Peak and peak-to-peak values of an a.c. signal

▸ **Root mean square (r.m.s.)** – the value of a direct current that would produce an equivalent heating effect as the alternating current. For a sine wave, the r.m.s. voltage (V_{rms}) is related to the peak voltage (V_{peak}) by

$$V_{rms} = \frac{1}{\sqrt{2}}V_{peak} \quad \text{or, in terms of current:} \quad I_{rms} = \frac{1}{\sqrt{2}}I_{peak}$$

▸ **Average** – the average of all the instantaneous measurements in one half-cycle. For a sine wave, the average voltage (V_{avg}) is related to the peak voltage (V_{peak}) by

$$V_{avg} = \frac{2}{\pi}V_{peak} \quad \text{or, in terms of current:} \quad I_{avg} = \frac{2}{\pi}I_{peak}$$

▸ **Instantaneous** – the value of voltage or current at a particular time instant during the sinusoidal cycle. Instantaneous voltage and current are zero when the waveform crosses the time axis where it changes polarity.

▸ **Form factor** – equal to the r.m.s. voltage (V_{rms}) divided by the average voltage (V_{avg}). For a sine wave, the form factor is a constant:

$$\text{form factor} = \frac{V_{rms}}{V_{avg}} = \frac{V_{peak}}{\sqrt{2}} \times \frac{\pi}{2\,V_{peak}} = \frac{\pi}{2\sqrt{2}} = 1.11$$

This relationship is also true for current.

Non-sinusoidal waveforms

Figure 1.97 shows square, triangular and sawtooth waves, which are all examples of non-sinusoidal waveforms.

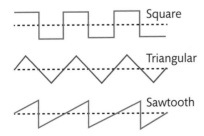

▸ **Figure 1.97** Some examples of non-sinusoidal waveforms

Case study

UK mains electrical supply

The majority of electricity in the UK is still generated by burning fossil fuels to drive steam turbines, which in turn drive generators. A two-pole generator supplies one complete a.c. cycle per rotation and so, when driven at 3000 rpm, it will generate 3000 cycles per minute or 50 cycles per second. It is extremely important for generating companies to control the rotational speed of their generators to maintain this 50 Hz frequency.

Domestic premises in the UK are supplied electricity through an extensive network of electrical supply cables and infrastructure, collectively known as the national grid. A series of local substations receive voltages, typically 11 kV 50 Hz a.c., from overhead supply lines and then use step-down transformers to output the 230 V 50 Hz a.c. standard UK domestic supply voltage. This is then fed to homes through underground cables.

Check your knowledge

1 A substation transformer has 4500 windings on its primary coil. Calculate the number of windings required on the secondary coil.

2 Explain two significant causes of energy loss that can occur in transformers and how these can be minimised.

3 Calculate the peak voltage of a 230 V a.c. sinusoidal waveform.

4 A 2.5 A current flows through an inductor with inductance 0.3 H and a resistance of 9 Ω when connected to a 240 V 50 Hz a.c. supply. Calculate the change in current that would occur if the mains frequency were to drop to 45 Hz.

Combining sinusoidal waveforms

The result of adding two sinusoidal voltages together can be determined either graphically or by the vector addition of phasors.

Graphical approach

In **Figure 1.98**, two waveforms are drawn on the same axes. At any point on the time axis, the resultant instantaneous voltage is obtained by adding together the individual instantaneous voltages at that point. The dotted line shows the resultant waveform generated in this way.

This approach can be used to combine any kinds of waveforms, irrespective of their frequency, amplitude or phase difference. The obvious downside to this approach, as with all manual graphical techniques, is that it is time-consuming and the accuracy of the results will be limited.

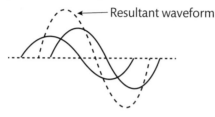

▶ **Figure 1.98** Graphical representation of adding two waveforms

Vector addition of phasors

This approach builds on the techniques explained in the sections on phasors and vector addition in **A2 Trigonometric methods**. It should be noted that it is valid only when applied to sinusoidal waveforms with the **same frequency**.

Consider two a.c. waveforms represented by

$$v_1 = 100 \sin (100 \omega t) \text{ and } v_2 = 200 \sin \left(100 \omega t - \frac{\pi}{6}\right)$$

If you compare these to the standard form for a sinusoidal phasor

$$v = V \sin (\omega t + \varPhi)$$

you can see that the magnitude and direction of the phasors are as follows:

▶ v_1 has magnitude 100 with no phase shift, so the direction of the phasor will be 0° (horizontal).

▶ v_2 has magnitude 200 with a phase shift of $-\frac{\pi}{6}$ rad, which means the direction of the phasor will be 30° below the horizontal.

These phasors can therefore be drawn as in **Figure 1.99**.

An accurate scale drawing will allow you to determine the magnitude of the resultant peak voltage (V_R) and phase angle ($\varPhi$) by direct measurement. It is often more convenient to convert the phase angle from radians to degrees to allow the use of a conventional protractor.

You could take a more analytical approach by redrawing the phasor diagram as shown in **Figure 1.100**.

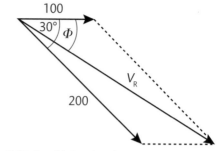

▶ **Figure 1.99** Combining the phasors of a.c. waveforms

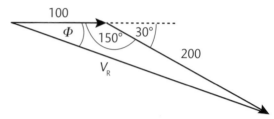

▶ **Figure 1.100** Vector addition of phasors

The magnitude of the resultant peak voltage (V_R) can be determined from the cosine rule:

$$a^2 = b^2 + c^2 - 2bc \cos A$$

So $V_R{}^2 = 200^2 + 100^2 - 2 \times 200 \times 100 \times \cos 150° = 84\,641$

$$V_R = 290.93 \text{ V}$$

You can find the resultant phase angle ($\varPhi$) by using the sine rule:

$$\frac{a}{\sin A} = \frac{b}{\sin B}$$

$$\frac{290.93}{\sin 150°} = \frac{200}{\sin \varPhi}$$

Rearranging and solving for $\varPhi$ gives

$$\varPhi = 20.10° \text{ or } 0.351 \text{ rad}$$

Then you can express the resultant waveform as

$$v_R \approx 291 \sin (\omega t - 0.35)$$

Remember that the phase angle must be in radians if it is to be used in the general form for a sinusoidal phasor.

Note that here the phase angle lags below the horizontal and so is considered negative.

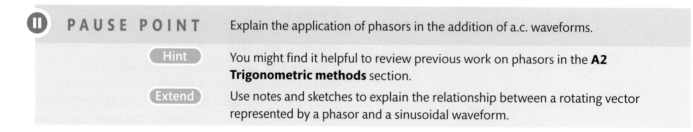

Ⅱ PAUSE POINT Explain the application of phasors in the addition of a.c. waveforms.

Hint You might find it helpful to review previous work on phasors in the **A2 Trigonometric methods** section.

Extend Use notes and sketches to explain the relationship between a rotating vector represented by a phasor and a sinusoidal waveform.

Impedance

A.c. circuits that contain capacitors and/or inductors are more complex to analyse than those containing purely ohmic resistance.

Impedance (Z) is the total opposition to the flow of electricity exhibited in an a.c. circuit.

The voltage and current in a purely resistive a.c. circuit are in phase, and the impedance of the circuit is equal to its resistance.

In circuits containing either capacitance or inductance, the impedance is made up of two parts: the **resistance** (R), which is independent of frequency, and the **reactance** (X), which varies with frequency. There are two forms of reactance, both dependent on frequency (f):

▸ **Capacitive reactance** $X_C = \dfrac{1}{2\pi fC}$, where C is the capacitance value in farads. Unit: ohm (Ω).

▸ **Inductive reactance** $X_L = 2\pi fL$, where L is the inductance value in henrys. Unit: ohm (Ω).

The total impedance (Z) is made up of the resistance (R) and the reactance (X). However, these cannot simply be added together, as phase changes between voltage and current waveforms must be taken into account. Impedance is therefore the vector sum of the ohmic resistance and the reactance present in a circuit.

▸ For a circuit with resistance R and inductive reactance X_L, the total impedance (Z) is given by

$$Z = \sqrt{X_L^2 + R^2}$$

▸ For a circuit with resistance R and capacitive reactance X_C, the total impedance (Z) is given by

$$Z = \sqrt{X_C^2 + R^2}$$

Alternating current rectification

Rectification is the conversion of an a.c. waveform into d.c. You looked at this in the section on diodes in **E2 Direct current circuit theory**.

Assessment practice 1.7

1 Two a.c. voltage waveforms are represented by $v_1 = 65\sin(80\omega t)$ and $v_2 = 90\sin(80\omega t - \frac{\pi}{5})$.

   a) Draw a sketch of the phasors that can be used to represent these waveforms. (2 marks)

   b) Use trigonometric methods to find the magnitude and phase angle of the resultant phasor when v_1 and v_2 are combined. (4 marks)

   c) State the resultant waveform in the standard form $v = V\sin(\omega t + \Phi)$. (1 mark)

   d) Calculate the frequency of the resultant waveform. (1 mark)

2 An engineer is testing an inductor with an inductance of 0.8 H and a resistance of 3 Ω. It is connected to a 240 V 50 Hz a.c. supply.

   Calculate the current drawn from the supply. (3 marks)

For each of the problems in this assessment practice, use the following stages to guide your progress through the task.

Plan
• Will a sketch help me to visualise the scenario given in the question?
• Is it clear to me how to approach this type of question? Will I have to refer back to my notes?

Do
• I can apply the methods taught in this unit to solve problems appropriately.
• I understand my own limitations and when I need to stop, revisit the textbook, look at my lesson notes or ask for help.

Review
• I can explain to others why I chose a particular method for solving this type of problem and how to apply it correctly.
• I can evaluate how effective my planning was in approaching this question and how I might make improvements for next time.

Assessment practice 1.8 synoptic question

A remote ski cabin relies on a diesel engine to generate its electricity. The system is depicted in **Figure 1.101**.

The diesel used to run the engine has a total energy content of 36 MJ l⁻¹ and is consumed at a rate of 3.2 l h⁻¹.

The generator driven by the diesel engine provides a 240 V electrical supply at 32 A.

Waste heat from the diesel engine is used to heat the cabin. This is transferred from circulating engine coolant via a heat exchanger into a glycol-filled heating system.

The diesel engine generates 15 kW of energy wasted as heat. The glycol in the heat circulation system has a specific heat capacity of 4.18 kJ kg⁻¹ K⁻¹ and a density of 1096 kg m⁻³. The flow rate of glycol through the heat exchanger is 12 l s⁻¹ and the temperature difference maintained between the input and output pipes is +0.2°C.

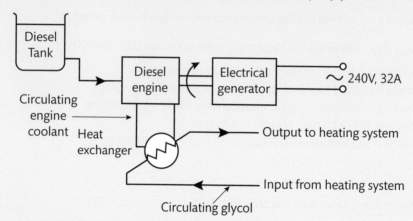

▶ **Figure 1.101** Electricity generator using a diesel engine

a) Calculate the thermal efficiency of the heat transfer between the diesel engine and the glycol heat circulation system.   (7 marks)

b) Calculate the overall system efficiency.   (7 marks)

THINK ▶FUTURE

Rob Taylor

Product designer for an industrial lighting company

I've been working as an industrial product designer as part of a multi-disciplinary design team for four years. In the industrial lighting products that we work on, function, reliability, safety and efficiency are all key factors that we must get right. Of course, aesthetics are also an important aspect of our work, especially when working with architects to design bespoke lighting fixtures for big projects.

I am lucky to work with colleagues who have a diverse range of electromechanical and electronic specialist knowledge and experience, and I am still learning new things every day. When designing products, I have access to a range of materials and manufacturing techniques in our UK factory. A single product might involve designing sheet metal parts, aluminium castings and plastic injection moulded components. Internal electrical wiring must then be specified and lamps and their electronic drive gear positioned inside the luminaire. Safe and reliable ways of suspending the complete luminaires from a high warehouse roof or integrating them as part of an office ceiling system also need to be considered. I have even worked to design curved aluminium reflectors to distribute light evenly throughout a work area. Every project brings something new, and with recent developments in high-output LED technology, which is revolutionising lighting, these are exciting times to be involved in the lighting industry.

Focusing your skills

Seeing the whole picture

When designing a complete electromechanical product, machine or system it is important to appreciate the principles that underpin its operation and how these can affect one another. Here are some example questions that a designer might ask:

- Does the wiring size specified have a current rating significantly above the levels expected during normal operation? What would happen if safe working loads were exceeded?
- What would be the effects of extreme high or low temperatures on electrical and mechanical systems?
- Are mechanical or electrical systems likely to lose energy as heat during operation? What might the main losses be and how could these be reduced?

- If cooling is required, how much heat energy would need to be removed? Which type of coolant would be appropriate, what flow rate would be needed, and how could the coolant be circulated?
- Are electrical devices likely to cause electromagnetic interference and affect the operation of nearby sensitive electronic systems? If there is no room for the devices to simply be moved away from each other, how else might such interference be prevented?
- Are all the components of sufficient mechanical strength to withstand the working stresses encountered during operational extremes?
- How can I minimise the amount of material used to achieve the required outcome? What solid shapes might be most useful?
- How will the static forces existing in mechanical components be affected by linear or rotational movement?

Getting ready for assessment

This section has been written to help you to do your best when you take the external examination. Read through it carefully and ask your tutor if there is anything you are not sure about.

About the test

This unit is externally assessed using an unseen paper-based examination. Pearson sets and marks the examination. The assessment must be taken under examination conditions.

As the guidelines for assessment can change, you should refer to the official assessment guidance on the Pearson Qualifications website for the latest definitive guidance.

Make sure you are familiar with the command words and know how to apply them within the context of the examination. The table below shows the command words for each Assessment outcome.

Assessment outcome	Command words
AO1	calculate, describe, explain
AO2	calculate, find, solve
AO3	find, calculate, describe, draw, explain
AO4	calculate, draw
AO5	calculate, draw, explain

Remember that all the questions are compulsory and you should attempt to answer each one.

Organise your time based on the marks available for each question. Set yourself a timetable for working through the test and then stick to it — do not spend ages on a short 1–2 mark question and then find you only have a few minutes for a longer 6–7 mark question.

Try answering all the simpler questions first, then come back to the harder questions. This should give you more time for the harder questions.

Sitting the test

Listen to and read carefully any instructions you are given. Lots of marks are often lost through not reading questions properly and misunderstanding what the question is asking.

Most questions contain command words. Understanding what these words mean will help you to understand what the question is asking you to do.

Command word	Definition – what it is asking you to do
Calculate	Learners judge the number or amount of something by using the information they already have, and add, subtract, multiply or divide numbers. For example, 'Calculate the reaction forces…'
Describe	Learners give a clear, objective account in their own words showing recall, and in some cases application, of the relevant features and information about a subject. For example, 'Describe the process of heat transfer…'
Draw	Learners make a graphical representation of data by hand (as in a diagram). For example, 'Draw a diagram to represent…'
Explain	Learners make something clear or easy to understand by describing or giving information about it. For example, 'Explain one factor affecting…'
Find	Learners discover the facts or truth about something. For example, 'Find the coordinates where…'
Identify	Learners recognise or establish as being a particular person or thing; verify the identity of. For example, 'Identify the energy loss…'
Label	Learners affix a label to; mark with a label. For example, 'Label the diagram to show…'
Solve	Learners find the answer or explanation to a problem. For example, 'Solve the equation to…'
State	Learners declare definitely or specifically. For example, 'State all three conditions for…'

Sample answers

For some of the questions you will be given background information on which the questions are based.

Look at the sample questions that follow and our tips on how to answer these well.

Answering applied mathematics calculation questions

- Identify the method required to solve the type of problem set.
- Explain each step and show all your workings.
- Express your answer in appropriate units.
- Check your answer.

Worked example

The velocity of a model rocket fired vertically upwards is given by the equation $v = 2t^2 + 5t - 11$.

Find, by use of the quadratic formula, the time when the rocket reached its highest point. (2 marks)

Answer

At its highest point the velocity of the rocket will have fallen to 0, so

$$2t^2 + 5t - 11 = 0$$

The quadratic formula takes the form

$$x = \frac{-b \pm \sqrt{b^2 - 4ac}}{2a}$$

In this problem:

$$a = 2, b = 5, c = -11$$

$$\text{so } t = \frac{-5 \pm \sqrt{5^2 - 4 \times 2 \times (-11)}}{2 \times 2} = \frac{-5 \pm 10.63}{4} = 1.41 \text{ or } -3.91$$

The negative value is not valid in the context of the question, so check the answer by substituting $t = 1.41$ into $v = 2t^2 + 5t - 11$:

$v = 3.96\ldots + 7.03\ldots - 11 = 0$ (remember to use unrounded values)

The highest point will be reached after 1.41 s.

Where calculations are involved, marks are awarded for 'knowing a method and attempting to apply it' at each stage. It is essential, therefore, to show your working and make clear which principle or formula is being applied. Further marks are awarded for accuracy, that is, for reaching the correct numerical answer, only if the relevant method marks have already been awarded.

Answering engineering principles calculation questions

- Identify the variables from the information given in the question.
- Write down the formula you need.
- Substitute numbers into the formula.
- Explain each step and show all your workings.
- Express your answer in appropriate units.
- Check your answer.

Worked example

An engineer is testing an inductor with an inductance of 0.5 H and a resistance of 8 Ω. It is connected to a 120 V 50 Hz a.c. supply.

Calculate the current drawn from the power supply. (3 marks)

Answer

The solution will require the calculation of I_{rms}.

The values given in the question are:

$L = 0.5\,H$, $R = 8\,\Omega$, $V_{rms} = 120\,V$, $f = 50\,Hz$

The inductive reactance is $X_L = 2\pi f L = 2\pi \times 50 \times 0.5 = 157.08\,\Omega$.

The total impedance is given by $Z = \sqrt{X_L^2 + R^2} = \sqrt{157.08^2 + 8^2} = 157.28\,\Omega$.

So the a.c. current is $I_{rms} = \dfrac{V_{rms}}{Z} = \dfrac{120}{157.28} = 0.763\,A$ or 763 mA.

Answering engineering principles short-answer questions – state

- Read the question carefully.
- Make sure that you make the same number of points as there are marks available in the question.

Worked example

State two sources of energy loss that affect the efficiency of electrical transformers. (2 marks)

Answer

Eddy currents. Hysteresis losses.

This answer provides two accurate points and full marks would be awarded. Command words such as 'give', 'state' or 'identify' can be answered in single words or brief statements. There is no need to write in full sentences because the examiner is only testing your ability to recall information.

Answering engineering principles short-answer questions – describe, explain

Worked example

Describe the process of heat transfer through conduction. (4 marks)

Answer

Heat transfer by direct contact between adjacent atoms:

Heat energy is a measure of the vibrational kinetic energy possessed by atoms. These vibrations are passed between atoms that are in direct contact with each other.

Heat transfer by the movement of electrons:

Heat energy can also be distributed through a material by the transfer of vibrational kinetic energy possessed by free-moving electrons as they move around within a material. This helps to explain why metals are in general good thermal (as well as electrical) conductors.

The mechanism of energy transfer through direct contact between atoms and the secondary mechanism involving energy transfer by free-moving electrons have been correctly identified. Each mechanism is then explained and expanded upon sufficiently for the award of the additional marks available.

Answering synoptic questions

Worked example

A heat engine driving a combined heat and power system outputs mechanical energy to drive a generator.

The energy is provided by a combustion process that uses air and a fuel with an energy content of $40\,MJ\,kg^{-1}$, which is supplied at a rate of $0.003\,kg\,s^{-1}$. The generator has a rotor that turns at 1500 rpm and has a torque of 255 N m. The output of the generator is 6.5 A at 400 V.

a) Explain how energy loss processes in both mechanical and electrical equipment affect the efficiency of the system. (4 marks)

b) Calculate the efficiency with which the heat engine provides mechanical work to the generator. (7 marks)

c) Calculate the overall system efficiency. (3 marks)

Synoptic questions, such as the example given above, are basically a combination of interrelated short-answer and calculation questions based around a common complex system or theme. As such they can be broken down into individual elements that should be approached in a similar way to the worked examples already discussed.

Synoptic questions will require you to make links between engineering principles from the full range of disciplines covered in the unit.

Delivery of Engineering Processes Safely as a Team

2

Getting to know your unit

Engineering processes, whether concerned with the manufacture of a product or the delivery of an engineering service, are the cornerstones of all modern industrial engineering. A single individual cannot carry out any complex industrial function effectively – often the coordinated efforts of hundreds or even thousands of people are required to manufacture a complex product such as a car or an aeroplane. This unit covers a range of practical and teamworking skills that are necessary when manufacturing a product or delivering a service safely as a team.

How you will be assessed

This unit will be assessed by a series of internally assessed tasks set by your tutor. Throughout this unit you will find assessment practice activities to help you work towards your assessment. Completing these activities will not mean that you have achieved a particular grade, but you will have carried out useful research or preparation that will be relevant when it comes to your final assignment.

In order for you to achieve the tasks in your assignment, it is important to check that you have met all of the assessment criteria. You can do this as you work your way through each assignment.

If you are hoping to gain a Merit or Distinction grade, you should also make sure that you present the information in your assignments in the style that is required by the relevant assessment criterion. For example, Merit criteria require you to analyse, and Distinction criteria require you to evaluate.

The assignments set by your tutor will consist of a number of tasks designed to meet the criteria in the table. They are likely to take the form of written reports, but may also include activities such as the following:

▶ Reviewing and analysing case studies based on the manufacture of an engineered product or delivery of a service in terms of the processes used and the influence of human factors.

▶ Creating engineering drawings using computer-aided design (CAD) software.

▶ Carrying out practical engineering processes both as a team leader and as a member of a team.

Assessment criteria

This table shows what you must do in order to achieve a **Pass**, **Merit** or **Distinction** grade, and where you can find activities to help you.

Pass	Merit	Distinction

Learning aim **A**   Examine common engineering processes to create products or deliver services safely and effectively as a team

Pass	Merit	Distinction
A.P1 Explain how three engineering processes are used safely when manufacturing a given product or when delivering a given service. **Assessment practice 2.1**	**A.M1** Analyse why three engineering processes are used to manufacture a product or to deliver a service and how human factors, as an individual and as a team, affect the performance of engineering processes. **Assessment practice 2.1**	**A.D1** Evaluate, using high quality written language, the effectiveness of using different engineering processes to manufacture a product or to deliver a service and how human factors, as an individual and as a team, affect the performance of engineering processes. **Assessment practice 2.1**
A.P2 Explain how human factors, as an individual or as a team, affect the performance of engineering processes. **Assessment practice 2.1**		

Learning aim **B**   Develop two-dimensional computer-aided drawings that can be used in engineering processes

Pass	Merit	Distinction
B.P3 Create an orthographic projection of a given component containing at least three different types of feature. **Assessment practice 2.2**	**B.M2** Produce, using layers, an accurate orthographic projection of a component containing at least three different types of feature and a circuit diagram containing at least six different component types that mainly meet an international standard. **Assessment practice 2.2**	**B.D2** Refine, using layers, an accurate orthographic projection of a component containing at least three different types of common feature and a circuit diagram containing at least six different component types to an international standard. **Assessment practice 2.2**
B.P4 Create a diagram of a given electrical circuit containing at least six different component types. **Assessment practice 2.2**		

Learning aim **C**   Carry out engineering processes safely to manufacture a product or to deliver a service effectively as a team

Pass	Merit	Distinction
C.P5 Manage own contributions to set up and organise a team in order to manufacture a product or deliver a service. **Assessment practice 2.3**	**C.M3** Manage own contributions safely and effectively using feedback from peers, as a team member and a team leader, to manufacture a product or to deliver a service. **Assessment practice 2.3**	**C.D3** Consistently manage own contributions effectively using feedback from peers, as a team member and a team leader, to set up, organise and manufacture a product or deliver a service safely, demonstrating forward thinking, adaptability or initiative. **Assessment practice 2.3**
C.P6 Produce, as an individual team member, a risk assessment of at least one engineering process. **Assessment practice 2.3**		
C.P7 Set up, as an individual team member, at least one process safely by interpreting technical documentation. **Assessment practice 2.3**		
C.P8 Manage own contributions safely, as a team member and a team leader, to manufacture a batch of an engineered product or to deliver a batch of an engineering service. **Assessment practice 2.3**		

Getting started

In a small group, make a list of situations where you have had to work as part of a team. Think about how teamworking compares with working alone and make a list of the advantages and disadvantages of the two approaches to completing a task.

A Examine common engineering processes to create products or deliver services safely and effectively as a team

Engineering projects are usually large and complex, requiring a range of specialist skills to complete. Some of the great feats of engineering from history, such as Concorde or the Channel Tunnel, involved tens of thousands of people working together over decades to bring the project to fruition. The manufacture of everything, from condensing boilers to cars, depends on multi-skilled **teams** of engineers and technicians working together to manufacture products quickly, efficiently and in the necessary numbers to satisfy demand. After manufacture, similar multi-skilled teams are relied upon to deliver the services that help to maintain and repair these complex products.

Key terms

Team – a group containing three or more individuals who have a common objective or shared goal.

Batch – three or more products manufactured or services delivered together.

A1 Common engineering processes

Preparation before product manufacture or product delivery

Generally engineering products or services start life as ideas on how to solve a particular problem or satisfy some other demand from a customer or the wider marketplace.

These ideas must be developed, evaluated and refined to achieve a viable solution. Once a solution is established, it must be communicated effectively to the people who will be asked to manufacture the product or deliver the service.

Preparation will include these documents:

▶ **Technical specifications** – define exactly what a product or service will do.

▶ **Engineering drawings** – define exactly what the individual components of a product will look like and how they should be assembled during manufacture to make the final product.

▶ **Scale of production** – defines the number of products that need to be manufactured or the number of times a service needs to be performed, hence dictating the approach to manufacture or service delivery; (see **Table 2.1**).

▶ **Work plans** – define a standard methodology that should be followed when manufacturing a product or delivering a service.

▶ **Quality control documents** – define the checks that should take place both during and after manufacturing a product or delivering a service.

▶ **Table 2.1** Characteristics of different scales of production

	One-off	Small batch	Mass or large batch	Continuous
Unit cost	high	medium	low	low
Tools and equipment	general	specialised	specialised and dedicated	dedicated
Initial investment	low	medium	high	high
Production efficiency	low	medium	high	very high
Labour type	skilled	skilled and semi-skilled	semi-skilled and unskilled	unskilled
Labour cost	high	medium	low	low

Standards and reference material

The importance of BS8888

The people who design a product and create its technical specifications are often not the same people who manufacture the product. In fact, they are increasingly likely to be thousands of miles apart in different countries and may not even share a common language.

It is vital, therefore, that the drawings and information that you generate comply with the widely adopted rules and conventions laid down in BS8888 for writing technical specifications. In this way you are adopting a common technical language that will be understood by engineering companies globally.

Reference charts

When preparing drawings and work plans it is often necessary to use information from reference charts or other sources of technical information. These can be found in engineering handbooks and are often displayed as posters in engineering workshops. The charts give information such as that in **Tables 2.2** and **2.3**.

Products and services

The main driver of economic activity through which wealth can be generated is the provision of products and services.

Products

Products are types of goods that are manufactured and then sold to customers. Upon purchase, customers take ownership of products, which can then be used when the customer needs them until they break down or wear out and have to be repaired or replaced. Products are physical items such as cars, washing machines and bicycles.

Services

A **service** can be described as a series of activities that provides benefit (value) to customers. Examples of services are the processing of a credit card transaction by a bank or an MOT inspection by a garage.

▶ **Table 2.2** Metric (course) tapping drill sizes

Thread size	Tap drill (mm)
M3 × 0.5	2.50
M4 × 0.7	3.30
M5 × 0.8	4.20
M6 × 1	5.00
M8 × 1.25	6.80
M10 × 15	8.50

▶ **Table 2.3** Cutting speeds for commonly used engineering metals (using high-speed steel tooling)

Material type	Cutting speed (m/min)
Mild steel	30–38
Cast iron	18–24
Carbon steel	21–40
Stainless steel	23–40
Aluminium	75–105
Brass	90–210

Key terms

Product – the final tangible outcome of a manufacturing process, often referred to in economics as 'goods' (e.g. a car, television or chair).

Service – activities that provide some intangible benefit to a customer (e.g. processing a credit card payment or performing an MOT inspection on a car).

⏸ PAUSE POINT   Working in small groups, list some services that you might associate with the engineering sector.

> **Hint**   Such services might involve disassembly, maintenance activities or part replacement.

> **Extend**   Choose one of the engineering services you identified and break it down into the separate steps that would be involved in its delivery.

Common processes used to manufacture engineered products

Bench fitting

Bench fitting is the general term used for a wide range of engineering workshop activities carried out at a bench, which needs to be substantial and rigid and usually has a steel working surface. The fitter's vice (see **Figure 2.1**) is possibly the most important piece of equipment in bench fitting. It is used to secure a workpiece or assembly in place while the engineer uses hand tools such as files, saws or taps.

▶ **Figure 2.1** A fitter's vice

It is usually better to work on individual components or sub-assemblies from larger products at a bench where you can securely hold and orient them as necessary to access various parts. For example, a motor vehicle technician wouldn't disassemble a disc brake calliper while it was still on a vehicle.

Files

You would use a double-cut forged carbon steel file to shape metals such as brass, aluminium and steel.

Files come in many shapes and sizes, with different tooth pitches suited to different applications. Rough and bastard files with a large pitch between the rows of teeth are most suited to rapid metal removal with reduced clogging, but they give a poor surface finish. Smooth files with a small pitch between the teeth give an excellent surface finish but cannot remove material quickly.

The general-purpose choice of file in most workshops is the second-cut file (see **Figure 2.2**), which has a tooth pitch between those of the rough and smooth files. Second-cut files are capable of reasonable rates of stock removal and also give surface finishes that are acceptable in most circumstances.

▶ **Figure 2.2** A second-cut flat file

Filing techniques

Files can be used in different ways to give different finishes:

▶ cross filing – used for rapid material removal.

▶ draw filing – used for finishing to improve surface finish.

▶ straight filing – used to flatten surfaces (this is a highly skilled **operation** that can take years to master).

As the engineer files material from the workpiece, the spaces between the file teeth will become clogged. This is called 'pinning'. You can minimise pinning by rubbing chalk into the teeth periodically as you work. When pinning starts to affect the way in which the file cuts or begins to mark the surface being worked on, the teeth should be cleaned with a special wire brush called a file card.

> **Key term**
>
> **Operation** – a single step in a manufacturing process. For example, this could be marking out the position of a hole when manufacturing a product or removing an access panel when delivering an engineering service.

Cutting

You usually use a hacksaw (**Figure 2.3**) to cut metal. High-quality hacksaw blades are constructed from high-speed steel (HSS) that has been hardened to enable the teeth to cut soft metals like brass and aluminium, as well as most steels, accurately and with ease.

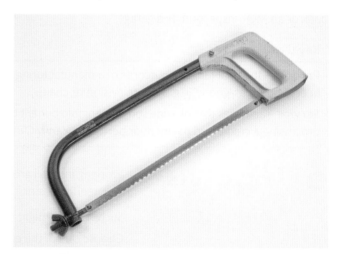

▶ **Figure 2.3** A hacksaw

HSS hacksaw blades are necessarily hard, but as a consequence they can also be brittle and have a tendency to snap when bent, twisted or used incorrectly. These problems can be overcome to an extent by using semi-flexible bi-metal blades. These have a strip of HSS teeth welded to a tougher and more flexible, but also softer, steel backing.

General-purpose carbon steel blades, which have been differentially hardened to make the teeth hard but leave the back tough and flexible, are also commonplace. Generally these are suitable only for occasional use or use with soft materials. They tend to wear quickly and become blunt when used with harder materials such as steel.

Like other saws, hacksaw blades are available with different teeth pitches. These are usually defined in terms of the number of teeth per inch (TPI). For general-purpose work, a 24 TPI bi-metal blade is ideal.

Cutting internal threads

Internal threads are cut by hand with taps held in a tap wrench (**Figure 2.4**).

▶ **Figure 2.4** A thread-cutting tap held in a tap wrench

Before using the tap to cut a thread, you need to drill a hole with the appropriate diameter for the thread required. This diameter is known as the tapping drill size and can be obtained from appropriate engineering workshop reference tables (such as **Table 2.2**).

Taps are generally provided in sets of three with progressively shorter leads. The first-cut (or taper) tap has a long lead-in to enable you to start the thread in the plain hole. You must ensure that the thread is aligned with the axis of the hole. The second-cut tap has less of a lead-in and can be used to increase the depth of the thread. The third-cut (or plug) tap can be used to finish the thread-cutting through the full thickness of the material or to the bottom of a blind hole.

Cutting external threads

You cut external threads by hand using a split button die secured in a die holder (**Figure 2.5**).

▶ **Figure 2.5** A split button die in a die holder

First, you need to prepare a rod of material of the correct diameter for the thread required. This is the outside diameter of the thread, so for an M8 thread you will need rod of 8.0 mm diameter.

Like taps, split button dies also have a lead-in, so you must similarly orient them correctly in the die holder. To enable easy starting on the first cut, you can spread the split die a little using the central screw on the die holder. Again, you need to take great care at this stage to ensure that the die is aligned properly so that you don't end up with what is known as a drunken thread. You then make the second and third cuts by progressively closing the gap in the split die.

Secondary machining

Types of workshop machine include pillar drills, lathes and milling machines. You typically use these to make holes, slots or other features in a single workpiece by removing the material required by the component design.

Drilling using a pillar drill

You primarily use a pillar drill (**Figure 2.6**) to cut circular holes in a range of materials. The pillar drill has the advantage over hand drills in that the rotating chuck moves vertically up and down by means of an operating handle and the depth of cut can be controlled using adjustable stops.

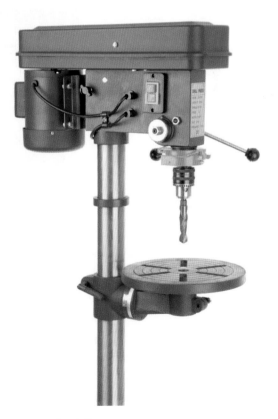

▶ **Figure 2.6** Pillar drill

- Features – through holes, blind holes, countersinking, flat-bottomed holes, counterbored holes.
- Tooling – straight-shanked twist drill bits (1–13 mm diameter), taper-shank twist drills (>13 mm diameter), centre drills, countersinks, counterbores, flat-bottomed drills.
- Tool holding – keyed or keyless Jacobs chuck.
- Work holding – machine vice, clamps.
- Parameters –
  - the spindle speed (N) in revolutions per minute (rpm) must be selected before drilling. This will depend on a range of factors, but is generally found using the formula $N = \dfrac{1000S}{\pi D}$, where D is the diameter (in millimetres) of the hole being drilled and S is the recommended cutting speed (in m/min) of the material being drilled, which can be obtained from general workshop data tables and charts.
  - You can adjust the spindle speed by moving the spindle drive belt between pulleys. The machine must be isolated from its power source while this procedure is carried out.

Turning using a centre lathe

You use a centre lathe (**Figure 2.7**) to perform turning operations to produce round features on cylindrical components. Unlike other machining processes that involve turning, the tool remains stationary while the workpiece is rotated.

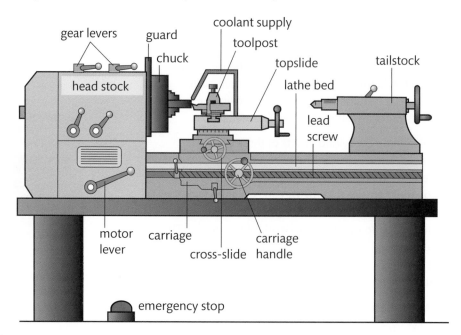

▶ **Figure 2.7** Schematic of centre lathe

- Features – flat faces, parallel diameters, stepped diameters, tapered diameters, chamfers, grooves, knurls, axial drilled holes.
- Tooling – turning tool, facing tool, parting tool, knurling tool, centre drills.
- Tool holding – turning tools are secured on an adjustable toolpost and must be set to the correct height and cutting angle when installed; a tailstock-mounted Jacobs chuck is used for drilling operations.
- Work holding – cylindrical workpieces are held in a three-jaw chuck; square or rectangular workpieces are held in a four-jaw chuck.

▶ Parameters –

▶ The cutting speed (N) in revolutions per minute (rpm) must be selected before turning. This will depend on a range of factors, but is generally found using the formula $N = \dfrac{1000S}{\pi D}$, where D is the diameter (in millimetres) of the workpiece being turned and S is the recommended cutting speed (in m/min) of the material, which can be obtained from general workshop data tables and charts. You can change the cutting speed by setting the gear change levers as required.

▶ The feed rate is the rate at which the carriage (and so the toolpost and tool) moves parallel to the workpiece. It is usually stated in millimetres moved per revolution (mm/rev). This will usually depend on the surface finish required because the lower the feed rate the better the finish. For deep roughing cuts, a high feed rate will enable lots of material to be removed quickly. When finishing a workpiece, shallow cuts at a low feed rate will give a good surface finish.

Milling using a vertical milling machine

Milling is a machining operation that uses a rotating cutter to remove material from a workpiece (**Figure 2.8**). During operation the position of the rotating cutter is fixed, and you move the machine bed to which the workpiece is clamped horizontally and vertically as required.

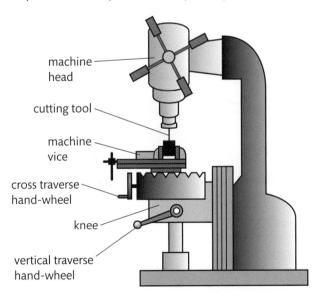

- machine head
- cutting tool
- machine vice
- cross traverse hand-wheel
- knee
- vertical traverse hand-wheel

▶ **Figure 2.8** Schematic of vertical milling machine

▶ Features – flat surfaces, holes, grooves, slots, steps.

▶ Tooling – end mill, slot drill, face mill.

▶ Tool holding – milling tools have standard shank diameters and are held in an appropriate collet which grips over a large surface area to prevent any tool movement or slippage while in use.

▶ Work holding – machine vice, clamps.

▶ Parameters –

▶ The spindle speed (N) in revolutions per minute (rpm) must be selected before milling. This will depend on a range of factors, but is generally found using the formula $N = \dfrac{1000S}{\pi D}$, where D is the diameter (in millimetres) of the milling cutter and S is the recommended cutting speed (in m/min) of the material, which can be obtained from general workshop data tables and charts.

▶ The vertical and horizontal feed rates generally depend on the surface finish required. A low feed rate will give a better finish but only remove small amounts of material.

Fabrication

Instead of removing material from a solid workpiece as with machining processes, fabrication involves the forming and joining of sheet metal to components in order to manufacture thin-walled products such as a toolbox. Some common fabrication processes, such as shearing, rolling and welding, are described below.

Shearing

You can use hand shears (sometimes referred to as tin snips) to cut thin sheet metal. These are available with straight or curved blades and in a range of sizes. A guillotine or press shear can be used to make longer, straighter and more accurate cuts in a wider range of material thicknesses than you can with hand shears.

Forming

You usually making tight and uniform folds in sheet material using a press-mounted V-block and blade or a box press.

Rolling

Rolling allows you to bend sheet material uniformly into large-diameter curves or even tubes by using a series of adjustable rollers.

PAUSE POINT Calculate a suitable spindle speed for drilling a 10 mm diameter hole in a mild steel workpiece.

Hint You will need to refer to the cutting speeds for commonly used engineering materials given in **Table 2.3**.

Extend Investigate the range of spindle speeds available on a pillar drill in your workshops.

In a general-purpose workshop these machines are usually manually operated using long handles or foot pedals. **Figure 2.9** shows an example of a small, hand-operated combined shear, roll and V-block and blade machine that gives three-in-one functionality.

▶ **Figure 2.9** General-purpose combined shear, roll and V-block and blade machine for the small workshop

MIG welding

Metal inert gas (MIG) welding (**Figure 2.10**) is a form of electrical resistance welding where a consumable welding wire is fed from a reel to the welding gun during operation. A high electrical current is fed through this wire and the workpiece, and a high-temperature electrical arc at the point of contact provides the heat necessary to form the weld. To prevent oxidation of the molten materials as the weld is formed, air is excluded from around the weld as it is formed by using a shield of inert gas (usually argon) fed from a cylinder to the tip of the welding gun. MIG welding is suitable for a variety of material thicknesses, from car body panels to heavy beams, because both the supply current and the wire feed rate can be quickly and easily adjusted.

▶ **Figure 2.10** MIG welding

Spot welding

Spot welding is a form of electrical resistance welding suitable only for joining thin sheet metal components. It works by passing an electrical current through a small area as sheet components are pressed together by the welding electrodes. The temperature in the material between the electrodes is sufficient to melt and fuse the sheets together.

Electrical wiring

Engineered products often contain electrical components that need to be wired up and connected together. Before assembly into a product, you can make a wiring loom, where all the necessary wires are cut to length and fitted with the necessary connectors to join with switches and other components. This saves time during installation and is common practice in the automotive and aerospace industries.

Soldering

You can make reliable and permanent electrical connections by soldering. Solder is a low-melting-point metal alloy (commonly 60% tin and 40% lead) that can be melted safely and easily using a soldering iron (**Figure 2.11**). Molten solder adheres extremely well to the surface of copper wires (and many copper alloys such as brass) and can be used to make low-resistance electrical connections, while also providing good mechanical strength after it has cooled and solidified.

▶ **Figure 2.11** A soldering iron

You solder components on printed circuit boards (PCBs) by passing the board's underside through the surface of a bath of molten solder so that hundreds of soldered joints are made in a single operation.

Electrical connectors

There are many situations in which it is either inconvenient or too time-consuming to make individual soldered connections. Sometimes there is a need for temporary connections, which need to be easily disconnected. You can use many different types of electrical connectors in these cases, including screw terminal blocks, crimped spade connectors and bullet connectors (**Figure 2.12**), as well as plugs and sockets in hundreds of different styles and sizes.

Primary forming

Primary forming is a term used to describe the reshaping of metals without the removal of material. Primary forming processes include casting, forging and moulding.

Casting

You use casting processes when large numbers of components are required or when the shape of a metal component is complex and difficult to machine. Processes take many forms, including sand casting (**Figure 2.13**), investment casting and die casting. All work on the basic principle of pouring molten metal into a die or mould that is pre-formed in the required shape. Once cooled, you can remove the cast component from the die or mould and either use it directly or use secondary machining processes to refine it.

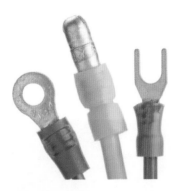

▶ **Figure 2.12** Crimp ring, bullet and spade terminal connectors

▶ **Figure 2.13** Pouring molten iron in sand casting

Link

Casting and forging processes are covered in more detail in Unit 25: Mechanical Behaviour of Metallic Materials.

Forging

Forging does not involve melting the metal, and so the internal grain structure of the material is different from that of cast products that have undergone complete recrystallisation in their new shape. Forging deforms and stretches the internal structure of the material, leaving it considerably stronger and crucially tougher than cast components.

It is possible to cold forge some materials, but you usually need heating to soften the metal and make it malleable so that it can be reshaped.

Traditionally forging was carried out in a blacksmith's forge, and the reshaping was done with an anvil and hammer. More modern forms of forging include drop and press forging (**Figure 2.14**), where material is forced into a shaped die using a series of blows or the application of extremely high pressing forces.

⏸ PAUSE POINT   What process is used in the manufacture of spanners?

   Hint   Inspect a spanner in your workshop.

   Extend   Why do you think this process is used for making a spanner in preference to alternative methods?

▶ **Figure 2.14** A freshly struck, and still hot, drop-forged component ready for removal from the die

Common processes used in engineering services

Disassembly, replacement and refitting

You will often need to disassemble an engineered product in order to maintain, service or repair it. You usually do this by using general tools such as screwdrivers, spanners or hexagon keys.

Make sure that you collect and label all the fixings (such as screws, clips and bolts) so that you can easily identify them during re-assembly.

Parts that have worn during normal operation (such as cam followers, push rods or bearings) should be replaced in the same position and orientations as they were in before removal, so you should also carefully record this information.

You will need specialist tools in certain disassembly and re-assembly operations. These might include security screw fasteners (which require a specific driver bit to remove), bearing pullers or specialist alignment tools.

Inspection

A visual inspection of the condition of certain components, such as those prone to wear like cutting edges on tools, lubricant levels or the condition of the coolant being used on a lathe, is often enough to prompt further investigation and component replacement, adjustment or repair.

Other signs which indicate that further investigation is required might be noisy operation, vibration, over-heating, a blown fuse or any other problem affecting the correct operation of the system.

Systems servicing

Most engineering systems require regular system servicing as part of a preventative maintenance plan. You are probably most familiar with this in relation to motor vehicles, which need to undergo regular oil and filter changes to help prevent premature engine wear. Waste engine oil has to be collected, safely disposed of and replaced with the correct grade and quantity of new oil.

There are hundreds of examples of industrial systems that require similar attention. Engineers in workshops or factories that use compressed air must regularly drain the water that accumulates in tanks and pipework to prevent corrosion. They need to top up lubrication fluids so that the compressed air feed to power tools carries sufficient oil to lubricate the working parts. To perform these procedures safely, you must depressurise the system, and isolate and lockout the compressor to prevent accidental operation while the system is being worked on.

Installation

The installation of new equipment or parts is often necessary in a range of circumstances. This might be the installation of a central heating boiler in a new house, a new machine, such as a laser cutter in a workshop, or a hydraulic winch on an off-road vehicle.

In all these circumstances, the manufacturer of the product to be installed will provide detailed installation instructions.

A2 Health and safety

It is both a legal and a moral responsibility of employers and their employees to take health and safety in the workplace seriously. There is a well-established legal and regulatory framework in the UK that helps to ensure the safety of workers. The following is a selection of some of the more important legislation and regulations that are commonly encountered in engineering.

Health and Safety at Work etc Act 1974

The Health and Safety at Work etc Act (HASAWA) is the cornerstone of British health and safety legislation. It establishes in law the legal responsibilities that an employer has towards their employees as well as visitors and members of the public while on their premises. As a consequence, if you are an employee, apprentice or on work experience in an engineering business, then the employer must provide you with:

▶ safe machinery and systems of work to help prevent accidental injury including the provision of appropriate personal protective equipment (PPE)

▶ a healthy workplace, which should include adequate lighting, heating (or air conditioning) and washing facilities

▶ a safe workplace with appropriate fire alarms, clearly defined exit routes and established emergency procedures

▶ safe methods of storing, transporting, handling, using and disposing of any substances or materials encountered as part of your employment

▶ appropriate training and instruction in how to carry out your duties safely.

Employees (and any other persons with permission to be on site) also have a legal responsibility to look after their own and their colleagues' health and safety, and to cooperate with their employer on health and safety issues.

Suppliers and manufacturers of materials and equipment used in the workplace also have a legal obligation under the Act to ensure these are safe. They must provide adequate information (e.g. data sheets, operating instructions) on the correct use of the products they supply.

HASAWA also established the Health and Safety Executive (HSE) that is responsible for developing national health and safety guidance and enforcing the provisions made in the Act. HSE inspectors have the right to enter any workplace where they believe dangerous working practices are being used. If a serious breach of health and safety law is discovered, notices can be served to halt dangerous work and/or prosecutions made against the company involved.

Reporting of Injuries, Diseases and Dangerous Occurrences Regulations 2013

All accidents and incidents (including minor ones and near misses) in the workplace should be recorded in the company accident book. These records will be used when reviewing risk assessments, and they often prompt the introduction of additional control measures.

Under the Reporting of Injuries, Diseases and Dangerous Occurrences Regulations (RIDDOR), an employer has a legal duty to report certain types of serious accident, incident or dangerous occurrence directly to the Health and Safety Executive (HSE). These reports are used by the HSE to compile statistics on the numbers of different types of accidents. If the HSE suspects that a company has failed to meet its legal responsibilities with regard to the health and safety of its employees, the reports are also likely to prompt an investigation.

An accident or incident at work becomes reportable under RIDDOR if certain types of injury result. These include, but are not restricted to:

▶ death

▶ the fracture of any bone, except a digit or toe

▶ amputation of any part of the body

▶ permanent or temporary loss of sight

▶ any crush that has damaged internal organs in the head or torso

▶ serious burns

▶ scalping that requires hospital treatment

▶ loss of consciousness resulting from a head injury or suffocation

▶ certain injuries that have resulted from working in an enclosed space.

Some occupational diseases must also be notified if contracted while at work, such as occupational dermatitis caused by contact with cement dust, oils or waste materials.

Dangerous occurrences that must be notified to the HSE include:

▶ failure of any crane, lift, hoist or derrick

▶ failure of a pressurised container, such as a tank on a compressor

▶ a forklift truck tipping over.

Personal Protective Equipment at Work Regulations 1992

Employers have an obligation to provide appropriate PPE when its use is deemed necessary by a risk assessment.

PPE must be fit for purpose in order to provide the required level of protection. For example, a boiler suit provided for use by a welder should be manufactured from a flame-retardant material, and dust masks must be of an appropriate type and sufficient quality to provide effective protection against the inhalation of dust or particulates.

All PPE must be provided free of charge to relevant employees, and it must be cleaned and maintained by the employer. Worn or damaged equipment that is no longer fit for purpose must be replaced.

Control of Substances Hazardous to Health Regulations 2002

The Control of Substances Hazardous to Health (COSHH) Regulations are designed to ensure the safe use and handling of the potentially hazardous substances that are often encountered in engineering. For instance, frequent and prolonged skin contact with some common fluids, such as petrol and oil, can cause long-term skin problems, which is why you should use barrier creams and gloves to minimise your exposure.

Some chemicals, such as those used as solvents in paints and glue, can cause more immediate damage or even death if inhaled in sufficient quantities.

Potentially harmful fumes, dust, liquids and chemical vapours can enter your body through a number of mechanisms, including being absorbed directly through the skin or eyes, entering the bloodstream through a cut or broken skin, and being inhaled or even ingested in contaminated food or drink.

Very few substances commonly encountered in engineering will cause immediate acute symptoms, and even fewer have the potential to pose an immediate threat to life in normal circumstances. However, prolonged exposure to a range of substances has been shown to be severely detrimental to long-term health. For instance, exposure to coal dust over the working life of a miner is known to cause chronic lung diseases, and exposure to asbestos, used for years as a heat-proof insulation material, is directly linked to an aggressive and incurable form of lung cancer.

When using or working with potentially hazardous chemicals you should:

▶ ensure that they are stored securely in appropriate and labelled containers
▶ read and understand the labels and/or the manufacturer's data sheets on the safe use of the substances, and follow the recommendations
▶ always use the correct type of PPE to protect you from contact with hazardous substances
▶ investigate the availability of safer alternatives.

Appropriate warnings and labels should be displayed where hazardous materials are present. Examples of internationally recognised signs are shown in **Figure 2.15**.

▶ **Figure 2.15** Examples of COSHH warning symbols that can be found on product labels

Manual Handling Operations Regulations 1992

About a third of all reported injuries at work are the result of people using incorrect manual handling methods when lifting or moving heavy objects. In engineering there is often the need to move heavy tooling, equipment, components or products, so you must be familiar with safe ways in which this can be achieved.

Employers should ensure that their staff avoid moving any items in the workplace in a way that could cause them injury. Where repetitive movement of heavy objects is required, some mechanical aid should be employed to lessen the reliance on manual handling. For instance, on a vehicle production line, heavy sub-assemblies such as doors are picked up and supported in position by mechanical hoists while they are bolted into place.

Pallet trucks, trolleys and conveyors are all widely used to lift and move objects around factories and production lines. If manual handling cannot be avoided, then it is important that staff be trained in the safest methods and techniques to minimise any risk of injury.

Research

Carry out your own research to find out more details about employer and employee responsibilities under the health and safety requirements mentioned above.

⏸ **PAUSE POINT**   Identify any substances in your workshop that are potentially hazardous.

Hint   To comply with COSHH regulations, these substances will be locked away, so you'll need a technician or tutor to join your investigation.

Extend   Investigate the availability of safer alternatives.

A3 Human factors

Human factors affecting productivity

The productivity of any organisation depends on the people carrying out the processes. Their skill level, training, experience, enthusiasm and conscientiousness all play a part.

Reliability

An important factor in maintaining productivity is that staff arrive at work on time and have a good attendance record. Absenteeism increases pressure on fellow workers as the absentee's duties need to be performed by others who perhaps lack the relevant knowledge and experience or who may become overloaded. This inevitably leads to a decrease in work output.

Quality

In modern factories, quality assurance systems go well beyond the traditional quality control checks carried out at the end of the manufacturing process. All staff are expected to perform continual quality checks as they work to ensure that any problems they encounter or perhaps even cause are not passed on to subsequent stages without being addressed. This requires a certain level of trust that is cultivated in a team-oriented environment where allocating blame is less important than addressing and solving problems together. It is vital that individuals are conscious of the importance of the quality of their own work and the work of their team and the potential effects on the wider organisation.

Safety

As noted in the 'Health and safety' section above, it is a legal requirement that all employees in an organisation manage their own health and safety and do nothing to jeopardise the safety of others. Safe working practices also help to maintain productivity. Stoppages while incidents and accidents are investigated or damage repaired can be significant and have a disastrous effect on productivity, so it is an employee's ethical duty to ensure that nothing harms their colleagues or equipment.

Human factors affecting the performance of individuals and teams

Professionalism

The regulatory body for the engineering profession in the UK is the Engineering Council. It is the role of such professional bodies both to set and to regulate the standards of technical competence, commitment and ethical behaviour of its members. These are laid out in the UK Standard for Professional Engineering Competence (UK-SPEC).

Some of the expectations of professional conduct for engineers are:

- to operate and act responsibly, avoiding negative environmental, social or economic impact when conducting work-related activities
- to accept responsibility for your own work and that of others
- to exercise personal responsibility – for example, managing a project through to completion to an agreed deadline
- to manage and apply safe systems of work – for example, carrying out risk assessments
- to take responsibility for identifying your own limitations and training requirements
- to carry out continuing professional development (CPD) activities to enhance operational knowledge and competence
- to act with due care, skill and diligence throughout your professional life.

Ethical principles

If an individual, team or larger organisation is seen to conduct themselves ethically with a strong sense of right and wrong, then they are perceived as trustworthy and reliable. Some of the qualities displayed by those considered to be acting ethically include:

- Rigour – approaching each task or project in a thorough and careful manner.
- Honesty – being truthful and straightforward in dealings with other people (and yourself).
- Integrity – able to recognise right from wrong and willing to do the right thing.
- Respect – displaying regard for the feelings and opinions of others, valuing their contribution.
- Responsibility – being accountable for your actions and, when things go wrong, willing to admit you were wrong and work to put things right.

Behaviours

Some personal characteristics and forms of behaviour can influence how individuals are perceived and affect their roles within a team. For instance:

- Strong values – maintaining your principles and standards of behaviour is important if you are to retain your self-respect and gain the respect of other people.
- Attitude – maintaining a positive outlook can help to bolster the morale of a team under pressure.
- Persuasion – getting people to embrace change or try something new is best achieved by encouragement, enthusiasm and reasonable argument. An individual who has given their support and agreement to a course of action through collaboration will be more committed and effective than if they were coerced.

- Coercion – an unacceptable form of forceful persuasion, which might include threats or emotional blackmail, that will cause long-term damage to relationships and morale.
- Rapport – establishing good relationships with groups or individuals by understanding their thought processes, cares and concerns and communicating clearly and effectively with them is a valuable skill.
- Authority – although the person in charge of a team or organisation has the right to make decisions and direct the activities of others, the way in which they choose to exercise their authority can be damaging if it is not done respectfully.

Limitations

There are both physical and mental limitations to the activities that can be undertaken by an individual or a team in a busy working environment. You must consider the following:

- Stress – a sense of mental and emotional strain or anxiety that can be brought about by the demands and pressure of work.
- Time pressure – having to complete tasks in a fixed time frame or at a defined rate.

- Fatigue – physical or mental tiredness.
- Memory – it is only possible to retain a limited number of facts and effectively process a limited amount of information at any one time.
- Capability – the extent of someone's ability to complete a task.
- Motivation – the reasons why people act in the ways they do; put simply, it is their desire to do certain things. People tend to be motivated by a variety of factors, and it shouldn't be assumed that all people are alike in what they find motivating.
- Knowledge – the depth of a person's technical knowledge must be sufficient for the task in hand.
- Experience – generally people become quicker and more effective at performing tasks or processes if they have done them before.
- Health – some physical tasks are unsuitable for those with particular health issues or physical limitations.

Other factors, such as the consumption of drugs or alcohol, can seriously affect an individual's ability to perform tasks safely.

II   PAUSE POINT   Which ethical and behavioural characteristics do you believe you already possess?

Hint   Go through the list one at a time and be honest about your assessment of yourself.

Extend   Ask a colleague to carry out the same procedure, but to look for the characteristics they recognise in you. Do the two lists match?

Assessment practice 2.1   A.P1   A.P2   A.M1   A.D1

You are working in a small engineering company, and your manager has asked you to take a look at a prototype of a screwdriver and determine the best way to manufacture five more as samples for potential clients. Your recommendation should be fully justified and include a comparison with alternative processes.

If the screwdriver eventually goes into production, it will be required in large numbers. Your manager has explained that when demand reaches a certain level, a fully automated manufacturing system will be introduced that will eliminate human involvement in the process almost entirely.

This would require a significant investment, and your company is keen to establish how a range of human factors could actually benefit the performance of engineering processes involved in manufacturing.

Your manager has asked you to write a technical report that will address the issues that she will present to the operations director of the company. Care must be taken to ensure that a high standard of written language is used throughout.

Plan
- What is the task? What am I being asked to do?
- How confident do I feel in my own abilities to complete this task?
- Are there any areas I think I may struggle with?

Do
- I am confident in what I'm doing and know what I should be achieving.
- I can identify where I've gone wrong and adjust my thinking/approach to get myself back on course.

Review
- I can explain what the task was and how I approached its execution.
- I can identify the parts of the task that I found most challenging and seek ways that will help me to overcome any difficulties.

B Develop 2D computer-aided drawings that can be used in engineering processes

> **Link**
>
> Unit 10: Computer-Aided Design in Engineering covers many of the techniques and concepts required by this learning aim. Unit 10 also includes additional background material that you will find helpful when you practise using CAD software to produce your own drawings

B1 Principles of engineering drawing

Orthographic projections

An orthographic projection is a method of positioning the different views of an object when representing it in an engineering drawing, such as that in **Figure 2.16**. The rules that dictate where each view goes in an orthographic projection can be confusing, so you may find the following rules of thumb helpful.

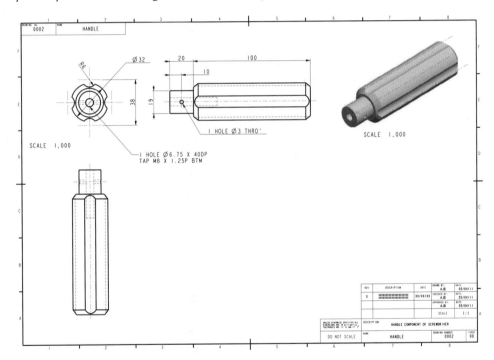

▶ **Figure 2.16** Third-angle orthographic projection of a screwdriver handle component

Third-angle orthographic projection

The standard symbol (**Figure 2.17**) that you will find on a drawing arranged in a third-angle projection looks very much like something that you are probably familiar with – a traffic cone. This will help remind you how to set out the drawing.

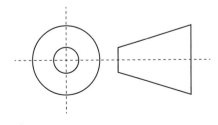

▶ **Figure 2.17** Symbol for third-angle projection

Worked Example

Draw a third-angle orthographic projection of a real traffic cone.

Solution

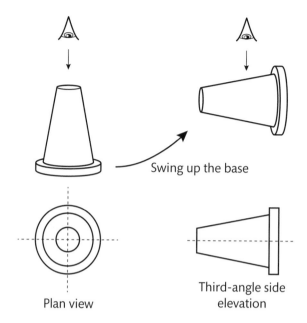

Swing up the base

Plan view

Third-angle side elevation

▶ **Figure 2.18** Rule of thumb to create a third-angle projected view

Step 1: Draw the plan first – this shows the view of the object from directly above. In this case the plan view consists of two concentric circles, representing the tapering section of the cone, inside the larger circular base (left-hand side of **Figure 2.18**).

Once this has been drawn, all the other views in the drawing are projected from the plan.

Step 2: Now imagine picking up the traffic cone by its narrow pointed end and swinging the bottom of the cone out in the direction of the next projected view to be generated (a side elevation). What you now see from above is the way this side of the cone should be drawn as a third-angle projection (right-hand side of **Figure 2.18**).

Step 3: Repeat this process to create as many views as are necessary to show all the features on an object.

First-angle orthographic projection

The standard symbol for a first-angle orthographic projection is shown in **Figure 2.19**.

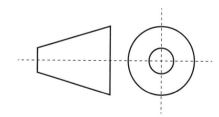

▶ **Figure 2.19** Symbol for first-angle projection

Worked Example

Draw a first-angle orthographic projection of a real traffic cone.

Solution

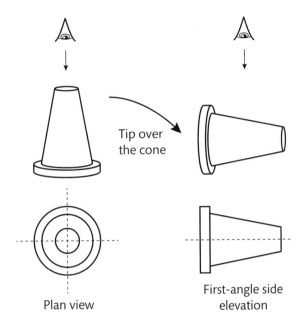

▶ **Figure 2.20** Rule of thumb to create a first-angle projected view

This follows a very similar method to that used above for a third-angle projection. The only difference lies in Step 2: instead of picking up the cone and swinging the base out, you just tip it over in the direction of the required projected view.

Dimensions

Dimensions are an essential element of any engineering drawing. It takes only one missing or poorly formatted dimension to cause unnecessary confusion and possible delays.

Dimensioning should:

▶ not interfere with the component drawing lines or any other features

▶ be neatly aligned and spaced out evenly

▶ be consistent in the use of font, text size and style

▶ be clear and easy to interpret

▶ comply with the requirements of an appropriate standard (such as BS8888).

Examples of good practice when drawing linear dimensions and the different terms used when referring to the features of dimensions are shown in **Figure 2.21**.

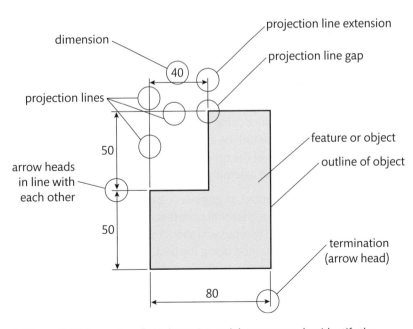

▶ **Figure 2.21** Examples of good practice and the terms used to identify the features of dimensions

Several acceptable methods can be used to dimension circles; these are shown in **Figure 2.22**. The actual method to use often comes down to personal preference, but the underlying principle here, as with all dimensioning, should be to choose the method that gives optimum clarity.

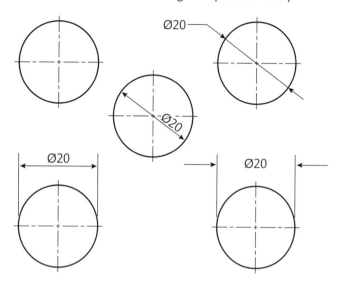

Tolerances

Tolerances define the allowable variation between the measured dimensions of a completed product and the design dimensions stated on an engineering drawing.

Generally speaking, a reduction in tolerance means an increase in associated manufacturing cost. As tolerances become smaller, costs can rise because the machinery used in manufacturing needs to be capable of working to the required accuracy and precision.

The tolerances required will depend on the product being manufactured. For manufacturing a sheet metal wheelbarrow, the tolerances are likely to be large (of the order of ±2 mm), because this level of variation will not detrimentally affect the functioning of the product. At the other extreme, a jet engine turbine assembly will require manufacturing tolerances of the order of ±0.005 mm to prevent vibration when the assembly rotates at high speed.

Tolerances can be represented on engineering drawings in a number of ways. Again, personal preference plays a part in which method you use, but you should stick to the principle of the clearest method being the best.

Research

Search the internet to find different ways in which tolerances can be represented on engineering drawings.

Material

An engineering drawing generally states the material from which the component is made. This should be specified as a particular type, grade or alloy.

Surface finish

The roughness of a finished surface is defined in terms of its roughness average (Ra) measured in micrometres (μm), which is the arithmetic mean of the sizes of the peaks and troughs present on a surface when viewed under a microscope.

You should state the Ra value for each surface shown on an engineering drawing. For example, a cast iron engine block has a relatively rough cast finish (Ra ≈ 20 μm) over most of its surface, but on bearing or mating surfaces, such as the cylinder bores or where the head is mounted, the surfaces are ground to a much lower roughness (Ra ≈ 0.2 μm).

Figure 2.23 shows the standard symbol for surface roughness. In this case Ra = 3.0 μm is specified. This is typical of the finish left by turning and milling operations.

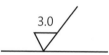

▶ **Figure 2.23** The standard symbol for surface roughness (indicating Ra = 3.0 μm)

Scale

It is important to select a suitable scale for any engineering drawing you are asked to complete. Few things will fit conveniently on a standard sheet of A4 or A3 paper. More often than not, objects are drawn either larger or smaller than life-size.

Small objects, such as the components of a mobile phone, would usually need to be drawn larger than in reality so that sufficient detail can be shown with the clarity necessary. When making small objects larger in a drawing, it is conventional to use one of the following scaling factors: 2 : 1, 5 : 1, 10 : 1, 20 : 1 or 50 : 1.

Large objects, such as a car door, would be too large to draw full size on a sheet of A3, so they should be drawn to a smaller scale. When making large objects smaller in a drawing, it is conventional to use one of the following scaling factors: 1 : 2, 1 : 5, 1 : 10, 1 : 50 or 1 : 100.

Drawing conventions

BS8888

The accepted conventions used in engineering drawings and other documentation for specifying a product are laid out in British Standard BS8888 *Technical Product Documentation and Specification*.

The topics covered in BS8888 relevant to this unit are: scales, dimensioning, tolerancing, surface finish, lines, arrows and lettering, projections and views, and symbols and abbreviations.

BS60617

A range of standard electrical and electronic symbols for use in wiring diagrams and circuit layouts are defined in BS60617.

General layout and title block

A standard layout should be used for all the drawings produced by any one organisation. Layouts will differ slightly between companies, but they should all be based on the conventions laid down in BS8888.

▶ A drawing's extent is defined by a border. This often includes vertical and horizontal divisions that form a coordinate system for identifying specific features in a drawing.

▶ The title block, usually positioned in the bottom right corner of a drawing, contains basic information about the drawing itself, including:

  ▶ drawing number – a unique reference number that can be used to identify the drawing

  ▶ projection symbol – a symbol that specifies whether the drawing views are arranged in first- or third-angle orthographic projection

  ▶ scale – the scale applicable to the component drawn

  ▶ units – the units in which the dimensions are stated

  ▶ general tolerances – the tolerances that apply where no specific tolerance is stated on a dimension

  ▶ name of author – provides traceability back to the person who created the drawing

  ▶ date – specifies when the drawing was completed.

▶ Parts referencing – when the drawing depicts an assembly, it is common for each element to be labelled with a balloon note and referenced in a table that contains additional information such as part numbers and a description.

Views

The different views portrayed in engineering drawings are:

▶ **Plan** – the view of an object from directly above.

▶ **Elevations** – the front, back and side views of an object or component. Where only a single side view of an object is required, this is sometimes referred to as the end view.

▶ **Section** – sometimes it is necessary to provide details of a cross-sectional slice through a component to reveal internal or otherwise hidden details. This is provided in a section view.

▶ **Hatching** – used to define areas that have been sectioned.

▶ **Auxiliary view** – if a component cannot be fully represented by the standard orthographic plan view and front, back and side elevations, then an additional auxiliary view might be required. This allows features that are not aligned with standard orthographic views to be shown and clearly dimensioned.

Line types

There are several standard line types that can be used to represent the features on an engineering drawing. These are shown in **Table 2.4**.

▶ **Table 2.4** Standard line types used in engineering drawings

Line type	Usage	Example
Continuous thick line	Visible outlines and edges	
Continuous thin line	Dimension projection and leader lines, hatching	
Chain thin line	Centre lines of symmetry	
Dashed thin line	Hidden outlines and edges	

> **Link**
>
> Unit 10, Figure 10.2, shows some other standard line types used in engineering drawings.

Circuit diagram symbols and components

Commonly used electrical and electronic components are represented on circuit diagrams using standard symbols, which include those shown in **Table 2.5**.

▶ **Table 2.5** Some component symbols used in engineering drawings

Component	Symbol
Cell	
Battery	
Switch – single pole single throw (SPST)	
Resistor	
Diode	
Capacitor (polarised)	

▶ **Table 2.5** Some component symbols used in engineering drawings – *continued*

Component	Symbol
Transistor (NPN)	
Integrated circuit	
Light emitting diode (LED)	
Motor	M
Buzzer	

> **Link**
>
> More information can be found in Unit 19: Electronic Devices and Circuits.

Abbreviations

A number of standard abbreviations can be used in engineering drawings. **Table 2.6** lists some of these.

▶ **Table 2.6** Some abbreviations used in engineering drawings

Abbreviation	Meaning
A/F	across flats
CHAM	chamfer
DIA	diameter
R	radius
PCD	pitch circle diameter

Lettering

It is vital that the lettering used in the title block, notes and dimensions are of a size and type that ensures clarity and legibility. To facilitate this, all lettering should be in upper case and additional formatting such as underlining should be avoided.

On drawings sized A2, A3 and A4 the lettering used in the title block for important information such as the drawing number and title should have a minimum height of 3.5 mm.

Lettering used in notes and annotations elsewhere on drawings of this size should have a minimum height of 2.5 mm.

General notes should be grouped together in one place on the drawing wherever sufficient space is available. However, notes relating directly to a specific view or feature may be located adjacent to that view or feature.

All notes should be aligned in the same direction, conventionally this will be parallel to the bottom of the drawing.

Common features

Often time and effort can be saved by using simplified representations of common features and components in engineering drawings, such as springs, threads and splines (see **Figure 2.24**).

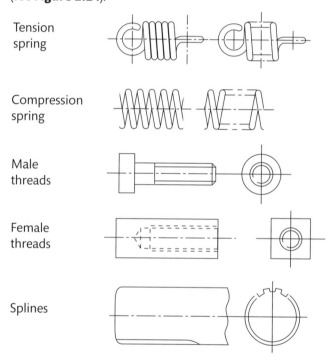

Tension spring

Compression spring

Male threads

Female threads

Splines

▶ **Figure 2.24** Simplified representations of springs, threads and splines used in engineering drawings

B2 Two-dimensional (2D) computer-aided drawing

Coordinates

A computer-aided design (CAD) package displays a system of coordinates on-screen. These coordinates allow you to navigate around a drawing and to specify accurately the positions of features such as the centre of a circle or the start and end points of a line.

There are three coordinate systems available for use in most CAD packages:

- **Absolute coordinates** – these define your position in the drawing using X and Y Cartesian coordinates with reference to a fixed datum point or origin, often the bottom left corner of the drawing.
- **Relative coordinates** – these let you choose a position as the origin of a Cartesian coordinate system. This is useful when you are drawing a number of different features on the same diagram. For example, if you define the starting point of a line as the origin of the coordinate system, then the end of the line can be specified more easily using X and Y coordinates.
- **Polar coordinates** – these are a type of relative coordinate system that is particular useful when positioning features in circular arrays. Instead of using distances along two perpendicular (X and Y) directions, polar coordinates use an angle and a distance from the origin to specify a position.

Templates

The starting point for most CAD drawings will be a basic template, which already contains a border and a partially completed title block. Using a standard template saves time and helps to ensure that all the drawings an organisation produces look similar and contain all the required information in the title block.

A template will probably be the first thing you create in CAD and will be used in all your subsequent drawings, so it is important to get it right.

Layers

Layers in a CAD drawing can be used to overlay different information on a common drawing outline. Each feature you add to a drawing will belong to a specified layer and can be moved to a different layer if needed. You can specify any number of layers in a drawing. You can give the layers names and colours and control them in different ways. For instance, if the drawing outline and dimensions are on separate layers, then you have the option to hide the dimension layer so that only the drawing outline is visible.

Common commands in CAD

- Line – you create straight lines by specifying a start and an end position. You can do this by using a mouse to click between grid snaps or by clicking with the mouse to start a line and entering the relative coordinates of its end point.
- Circle – you can define circles in a number of ways. A common method is to define the centre of the circle (in the same way as you define the starting point of a line) and then input its diameter.
- Arc – this is part of the circumference of a circle. A common method of drawing an arc is to define the centre of the circle and the start and end points of the arc.

- Polygon – regular polygons include shapes such as squares, rectangles and hexagons, which are all commonly required in engineering drawings.
- Chamfer – this describes the replacement of a sharp (90°) corner with an edge of a specified angle.
- Fillet – this describes the replacement of a sharp (90°) corner with a curve of a specified radius.
- Grid – a system of on-screen dots that help you lay out the drawing. You can specify different grid spacings appropriate to the size and scale of the drawing you are producing.
- Snap – this allows you to use the mouse pointer to specify an exact position in a drawing. In its simplest form, when you click the mouse button, a snap will jump to the grid point nearest to the mouse pointer. Snapping is useful when you want to make a seamless join with the end point or midpoint of a line, or with intersections and numerous other entities on your drawing.
- Copy – replicating an **entity** or group of entities that make up a component on a drawing.

> **Key term**
>
> **Entity** – a discrete element of a drawing, such as a line or circle.

- Rotate – turning an entity or group of entities around a given centre by a specified angle.
- Erase – deleting an entity or group of entities from the drawing.
- Stretch – resizing an entity or group of entities by specified amounts in the X and Y directions.
- Trim – removing specified sections of overlapping lines.
- Scale – establishes the scale of the engineering drawing being produced, letting you use the actual dimensions of an object when creating the drawing.
- Dimensioning – methods of dimensioning the features of your drawing will vary depending on the particular CAD package being used. Default dimension styles and formats are generally pre-set in the program preferences so that they will be used consistently. The actual value for a dimension is usually generated automatically according to the exact size of the drawing you have produced, but positioning will be up to you.
- Text – text boxes are frequently used in drawings to provide additional information about a component. Again, the default style and format used will generally be pre-set in the program preferences so that the text will have a consistent appearance.

- Pan – moving a particular area of a drawing so that it is more conveniently positioned on the screen. It is often used in conjunction with the zoom function.
- Zoom – you zoom in to enlarge a particular area of a drawing and zoom out to display more of the drawing.
- Cross-hatching – most CAD packages will have a number of pre-defined hatch patterns that are usually used to denote areas of cross-sectioning. Methods of applying cross-hatching differ between CAD packages, but generally cross-hatching can be applied only to an area that is fully bounded by a continuous line.

- Importing standard components or symbols – one of the advantages of CAD is that there are libraries containing thousands of standard pre-drawn components or symbols that can be imported into your drawings. This can save a great deal of time and effort when producing circuit diagrams.

> **Link**
>
> See Unit 10 for more details on computer-aided design in engineering.

❚❚ PAUSE POINT Experiment by exploring the capabilities and commands available in the 2D CAD package you will be using.

Hint Start by drawing simple entities such as lines and circles before attempting to join them together.

Extend Find different ways of drawing similar features. How many ways can you draw a circle in the CAD package you're using?

Assessment practice 2.2 · B.P3 · B.P4 · B.M2 · B.D2

A colleague in technical sales has been working with a client on a new product and has brought in some hand sketches they have drawn up during a recent meeting.

The first sketch is of a stepped drive shaft with a milled keyway cut in one end.

The second is of an electronic circuit that monitors the temperature of the oil in the gearbox that drives the shaft.

Your colleague has asked you to use the information in the sketches to produce an engineering drawing for the shaft and a circuit diagram for the temperature sensor. Use a 2D CAD package to complete these drawings, ensuring that they both comply with appropriate international standards.

Plan
- What is the task? What am I being asked to do?
- What tools will I need in order to complete the task?
- How confident do I feel in my own abilities in using the tools? Are there any areas I think I may struggle with?

Do
- I know what I'm doing and what I am aiming to achieve.
- I can identify the parts of the task that I find most difficult and devise strategies to overcome those difficulties.

Review
- I can explain what the task involved and how I approached the work.
- I can explain how I would deal with the hard elements differently next time.

C Carry out engineering processes safely to manufacture a product or to deliver a service effectively as a team

C1 Principles of effective teams

If teams are going to work together to achieve a common goal, then every member of the team needs to understand what that goal is, what it will look like when it is achieved, the part they are expected to play and how this fits with the activities of other members of the team.

Good communication

Clear communication within teams and by team members to external stakeholders and interested parties is vital if the project is to be effective. Communication skills are important and can take many forms:

▶ Verbal – The language you use and the clarity with which you are able to get your message across are extremely important when talking to people. You are likely to need to enunciate your words carefully and talk more loudly during a formal presentation. You should always remain conscious of the reaction of your audience – they will usually make you aware if they are unable to hear or understand you properly.

▶ Written – The quality and clarity of written communications must always be high and should be tailored to suit different readers. A technical report will use appropriate technical terms, impersonal objective language and a structure appropriate to its readership. In contrast, a letter requesting a meeting to introduce a new product to a potential customer will be formal but friendly, because it seeks to persuade the reader to consider using your services. Use of correct spelling and grammar may not be sufficient to make a good impression, but repeated mistakes in a letter or report will always leave a poor one.

▶ Effective listening – Other team members can only make effective contributions if they are encouraged to express their thoughts during meetings and discussions. This means giving them the opportunity to contribute. You must listen carefully to what they have to say without curtailment or interruption. Ask questions to make sure you really understand what they are saying – it could be crucial to the success of your project. Maintain eye contact and provide encouragement by using non-verbal cues such as nodding as people speak.

▶ Respect for others' opinions – Although you may not agree with all that someone else brings to a discussion, it is important that they have the opportunity to express themselves and are treated with respect. This will encourage them to keep contributing their ideas, some of which might prove valuable.

▶ Negotiation – Negotiation is a process used to reconcile differences between parties. This could mean haggling over the price of a contract or trying to agree how the workload should be distributed within a team. Negotiation should be carried out in a reasonable and fair way, with the aim of establishing a compromise that is acceptable to all parties. Nobody should walk away from a negotiated settlement feeling that they have been treated unfairly or unreasonably.

▶ Assertiveness – A team leader must have a certain level of assertiveness to establish and maintain the direction in which the team should proceed. But don't confuse assertiveness with being dictatorial. Being assertive can help to build loyalty and maintain the focus of the team as they move forward with you. Being dictatorial and imposing your opinions without establishing a general consensus breeds resentment and disquiet within a team.

▶ Body language (non-verbal actions) – How you present yourself can often have an unconscious effect on how others perceive you. Maintain eye contact when you are talking to people; smile and shake hands when being introduced to colleagues; use your arms to help emphasise what you are saying; and definitely keep your hands out of your pockets!

Planning

The following are important when planning a project:

▶ Setting targets – A team needs to have a clearly defined target that they are working together to attain. It is important that all members of the team understand what the final outcome of their activities should be and the consequences to the organisation of missing the deadline for their target.

▶ Considering alternative approaches – The full team should meet early in the project to brainstorm ways by which their target might be reached. If everyone is involved in this way from the very start, team members will feel more involved in the project and are more likely to buy in to the decisions reached and the processes chosen.

▶ Organisation – see Section C2 'Team set-up and organisation'.

⏸ **PAUSE POINT** In small groups, take it in turns to express your mood using body language only.

Hint Try the easy ones first, such as happiness, sadness and excitement.

Extend Try to express more complex emotions, such as boredom, frustration or disbelief, and see if your colleagues interpret them correctly.

Motivation

Maintaining the motivation of individuals and teams is essential if they are to function effectively. There are many factors that can influence motivation.

▶ Shared goals – A sense of shared enterprise towards established and achievable goals will help to sustain the motivation of individuals in a team.

▶ Collaboration – People generally enjoy working collaboratively with their colleagues in situations where their opinions and suggestions are valued and they feel they are making a valuable contribution to a larger process.

▶ Reaching agreements – Effective negotiation will leave all parties feeling fairy treated when agreeing workloads or allocating responsibilities within a team. People generally resent having decisions imposed upon them without consultation or discussion. Change is best implemented by reaching mutual agreement where possible.

▶ Fairness – A team leader must act impartially in all matters. Any hint of favouritism should be avoided because it will have a negative impact on the morale of other team members.

▶ Opportunities to take responsibility – Encourage team members who are unused to managerial roles or positions of responsibility to take charge of parts of a project in which they have valuable experience or specialist knowledge. For instance, the setter who has worked with and maintained a specialist machine will have important knowledge to organise its transfer and re-installation.

▶ Constructive feedback – Comments relating to the work carried out by members of a team should be both timely and constructive. When things go wrong, other members of the team need to know quickly. Any issues raised should be worked on and resolved collaboratively with colleagues that have been affected by the problem.

In general, the allocation of blame will neither help solve a problem nor improve the performance of a team or individual. Working together to solve issues will help to cement relationships and demonstrate to those involved the importance of their part in the bigger picture and the potential effects their actions can have on the work of others.

Remember the rule of thumb that ninety per cent of feedback should be positive. The motivating effect of thanking someone for a job well done and confirming that you appreciate the efforts being made by others is invaluable in raising morale and maintaining motivation – but be careful not to give false praise because this can mislead.

Working with others

There are several aspects of being able to work successfully as a team.

▶ Being a team player – To work successfully in a team, you need to open up to the idea that your colleagues might actually have some useful input, which could improve your way of working. In a multi-disciplinary team, for example, some of the tasks you currently perform might be done more effectively or efficiently by a colleague. Sometimes being part of a team means putting your feelings and opinions to one side and doing what is best for the team as a whole.

▶ Flexibility/adaptability – Your willingness and ability to carry out multiple roles within a team will make you a valuable asset to any organisation. Working together is often about concentrating greatest effort where it is needed most at any one time. This might be in an area that is outside your core competency or specialism but in which you could effectively contribute if you are prepared to adapt.

▶ Social skills – It is important to remember that you must practise good language and behavioural skills at all times in the workplace so that you don't cause embarrassment or offence. Some colleagues or team members might become close personal friends over time.

▶ Supporting others – You will work closely with others in a team. A sense of loyalty and camaraderie among team members is important. Mentoring and supporting colleagues through difficulties both inside and outside of work help to cement these relationships.

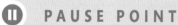 **PAUSE POINT**   What are the main benefits of working in a team?

    Hint   Try to think of an example from your own personal experience.

    Extend   What, if any, are the potential pitfalls of being part of a team and working closely with others?

The working environment

Here are some attributes of a working environment that are conducive to successful outcomes.

▶ Safe – Health and safety at work is vitally important to you and your team members, and everybody should take responsibility to ensure that safe working practices are rigidly adhered to, even if this means that a particular process might take a little longer to complete.

▶ Supportive – A supportive working environment might mean more than just the friendship of your colleagues. Your workplace might provide on-site childcare facilities to support working parents or agree to you working from home or becoming a part-time employee if you need to establish a better work–life balance. Treating employees as real people and appreciating that an employee's life outside work can and will affect their contribution inside is likely to increase staff loyalty and the long-term retention of staff.

▶ Challenging – Working within your comfort zone on the same kind of work day after day can lead to boredom, complacency and a lack of fulfilment. Most people need a little challenge in their working lives. Boring, unfulfilling jobs lead to bored and unfulfilled employees, and this will in turn harm productivity and the effectiveness of an organisation.

▶ Opportunities to show initiative and leadership – Providing opportunities for employees to show initiative and leadership encourages them to push themselves, develop and achieve their full potential. Effective leaders are rarely born, but are developed in organisations that provide sufficient training, challenge and progression opportunities to their staff.

Case study

The Toyota Production System (TPS)

▶ **Figure 2.25** Car being manufactured on a production line

The Toyota Production System developed by the Japanese car manufacturing giant is the forerunner of what has come to be known as the 'lean philosophy'. Since its development in the latter part of the twentieth century, the lean philosophy has been applied to everything from the delivery of services in hospitals to the manufacture of aircraft.

A cornerstone of TPS is the empowerment of the employees who actually carry out manufacturing operations. Any one of them can stop the production line in order to address a quality problem they have encountered. The changes needed to solve the problem can be made by the team itself immediately, without having to seek further guidance – this minimises delay. Taking collaborative ownership of a process in this way and exercising both individual and joint responsibility for the quality of the work leaving their production cell has meant that TPS provides an environment where near-perfect quality is the norm.

Check your knowledge

How do you think that the TPS approach benefits the employees and the company as a whole?

C2 Team set-up and organisation

Team competencies and development

When establishing a team it is vital to recruit team members who have the correct mix of skills to complete a project successfully.

Professional team members will have a diverse mix of experience that can be passed on to other members to develop their abilities and effectiveness. These development activities help to enhance the flexibility and capability of individuals and of the team as a whole.

Constructive peer feedback

Most multi-disciplinary teams have a fairly flat hierarchical structure, often with a single project manager responsible for coordinating the efforts of the whole team. As such, most opportunities for receiving feedback on your work will come from your colleagues, who might be affected by issues that will benefit from your improvement. Peer feedback can be a very valuable and powerful developmental tool to ensure the smooth running of a project. Problems tend to be exposed quickly and can be addressed immediately.

Roles and responsibilities

The allocation of roles within a team is often a matter of common sense because team members will usually be selected on the basis of their skills or expertise in a specific area. It is important to ensure that sufficient staff are allocated to each task. Core responsibilities will be clearly laid out for each member of the team, but there is usually some flexibility.

Timescales and planning

The project manager usually plans a project and coordinates the efforts of the team involved in its delivery. The schedule can be extremely complex for a large project.

In practice, planning the delivery of projects often moves backwards from a required completion date, and staff and resources are determined so that the project can be completed on time and within budget.

At the heart of the plan is a detailed list of all the activities that need to take place to complete the project and the order in which they should be carried out. Some activities can be performed concurrently, with different parts of the team working on different elements of the project at the same time. Other activities cannot be started until the completion of some previous task or tasks.

You can use flow charts, tabulated data, Gantt charts and/or network diagrams to visualise project planning activities.

Objectives and target setting

It is important that the scope of any project is properly established and its success criteria are well defined. In other words, you must state exactly what it is you are trying to achieve and define clearly how you will know when you have achieved it.

For example, suppose the scope of a project is described as:

'Change the coolant on a centre lathe.'

This is not sufficiently clear – will the project include cleaning out the coolant tank after it has been drained or safe disposal of the old fluid? When is the coolant change going to take place? What will define completion of the project?

A better description might be:

'Change the coolant on a centre lathe. Machine to be taken out of service and handed over to maintenance team at 17.00 on 04/05/16. Maintenance team to complete the following: drain old fluid, clean out storage tank, flush pipework, transfer waste to waste fluid drums (in preparation for collection), refill system with new coolant to equipment manufacturer's specification, test operation of coolant pump and delivery system. Machine to be returned to service and handed back to manufacturing team at 19.00 on 04/05/16.'

During a project or process, a number of milestones or gateways may also be established in the plan to enable the project manager to measure the progress of the project. At these key points in the project, the team might be gathered together to review the work completed so far and suggest any improvements or changes that might be required in the next phase of the project.

C3 Health and safety risk assessment

Risk assessment in an engineering workshop

The Health and Safety Executive (HSE) gives clear guidance (www.hse.gov.uk/pubns/indg163.pdf) on how to conduct a risk assessment in five steps:

1 Identify the hazards.
2 Decide who might be harmed and how.
3 Evaluate the risks and decide on precautions.
4 Record your significant findings.
5 Review your assessment and update if necessary.

Identifying hazards

Hazards are best described as anything that has the potential to cause harm. You can identify hazards by inspecting your working environment and the equipment available to you. In an engineering environment, hazards might include things such as:

▸ having an untidy workshop
▸ using hand tools, such as a hacksaw
▸ using machine tools, such as pillar drill
▸ exposure to chemical substances, such as lathe coolant
▸ using hot processes, such as welding.

Deciding who might be harmed and how

For each of the hazards identified, you need to determine how an injury might happen and who might be injured. For example, if you consider the hazard of 'having an untidy workshop', someone might be injured by tripping over or slipping on something left on the floor. The consequences of a fall can be serious and could lead to head or back injury, cuts or broken bones. Those at risk of the hazard include anyone who has access to the workshop, such as technicians, supervisors, managers and site visitors.

Evaluating risk and adopting control measures

Risk is related to the likelihood of a hazard actually causing an injury and the likely severity of that injury. It is sometimes useful to establish a rating for each risk so that the risks can be compared and categorised as trivial, acceptable, or unacceptable and requiring immediate action.

A risk rating is the product of a likelihood score, which represents how likely it is that an injury will occur, and a severity score, which represents the potential seriousness of any injury. The calculated risk ratings can be displayed in a matrix as shown in **Figure 2.26**.

Severity

Likelihood			Minor injury (First aid) 1	Moderate injury (lost time) 2	Serious injury (RIDDOR reportable) 3	Major injury (RIDDOR reportable) 4	Fatality (RIDDOR reportable) 5
Extremely unlikely	1		1	2	3	4	5
Unlikely	2		2	4	6	8	10
Likely	3		3	6	9	12	15
Extremely likely	4		4	8	12	16	20
Almost certain	5		5	10	15	20	25

▶ **Figure 2.26** Example of a risk rating matrix

In this example, where a risk rating is 4 or less (the green cells in **Figure 2.26**), the risk is considered trivial and no additional control measures need to be put in place. Risks with a rating of 5–10 (amber cells in **Figure 2.26**) are accepted, but any measures put in place to limit the risk should be reviewed regularly and efforts should be made to reduce the risk rating further. Immediate action must be taken to reduce the risk if its rating is 12 or above (red cells in **Figure 2.26**).

You can use a variety of techniques to control risk – each has a different level of effectiveness (see **Table 2.7**).

▶ **Table 2.7** A hierarchy of effectiveness for control measures

Control measure (in order of decreasing effectiveness)	Examples and comments
Eliminate the hazard	This might mean stopping use of a particular process, material or piece of equipment because it is unacceptably dangerous. For example, exposure to lead fumes during soldering operations in the electronics industry was extremely difficult to control effectively. As a consequence manufacturers changed the type of solder used to one which no longer contained lead, thus eliminating the problem.
Reduce the severity of the hazard	Building sites use electrical tools that operate on a reduced, and therefore safer, voltage of 110 V instead of the UK standard supply voltage of 240 V.
Isolate personnel from the hazard	Computer numerical control (CNC) machine tools, such as lathes and milling machines, are completely enclosed during operation.
Limit the extent of exposure or contact with the hazard	A residual current device (RCD) will limit exposure to electric current if an electrical device develops a fault.

Welding curtains are used to limit the exposure to UV radiation of workers near where welding operations are being carried out. |
| Personal protective equipment | Goggles provide eye protection from flying debris that might otherwise cause eye damage. |
| Discipline and safe working practices | Instruction and training on safe working practices are important but rely on good discipline, concentration and responsible behaviour, which cannot always be guaranteed. |

Other means of reducing risk in the workplace include:

▶ good design – for example, improving the operational safety of equipment; designating exit routes from workplaces.

▶ permit-to-work system and guards – allowing only designated personnel to work in hazardous areas or on hazardous jobs; physical barriers to keep visitors or unauthorised people away from dangerous equipment.

▶ testing and maintenance – checking that equipment is in good working order and safe; revealing potentially dangerous faults.

Recording significant findings

Carrying out a risk assessment and putting in place the necessary control measures are insufficient on their own. Keeping proper records is also extremely important because these prove that you have carried out your legal obligations to provide a safe place of work.

HSE provides pro forma documentation to enable record-keeping to be done effectively. Prompts are provided to ensure that important steps are not missed (see **Figure 2.27**).

Company name:						
What are the hazards?	Who might be harmed and how?	What are you already doing?	What further action is necessary?	Action by who?	Action by when?	Done
Slips and trips	Staff and visitors may be injured if they trip over objects or slip on spillages.	• General good housekeepoing. • All areas well lit, including stairs. • No trailing leads or cables. • Staff keep work areas clear, e.g. no boxes left in walkways, deliveries stored immediately, offices cleared each evening.	• Better housekeeping in staff kitchen needed, e.g. on spills. • Arrange for loose carpet tile on second floor to be repaired/replaced.	All staff, supervisor to monitor Manager	From now on 01/10/10	01/10/10 01/10/10

▶ **Figure 2.27** Example of an HSE-approved risk assessment template

Reviewing the assessment

Work methods, tools and machinery tend to change over time, so it very important that risk assessments are reviewed periodically to ensure that they are still relevant and effective.

⏸ PAUSE POINT Working with a small group of colleagues, carry out a risk assessment on the use of a pillar drill.

Hint Follow the procedures laid out in the five steps to risk assessment.

Extend How does your risk assessment compare with the work carried out by other groups?

Case study

Slippery workshop floors

An employee slipped on the dusty concrete floors of a workshop during a routine visit; luckily he wasn't hurt. However, this incident did spur the company into action to investigate methods of making the workshop safer.

The nature of the work being carried out meant that graphite dust was constantly being produced, and despite using an extraction system, some of this dust found its way onto the floor.

The company decided to take a two-pronged approach to reducing the risk of slipping by both upgrading the extraction and filtration system to capture more dust and installing textured resin flooring in the workshop.

Check your knowledge

What might have been the consequences for the company if the employee had slipped and been seriously injured?

C4 Preparation activities for batch manufacture or batch service delivery

Health and safety factors

In all branches of engineering, health and safety needs to be considered during the planning and preparation for any new manufacturing or process task.

In both manufacturing and service processes, risk assessments and safe working prractices need to be eastablished prior to the commencement of any work.

On completion of the planning and preparation activities, a comprehensive risk assessment on the manufacturing or service delivery plan you have developed is required. This should be carried out according to the approach explained in Section C3.

Defining the required outcomes

When planning the manufacture of a product, you will need a materials list, relevant engineering drawings for the manufactured components and a general assembly drawing detailing how the finished product is put together (if necessary). These documents will contain a complete definition of the product being manufactured and all of its component parts.

When planning to provide an engineering service, you will need a full description of the required outcomes that must be achieved. For instance, to perform the renewal of the coolant in a centre lathe, the outcomes might be defined as: supply and renewal of coolant; removal and safe disposal of the used fluid; cleaning out the tank; testing the fluid flow.

Determining the processes required

For manufacturing a product, this phase of planning and preparation will require careful inspection of the specification and component drawings in order to determine suitable and efficient methods of manufacturing. This will help you select appropriate manufacturing processes for each step. A single product may require an understanding of several different process types. For example, the manufacture of a light fitting might require: mechanical processes to create formed metal components, assembly processes to fix components together, electrical processes to install wiring and connect electrical components, and testing processes to ensure the product is electrically safe and functioning properly.

For an engineering service, you need to take a close look at the requirements of the service being planned and determine the best way to achieve each of the required outcomes. This will require research into the equipment you will be working on (where necessary), and might involve consideration of a number of disassembly, replacement and test processes. This will help you select appropriate engineering servicing processes for each step. A service may require an understanding of several different process types. For example, changing the coolant on a lathe will require: effective use of safe working practices such as electrical isolation of the lathe to prevent accidental activation, safe material handling and disposal, disassembly processes using hand tools to gain access to the coolant tank, safe cleaning and fluid replacement processes, and testing processes on the coolant pump, coolant flow and checking for leaks or blockages.

Sequence of operations

As part of the preparation, define the sequence of operations that need to be carried out. Identify the operations that must be performed sequentially and those that can be carried out concurrently. This will form the basis of your manufacturing plan (for a product) or delivery plan (for a service), which defines what must be done, how, when and by whom.

Equipment

Determine the equipment that will be needed to complete each operation in the manufacturing plan or service delivery plan.

Quality control

Define when and how checks will be carried out to ensure that the product has been manufactured or the service delivered as originally specified. For a product, this might include taking measurements, assessing the fit and finish, and testing the finished product. For a service, this might include functional testing and ensuring that all fixings have been replaced and tightened to the required torque.

Materials and parts

For a product, gather together the materials and parts, such as screws or washers, that will be required to complete all the manufactured components and assemble a finished product. For a service, gather together the materials and replacement parts that will be required to complete the engineering service.

C5 Delivery of manufacturing or service engineering processes

Manufacturing engineered products

Many manufactured products are suitable for consideration in this unit – for example, a screwdriver, a toolmaker's clamp, a sheet metal toolbox, callipers, a ball joint splitter and a wiring loom.

You need to consider a number of basic workshop processes when manufacturing an engineered product.

▶ Marking-out processes are used to indicate the positions of features on a workpiece clearly so that they can be created accurately. This might mean using a rule, square and scriber to mark the position of a saw cut or using a centre punch to mark the position of a drilled hole. This stage must be done precisely, and there are a range of specialist tools that will help you achieve the required accuracy – for example, a scriber, rule/tape, punch, square, vernier height gauge or marking-out medium.

▶ Manual processes are usually used for one-off production because they are versatile. The down side is that they are time-consuming and usually require a high degree of skill to use proficiently and achieve a good finish. Tools used in manual processes might include shears, punch, guillotine, bender, saw, tap, die or file.

▶ Machining processes tend to be used for the quick and accurate removal of larger amounts of material than would be practicable with hand tools. Typically these might include a pillar drill, lathe or milling machine.

▶ Assembly processes complete the final product once all its component parts have been manufactured. There are numerous joining methods used in assembly, which might include the use of adhesives, mechanical fasteners such as screws, rivets or clips, and cable ties or wiring connectors such as bullets, spades or terminal blocks.

▶ Quantity production techniques are used to semi-automate processes that need to be repeated on a large number of products. Manufacturing aids, which help to cut down production times, might include bespoke form tools, templates, jigs, moulds, fixtures or stops.

▶ Measuring processes must be employed to ensure that work is marked out and manufactured accurately. This is likely to require the use of specialised engineering measuring devices such as micrometers, vernier or digital callipers and comparators.

Engineering services

Many engineering services are suitable for consideration in this unit – for example, refurbishing an alternator (including worn part replacement and testing), modification of pipework (including the connection of valves and operational checks), modifying and rewiring electrical switch panels, and performing a service on a centre lathe (including coolant renewal).

Delivering engineering services

Engineering services will not usually require the use of heavy workshop machine tools. Instead, they concentrate not on the manufacture of new parts but on the refurbishment or repair of existing products. Again, there are a number of workshop processes that will need to be considered when carrying out engineering services.

▶ Disassembly, removal and strip processes all involve the removal of parts. You will usually do this because the parts need to be checked or they are simply in the way and preventing the removal of other components that need to be checked. You will need a range of tools to remove the fasteners that hold a product together, such as screwdrivers, wrenches, spanners, sockets, pliers/grips and hexagon keys.

▶ A manual process may be needed in removing, repairing, refurbishing or refitting components, such as when removing burrs, cleaning, trimming pipes to length or cutting gasket material. You may need to use other hand tools such as snips, cutters, knives, punches, saws, files or hammers.

▶ Assembly processes are those required to reinstate the product into working condition. It is fundamentally the opposite of disassembly, but some of the electrical connections or mechanical fixings may need to be totally replaced. These processes might include using a soldering iron to connect wiring, refitting or replacing mechanical fasteners, using a torque wrench to ensure that machine screws are correctly tightened, fitting new wiring crimp connectors, and using pneumatic tools or clamps to hold components in place while fixings are refitted.

▶ You will use inspection/testing processes to verify the function of the product once it has undergone a service and to ensure that components have been assembled correctly. This might mean using a multimeter to check continuity of electrical connections, flow meters to verify that fluid input and output are as required, or a pressure sensor/gauge to ensure that appropriate system pressures are maintained and there are no leaks.

In your role as a shop floor supervisor in a small engineering company, you have been tasked with setting up, leading and working alongside a small team of two manufacturing technicians to make a batch of five sample screwdrivers for approval by a client.

- Discuss your role as an effective team leader in manufacturing the screwdrivers.
- Discuss the requirements in setting up the manufacturing process.

Ensure that you include safety, efficiency and quality aspects in your answers. You will be expected to produce a risk assessment for at least one engineering process and discuss any necessary set-up activities.

Plan

- What is the task? What am I being asked to do?
- What issues do I need to consider and what research should I undertake in completing this task?
- Are there any areas I think I may struggle with?

Do

- I know what I'm doing and am confident that I can achieve my goals.
- I can identify the challenging aspects of the task and refine my thinking/approach to overcome these difficulties.

Review

- I can explain what the task involved and the steps I took to complete it.
- I can identify the parts of my knowledge and understanding that require further development.

THINK ▶FUTURE

Patrick Makin

Apprentice engineering technician

I'm now in my third year as an apprentice engineering technician with a multinational manufacturing company. At school I completed a BTEC National in Engineering, and as part of the course we did some project work with my current employer and worked with them to solve some real-world engineering problems. Well, I must have done something right because they encouraged me to apply for one of their apprenticeships. At the start of Year 13 I did just that and sent off my application. After an interview I was accepted! I joined the programme immediately after leaving school (with a Distinction in my engineering course) and couldn't wait to get stuck in.

I am now settled in and really enjoying myself. I have been supported in loads of different ways by the company and am considered as part of the team by my work colleagues who I really enjoy working alongside.

Focusing your skills

Becoming established in a new environment

It can be tough when you start a new job or join an unfamiliar team. Here are a few tips to help with the transition:

- Listen – starting a new role will mean having to absorb of information. You need to pay attention to what you are told. People will generally cut you a bit of slack in the first few weeks, but don't expect that to last.

- Take a few notes – note down important information such as start, finish and break times.

- Address people by name – this helps to establish working relationships more quickly. Try to remember names. You could even make a note of people's names as you are introduced to them.

- Be aware of what your body language is saying – you might be intently listening to a presentation, but if you're leaning back in your chair with your eyes closed it will look like you are not.

- Stay safe – during your induction and initial tour of the facilities make sure you pay attention to the emergency procedures, where to find the exits, how to contact a first-aider and all that stuff you hope you'll never need.

- Ask questions – don't be afraid to ask questions, but make sure that you don't keep asking the same ones of the same people.

- Learn how the equipment works – you need to know how the equipment you will be using works and if there are any dos and don'ts that do not appear in the manuals. Obviously don't use a piece of equipment unless you have been trained to do so and are familiar with any relevant safety requirements.

- Ask for help – if you are lost (this can happen in a big facility!), confused about something or simply need direction to the nearest toilet, then don't be afraid to ask. Take the opportunity to introduce yourself at the same time.

- Be punctual – be on time in the morning, don't be late back from breaks and don't be the first to rush out of the door at the end of the day.

- Be friendly – if you are friendly towards your colleagues, then they will usually reciprocate. If you're in a bad mood then bite your lip and don't take it out on your colleagues.

- Maintain your sense of humour – smile!

Getting ready for assessment

Hazeem is working towards a BTEC National in Engineering. He was given an assignment for Learning aim A that asked him to evaluate the manufacturing processes used to make a screwdriver and the impact that human factors have on manufacturing operations.

His findings had to be written in a technical report that would be passed on to the operations director of the company in the assessment scenario.

The report had to:

▶ justify the use of at least three manufacturing processes used to manufacture the screwdriver, comparing them with possible alternatives

▶ evaluate how human factors and characteristics affect the performance of individuals and teams when carrying out manufacturing processes.

Hazeem shares his experience below.

How I got started

First, I had a really good look at the screwdriver we were given to use as the basis of the assignment. From my notes on manufacturing processes I was able to recognise the tools and techniques that could be used to make the different parts. What really helped was drawing pictures of the different components on A3 sheets of plain paper and then adding annotations to record my thoughts on the different ways each feature could be made. The assessment criteria said I had to consider at least three different processes, but I made sure I had plenty to choose from.

The second part of the assignment on the effects of human factors on individual and team performance was harder to get my head around. I started by reviewing my notes and making a mind map of all the factors that I thought were most important. This helped me visualise what I should write about and helped me to structure my written report.

How I brought it all together

I wanted my report to look professional because it was going to be read by a senior business manager, so I used a simple Arial font and included a footer on each sheet containing my name and page numbers. I added a title page, which included a photograph of the screwdriver, and then wrote a short introduction. I included lots of relevant photographs to support what I was saying in the text and to add some visual interest.

While I was writing the report, I ticked off the points I had covered on the annotated sketches and mind maps I had created previously in my preparation for the assignment.

What I learned from the experience

I think my assignment went really well on the whole. Keeping good notes of the things we covered in class really helped later on. I am going to buy a notebook to keep all my class notes together though. Trying to keep all my loose sheets of paper in order is a pain.

If I were going to do this again, I would make sure that I got hold of a copy of the unit specification so I could look at the 'Essential information for assessment decisions' page while I was writing the assignment. This would have been really useful because it explains what the assessment criteria actually mean in practice, and tutors use this when assessing the work.

Think about it

▶ Have you planned what you need to do in the time available for completing the assignment to make sure that you meet the submission date?

▶ Do you have your class notes on manufacturing processes and the influence that human factors can have?

▶ Is your information written in your own words and referenced clearly where you have used quotations or information from a book or website?

Getting to know your unit

Engineering product design and manufacture is the process of transforming a user- or market-driven need for a new or revised design into a commercial product that addresses this need. To turn design ideas into viable products, engineers need to consider the requirements of the user and any relevant regulatory standards to generate concepts, and then use their knowledge of materials, components and engineering principles to produce effective solutions.

This unit covers the triggers that motivate the need for new designs, potential challenges to and constraints on your design, how to specify what you want from your product, the use of iterative processes to develop an effective solution, and how to analyse, validate and communicate your work.

How you will be assessed

This unit will be assessed under supervised conditions. Before the supervised assessment period, you will be provided with an externally set case study. You will be allowed to carry out independent preparatory research on the case study and then use this information in the assessment.

During the supervised assessment period you will complete a task in which you will be expected to follow a standard development process of interpreting a brief, scoping initial design ideas, preparing a design proposal and evaluating your proposal.

Your completed assessment will then be forwarded to Pearson for marking.

As the guidelines for assessment can change, you should refer to the official assessment guidance on the Pearson Qualifications website for the latest definitive guidance.

To achieve a Pass, a learner is expected to demonstrate these attributes across the essential content of the unit:

▶ You demonstrate knowledge and understanding of iterative design methodologies, processes, features and procedures and their application to engineering products.

▶ You can interpret a design brief to generate ideas and use a range of skills and techniques to develop modified products in context.

▶ You can use research and analytical skills to create a product design specification to meet the requirements of the brief.

▶ You can make recommendations and proposals relevant to familiar and unfamiliar situations, taking into consideration sustainability and safety issues.

▶ You are able to make an evaluation of your design proposal and provide technical justifications in validation of your design solution.

For an award of Distinction you will need to show:

▶ A thorough knowledge and understanding of iterative design methodologies, processes, features and procedures and their application to engineering products in context.

▶ You can interpret a design brief to generate complex design ideas and can apply a range of skills and techniques to develop modified products in context, with justification for the design decisions.

▶ You can use comprehensive research and analytical skills to create a product design specification that fully and effectively meets the requirements of the brief.

▶ You can make justified recommendations and proposals for familiar and unfamiliar situations, taking into consideration sustainability and safety issues.

▶ You are able to select appropriate techniques and processes to design ideas and can justify applications in arriving at creative, feasible and optimised solutions.

▶ You are able to make a robust evaluation of your design proposal and provide detailed technical justifications in validation of your design solution.

This table outlines the skills and knowledge you will need to demonstrate through the assessment.

Assessment outcomes	
AO1	Demonstrate knowledge and understanding of engineering products and design
AO2	Apply knowledge and understanding of engineering methodologies, processes, features and procedures to iterative design
AO3	Analyse data and information and make connections between engineering concepts, processes, features, procedures, materials, standards and regulatory requirements
AO4	Evaluate engineering product design ideas, manufacturing processes and other design choices
AO5	Be able to develop and communicate reasoned design solutions with appropriate justification

The following table outlines the areas of essential content with which you will need to be familiar before you take the assessment.

Essential content	
A Design triggers, challenges, constraints and opportunities, and materials and processes	
A1	Design triggers
A2	Design challenges
A3	Equipment-level and system-level constraints and opportunities
A4	Material properties
A5	Mechanical power transmission
A6	Manufacturing processes
B Interpreting a brief into operational requirements and analysing existing products	
B1	Design for a customer
B2	Regulatory constraints and opportunities
B3	Market analysis
B4	Performance analysis
B5	Manufacturing analysis
C Using an iterative process to design ideas and develop a modified product proposal	
C1	Design proposals
C2	Communicating designs
C3	Iterative development process
D Technical justification and validation of the design solution	
D1	Statistical methods
D2	Validating designs

Getting started

Work in groups to list all the things you might want from a particular product. Write down what you think these requirements might mean in terms of the decisions the designer would make – consider the cost of manufacturing the product, the materials used, the product's strength and durability, its sustainability and reliability, and so on.

A Design triggers, challenges, constraints and opportunities, and materials and processes

A1 Design triggers

The initiation of a new product does not normally happen spontaneously. Generally, there will be some sort of trigger that motivates a new design activity.

Market pull or technology push

Market pull

Market pull refers to the development of new designs based on a need for either a completely new product to meet some demand or an improved product to address customer dissatisfaction with existing products. In the automotive market, for example, customers are always looking for better fuel economy, greater comfort and improved safety; as a result, manufacturers tend to focus on these areas.

The identification of these needs may be through informal customer feedback, such as sales agents chatting with customers, or more formal market research. Market pull can also arise from the pressure to upgrade a product in response to competitor developments.

Customer feedback and market research can be used to identify areas of the market where a new product with the correct characteristics could make an impact. By definition they indicate what customers want, so if this can be delivered as required and on time, a commercially successful product should result.

Market pull does, however, have its limitations. Customers may be able to identify a need, but the technology to turn this need into a real product may still be some way off. For example, there may be market demand for a hoverboard or teleportation device, but if it cannot be produced with current technology, then it will not become reality. Designers also have to carefully consider the timeliness of the need – customer demands may change or even be completely superseded over the product's development cycle, so that the engineering company may end up giving the customer what they wanted three years ago, not what they want now.

Technology push

Technology push refers to the development of new designs made possible by the introduction of new processes or technologies. In the case of manufacturing, it may mean that products can be produced much more cost-effectively, reliably or to a greater precision than was previously possible. Technological advances may also allow new features and functions to be incorporated into a product. For example, advances in electronics, sensors and touchscreens have led to the emergence of a completely new market in smartphones and tablets since 2010.

Technology push products are often highly innovative and can mark a step change in a particular area of the market or even stimulate the creation of brand new market sectors.

However, care does have to be taken. Developing a new technology into a product does not guarantee commercial success in the market. 3D television is an example of technology push where an upgrading of television design and broadcasting capabilities made the innovation possible, but customer demand did not match expectations.

> **Reflect**
>
> Imagine that your company has become aware of a new sponge-like material which has the ability to draw in powder and dust rather than liquids. It comes in sheet or strip form and is 1 mm thick. It can be cut with scissors and can also be rolled up and unrolled without being damaged.
> - How might you try to collect tiny particles using this material?
> - List as many possible uses as you can think of for the material.
> - Pick one of your ideas and think how you might develop the idea further. Are there any limitations to your idea? Could these be resolved?

Demand

You want to create a product for which there is a market and customer demand. The total demand and your company's anticipated share of the market will give a predicted value for sales, and this can help you decide whether to start developing the product and how much time and money to invest in its development.

You also need to consider demand, not just for now but for the life of the product. Is this a product you will produce as a one-off or a special order for a single customer, or is it something that will be for sale over a number of years on an open market? How do you think demand for the product might change over time? Is the total market for this type of product rising, falling or steady? How will your company's market share change over time? **Figure 3.1** illustrates what different levels of demand and profitability mean in terms of whether it is worthwhile to develop a new product.

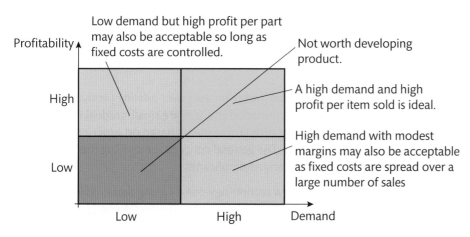

▶ **Figure 3.1** Profitability against demand chart

Profitability

For any planned product, you must consider whether the potential sales revenue will cover the costs of designing the product, making it and any **overheads**. In addition, you will want to make enough profit so that your organisation will be able to invest

> **Key term**
>
> **Overheads** – costs that relate to extra services involved in the manufacture of a product, such as equipment costs and heating or lighting; they are not directly related to the cost of labour or materials.

in future products and plant as well as pay workers and shareholders. **Figure 3.2** summarises some of the financial items you need to consider when planning the development of a new product.

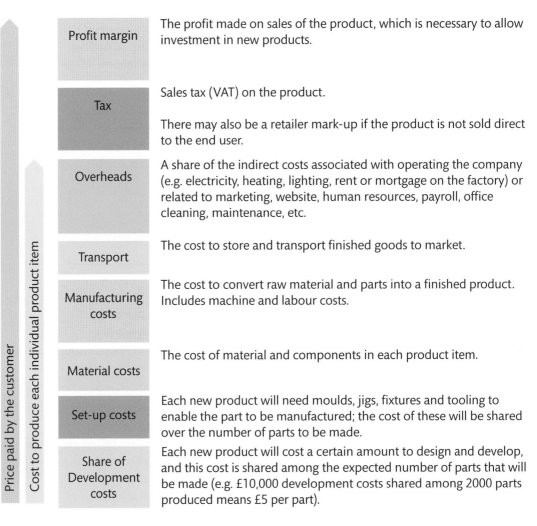

Profit margin	The profit made on sales of the product, which is necessary to allow investment in new products.
Tax	Sales tax (VAT) on the product. There may also be a retailer mark-up if the product is not sold direct to the end user.
Overheads	A share of the indirect costs associated with operating the company (e.g. electricity, heating, lighting, rent or mortgage on the factory) or related to marketing, website, human resources, payroll, office cleaning, maintenance, etc.
Transport	The cost to store and transport finished goods to market.
Manufacturing costs	The cost to convert raw material and parts into a finished product. Includes machine and labour costs.
Material costs	The cost of material and components in each product item.
Set-up costs	Each new product will need moulds, jigs, fixtures and tooling to enable the part to be manufactured; the cost of these will be shared over the number of parts to be made.
Share of Development costs	Each new product will cost a certain amount to design and develop, and this cost is shared among the expected number of parts that will be made (e.g. £10,000 development costs shared among 2000 parts produced means £5 per part).

(Left axis, upper arrow: *Price paid by the customer*; lower arrow: *Cost to produce each individual product item*)

▶ **Figure 3.2** Profit and costs

Innovation

When developing a new product, the improvements over existing products could be either *incremental* or *innovative*. Incremental changes are small, gradual improvements over the existing product. These are often refinements based on experience with previous **iterations** of the product. The general concepts underlying the operation of the product are unlikely to have changed significantly.

Innovation may happen when a new technology comes along that enables the production of items that were impossible before – see the section on technology push above. Innovation can also occur in situations where the way of achieving a given product's functionality is examined from a non-conventional perspective, or where technologies or approaches from another discipline are brought in. As a classic example of the latter, Owen Finlay Maclaren, a designer involved in creating the folding undercarriages of Spitfire fighter aircraft in the run-up to the Second World War, later used his knowledge of lightweight fabrication and folding mechanisms to create collapsible pushchairs and baby buggies.

Innovation in your product's design does offer the possibility of creating a clear difference between yourself and your competitors. For commercial success, the innovation must not simply be a novelty but has to be something that the customer values, so understanding what your end users want from your product is crucial.

Market research

Understanding your market is essential to the creation of successful products. Market research covers a range of activities that help you to understand what your customers want and what competitors might be planning now and in the future.

Market research can be carried out in a variety of ways, depending on the size of your company, your relationship with your end users and the nature of the products. If your company is a small, highly specialised organisation selling directly to a limited number of customers, your sales engineers may carry out the market research themselves. They will aim to find out how many products your customers are likely to buy over the coming years, the limitations of your (and your rivals') current products, and any anticipated changes to customers' expectations of what they want or need from your products. For example, is price going to be more important than weight? Are there any new features customers would like to see, or are there changes in their operations that might make your current products obsolete? A drawback to this approach is that it limits you to your current customer base and product areas and may not help if you wish to move into new product areas.

For larger companies with wider customer bases, such research is likely to be carried out by specialist teams employed either directly by the company or on a consultancy basis.

Market research can also help during the development of new products. For example, you may have a range of concepts for a new product but need some guidance on which would be most likely to appeal; or you may have a technology push innovation that, while technically interesting, may or may not provide much benefit to the end users of your product. By running your ideas past groups of potential users, you can collect data to help you select the ideas most likely to succeed.

Performance issues

Dissatisfaction with products currently available on the market will always be a trigger to design improved versions. Existing products may be too costly, too unreliable, require too much servicing, have too short a life, be too costly to run, be too heavy or have too large

a carbon footprint. Knowing these limitations, or using market research to pinpoint them, can act as a trigger to develop new products to address the unsatisfactory issues. However, care must be taken not to simply swap one weakness for another. For example, it might seem straightforward to solve problems with battery life in a mobile tablet, but if this is done by putting in a bigger battery, then customer satisfaction is likely to go down because of the increased cost and weight.

Discussion

Choose a product from your workshop or home, and in groups discuss what makes it effective for the purpose it was designed for, and also where it is less effective. For example, think about whether you like using the product. Is it too heavy? Does it run out of power too quickly? Is it tricky to set up or to put away?

If you were asked to improve the product, how important would you consider each of the limitations to be? Would you be prepared to pay more for the product if the current issues were addressed? Would you perhaps buy another product with different limitations if it didn't have those issues?

Can you propose a no-cost solution (in terms of money, weight or functionality) to one of your concerns?

Sustainability

Making a product more sustainable is both a commercial and an environmental driver in the creation of new designs. There may be an immediate pressure for existing products to be replaced or removed from sale, due to them no longer meeting environmental targets or because products associated with their use or operation will become unavailable in the future. This offers an opportunity for new products to be developed, and can even lead to significant early sales if products already in use must be replaced immediately.

Designing out risk

Naturally, your product needs to be safe to use. Risks to the user, or anyone coming into contact with the product, must be eliminated or minimised. To evaluate risk, you normally consider the likelihood of an accident happening and the severity of the consequence if it does. The combination of these factors gives a measure of the overall risk.

For example, you might be designing a product that runs a little warm and may cause some discomfort if you were to touch it for a long time, but it would not cause burning of the skin. If your product were operating where people were unlikely to come into contact with it, then the likelihood of anyone touching it would be low. The consequences would also be low (some discomfort), so the overall risk would be low and you may decide that the cost involved in completely eliminating the problem would not be justifiable. If, however, the likelihood of a safety issue was higher (e.g. if people would be exposed to the heat on a regular basis or if the heat was sufficient to cause burning), then some form of redesign might be needed to move the hot parts of the product away from the user, to reduce the surface temperature or to shield the product.

> **Key term**
>
> **Decommission** – to remove or withdraw from service.

When designing a product, it is always good to envisage risks associated with the installation, use and **decommissioning** of your product. If the likelihood of a particular risk is not low, or if the consequence is non-trivial, then you need to address the risk.

> **Link**
>
> Risks and risk management are discussed further in Unit 5.

A2 Design challenges

There are many commercial, regulatory and public policy-related trends that present challenges to engineering design.

Reduction of energy wasted during design

There is always pressure to bring products to market in a timely matter. Spending an extra six months developing a product uses additional resources and also means that competitors may steal a lead on the market. The use of tools to speed the development process is therefore becoming increasingly important. Advances in design tools and products offer ways to achieve time and energy savings during the development cycle. For example, 3D printer technology can be used to create early-stage prototypes much faster than traditional methods. Moreover, in many cases, engineers can quickly simulate different design scenarios using a variety of computer-aided techniques. These can range from fairly simple numerical models in a spreadsheet (e.g. to simulate different permutations of strength, volume, weight and cost for simple parts) to more complex simulations modelled with computer-aided design (CAD) packages.

Reduction of energy wasted during operation

Reducing the energy used during the operation of an engineered product is often a key driver and a serious challenge in the creation of new products. This goal of reduced energy usage can be driven by consumers' desire to have more environmentally friendly products or by the company's need to lower operating costs. Thus there is continual pressure for consumer products such as motor vehicles to have improved fuel economy, or central-heating boilers to have increased efficiency. In addition, legislation is requiring ever greater energy efficiency of an ever wider range of products. For example, since 2007 traditional incandescent filament light bulbs have been phased out in favour of more efficient LED and fluorescent bulbs.

Reduction of physical dimensions

Reduction of physical dimensions is often a design constraint. This may be related to customer demand for more compact products, for portability or aesthetic reasons.

There can also be pressure to reduce the physical dimensions of some products that are parts of larger systems, in order to free up space elsewhere. For example, reducing the physical depth of the backrest in aircraft or car seats can leave more space and hence give more comfort to passengers in the seat behind.

Reduction of mass

Reducing the mass of products is again a common challenge in product design or revision. This may be done for practical reasons – for example, where a product such as a hand tool needs to be portable and generally easier to manipulate.

Reducing mass also reduces the energy required to move products. A lighter product will be much more responsive and so easier to bring up to working speed – or to decelerate to a halt. Doing so will expend less energy, which means that the product will be more efficient. An example of this is the elimination of spare wheels in many new cars and the move towards electronic rather than mechanical hand brakes, both of which reduce the weight of the car.

It is particularly important to reduce the weight of key components, as this can offer compound benefits, with further weight savings elsewhere. Reducing weight in the body and interior of a car, for example, results in less load placed on the engine, gearbox, suspension and brakes. This means that there may be scope to reduce the weight of these parts too.

⏸ PAUSE POINT

Estimate the weight saving if no spare wheel is supplied in a family hatchback car. How does this compare to the overall weight of the car? Where else could weight be reduced?

Hint   Try looking at car brochures or websites. Some car models offer the spare wheel as an option.

Extend   Investigate how car performance and weight are related. Do you think it is a simple proportional relationship?

Increase in component efficiency

Increasing component efficiency can mean the individual elements within a product having their mass, size or energy efficiency improved – but in a design sense it also refers to enhancing the value and functionality that can be gained from a single component.

For example, if a moulded or cast part is to be produced, it is often possible to combine a number of functions into a single part, thus reducing the need for more parts or more complex secondary manufacturing processes. A box to contain electronics may take the form of the box itself, guides for holding the circuit boards and mounting holes for screwing the box to a baseplate, but it may be possible to achieve all of these functions in a single well-designed moulding. This potentially lowers the cost of the product by reducing the part count and assembly time.

Energy recovery features

Energy recovery means taking energy that might have been wasted during the operation of a product and returning this as useful energy. For example, combined heat and power plants produce both electricity and hot water, where the water is heated by extracting excess heat energy from the electricity generation process. Another example is regenerative braking in wholly or partly electrically driven vehicles. In this system, under braking conditions the connection to the drive motor is altered, allowing the motor to act as a generator, and the power generated is then returned to the vehicle batteries for later use.

Sustainability issues and reduced product life cycle costs

Another challenge for designers is to reduce product life cycle costs. While the focus in product design has traditionally been to reduce manufacturing costs, whole life cycle costs must now be considered at all stages of the design process. This includes considering sustainability as well as operating costs throughout the life of the product. Having more power-efficient products is obviously important, but you also need to take into account the financial and environmental costs of servicing your product, particularly where this includes, for example, the appropriate disposal of oils or lubricants or the use of solvents in cleaning.

You must also consider end-of-life disposal costs. A product that can readily be stripped to recyclable constituent materials will have much lower end-of-life financial and environmental costs than one that has to be placed in landfill or which needs to have toxic components treated before final disposal.

Integration of different power sources for vehicles

A special case exists currently with vehicles that are increasingly combining different power sources, with the aim of balancing the positive and negative attributes of each power source. Entirely electric power gives off no pollutants at point of use and offers the possibility of energy recovery through regenerative braking, but it relies on batteries, which limits the range of the vehicle. Traditional internal combustion engines give good range but discharge pollutants into the atmosphere. Hybrid vehicles typically use both power sources, and these can be balanced to try to exploit the advantages of each.

Integration of the two systems in such a way that the user will have a seamless experience poses a challenge to the designer, and a number of approaches are used. In the parallel approach, both an electric motor and a conventional engine are used, each capable of driving the wheels, and the systems can work in isolation or together, depending on road conditions. In a serial system, the combustion engine has no mechanical connection to the wheels and instead acts as a generator to ensure that the batteries are topped up to drive the wheels via electric motors. Computer controllers, which continually monitor road conditions, battery power and driver demand, are used to optimise the use of these systems.

Reduced use of resources in high-value manufacturing

High-value manufacturing is a term used to describe manufacturing that achieves value by producing items not only in a more cost-effective way but also in ways that are driven by high levels of knowledge and skills. It is normally characterised by the use of new technologies, materials and processes, and is associated with sectors such as aerospace, microelectronics, photonics and medical product design. This can often mean that there is a particular need for efficiency to optimise the use of resources due to the exotic materials or processes needed and to offset the high levels of investment in equipment and research.

Designing out risk

As mentioned earlier, it is important to review the possible risks associated with your product. Are there any areas which are not as safe as they could be – such as sharp edges or corners, parts that run hot or cold enough to cause burns, or potential electrical faults caused by cables working free under use? Where risks are likely or may have significant consequences, they must be addressed. Trying to do this as part of the overall design will generally be more satisfactory and cost-effective. For example, designing a folding baby buggy such that fingers cannot get trapped in the first place is better than having to fit a range of guards to the device to prevent injury.

Another consideration is commercial risk and the need to be aware of customer needs and the state of the market throughout the production life. Adding an extra feature to your product may present a risk if it adds more cost than value for the customer, and in this case you could see a marked drop in sales. Similarly, you need to think about risks due to external factors, which could mean that your product becomes unviable sooner than you expected. For example, your product might become obsolete because of new environmental regulations, or it may rely on a key component available only from a single supplier who then stops making the part.

A3 Equipment-level and system-level constraints and opportunities

Selecting different solutions for equipment interfaces

Very often, when designing a product, you will find yourself building a system that is more a collection of standard mechanical, electrical, hydraulic and pneumatic parts than components designed from scratch. An example of this would be an automated production line.

Such a system often consists of specialised mechanical components (such as jaws designed to hold specific parts), customised modular parts (such as the physical structure of the production line, which may be built from extruded sections that are cut to length and modified as required), electrical drive components, controllers and sensors and pneumatic components (which may be used to lift and carry the parts).

For the system to be efficient, it is important to think carefully about how the components will link up and how to ensure consistency within the system. The aim is to design a system that is efficient in use but also easy to extend or develop. You should avoid running parallel, incompatible systems that could lead to duplication or redundancy and would be dependent on a wide range of specialist tooling for construction and servicing, as well as specialist staff with the necessary skills and knowledge to cover the different systems.

- For electrical products:
  - What power supply is required?
  - Is this a.c., d.c. or three-phase?
  - Can you have a single power source for all elements, or would you need a low-voltage, low-current supply for control circuitry and a separate system for high-power items such as motors or heaters?

- For signal transmission:
  - What voltage is to be used?
  - What is the communication protocol – what sort of coding do sensors and actuators use to get signals to and from the controller?
  - Is the communication wired or wireless?

- For pneumatic and hydraulic products:
  - What connectors are to be used?
  - What pressure are they to operate at?
  - What ancillaries (e.g. filters, regulators, lubricators) are needed?

- For mechanical products:
  - Do you need to work to standards?
  - What fixing types (e.g. screw sizes) are to be used?
  - For modular systems – will you use standard distances between fixing holes to enable easy addition and removal of parts?

- For control systems:
  - What type of sensors are you using?
  - What types of actuators, motors and drives are you using?
  - What type of controller is most appropriate (e.g. programmable logic controller, PC, microcontroller)?
  - Are all these parts compatible?
  - What software does the control system use?

In some cases the decision might be one you could make entirely independently. However, factors that could influence your decisions may include your customers' infrastructure, skills and experience, your own organisation's experience and the likelihood of having to modify or upgrade the product or system.

Systems integration compromises

When developing a system it is likely that you will need to make compromises, as you have to balance the demands of individual elements to create the strongest overall system.

Many of these compromises will relate to physical packaging issues and how components are physically placed relative to each other. For example, imagine you are designing a high-quality drone aircraft (like the one in **Figure 3.3**) for use by surveyors mapping out land.

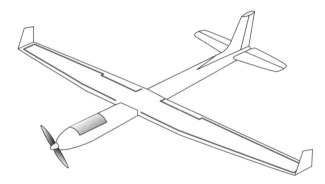

▶ **Figure 3.3** A drone aircraft

The various components within the drone must work in harmony. For efficiency, the electrics should be consistent in the voltage they require, to avoid the need for multiple power packs. This may mean a compromise if the ideal circuitry for the flight controls does not match that of the surveillance camera and sensors. The placement of components may also need to be compromised on. For example, if the drone will have a petrol engine, the logical location for the fuel tank is at the front close to the engine – this decreases the length of fuel pipe needed, reduces the likelihood of needing a fuel pump and allows for modular assembly. From a stability point of view, however, a central location makes more sense because as fuel is used the aircraft will remain in balance. With a nose-mounted fuel tank the aircraft would gradually become light-nosed as the fuel tank empties.

Similar sorts of design trade-offs would need to be made in other products. It often makes sense to place a certain component centrally within a product, but cooling or servicing requirements, for instance, might suggest a position closer to the outer surface for improved ventilation and access.

Equipment product design specification

When designing a product made from multiple subsystems and components, it is not uncommon to draw up a **product design specification** (**PDS**), not only for the entire finished product but also for each subsystem and even each component. These specifications can be used to help ensure that the finished product meets requirements in terms of function, cost, weight, serviceability, etc.

> **Key term**
>
> **Product design specification** (**PDS**) – a document detailing the requirements in various areas that a product, service, system or process must meet.

For example, a product might have a target weight. This can then be broken down into target weights of subsystems and individual components. Should a particular component or subsystem not be able to meet its target weight (e.g. motors with the required torque are all heavier than the target weight), then this issue can be flagged up early in the design stage and the system re-evaluated to see if the extra weight could be offset elsewhere in the design.

There may be other specifications you need to place on your components. You might want to avoid a complex cooling system, so components must be specified to avoid generating excess heat. Similarly, electrical elements will need to have a particular electromagnetic compatibility to ensure that they do not cause interference with other elements in the product and on the outside.

> **Link**
>
> Product design specifications and technical specifications are also covered in Unit 5.

Cost-effective manufacturing

Your product needs to be cost-effective to manufacture, to ensure that it covers its costs and generates a profit over the course of its production life.

For one-offs and short-run parts, the use of custom fixtures, moulds and tools will normally be kept to a minimum, and flexible machine tools such as traditional lathes, mills and computer numerical control (CNC) machines are likely to be used. For medium to large volumes, it is likely that you will factor in the cost of specialist tools to streamline the manufacturing process. These enable fast and cost-effective production of components and assemblies but require both capital investment and set-up costs each time a tool is used.

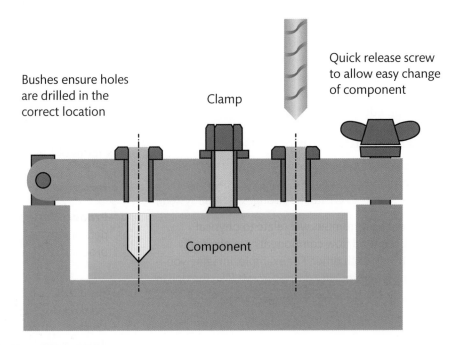

Bushes ensure holes are drilled in the correct location

Clamp

Quick release screw to allow easy change of component

Component

▶ **Figure 3.4** A drill jig

Another approach to speeding up manufacturing operations is to produce specialist fixtures that make it quick and easy to clamp parts, or jigs to aid the drilling of holes in the correct location (**Figure 3.4**). Again, for a single part, making a jig is likely to be much more expensive than making the part directly; however, if production numbers are higher, the time saved in drilling the holes at the correct locations is likely to return the cost of the jig.

With any tooling it is important also to factor in set-up costs, which are a fixed cost each time you manufacture a batch of products. For example, before parts are produced by moulding, the mould needs to be loaded with the correct polymer; it then needs to be fitted to the machine and possibly have some trial components run off to check that the device is operating at the correct temperatures and pressures.

A4 Material properties

Mechanical properties

Many of the mechanical properties of a material can be determined by what is known as a tensile test, in which a sample of the material is held between two jaws and gradually pulled apart. As this happens, the force being exerted and the amount by which the specimen is stretched are recorded.

This force and deflection data is generally converted into stress and strain values.

▶ The stress (σ) is the force applied divided by the cross-sectional area of the specimen:

$\sigma = \dfrac{\text{force}}{\text{area}}$, with units of N m^{-2} or pascal (Pa).

▶ The strain (ε) is a measure of the deformation of a component under a given load:

▶ $\varepsilon = \dfrac{\text{change in length}}{\text{original length}}$, which is dimensionless, but sometimes values are given as multiples of 10^{-6} because they tend to be very small.

The stress and strain can be plotted on a graph (see **Figure 3.5**) to find quantities such as the Young's modulus, yield strength and ultimate tensile strength.

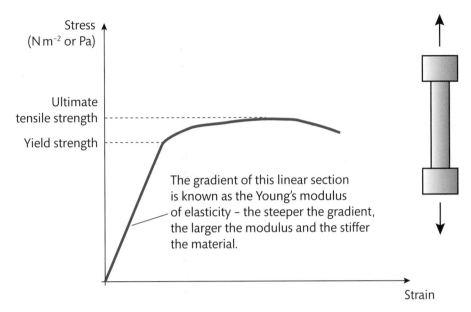

Stress
(N m⁻² or Pa)

Ultimate tensile strength

Yield strength

The gradient of this linear section is known as the Young's modulus of elasticity – the steeper the gradient, the larger the modulus and the stiffer the material.

Strain

▶ **Figure 3.5** Tensile testing of materials helps us to understand their strength and stiffness

For most materials, when a load is applied, the sample will stretch proportionally to the load. This behaviour corresponds to the straight-line portion of the graph in **Figure 3.5** and is known as the linear *elastic* phase. After the load is removed, the specimen would be undamaged and return to its original length. The gradient of this linear region is called the modulus of elasticity or Young's modulus, usually denoted by *E*, and is a measure of the stiffness of the material. The steeper the gradient of the linear elastic phase, the greater the modulus of elasticity and the more rigid the material.

As the load is increased further, a point will be reached where the material moves into a phase of *plastic* deformation. The end point of the linear region of the stress–strain graph is called the yield strength of the material (also known as yield point), and once this is exceeded the material will become permanently deformed. If the load is released during the plastic phase, the test specimen will shrink back following a line parallel to that of the linear elastic phase but will remain stretched beyond its original length.

During the plastic phase the specimen is likely to experience its maximum load – this corresponds to the ultimate tensile strength of the material (the maximum point on the graph in **Figure 3.5**). As the material continues to be stretched, it will eventually break, and the strain at this point is an indicator of the material's ductility.

Table 3.1 shows the yield strength, ultimate tensile strength, modulus of elasticity and ductility of four materials. The values given are indicative and will vary depending on the precise grade of material, the temperature and other factors. As a designer, you want to design your product so that the stresses within it stay well below the yield strength, thereby avoiding permanent damage to the product.

> **Link**
>
> For more details on stress, strain and modulus of elasticity, see Unit 1 (Section B2) and Unit 25.

▶ **Table 3.1** Mechanical properties of four materials

Material	Yield strength in tension (MPa)	Ultimate tensile strength in tension (MPa)	Modulus of elasticity (GPa)	Ductility (percentage elongation of 50 mm sample)
Structural steel	250	400	200	21
Aluminium alloy	300	500	70	19
Cast iron	180	–	67	–
ABS polymer	42	–	2.3	25

Physical, thermal, electrical and magnetic properties

In addition to mechanical properties, there are a number of physical properties that are of interest to the designer. **Table 3.2** summarises some of the thermal, electrical and magnetic properties that an engineer may need to consider when designing products.

▶ **Table 3.2** Physical properties of materials

Property	Symbol	Unit	Description	Example design application
Shear modulus	G	GPa	The stiffness of a material in shear or torsion	Ensuring that a driveshaft in a gearbox does not twist excessively
Poisson's ratio	ν	none	A ratio that measures how much a material thins as it is stretched	Understanding material behaviour in 2D and 3D situations
Density	ρ	$kg\,m^{-3}$	The mass of material per unit volume	Selecting material for a lightweight sports product
Fracture toughness	K_{IC}	$Pa\,m^{1/2}$	The resistance of a material to fracture	Designing a product so that it will be less likely to fracture in areas of stress concentration
Coefficient of thermal expansion	α_L	$m\,(m\,°C)^{-1}$ or $m\,(m\,K)^{-1}$	How much a material expands (or contracts) as the temperature changes	Making allowances for parts to expand with temperature and not cause a product to malfunction
Coefficient of thermal conductivity	k	$W\,(m\,K)^{-1}$	How well a material conducts heat	Designing an air-cooled engine
Melting point		°C or K	Temperature at which a material changes from solid to liquid form	Planning injection moulding or die casting
Specific heat	C	$J\,K^{-1}$	How quickly a material warms when subject to thermal energy	Estimating how quickly a product will take to reach an operating temperature
Electrical resistivity	ρ	$\Omega\,m$	The resistance of a material to the passage of electricity	Selecting a cover material for a battery terminal connector
Electrical conductivity	σ or κ	$S\,m^{-1}$	The reciprocal of resistivity $\left(\frac{1}{\rho}\right)$; measures a material's ability to conduct electric current	Selecting the appropriate size of electric cable for high-power applications
Permeability	μ	$H\,m^{-1}$	The ability of a material to form a magnetic field	Selecting appropriate materials for magnetic couplings

Advanced materials

Besides traditional materials, a range of new materials has become available over recent years.

Biomaterials

Biomaterials are materials that can be used in medical devices and implants. As they are designed to interact with living tissue either *in vivo* (in the body) or *in vitro* (in glass – in the lab), they must have a high degree of biocompatibility to avoid damaging the cells.

Traditional biomaterials include the ceramics, amalgam and cements used in dentistry.

Common applications of biomaterials include replacement joints, which can be made from a variety of metallic (e.g. stainless steel, titanium), ceramic and polymer (e.g. polythene) materials. These materials must not only have the appropriate strength and wear characteristics but also be biocompatible so that the body will not attempt to reject the implant.

More advanced biomaterials include those used to create miniature scaffolds, which can then be used to grow living tissue for research and medical uses. These scaffolds may be made from ceramics for bone-type tissue engineering or hydrogels for softer tissue. Scaffolds may also be designed to biodegrade as the tissue grows and becomes dominant.

Smart alloys

Smart alloys, also called shape memory alloys, are special materials made of either copper–aluminium–nickel compounds or nickel–titanium alloys. A straight smart alloy wire can be bent into a particular form, but when heated above a certain temperature – typically by passing an electrical current through it – it will return to the original straight form.

Smart alloys can be used as actuators, where they offer the potential to create much simpler and lighter systems than those involving motors and gears. These alloys also exhibit super-elastic properties, meaning that parts made from them can be deformed considerably more than traditional metals without incurring permanent damage. They are, however, expensive in bulk and have other limitations, so they are not practical to use for large parts.

> **Link**
>
> Shape memory alloys are also discussed in Unit 25, Section A2.

Nanoengineered materials

A relatively recent advance in materials science is the development and application of nanoengineered materials. These are materials whose fundamental particles have at least one dimension smaller than 100 nm (a nanometre is 10^{-9} m, so 100 nm = 10^{-7} m). Nanomaterials exist naturally, and new types of nanoengineered materials are often developed directly from the molecular or crystalline structure of the source material. Such developments can impart novel properties in terms of the strength, purity or shape of the material structure, which in turn may offer benefits not available from more traditional materials.

One of the best-known families of nanomaterials is the family of fullerenes, which consist of carbon molecules formed into tubes or ball-like particles. These particles can be very strong and are also very effective conductors of heat and electricity. Nanoengineering of materials is, however, a very new technology, and the creation, processing and applications of these materials are constantly evolving.

Modes of failure

When designing a product, you need to think about how it might physically fail. Failure could occur under tension, compression or bending, for example. These three kinds of loading are illustrated in **Figure 3.6**.

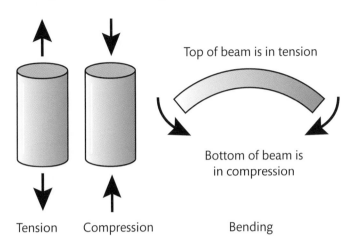

▸ **Figure 3.6** Loading of a component under tension, compression and bending

When a component carries a load in tension, the stresses in the component increase with the force applied. If these stresses exceed the yield strength of the material, the component will be permanently stretched. After releasing the load, the component will be longer than it originally was and is also likely to show signs of thinning. If the loading continues to be increased beyond the yield strength, the component will eventually break. As a designer, you want to ensure that the yield strength of your materials will never be exceeded under normal operating conditions.

When a material is compressed under a load, it will experience a similar effect as in tension. Bulk or bearing-type failures can occur when the yield strength of the material is reached.

For many metals and plastics, the stresses at which a part might fail in tension and under compression are similar. However, for brittle materials such as concrete or cast iron, the load that can be carried under compression is many times higher than under tension.

Bending is a very common mode of failure. A component that is subject to bending will normally be under tension on one face, be under compression on the opposite face and have a neutral axis somewhere in the middle which is under neither tension nor compression. Stresses in bending can often be very high, even if the loads applied are modest. If your product is not designed correctly, failure will be likely.

Consider a conventional metal paperclip. How easy is it to unfold this into an approximately straight wire? How easy is it to then take that wire and pull it until it stretches? Why is this?

Hint What sort of loading is occurring in each case? What is happening to the material as you manipulate it?

Extend What happens if you keep bending and straightening the clip at the same point? Why do you think that happened?

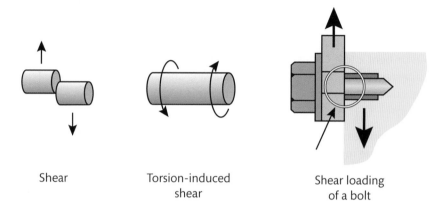

Shear   Torsion-induced   Shear loading
                    shear   of a bolt

▶ **Figure 3.7** Shear loading

Components can also fail under *shear*. In tension or compression the material is being pulled apart or crushed. Shear occurs when the material is forced to slide over itself (see **Figure 3.7**). A very common form of shear loading occurs in a shaft under torsional loading. In general, the allowable stress in shear is half that in tension.

Components can also fail due to *fatigue*. This happens when products are subjected to cyclic loading, such as when a load varies from tension to compression and back again many times. This can cause microscopic defects to form at loads much lower than the yield strength of the material. Fatigue can be more prominent in certain materials and is also associated with stress concentration factors, such as sudden changes in diameter of shafts, around holes, or in sharp internal corners.

Buckling is a particular type of failure that can occur when slender components are forced to carry a compressive load. This could happen in slender struts or columns and also in thin-walled structures, for example when you squash a soft drinks can. Buckling happens because slender structures are likely to bow under loading, and what was simple compression can then become a combination of compression and bending, which causes instability. The forces that can cause failure by buckling are typically much lower than those associated with conventional yielding.

Surface treatments

While bulk material properties will provide many of the characteristics required for your product, you may choose to coat or plate the surface for functional or cosmetic reasons.

Painted coatings

Paint can provide some protection against corrosion, as well as cosmetic benefits. Appropriate paints and coatings can be applied to most surfaces and can also be applied in layers, with each different coating adding a different function – for example, one layer might provide corrosion resistance, another could act as a base for a cosmetic coat, and the final layer could be a lacquer over a colour or metallic coat.

Paint may be applied using conventional brushes or sprayed on.

Powder coating

Powder coating offers many of the same benefits as conventional paint but can result in a tougher surface. The paint is applied as a powder, usually to metal parts, and is attracted to the surface of the parts by an electrostatic charge. The powder is then cured by placing the parts in an oven to yield a hard even surface.

Galvanising

Galvanising consists of zinc plating and is a form of coating applied to iron or steel parts. It often gives a characteristic crystalline 'spangled' surface finish, which can be seen on galvanised steel buckets, for example. Galvanising offers sacrificial protection in that the zinc will corrode before the underlying steel, and it will still provide protection even where the coating has been scratched through to the steel.

Anodising

Anodising is a technique commonly used to protect aluminium parts. It artificially increases the thickness of the layer of aluminium oxide that naturally forms around the parts, thus reducing the likelihood of corrosion of the aluminium underneath. Anodising also offers the opportunity to add colourings to give a decorative finish. Bicycle brake levers, for example, are commonly given coloured anodising in red, gold or blue tints.

Electroplating

Electroplating is an electrochemical process by which a base material is given a surface coating designed to improve corrosion resistance, wear or cosmetic characteristics. Probably the best-known electroplating process is chrome plating, which can give both decorative and wear-resistant surfaces.

Electroless plating

This technique can be used to plate nickel-based compounds onto materials to improve wear and corrosion resistance.

Lubrication

Lubrication is essential where there are parts that move each other – for example, the rotating hub of a motor, a cylinder moving in its bore or a control cable moving inside its casing.

Reasons for suitable lubrication

▸ Reduction of friction – By applying a thin film of lubricant over parts, the components do not come into direct contact with each other, and this reduces the effective coefficient of friction.

▸ Reduction in wear – Due to the film of lubricant, the moving parts do not directly contact each other and so wear can be much reduced.

▸ Reduction in heat build-up – As friction is reduced, so too is the heat associated with friction.

▸ Removal of debris – In systems where the lubricant is flowing or recirculating (such as in a car engine), it can help to wash away any dirt or debris from the moving parts. This debris can then be collected in a filter.

▸ Corrosion protection – Liquid or grease-based lubricants can act as a barrier to moisture, thus helping to prevent corrosion.

Types of lubricant

The lubricant can be a liquid, a grease or a solid.

A liquid lubricant may be a traditional oil or may contain additives such as PTFE to reduce friction further. The lubricant may be circulating, such as in a car engine, or it may be sprayed or dripped on. In the case of circulating lubricants, it is necessary to follow a regular regime of changing the lubricant and filter. In systems where the lubricant is applied directly, it will also need to be replaced on a regular basis. As the original lubricant is likely to be contaminated with dirt, application of new lubricant is normally preceded by a degreasing and cleaning process.

Grease-based lubricant may be packed around moving parts. For some components, the grease may be sealed in and designed to last the life of the part, but for others re-application may be needed, and a grease gun can be used to apply fresh grease as required.

Solid lubricants are commonly based on graphite and can operate at high temperatures, which tend to degrade liquid lubricants.

A5 Mechanical power transmission

Mechanical motion

It is important to appreciate the different types of motion you might have to take into consideration as a design engineer. **Table 3.3** shows four main types of motion, along with examples of each.

▸ **Table 3.3** Main types of mechanical motion

Type of motion	Examples
Linear	• A hydraulic ram used in earth-moving equipment • A shock absorber in car suspension • The movement of a seat post to adjust seat height on a bicycle
Rotary	• The main shaft of a jet engine • A drill tip when in use • The rotation of a vehicle wheel
Reciprocating	• A piston in a car engine • A bicycle pump
Oscillating	• A mass on a spring • A pendulum

Linkages

In many design scenarios you need to link the motion of different mechanical components, for either power transmission or control purposes.

Perhaps one of the most common forms of linkage is the lever (**Figure 3.8**). This gives increased mechanical advantage in a system, allowing a relatively small force to generate a proportionately larger force, depending on the lever proportions.

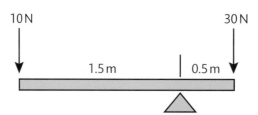

Applying a 10 N force at a distance, 1.5 m from the pivot will generate 10 × (1.5/0.5) = 30 N at the other end of the lever.

▶ **Figure 3.8** An example of leverage

Another common type of linkage or transmission is by transferring rotary motion from a drive source, such as an engine or a motor, to wheels or other outputs. This is commonly done via gears, a chain or belt drives (see **Figure 3.9**).

Gears are well suited to power applications. Often, motors and engines run fast and have too little torque to be of direct use. By using gears, the speed of rotation can be reduced and the torque increased. This is effectively another form of leverage. For example, if an 18-tooth pinion running at 3000 rpm is engaged with a 54-tooth gear, the shaft on this gear will run at $\frac{18}{54}$ × 3000 = 1000 rpm, while the torque will increase, nominally, by a factor of three, though in reality a bit less. Note that connecting gears will rotate in opposite directions.

Belt and chain drives also allow speed changes between input and output shafts, based on the number of teeth on their sprockets. Chain drives are commonly used on bicycles. Both chain drives and belt drives feature in car engines to drive cams and ancillary devices, such as alternators and water pumps.

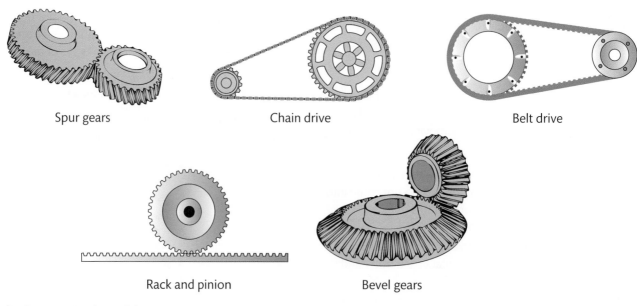

Spur gears   Chain drive   Belt drive

Rack and pinion   Bevel gears

▶ **Figure 3.9** Mechanical drive components

Gears can also be used to convert rotary to linear motion via rack and pinions, or to change the orientation of shafts via bevel gears.

> **Discussion**
>
> Imagine you are designing a single-speed folding commuter bicycle. Discuss the merits of using a conventional chain drive, a belt drive or a system based on two pairs of bevelled gears and a drive shaft.
>
> Think about what a user might want from a folding bike. Is performance everything or might practical considerations be more important? Would this influence your choice of drive?
>
> Discuss how you would determine the ratio between the drive and driven sprockets in a single-speed system. If you decided to offer both a single-speed and a five-speed system, might certain drive options limit your choice of gear system?

Other common linkages include four-bar linkages, slider-type linkages and cams, examples of which are shown in **Figure 3.10**.

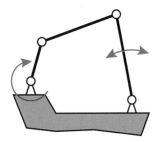

Four-bar linkages normally consist of three moving links, with the fourth link being a base. This example shows a windscreen wiper-type mechanism. The short link on the left would be driven by a motor and rotate through a full circle, while the link on the right would oscillate left and right through an arc.

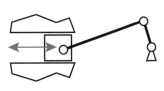

Slider-type linkages are also common. This example is similar to the mechanism of a reciprocating piston in a car engine driving a rotating crankshaft. The same sort of mechanism could run the other way – with a motor driving a rotary crankshaft, which then causes a piston to reciprocate – and could form the basis of a pump or air compressor.

Cams usually take rotary motion and convert it into oscillating linear motion. The cam itself is a specially profiled section of the rotating shaft; the profile determines the nature of the reciprocation.

▶ **Figure 3.10** Some other common types of mechanical linkage

Power sources

When designing many products you need to consider how the device will be powered. **Table 3.4** summarises some common forms of power sources and their advantages and disadvantages.

Source	Advantages	Disadvantages
Fossil fuels	Readily available (currently) Portable	Not sustainable Give off particulates and emissions when used
Biofuels	Becoming increasingly available Portable	May still give off emissions when combusted
Mechanical	Have a long history of use – these include flywheels, clockwork and spring-based energy storage, and gravitationally based systems such as pendulums and falling weights	Flywheels have been applied to some specialist transport applications, but these, like springs and gravitational energy-based systems, generally need to be coupled to an additional energy source for effective use
Hydrogen	Can be used in fuel cells, which can then generate electricity Portable and low-emissions (the only emission is water vapour)	There is currently limited infrastructure to supply hydrogen in an accessible form
Electricity – mains	No emissions at point of use Easily accessible	The generation of electricity may produce emissions via the combustion of gas or oil Not portable
Electricity – battery	Portable and easily applied Batteries may be recharged	Batteries may have limited life between recharging cycles Batteries may require special disposal at end of life
Wind and solar power	Renewable source with no emissions Offers electrical power in locations where mains power may not be available	Wind power and solar power are not dependable and may require battery back-up to provide consistent service

Control of power transmission

To help control power transmission, systems that include sensors, controllers and actuators are often developed.

▶ Sensors are used to monitor what is happening in a system. Examples include:
  ▶ pressure sensors
  ▶ temperature sensors
  ▶ displacement and position sensors, including potentiometers and linear variable differential transformers (LVDTs)
  ▶ load cells – which measure forces
  ▶ strain gauges – which measure part deformation and small movements
  ▶ limit switches.
▶ Controllers decide what signals need to be executed. Commercial controllers include:
  ▶ programmable logic controllers (PLCs) – specialist industrial controllers which are common in production systems
  ▶ computers with data acquisition add-ons
  ▶ microcontrollers – small compact chip-based devices that are easily embeddable in products (see Unit 6).
▶ Actuators turn responses into actions. Examples include:
  ▶ motors – which actuate pulleys, drive wheels, pumps, leadscrews, etc.

  ▶ servomotors – which provide a displacement output proportional to the control signal
  ▶ solenoid valves – electrically controlled pneumatic or hydraulic valves which can be used to operate rams, suction cups etc.

For example, a commercial greenhouse may have sensors that monitor the temperature inside, the amount of direct sunlight and moisture levels in the soil. The controller monitors these variables and, where necessary, sends signals to turn on heaters, activate rams to open windows, trigger motors to close blinds, or switch on pumps to supply water to the soil.

A6 Manufacturing processes

Understanding how products can be manufactured is a key aspect of any design process. This enables you to plan for parts to be made in the most cost-effective way.

Moulding and casting

A major family of manufacturing processes are those involving moulding or casting, summarised in **Table 3.5**. These processes can be used to produce relatively complex geometries in metals, plastics and ceramics, and normally involve placing a molten or other liquid form of the desired material into a mould and allowing the part to solidify.

▶ **Table 3.5** The manufacturing processes of moulding and casting

Process	Method	Typical materials	Notes
Moulding	Injection moulding	Polymers (e.g. ABS, polystyrene)	Used for a wide variety of plastic parts. Molten plastic is injected into a metal mould where it solidifies. A single mould may contain multiples of a single part or a set of parts for a single product. Two-shot moulding may be possible, where two materials are injected sequentially; this allows for soft rubber-type handles to be moulded onto hard plastic bases in tools or toothbrushes, for example.
	Blow moulding	Polymers (e.g. polythene)	Commonly used for hollow plastic products such as milk bottles or children's garden toys. A tube of molten plastic has high-pressure air blown down its centre, pushing the plastic against the inside of a metal mould, where it solidifies.
Casting	Die casting	Zinc, aluminium, magnesium	Similar in concept to injection moulding but used for producing metal rather than plastic parts. Can give very smooth and precise results, with parts not necessarily needing secondary machining. Tooling, however, can be expensive, so this technique tends to be limited to small-size, high-volume parts.
	Sand casting	Aluminium, iron, steel	Wooden, plastic or metal patterns of the desired parts are created. Moulds are formed by packing clay (a mix of sand, water and resins) around the patterns and then removing the patterns to create a cavity. 'Cores' can also be used to make hollows and internal features. Molten material is poured into the mould and, once solidified, the part can be recovered by breaking it out of the sand mould. As the patterns can be made from most materials, and no specialist casting machinery is required beyond a furnace to melt the material, the process can be affordable for short production runs. It does, however, leave parts with a rough texture, which would then normally require machining.
	Investment casting	Bronze, brass, steel, aluminium, precious metals	Uses wax patterns, which may be created directly or by using moulds themselves. The wax pattern is coated in layers of ceramic slurry, and the wax is melted out to leave a hard, hollow shell that forms a mould into which the material can be poured. The casing can then be broken off to recover the part. Parts produced by this process can be of very high quality and may need no further machining. The multi-stage process of first creating wax patterns, then the mould and then the part can make it relatively expensive, but it is well suited to small, high-value, complex components, such as turbine parts.
	Ceramic casting	Ceramics, pottery	Ceramic products are commonly produced using moulding techniques, most commonly slip casting. A plaster mould is created from patterns. A liquid containing ceramic elements is then placed inside the mould. The liquid dries to leave a ceramic part. This will generally need firing in a kiln to become usable. The technique is used for cups, pots and sinks but can also be used for engineered parts.

While each moulding or casting technique has its own unique requirements, there are several general guidelines for the design of moulded or cast parts (see **Figure 3.11**), which apply regardless of the technique.

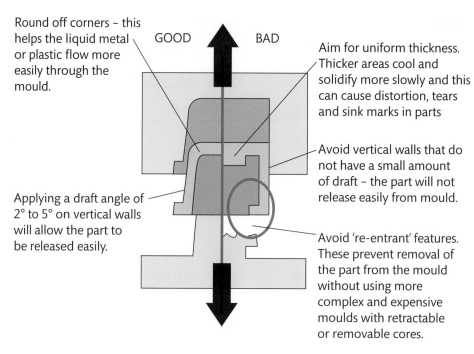

Round off corners – this helps the liquid metal or plastic flow more easily through the mould.

GOOD   BAD

Aim for uniform thickness. Thicker areas cool and solidify more slowly and this can cause distortion, tears and sink marks in parts

Applying a draft angle of 2° to 5° on vertical walls will allow the part to be released easily.

Avoid vertical walls that do not have a small amount of draft – the part will not release easily from mould.

Avoid 're-entrant' features. These prevent removal of the part from the mould without using more complex and expensive moulds with retractable or removable cores.

▶ **Figure 3.11** Some basic dos and don'ts for moulding and casting

Machining

Machining refers to a range of techniques where components are produced by cutting material from blocks or cylinder stock, or it can refer to a finishing operation on castings. Machining is highly flexible and can be applied to many materials, including most metals, polymers and woods.

Link

For more details on machining see Unit 2, Section A1.

Specific machining operations include drilling, turning and milling. These operations may be carried out on a manual machine, with the operator directly determining the cuts made, or via a CNC machine, where the cuts are made under computer control. Such machines typically have the ability to create threaded features via taps or thread-cutting tools.

Additive manufacturing

Additive manufacturing, also known as 3D printing or rapid prototyping, refers to a range of techniques that have emerged over recent years. Specific methods include stereolithography, fused deposition modelling and selective laser sintering. These techniques vary in the materials they can work with and in the accuracy they can achieve, but all are based on a 3D computer model of the part, sliced into a series of layers each a fraction of a millimetre thick. The part is then built up by depositing layers of material sequentially to recreate the model.

Additive manufacturing is attractive as it allows parts to be made directly from CAD files and also enables the creation of parts with geometries that would be difficult or impossible to make by machining or casting. The technology is relatively young and so developments are continually emerging – most notably the reduction in cost of equipment and the expansion of the range of materials that can be used.

Link

Additive manufacturing is also discussed in Unit 5.

Additive manufacturing techniques are, however, relatively slow, with parts typically taking many hours to produce. Also, manual clean-up of support material is usually necessary following part creation. Therefore, these techniques are currently best-suited to producing prototypes, one-offs or very small runs of products.

Assessment practice 3.1

Additive manufacture refers to a range of techniques where 3D parts are produced by adding material layer by layer to build an object based on a computer model.

- Identify at least four different forms of additive manufacture.
- List the materials that can be used with each technique.
- Summarise the pros and cons of each technique. (Hint: consider the cost of equipment and consumables, the precision of results, and how overhanging features are supported.)
- For each of the prototype applications below, identify which additive manufacturing method or methods, if any, would be suited to the application. If no method is suitable, suggest how you might prototype the part to evaluate the function.
  - Prototype to check the comfort of a games controller casing.
  - Prototype to check the fit of a casting in a pneumatic cylinder before committing to the manufacture of a mould.
  - Prototype to estimate the degree of flex in a new design of a spanner.
  - Prototype to evaluate a model aircraft wing profile in a wind tunnel.
  - Prototype to estimate the heat conduction in a proposed motorcycle cylinder head.

Plan
- What is the task? What am I being asked to do?
- What information will I need to collect? What resources can I use to help me complete the task?
- How confident do I feel in my own abilities to complete this task? Are there any areas I think I may struggle with?

Do
- I know what I'm doing and what I am aiming to achieve in this task.
- I can identify the parts of the task that I find most challenging and devise strategies to overcome those difficulties.

Review
- I can explain what the task involved and how I approached its completion.
- I can explain how I would deal with the hard elements differently next time.
- I can identify the parts of my knowledge and skills that could benefit from further development.

Joining

Few products are made from a single component, and it is often necessary to use joining techniques in constructing a product. Some common examples are shown in **Table 3.6**.

▶ **Table 3.6** Joining methods

Method	Uses, advantages and disadvantages
Screws, nuts	Easily available. Wide range of standard and special types. Allow for disassembly. While screws and nuts are generally metallic, they can be used to join any material.
Rivets	Easily available. Not readily disassembled.
Snap-fit	Allows fast assembly and eliminates the need for separate fasteners. May be difficult to disassemble.
Adhesives	Permanent. May also offer sealant function. Modern adhesives can be particularly strong. Difficult to disassemble.
Welding	Permanent – host material is melted to create join. As strong as host material. Range of different forms of welding available to suit different materials and applications.
Brazing, soldering	Metallic-based connection, but component material is not melted. Variants of these processes can be used for both mechanical and electrical connections.
Interference fits	A component has a hole in it with diameter very slightly smaller than the shaft it will fit on. Using an appropriate press or by heating the component to increase the hole diameter temporarily, the parts can be made to fit and will bind together due to the interference between them.

Look around your classroom, workshop or home. Pick two or three machines, tools, chairs or benches and think about how they might have been assembled. Why do you think that particular method was used?

Was the product assembled entirely in a factory or supplied partially assembled and then finished by the user? Does it need to be disassembled or partly stripped for cleaning or servicing?

Try to think of all the alternative ways the chosen product could be assembled, and identify the advantages and disadvantages of each method.

Other specialist techniques

Extrusion

In this process a metal, plastic or ceramic is forced through a die at high pressure and possibly also at high temperature to create **prismatic** lengths of material. These could be simple tubes but may also have much more complex forms.

Key term

Prismatic – having the form of a prism, that is, a 3D shape with identical ends and the same cross-sections all along its length.

Forging

Forging is the shaping of metal under high pressure and often at high temperature. This may be done by hand (blacksmithing), but for engineered parts it is normally done by placing the material between two dies to make the correct shape.

Pressing and punching

Parts are stamped from sheet material, using specialist dies. They may also be formed into 3D shapes using multi-stage presses. Car body panels are generally made using this process.

Thermoforming

Thermoforming involves the heating of sheet polymer until it is soft, at which point it is passed over an appropriately shaped mould, and air pressure or a vacuum is then used to draw the sheet into the mould where it can cool. The part can then be trimmed from the sheet.

Powder metallurgy

Metallic material in powder form is combined with additives and compacted in an appropriately shaped mould before sintering (heating) to create a solid part.

Composite manufacture

Composites are mixtures of more than one material without the constituent materials chemically combining. A particular type is fibre-reinforced composite, which consists of fibre strands or matting bonded in a resin. Such materials include fibreglass and carbon fibre. These materials can be turned into parts through a process known as layup, in which a mould of the part has layers of fibre matting placed into it. The fibre may be applied either infused with liquid resin or dry, with the resin applied later. The part is then cured. This may be done simply in air or more quickly in an oven. The parts are commonly placed in vacuum bags before curing. Removing the air helps to compact the laminates of fibre and gives a more predictable result. Once cured, the part can be removed from the mould and any excess can be trimmed away. Automated tow placement is an advanced variant of this process and uses specialist machinery to wind bundles of fibre over shaped mandrels. This ensures a high level of consistency and is used in applications such as the manufacture of aircraft parts or wind turbine blades.

Effects of processing

When processing any material, it is important to be aware that the processing may alter the nature of the material.

When casting metallic materials, the transition from molten to solid metal will cause the material to recrystallise, and the nature of this recrystallisation may not be predictable. By contrast, in forging operations the material remains solid – the hammering and squeezing causes the grains of crystalline material to flow into the required shape. This creates continuity in the material, producing parts that are stronger than if they were cast or machined.

Base materials may be processed to alter their characteristics in particular ways, such as making them stronger, easier to machine or more resistant to corrosion. For metallic materials this is most commonly done by alloying. For example, steel has iron as its base material, and has up to 2% carbon added along with small amounts of other materials. The carbon atoms sit in the iron crystal lattice, helping to prevent dislocation under load and resulting in a much harder and stronger material than the base iron. Adding in further elements can confer additional benefits – for example, adding chromium as an alloying element creates stainless steel, which has a hard, corrosion-resistant surface. Like iron, aluminium is rarely used by itself, but rather occurs in a wide range of alloy forms, where silicon, copper and other materials are added to achieve strength, castability, machinability and other beneficial properties. Similar blending of base materials can also be done with polymers – for example, blends of

ABS and polycarbonate are commonly used for injection-moulded parts.

Various process parameters can also alter the properties of the material. Machining, for example, can change the nature of the material. It may simply affect the surface finish – deep cuts will generally give a coarser finish than shallow cuts – but machining can also cause work hardening in certain materials, including stainless steel, making further machining difficult. It is therefore important to consult standard set-up tables to ensure that any processing is done within the recommended limits. For most conventional processes – such as casting, welding, brazing, and so on – there are recommended sets of process parameters to use so that you can be confident that the material will perform as expected.

Link

The crystalline grain structure of metallic materials and the effects of processing on the mechanical properties of metallic materials (including issues relating to recrystallisation, alloying and work hardening) are covered in detail in Unit 25 (Sections A4 and A5).

Scales of manufacture

The quantities of product you are planning to produce will influence your design. In general, as the total volume to be manufactured rises, it becomes more viable to use larger numbers of specialist jigs and moulds or even entire custom systems to make the part. If the total volume is very small, the cost of product-specific tooling will be too great and more general manufacturing equipment should be used.

One-off

Examples of one-off products include large infrastructural projects, such as oil rigs or ships, but may also be smaller specialist items, such as a piece of research equipment or even the jigs and moulds that are used in high-volume manufacturing. The manufacture of these sorts of parts will generally involve little automation, and the use of part-specific tooling will normally be kept to a minimum.

Small batch

Small-batch products are manufactured in small amounts at any one time, although they may achieve high total volumes over their production life. For example, you may produce 100 of a specific product each week, with this production taking place on a single afternoon. If production lasts two years, then around 10 000 items will be produced over the life of the product. An advantage of small-batch production is that, assuming your customers give you a fairly steady number of orders per week, you do not need to store lots of finished items and can quickly get the money back on your parts. By holding only a small volume of stock at any time, you are also protected if the order book dries up. The downside is that if you need to use specialist jigs and fixtures, then you will have costs associated with set-up as you swap tooling from one product to the next. For example, for a batch of 100 parts which takes a four-hour afternoon to complete, two hours of this may be spent putting in, aligning, testing and then taking out the specialist jigs or moulds, with only two hours available for actually making the parts.

Large batch

With large-batch manufacture, more parts are made at any one time, and the time spent on changing over tooling becomes less significant. Returning to the above example, if you were to spend an eight-hour day making parts, you would still need to spend two hours configuring and testing the tooling but would then have six hours rather than two hours of actual production and so could produce three times the number of parts for twice the manufacturing cost.

The downside of large-batch manufacture is that you would need to store many more finished products and it will take longer for you to sell the finished parts and get your money back.

Small batch versus large batch

Small batch is a nimble system and can be cost-effective if tool changeover time could be minimised. To try to reduce set-up costs, a lot of work has gone into developing systems that allow quick changeover of tooling or processes that make use of CNC machines and similar flexible systems capable of producing or assembling a range of parts without the need for physical tooling changes.

Mass production

Mass production is where large numbers of similar or identical products are produced, normally on a production line. Production machinery is often very specialised and designed specifically for a given product. The process can be highly cost-effective, but it can be difficult to introduce the manufacture of even slightly different products into such a system at a later stage.

Continuous production

Continuous production is a flow-type production system. Common in the chemical, petrochemical and raw materials processing industries, it is characterised by very specialist equipment with a high degree of automation.

How might you decide on what scale to manufacture your product? What factors would you need to consider?

Hint   Is there a way you could estimate the cost? Could you develop a numerical model to estimate an optimum batch size?

Extend   What practical factors would you also need to consider? What else might the machines be used for? How might you work out batch sizes for different components within the same product?

B   Interpreting a brief into operational requirements and analysing existing products

B1 Design for a customer

As a product design engineer, you need to be able to design products people want. Carefully understanding the needs of your customers is key to the success of your products. The customer will normally have a choice – if your product is too expensive, difficult to use, unreliable or lacks some of the key features expected and which are available in rival products, then it is likely your product will struggle in the market.

Types of customer

When thinking of a customer in the product design context, the end user might come to mind first, but this may not always be the case.

Internal customer

For complex products, such as cars and aircraft, while there will be an overall team looking at the design and manufacture of the entire system, the details will normally be delegated to specialist teams. These smaller teams, in turn, may break down the area they are responsible for into even more specific elements that are assigned to other teams.

In the case of a car, for example, you may be working in an ancillaries team, designing and developing water and oil pumps, and have the engine team as your customer, who in turn have the drivetrain team as their customer, who in turn have the overall new car team as their customer (**Figure 3.12**).

Very often, with an internal customer, you will be provided with a quite detailed specification as to what is required and will be expected to adhere to this. For the most effective outcomes you would, however, work closely with the customer throughout the design process.

External customer

An external customer is one from outside your organisation, who may often – but not always – be the end user. In contrast to the procedure for an internal customer, you will generally be putting a product onto the market, by basing your design decisions on your understanding of the types of product in demand. Your data will come from previous experience, market research and focus groups, but at the design stage you are unlikely to be working with specific customers. Understanding the market, the types of user and what they will want from the product is therefore crucial.

Product and service requirements

Before beginning to design a product, it is essential to fully understand the requirements associated with its use.

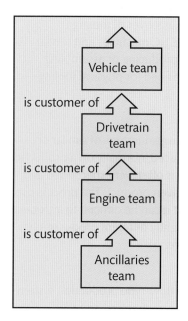

▶ **Figure 3.12** Example of internal customer relationships

Performance specifications

The performance specifications of a product set out what the user expects the product to be able to do. Consider an engine water pump. This would be expected to pump a certain number of litres of water per hour, which is the prime performance expectation. There are likely to be other performance requirements too, such as the temperature range over which the product would be expected to operate, the length of time for which the pump could be expected to work continuously, the time interval between servicing and the pump's oil consumption.

Compliance to operating standards

Most products will be expected to adhere to specific industry standards. These have been put in place for safety or standardisation reasons, and it is important to check first what standards might apply to your product and then adhere to them. For example, if you wish to bring out a new toy for children, then you must follow the 'Toy Safety Directive'. This obliges manufacturers to, among other things, make sure that the toy has been designed and manufactured to comply with the essential safety requirements during its foreseeable and normal period of use, carry out a safety assessment of the toy and ensure that the toy is accompanied by instructions for safe use.

Manufacturing quantities

Understanding the size of the market for your product, and what your product's share of the market is likely to be, will give you a planned manufacturing quantity. This is important in the design process and has both practical manufacturing and commercial implications. Certain manufacturing techniques are best suited to particular manufacturing quantities. For example, high-quality additive manufacturing (such as 3D printing) is often ideal for one-off or very small production runs, as no specialist moulds or fixtures are needed to create a new part; however, the actual manufacture of each part is slow. Injection moulding is a much faster process, but because a custom tool costing many thousands of pounds must be made for each component, it needs high manufacturing quantities to spread the cost of the tool.

Reliability and product support

You should also consider the requirements for your product in terms of reliability and support. You need to understand how your customers will use your product and what they expect from their relationship with it. For example:

▸ What is the consequence of your product failing while in use?

▸ Is there a danger to life should your product fail? (A car brake pedal? An artificial heart valve?)

▸ Is there a commercial consequence of your product failing? (Would a factory production line have to be stopped to fix the problem?)

▸ Should you 'life' your product to ensure it is replaced at a point before failure is likely?

▸ What support is needed for your product?

▸ What do you need to provide in terms of documentation to cover the set-up and use of your product?

▸ Does your product need servicing? (For example, lubrication, changing of filters, inspection routines?)

▸ Do you need to make sure there will be availability of serviceable parts?

▸ Can the user do the servicing themselves or do you need to ensure that specialist service technicians are available?

Product life cycle

Products have life cycles as they move through the stages of initial conception, manufacture, distribution and sale, operational use, reuse, recycling and disposal. As a designer, it is important to think about not only the initial user of your product but also whether your product can be reused in its original form, repurposed, recycled or otherwise disposed of.

As an example, consider car tyres.

▸ Original use – Can careful design and new compounds extend life?

▸ Retread – A partly worn tyre may be able to have a new tread moulded on to extend its life.

▸ Repurpose – A tyre no longer suitable for automotive use could be repurposed as a barrier for motorsports.

▸ Recycle – A tyre can be broken down into chips, the metal extracted and the rubber reused in specialist play surfaces.

Usability

Products need to be easy-to-use to be successful.

▸ How easy is it to operate the product?

▸ Is it intuitive, with functions that are laid out and operated as a user might expect?

▸ Are there conventions in place? (For example, indicator stalks tend to be on the left-hand side in most cars sold in the UK.)

▸ Is it easy to carry out basic servicing operations? (For example, battery changes, filter changes or lubrication.)

▸ Is it easy to clean?

▸ How well does it work with other products a user might be using at the same time?

Anthropometrics

This means understanding the physical body measurements of your product's users and is a key factor in making your products more usable. Designers can use anthropometric data, such as average height, eyeline position, weight, arm length, leg length and grasp details, together with parameters such as reach or strength, to design products to satisfy their customers. Typically, this kind of data is extracted from tables or databases and then converted into either computer or physical models of users. As a designer, you want your product to suit as much of the market population as possible, so it is common to aim the physical design at people between the 5th and 95th percentiles – 90% of the total population. Take care to select data relevant to your product's target population – designing a product for pregnant women will be different from designing a product for professional basketball players.

Product design specification or criteria

Before beginning to design a product you will need to draw up – or you may be given – a product design specification (PDS). This systematically lays down what is required from your product in terms of functionality, as well as any commercial constraints that apply. Creating a PDS ensures that you have formally considered everything required from your product, and you should refer to the PDS as you develop your product to check that it meets all elements of the specification.

Table 3.7 outlines many of the criteria that might appear in a PDS.

▶ **Table 3.7** Design criteria

Usage	What is the product supposed to do?
Performance	Are there specific target metrics of performance (e.g. speed, fuel consumption, emissions levels)?
Cost	What is the price at which you can sell each product to guarantee enough margin to cover product manufacturing and development costs and be sure the customer will still buy it?
Quantity	How many units of your product do you think customers will buy? This may affect how much you can afford to spend on development and will also influence the manufacturing methods available to you.
Maintenance	What level of maintenance will your product require? Will this be carried out by the user or require specialist support?
Finish	What, if any, particular finish is needed for aesthetic or protective reasons? This could be paint, coating, plating or polishing.
Materials	Is there a requirement to use particular materials due to market, regulatory, customer or manufacturing reasons?
Weight	What is the target weight for your product?
Aesthetics	What should the product look like? Must its look convey a certain image (e.g. high quality, robustness, professionalism)?
Life cycle	How will the product be used through its life? Will retailers expect a certain type of packaging? What happens when the product is no longer needed – how will it be reused, recycled or disposed of?
Sustainability	How can you ensure that the product is sustainable? Do you need to consider the materials and energy used during manufacture and use?
Carbon footprint	What is the target amount of carbon emissions associated with the product's design, development, manufacture, use and disposal?
Reliability	What level of failure will the customer tolerate? What aftercare and support is required?
Safety	What would the consequences be of your product failing? What is needed to minimise the risk of failure?
Testing	What tests, both regulatory and self-imposed, must the product pass to ensure viability?
Ergonomics	Who are you designing the product for? What key ergonomic and anthropometric data must be considered to design a successful product?
Usability	What will make your product usable? Do you need to factor in cleaning, battery replacement, storage, and so on?
Competition	What features must the product have to match or beat rival products? Are rival products likely to come out at the same time?
Manufacturing facility	Where will the product be made?
Manufacturing processes	Are there limits on the materials or processes you can use in order to match your company's current capabilities and tooling?

Commercial protection

Developing new products is expensive and time-consuming. A new product may feature design innovations in the way it has been manufactured, the materials used or the way it looks. These innovations are known as intellectual property. Protecting intellectual property is important to ensure that rivals cannot simply replicate your product.

Patents can be used to protect technical innovations. These may be new manufacturing processes, new products or new features in products, but they must genuinely be novel, be non-obvious, show inventiveness and have use. Designing a tennis racket with a slightly different shaped head is not likely to be patentable. However, fitting it with a genuinely novel device to maintain string tension might be. Holding a patent helps to prevent others from directly copying your innovation and allows you to license your idea for a fee.

While you may find yourself the inventor of a newly patented technology, you may also want to use a new technology, process or material that is protected by someone else's patent. In this case you could either enter negotiation with the inventors and strike a deal to use their technology, or try to achieve the same results yourself but by tackling the problem from a different angle to avoid infringing their patent.

Patents are normally region-specific and must be filed with a regional patent office (such as in Europe or the USA). If granted, they will apply in that region only. This will protect your rights over your innovation for up to 20 years. It is, however, important not to disclose your idea before filing, as once it is in the public domain the idea can no longer be patented and anyone will be free to use it.

Design rights are similar but are focused on how a product looks. An example might be the classic glass Coca-Cola bottle, which has a highly distinctive shape – although this does not necessarily improve its function as a bottle to hold a soft drink.

> **Link**
>
> Intellectual property is also discussed in Unit 4, Section A2.

A trademark is conventionally your company's name, logo, brand or slogan. So the Nike swoosh and 'Just Do It' slogan are trademarks and have been registered to avoid unauthorised use.

> **Reflect**
>
> Suppose you work as a designer in a company that makes products for the elderly. You have come up with a new process for making cushioning foam, which would also be ideal for use in trainers. The foam requires a unique chemical formulation and specialist machines to create it, both of which you have invented.
>
> How would you go about protecting and maximising the commercial return on your ideas if your goal is to work with an existing sports products supplier?
>
> How might this be different if you were to try to start up your own sports shoe company?

B2 Regulatory constraints and opportunities

Although you will have a specification as to what you – and the customer – want from your product, there may also be a range of external constraints that will apply to your product and manufacturing processes.

Legislation, standards, codes of practice and certification

Before you put a product on sale, it is essential to check that it meets any relevant regulatory and certification requirements. Additionally, there may be extra voluntary standards that you should aim to achieve.

▶ **Figure 3.13** The CE mark is used to verify that a product meets all appropriate standards

Look at any commercial product in your home, such as a television, radio or vacuum cleaner, and you will probably see the CE mark (**Figure 3.13**). This is the manufacturer's way of confirming that the product meets all the relevant standards.

In reality this symbol is a catch-all and, as a designer, you need to investigate which specific standards or directives would apply to your product. Some may be specific to a particular product type, such as child safety seats; others may relate to the ability of the product to operate in certain environments, such as explosive atmospheres.

A few examples of the standards that might apply to your engineering product are shown in **Table 3.8**. However, the number of standards is extensive, and they are under continual review – do not assume that the standards which were in place last year still apply now.

▶ **Table 3.8** Examples of safety standards

Product or requirement	Standard
Child car seats	ECE R44.03/04
Cycle helmets	BSEN 1078
Safety shoes	ISO 20345:2011
Items to be used in explosive atmospheres (e.g. petrochemical plants, grain silos)	ATEX 2014/34/EU
Electromagnetic compatibility	EMC 2014/30/EU
Boilers and water heater efficiency	EU Directive 2009/125/EC

In the past, standards and regulations were often tied to local requirements and the domestic standards agency. In the UK this was most commonly the British Standards Institution (BSI). Increasingly, however, national organisations are working together to harmonise regulations across international boundaries.

Environmental constraints

While issues such as safety and reliability drove many of the standards associated with the regulation of products, environmental concerns are now an increasing push in the creation of new products.

In addition to making sure that products meet relevant environmental standards, it is often a condition of satisfying the standards that this information is conveyed to users. For example, washing machines and fridges must have energy labels (**Figure 3.14**) that show the efficiency with which they use water and power. These labels can also cover other factors such as compliance with noise and electromagnetic regulations.

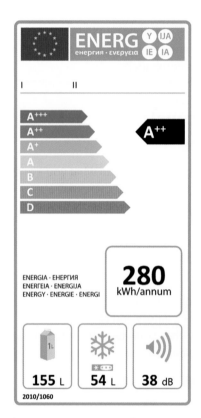

▶ **Figure 3.14** Energy label

In complying with these regulations a manufacturer measures the carbon footprint of a product while in use and throughout its life. Further regulations may cover disposal of products at the end of their life or the use of certain materials that may have environmentally detrimental effects in either their production or their disposal – for example, the banning of certain refrigerants.

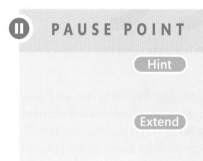

PAUSE POINT

You have been tasked to design an electric lawnmower. Find specific standards and regulations that might apply to this product.

Hint

How would you go about this? Can you source information from existing products either through product literature or by visiting garden centres? (Note: it may be difficult to access some standards as these may require subscriptions or payment.)

Extend

If you were to devise a performance label similar to the one for washing machines but applicable to lawnmowers, what might it look like? What would you consider to be important? How might you measure the data?

Health and safety

It is your responsibility as a designer to ensure that your products are safe to install, use and dispose of. Even if a product meets all the relevant standards and directives, it could still be unsafe.

Think carefully and ask yourself these questions about the product:

- Is it safe for the user under normal use?
- Could it be dangerous to others?
- Are there features that will run hot, or are sharp?
- Is it excessively noisy?
- Will it give off fumes in operation?
- Are there any other aspects of the design that are not as safe as they could be?
- Can you use guards or **interlocks** to protect the user from any dangerous features?
- Is it safe to install – is it heavy or does it have exposed electrical or cutting elements?
- Is it easy to clean and maintain?
- What instructions should you provide to instruct the user on safe use?

> **Key term**
>
> **Interlock** – a physical or electronic lock that prevents something from operating until guards are in place. For example, a CNC lathe will usually have an interlock to prevent the spindle from operating until the access door is closed.

Security

There may also be a need to consider data protection and data security. This is certainly the case if you are developing devices that a user will interact with to gain access to personal accounts (e.g. cash machines, card readers) or where personal data (such as medical information) may be stored. Such devices need to be designed so as to minimise the risk of interception or capture of data by third parties.

You may need to think about how data gathered from users might be used – in either planned or unintended ways. For example, the roll-out of smart energy meters to all households in the Netherlands was abandoned after it was realised that the data from these meters could be used not only to monitor energy consumption but also to analyse the comings and goings of residents in each home, thus infringing personal freedom and posing a security risk.

B3 Market analysis

When designing engineering products, keep in mind that you are working in a competitive market.

Bespoke products for a specific customer

In some cases a customer may commission your company to carry out a particular design for them, or the process may be put out to tender where a number of companies bid to do the work.

These types of commission are particularly common for high-value one-off products such as ships, oil rigs and buildings. Alternatively, the product may be of lower value and produced in modest quantities, but be of such a specialist nature that it is unlikely to be of interest to other customers.

The awarding of these jobs to particular design teams is based on adherence to the customer requirements, price and company reputation.

Products for sale on the open market

For products designed for sale on the open market, a different set of dynamics is in play. The customer will have a choice, and if your product is more expensive, less reliable, less stylish or harder to use than the competition then your sales will be limited.

A conventional approach to setting your product a step ahead of the competition is to have a unique selling point (USP). This could simply be a low price but is more commonly a feature or level of performance that your competitors cannot match. The USP can then be used by your company's marketing team to make your product seem more attractive to the customer. **Figure 3.15** shows examples of possible USPs for a new ladder design.

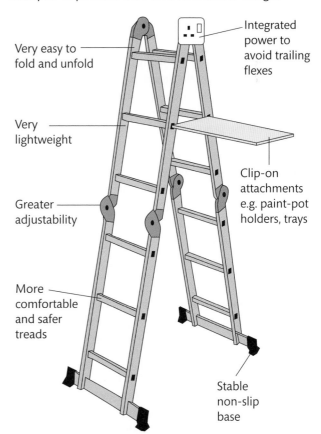

Very easy to fold and unfold

Very lightweight

Greater adjustability

More comfortable and safer treads

Integrated power to avoid trailing flexes

Clip-on attachments e.g. paint-pot holders, trays

Stable non-slip base

▶ **Figure 3.15** Potential unique selling points for a folding ladder – what else could be considered?

Of course you have to be careful. Your USP must be:

▶ justified – You cannot market your product as being the lightest if is not.

▶ something a customer would appreciate – Adding an extra feature may make your product unique, but if the customer does not value such a feature, it will simply add unnecessary cost.

▶ capable of being protected – You might have added a desirable extra feature, or your product may be much more efficient than similar products of your rivals, but if the design can be copied and you have not protected your intellectual property, your gains might be very short-lived as your competitors can replicate your innovations.

Benefits of the design

While a USP can help a product get attention and mark it out from otherwise very similar products, the overall performance of the product is most important in ensuring its success in the market. Sometimes there may not be very big differences between the individual elements of your design and those of competitors' products, but, as a whole, these incremental benefits may add up to give a more effective and balanced product. Do not neglect the details at the core of your product in pursuit of excellence in one particular area.

Obsolescence

Any particular design will normally have a period of time during which it can remain successful in the market before competition, regulation or disruption to the market makes the design obsolete. The lifespan of a product varies depending on the product and the market.

The Bic® Cristal® ballpoint pen, for example, has remained largely unchanged over nearly 70 years and is still a leading seller within its target market. Family cars usually have a production life of six years for a given model. By contrast, mobile phones will typically be superseded by updated versions within a year.

Obsolescence of product lines can come about for a number of reasons.

▶ A product may no longer be allowed to be sold due to changes in regulations. For example, tighter emissions rules have meant that a number of car engines are no longer viable.

▶ There may be a step change in the market (disruption), which may mean that products of a given type are no longer viable. For example, film-based cameras have largely disappeared in favour of digital cameras, which themselves, at least at the lower end of the market, face competition from camera-equipped mobile phones.

▶ Third-party products needed to support the operation of the product, such as specialist batteries or spare parts, may no longer be available.

▶ Developments in materials and technology may enable rivals to put out products with features and performance that exceed those of your product.

▶ New materials, technology or manufacturing techniques may allow rival products to be produced and sold at a lower price than you can match with your existing product.

▶ You will normally have your own development cycle to bring out improved replacement products on a regular basis. Such a planned cycle ensures that your product range will remain competitive but may also persuade regular customers to upgrade, rendering your older products obsolete.

It is always important to understand your customer. As soon as they are released, new high-end mobile devices and gadgets will be snapped up by a core of users. These users will not be concerned about replacing the product after a year or two. In contrast, a refrigerator, washing machine or cooker will often only be replaced when it fails, regardless of innovations in the intervening time period. Appreciating what your customer expects from your product is essential if you are to avoid costly mistakes.

B4 Performance analysis

Product form

The form of a product is its physical shape, size and appearance. Depending on the nature of the product, the product form is often tied very closely to the product function. For basic hand tools, such as a spanner or garden spade, the function very much determines the physical form. At a much more complex level, the shape of a Formula One car is largely dictated by aerodynamics. For devices whose functionality is primarily electronic or IT-based, there may be more scope to look at the form of the casings for the electronics, focusing on ergonomics, practicality or style, for example, based on customer expectations for the product.

Product functionality

Product functionality relates to how the product is expected to work.

What is your product expected to do?

For example, suppose you are designing a stackable chair of the type used in schools and waiting rooms. The functionality requirements might be that the product must:

- be able to support the weight of users
  - provide adequate back support
  - be such that the chairs can be stored efficiently
  - be easily carried
  - be such that the chair supports do not damage flooring
  - be able to be connected to other chairs to form rows
  - be easy to clean.

Note that these functions cover both the direct operation of the chair when someone is sitting on it and the storage and setting out of the chairs. These statements are quite general. It is the designer's job to develop them further, and the developed statements will then form part of the product design specification. For example, regarding the first point, about supporting the weight of the user, the designer would consider who the users will be, which percentiles of the user population to focus on and what **factors of safety** to use.

> **Key term**
>
> **Factor of safety** – describes how much stronger a system is than it needs to be for an expected load.

Technical considerations

Your product may also have certain specific technical requirements. These might include the output power of the device, the loads it should carry, its consumption of fuel or electricity, its maximum operating temperature and its weight.

Choice of materials and components

The choice of materials to use is a vital part of any design. Each design scenario will have its own specific demands, and balancing the characteristics of a range of different materials will be important in any successful design.

Strength

You do not want your product to break apart or be bent out of shape while in use. When designing a product, you need to have a good feel for the levels of stress that will be imposed on it and make sure that you will always be working below the yield strength of the material you are planning to use. You will usually want to implement factors of safety. These factors are put in place to help deal with unexpected variations in the loading on or behaviour of your product, which could result in the actual stresses being higher than those calculated. So if you were making a product from mild steel with a yield strength of 240 MPa, and working to a factor of safety of 3, then you would design the product such that anticipated loads would not exceed 80 MPa.

Stiffness

When a component is subject to a load it will flex – so how rigid does your design need to be? The rigidity of a structure is a function of the material and the geometry, with the material rigidity being governed by Young's modulus of elasticity for axial loading, or the shear modulus for torsional situations. Aluminium and its alloys typically have moduli of around a third of those for steel, so if you compare aluminium and steel bicycle frames you will find that aluminium frames typically use larger-diameter tubing to give an acceptable level of frame rigidity.

In many cases you will want a high degree of rigidity, so that the geometry of your design remains consistent and components stay aligned. In other cases, a bit of elasticity is desirable to give a degree of suspension or cushioning.

Fatigue characteristics

Is your product likely to be subject to fluctuating loads as a result of vibration, cyclic operation or rotational motion (such as where a rotating shaft is subject to an end load, like a car axle)? In such cases, to avoid failure you may need to consider the material's fatigue behaviour, even when the load is well below the yield strength. Prevention of fatigue failure can be managed by careful design and selection of material.

Thermal conductivity

If your product is likely to operate over a range of temperatures or generate its own heat, such as might happen in an engine, thermal conductivity could be important. You probably want a material with a high thermal conductivity to allow heat to be conducted quickly away from areas prone to overheating and then be distributed into the atmosphere. Aluminium and copper are good thermal conductors and are commonly used as heat sinks to carry heat away from the processors of computers.

Thermal expansion

Materials expand when heated, and this can cause products to distort as they warm up. This is why bridges, for example, commonly have expansion joints – these are gaps designed to allow road sections to expand into them as the temperature rises.

Density

Are you concerned about the weight of your product? If so, then the density – the mass per unit volume – of the material will be an important parameter. But be careful: if a material has low density but also low strength and rigidity, then you may need to use more of the material, possibly resulting in a higher overall product weight, to achieve the desired performance than if you had chosen a denser but stronger and more rigid material.

Ductility

Do you need to form your product? If you need to fold, stamp or press sheet material, you would want to use a fairly ductile or malleable material. Thin sheet mild steel, for example, can be bent by folding machines without problems, whereas brittle, high-carbon steel would simply snap.

Machinability

Some materials, such as aluminium and mild steel, machine easily, but stainless steel can be difficult because of work hardening. The vibrations generated by the machining will harden the material surface, making subsequent passes difficult.

Manufacturing suitability

If your product's geometry dictates that a particular process be used, such as casting, moulding, machining or welding, you will generally need to choose from a certain range of materials that are suited to that technique.

Corrosion resistance

Does your material work reliably in an environment where corrosion is likely to be an issue? Does it have natural corrosion resistance? Can you improve the corrosion resistance by coating or plating its surface? Is coating more practical or cost-effective than going for a corrosion-resistant material?

Electrical conductivity

Is the ability to conduct electricity necessary for your product? For example, a car chassis is commonly used as an earth return for the vehicle's electrics, as this reduces the amount of wiring needed. In other situations you may want your product to have electrical insulation properties. For example, plastic switches provide an extra safeguard against electric shocks to a user.

Price

How much does the material cost? Can you get away with fairly cost-effective and readily available materials or do you need to go for something more exotic? Take care when comparing costs in the selection of materials, as the prices of different materials may be quoted by weight, by length or by volume, and you need to relate this information to your design when analysing the overall cost of producing your design using different materials.

Choice of components

When choosing components, such as motors, connectors, bearings or switches, you will need to draw up a specification for what you require in terms of performance, size, weight, price and compatibility with the rest of the system. This will then be matched to the data provided by manufacturers and suppliers.

Environmental sustainability

Is your material sustainable? Is it from a renewable source? Can it be sourced from recycled materials? What amount of resources does a product use when in operation? Can you make it more efficient and reduce its energy consumption? How will the product be disposed of at the end of its life? Can it be stripped down easily into its component parts for recycling? Are there any elements within your product that need special handling, such as batteries or chemicals?

Interaction with other areas or components

Consider how your product will interact with other devices, components or the surrounding environment.

Ergonomic concerns

For example, if you are designing a steering wheel, you will need to consider the positioning of the indicator and wiper stalks and the wheel's placement in relation to the instrument panel.

Physical concerns

Suppose you are designing a fuel pump for an engine. How much space is available for the pump? Where will the connecting pipes come in? Once the engine is installed, will you have access to enable changing of the filter or removal of the pump? Do you need to use particular fasteners to be compatible with the overall system and to minimise the number of tools needed during assembly or servicing?

Electrical concerns

Does the product need to operate at a particular voltage to be compatible with the rest of the system? Does it need to use a particular digital communication protocol to talk to other parts of the system?

Thermal and environmental concerns

Is your product likely to be subject to heat, cold or vibration generated by other parts of the system? Do you need to design your product to cope with these factors, either directly or by using some form of shielding or insulation? Does your product generate heat or vibration that might be detrimental to other parts of the system? How might you design the product to limit these effects?

Likelihood of wear and failure

You need to anticipate that your product will wear through use. How long do you think your customers will expect your product to last? If the product contains moving parts, or parts subject to particular wear, do you plan to make these last the life of the product or do you anticipate that they will be replaced at various points in the product's life?

If the latter, do you need to define schedules or inspection processes to ensure that parts are replaced before failure? Do you need to make sure your design allows for easy replacement of worn parts?

B5 Manufacturing analysis

There are a number of technical aspects you need to consider if you are to design a product that can be made as easily, cost-effectively, sustainably and safely as possible.

Processes for manufacturing and assembly and manufacturing requirements

You should always design a product with a clear understanding of the manufacturing processes to be used. Each manufacturing process has its own constraints, in terms of the materials that can be used, the types of shape that can be made, the precision of the process, the specialist tooling costs and the manufacturing cost per part. In addition to these constraints, there may be a limit on the processes available to you, given your company's capabilities. To achieve a cost-effective and optimised design, it is important to select the appropriate manufacturing process and design your product accordingly. Refer to Section A6 for more details on various manufacturing techniques.

Quality indicators

When any design goes into production, the parts produced and the overall product will contain variations – no two items of output will be exactly the same. For example, the size and location of holes or ribs in a part will vary within measurement and processing tolerances; similarly for surface finishes, tool marks and assembly details. It is also the case that bought-in parts, whether stock materials or nuts, bolts, screws, seals, bearings, etc., will vary subject to tolerances.

It is important to consider which features need very tight control for the product to be acceptable and which can be produced with wider tolerances. Generally, if you set tighter tolerances or are more specific about the finishes required, then production will be more expensive. These parts may need more skilled production operators, more advanced machinery and secondary finishing processes, or they may suffer higher rates of rejection to meet the desired standards. Conversely, if parts are allowed to have features with wider tolerances where appropriate, it is likely that savings in time and money can be made.

In a given product it is likely that some features need to have tight tolerances and others much looser tolerances. An example is shown in **Figure 3.16**.

On an injection-moulded plastic part it might be perfectly acceptable to have ejector pin marks on the interior surfaces that are visible to the customer, but not on exterior surfaces. On sand-cast parts the natural texture may be acceptable in some areas, but where the part has to interact with other components it is likely that some form of machining or other finishing process will be needed.

These tolerances and manufacturing requirements would normally be recorded in the detailed part drawings and assembly drawings for the product. The information could then be used to draw up quality control documents and inspection check sheets to verify that components are being made to an acceptable standard. This helps to drive quality assurance, with the aim of minimising the number of parts produced that fail checks.

Similar quality considerations apply to the raw materials and standard parts bought in, as well as to the overall finished product.

Environmental sustainability

In terms of manufacturing processes there may be several areas to consider with regard to environmental impact and sustainability.

This hole locates a bearing which is held in place via an interference fit. Its diameter needs to be toleranced tightly (to 0.01 mm) or there is a risk that either the bearing will be too loose in the hole or the hole will be too small to fit the bearing.

This hole is simply an eyelet to help lift the motor. Its diameter can have a tolerance of ±1 mm without affecting its ability to work as a lifting eye.

▶ **Figure 3.16** Different hole diameter tolerances on a motor – what would be the consequence if both diameters were specified to ±1 mm, or both to ±0.1 mm?

- How sustainable are the materials you are planning to use?
- Are the materials renewable (such as wood), or is there a finite supply (such as copper)?
- Can you use materials that are wholly or partly recycled?
- What happens to waste material – machining swarf, moulding or casting sprues and runners, reject parts?
- How much energy is used in production? Can this be reduced?
- Is water used for cooling or cleaning during the manufacturing process? How is it disposed of? Can it be cleaned before disposal?
- Are there any by-products or other materials associated with production, such as solvents, cleaners or acids? How are these disposed of?

Design for manufacture

It is essential to always keep in mind how the product you are designing and the individual parts within it are to be manufactured.

Each manufacturing process has materials that suit it, particular shapes that can be made, tolerances and finishes that can be produced, and costs associated both with part-specific tooling and moulds and with the actual production of each part. In addition to these constraints, you also need to consider your company's capabilities and processes.

The example in **Figure 3.17** shows some alternative concepts for a simple component. Note that the geometry and detail design of the part are dependent on the manufacturing technique used. Which approach is finally chosen would depend on the materials that are most appropriate and available, the manufacturing capability of the company and the volume of parts to be produced.

Initial concept
The design specification calls for an eyelet mounted on a backing plate

Welded fabrication
The part is fabricated by two flat laser- or plasma-cut sheets welded together.

Positives:
low set-up costs with minimal specialist tooling; little waste of material

Negatives:
may lack precision due to thermal distortion;
welding method limits material choice (but as an alternative plastics could be bonded using adhesive)

Machined and assembled fabrication
The part is fabricated from two machined elements screwed together.

Positives:
low set-up costs with minimal specialist tooling;
wide choice of materials;
simple assembly of components

Negatives:
could be wasteful of material if machining from stock;
requires assembly process

Casting or moulding
The part is cast or moulded as a single piece.

Positives:
no assembly needed;
little waste of material;
some features can be moulded in to reduce secondary operations

Negatives:
cost of mould may be significant;
part needs draft angles to allow removal from mould;
limited choice of material

Sheet fabrication
The part is fabricated by folding flat laser- or plasma-cut sheets.

Positives:
low set-up costs with minimal specialist tooling;
little waste of material if parts are nested on sheet

Negatives:
may lack precision due to fold operations;
material choices limited (e.g. thin sheet steel, but may be possible to use plastics with heat-assisted folding)

▶ **Figure 3.17** The same component produced by several different manufacturing techniques

PAUSE POINT Consider the plastic case for a computer keyboard. What material do you think it is made of? Where does the raw material originate?

> **Hint** Ask your tutor if the college has any old or broken keyboards for you to examine. These will often have marks on the moulding that indicate that the material is for recycling purposes (although the marks will normally be on an inside surface and may be difficult to see without disassembly).

> **Extend** How recyclable is the material? Are there other materials that might be more sustainable? Could the keyboard case be designed to use less material?

Assessment practice 3.2

You have been asked to design a 'quick lift' jack for use in professional motorsport events. The function of the jack is to raise a car off the ground during pit stops, to allow tyres to be changed during the race.

Develop a product design specification for the jack. This should cover the key requirements of its operational use, together with other demands, expectations and constraints on the design.

Note that you are not being asked to do a full design of the jack. Rather, you will demonstrate through this task that you can take a customer requirement and turn it into a product design specification.

You must show that you understand what the customer wants and also what constraints there may be in terms of safety, reliability, price, anthropometrics, and so on.

For some of these factors, such as cost, you may need to investigate the price of comparable products to inform your thinking.

Consider carefully how the product will be used, stored etc.

Plan
- What is the task? What am I being asked to do?
- What information will I need to collect? What resources can I use to help me complete the task?
- How confident do I feel in my own abilities to complete this task? Are there any areas I think I may struggle with?

Do
- I know what I'm doing and what I am aiming to achieve in this task.
- I can identify the parts of the task that I find most challenging and devise strategies to overcome those difficulties.

Review
- I can explain what the task involved and how I approached its completion.
- I can explain how I would deal with the hard elements differently next time.
- I can identify the parts of my knowledge and skills that could benefit from further development.

C Using an iterative process to design ideas and develop a modified product proposal

C1 Design proposals

Initial and developed propositions to improve an engineering product

While sometimes you may be designing a brand new product to satisfy a customer's needs, at other times you may be asked to modify, refine or develop an existing design. Whether a product is brand new or an evolution of an existing design, many of the processes used will be similar, although the constraints may be different.

Table 3.9 lists some engineering design activities you may be involved in.

Develop an initial product proposal	You may be asked to develop a new and as yet unresolved product concept. You will be considering the materials, manufacturing processes, technical issues and other factors needed to turn the basic concept into a viable design.
Redesign an existing product to resolve a fault	An existing product may be failing in use. It will be your job to identify the underlying problem and propose an effective alternative design.
Reduce manufacturing costs of a product	Your company may wish to reduce the cost of manufacturing a product. This might be to ensure it can be sold at a competitive price or to increase profitability. Can the design be modified so that the product will be cheaper to make, easier to assemble or use fewer materials?
Develop a specialist variant	Your company may wish to produce a variant of an existing product with improved functionality (e.g. increased power, lighter weight, better corrosion resistance) or so that it can be used in new sectors such as food or medical applications.
Adapt to changing regulations	You may be unable to sell an existing product because new regulations or standards have made it obsolete. Can the product be modified to comply with the different requirements, or will you need to design an entirely new product?

Technical design criteria

When designing or redesigning any product, it is essential to understand fully what is expected of the product. See the sections on technical and product design specifications earlier in this unit (Sections A3 and B1) and in Unit 5.

Idea generation

In the design process you will usually generate a number of possible solutions based on the technical design criteria. At this stage it is beneficial to be open to new ideas. Do not be overly critical of any ideas that come up, as this is likely to slow the design process down. Just focus on generating as many ideas as possible.

If you are redesigning an existing product, questions that might help you generate ideas to give you a valid proposal include:

▸ Can you use different materials?

▸ Can you use fewer components?

▸ Can you use different manufacturing processes?

▸ Can you make your product bigger or smaller?

▸ Can you think of a completely different solution?

Morphological analysis

One approach to generating ideas involves using morphological analysis, where specific requirements of the product are identified and then a range of solutions is proposed for each item. This generates an array of possible product options.

For example, consider a cycle helmet, like the one shown in **Figure 3.18**.

▸ **Figure 3.18** What does a user require of a cycle helmet?

In **Table 3.10** the requirements for a cycle helmet are listed, and for each requirement ideas for possible solutions are generated.

▸ **Table 3.10** Morphological analysis of desired features of a cycle helmet

Requirement	**Possible solutions**			
Protects head from penetrating injury	Hard shell	Thick layer of protection	...	...
Cushions impact	Use of sponge-like foam	Use of expanded polystyrene-type material	Use of suspension straps	...
Suits a range of users	Helmet produced in wide range of sizes	Adjustable through tightening of retention straps	Ratchet-type adjuster	...
Can be secured at chinstrap	Buckle	Snap-fit	Stud	...
Keeps head cool	Vents	Active cooling (fan)	Thermally conductive material	...
Shields eyes from sun	Fixed peak	Detachable peak	Tinted eye shield	...
...	...	...	...	...

This kind of analysis can be used to drive the creation of ideas. Different features for the various requirements can be combined to generate a range of concepts, such as a hard shell with suspension straps, vents and a tinted eye shield.

PAUSE POINT   Carry out a morphological analysis, similar to the one above, for a professional builder's wheelbarrow.

Hint   Break down the operation of taking some rubble from a yard to a skip using the wheelbarrow. From this, list a range of requirements for the wheelbarrow.

Extend   How do you think the matrix would be different if the wheelbarrow were aimed at elderly leisure gardeners?

Initial design ideas

It will be necessary to select which of the initial ideas to take further. Review each one against the design criteria. Which concepts are fit for purpose and will do the job intended for them? What constraints are there on your design? Will some designs be too expensive, be too difficult to make, require specialist manufacturing equipment, or need complicated aftercare and servicing? Can any of your initial ideas be adjusted to enable them to meet these requirements? Can elements of different ideas be combined to give a better overall solution?

A common method to narrow down ideas is to use an evaluation matrix. A shortlist of possible concepts is drawn up, and each of these concepts is scored against the requirements in a weighted matrix. **Table 3.11** shows how this method might be applied to the cycle helmet example.

> Each concept is given a mark out of 10 to describe how well it meets the requirement.

▶ **Table 3.11** Evaluation matrix for the cycle helmet design

Evaluation criteria		Score (out of 10)		
Requirement	**Weighting**	**Concept 1**	**Concept 2**	**Concept 3**
Safety	0.30	8	6	9
Comfort	0.30	5	6	6
Style	0.20	5	3	4
Weight	0.10	6	5	5
Cost	0.10	3	4	5
Total	1.00	5.8	5.1	6.3

> Concept 3 scores highest overall.

> The requirements are weighted in proportion to their importance to the design.

> The total for each concept is the sum of the scores times the associated weightings. For concept 1: $(8 \times 0.3) + (5 \times 0.3) + (5 \times 0.2) + (6 \times 0.1) + (3 \times 0.1) = 5.8$

In this example, concept 3 had the highest score, so it *may* be the best design to take forward. However, you should always review the design to see if it really is the best approach. The criteria and weightings used in the table can be quite subjective, and small changes to these can often change the outcome, so use with caution.

Development of designs

Although the idea chosen may fundamentally do what is required, it is likely that the concept will need further development before it is fit for production. The following are some areas that you should consider in developing your design.

Aesthetics

What should your product look like? Will your product need to fit in with its environment or with convention? (You do not see many purple fridges.) Does the product need to convey an impression of quality or strength? Are you aiming for a 'family' look? How strongly do you want the product to represent your company's brand? (For example, all BMW cars share common design features.)

Ergonomics

Who are your product's users? How will they interact with the product? What do the size, strength, weight, reach and dexterity of your users mean for your design? How can you make the product easy to use? Do you need to think about safety issues (such as making sure gaps in folding mechanisms are either too big or too small to prevent fingers getting trapped)?

Size

Does your product have to be a particular size? Washing machines, for example, are generally 850 mm high and 600 mm wide to fit into a standard kitchen. If you are designing a replacement water pump for a car engine, you will need to take into account standard mounting points, coolant inlet and exit locations and maximum overall dimensions to avoid fouling on other parts.

Mechanical and electronic principles

How is your design going to work? Will it be strong enough? How do the various parts move relative to each other? How are components connected together? Does the product need servicing – for example, would it need lubrication points? What electronics are required? How will they link up? What do you need them to do? How will they be powered?

Materials

What strength and rigidity is the product required to have and consequently what materials will you use? Will your choice of materials influence the manufacturing techniques available to you? Will the environment in which your product is to be used affect your choice of material? Is the product going to have to work in a damp or corrosive atmosphere? Will it have to work in hot, cold or cyclic temperatures? Is the base material adequate for the job, or does it need coating, plating or heat treatment to give the required performance?

Manufacture

How are you proposing to manufacture your design? What is the expected production volume? Can the product's proposed geometry be produced by the planned manufacturing technique? Is the proposed method cost-effective?

Assembly

How will your product be put together? Is this going to be easy? Can parts be pre-combined to reduce assembly costs? Will it be difficult to line up parts for assembly and, if so, how can this be addressed? Can the product be assembled by a single operator with simple tools, or will it require specialist tools or machinery? How will it be held together? Will it be welded, glued, screwed, riveted or snapped together?

Costings

Can you estimate the cost of your product? The total cost will include the costs of raw materials, bought-in parts, processing and assembly. Is the cost more than the market is prepared to pay? Can any unnecessary costs be taken out by careful design?

Factors of safety

What factor of safety is needed? How much will you add to the expected loading to allow for misuse, unexpected loads or variability in the end product? What are the consequences of failure in terms of impact on individuals, cost of disruption and replacement, and cost to reputation? How likely is it that the product's normal terms of use will be exceeded? Do you need to add in a degree of redundancy? For example, car brakes are designed with dual circuits so that if one fails, the second should still give some stopping power.

Which components to buy in and how to choose these

For many products, you may need to buy in or subcontract the manufacture of some parts. Typically, these will be specialist parts that your company will have neither the in-house expertise nor the equipment to produce in a cost-effective way; they may include motors, seals, springs, nuts, bolts, washers, electronic components, sensors or switches. To select the parts to buy in, you generally first determine what you want from the item and then access manufacturer or standards data to identify the correct parts. For example, if you need to select a spring, you might go to a manufacturer's website and enter information such as whether your required spring should be tensile or compressive, how stiff it should be and what size it needs to be. The database would then return a recommendation for a particular spring.

Use of information sources

To support your design, it is important to consult with a range of external information sources.

▶ Materials data – Information on a material's strength, elasticity, density, corrosion resistance, electrical and thermal conductivity, processability, and so on will be available from suppliers or reference books. The data may be provided in tabular or graphical form. Computerised databases also allow for searching based on the properties required of a material.

▶ Manufacturer data – For the manufacture of many products you are likely to need a range of standard components. These might be items such as motors, drive belts, chains, gears, seals, bearings, switches, sensors or control boards. A manufacturer will provide a datasheet for each of their products, giving details of the dimensions, fitting instructions, performance and limitations.

▶ Standards – There are tables and standards governing the geometry of standard components, such as nuts, bolts and screws. For example, by referring to these you can find that an M6 screw will have a pitch of 1 mm and a head dimension of 10 mm across the flats.

Depending on your application, there may be many other sources to consider, including data provided by the suppliers of the machine tools needed to make your parts, regulatory documentation and ergonomic data sets.

Besides looking for written and online information, always consider asking others. If you are considering using an unusual material or a manufacturing process outside your company's normal experience, speak to the suppliers, who will usually have technical support experts available to advise you.

C2 Communicating designs

In practice, the design of a new or revised product is normally a team activity. For your design to be successful, it is essential to be able to communicate your ideas clearly to your engineering colleagues, subcontractors, customers and marketing teams.

While written technical communication is important, the nature of engineering design is such that sketches, drawings or renderings supported by notes are often the most effective means to get information across.

Freehand sketching is one of the most useful tools you can use to express your ideas. It is particularly powerful in the early stages of the design process, when new ideas can be quickly captured on paper. Aim to use sketching to create images that clearly communicate the form and function of your product. For this purpose, your sketches do not need to have any particular artistic quality. Adding annotations will also help to explain your ideas (see **Figure 3.19**).

Feel free to use both 2D and 3D effects in your sketches, as you think appropriate – 3D sketches are often best for showing the overall form of a product and the relationships between different elements, whereas 2D drawings can show technical detail and mechanisms more clearly.

You will need to use more formal forms of graphical communication once you are ready to turn your idea into a final design and then a physical product.

▶ Orthographic detail and assembly drawings – These provide the geometrical and dimensional information needed to create a physical component or assembly. They may be created by hand or generated with the help of CAD software.

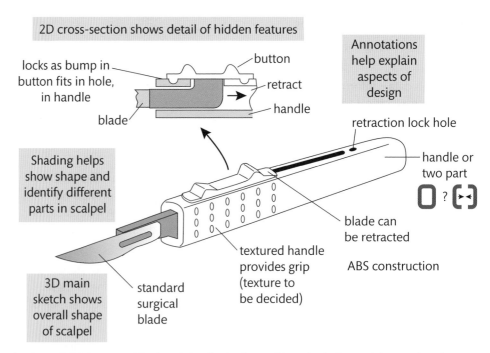

2D cross-section shows detail of hidden features

locks as bump in button fits in hole, in handle

button

retract

blade

handle

Annotations help explain aspects of design

retraction lock hole

handle or two part

Shading helps show shape and identify different parts in scalpel

blade can be retracted

textured handle provides grip (texture to be decided)

ABS construction

3D main sketch shows overall shape of scalpel

standard surgical blade

▶ **Figure 3.19** Annotated freehand sketch of a disposable scalpel – how might you show how the blade is retracted into the handle?

▶ Circuit diagrams (electrical, pneumatic or hydraulic) – These diagrams symbolically represent the layout and operation of different forms of control or power circuits within your product.

▶ Block diagrams – In these diagrams the main parts or sub-functions of a product or system are represented by blocks, and lines are drawn between the blocks to show the relationships between them.

▶ Flow charts – These are commonly used to describe sequences and decisions in the systems controlling the operation of your product.

In addition to graphical communication, you need to produce written documentation to describe your ideas and concepts fully.

▶ Technical report – a formally written report that describes the development of a product or the results of trials or tests. The tone should be formal and in the past tense. Be clear and concise and take care to use the correct technical terms.

▶ Design log – a detailed record of the design process with notes, sketches, block diagrams, test and market data gathered, and so on. The design log should be updated as concepts are generated, ideas proposed and design decisions made, and you should note the reasoning behind the decisions.

▶ Design specification – lists the requirements of your product in terms of functionality, cost, weight and any other needs.

▶ Parts list or bill of materials (BOM) – lists the materials, components to be manufactured and parts to be bought in to create your product. The list will normally include the quantity of each part, the supplier, the part reference number and possibly also costs.

Assessment practice 3.3

Suppose you work for a manufacturer of ride-on electric lawnmowers, and your company is planning to introduce a range of pull-along accessories such as trailers to increase the usefulness of their products.

To achieve this, you need to design a pair of couplings, one of which will be fitted to the rear of the ride-on lawnmower, with the mating part attached to the trailer (see **Figure 3.20**).

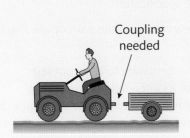
Coupling needed

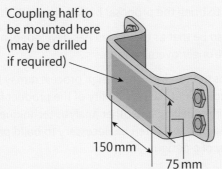
Coupling half to be mounted here (may be drilled if required)

150 mm

75 mm

▶ **Figure 3.20** Sketch of lawnmower trailer coupling – can you come up with an effective solution?

Both the lawnmower and the pull-along devices will have a steel plate that is 5 mm thick, 75 mm high and 150 mm broad, on which the couplings can be mounted.

The coupling should be strong enough to pull an 80 kg load over soft ground and be easy to attach and detach.

Your company also wishes to make a variant of the coupling that will allow electric power to be transferred to the trailer unit to power lights, leaf-blower units etc. The power connection will be rated at 24 V 10 A as a minimum.

Generate a proposal for the design of the coupling, which should include design variants for both the basic version and the version allowing electrical transfer.

Your proposal pack should take the form of a design log, which should include:

- a product design specification for the coupling
- a range of design ideas generated with the aid of a morphological analysis
- a systematic evaluation of the design ideas to show how you converged on a chosen solution – this should consider the key specification requirements
- an iteration of this evaluation and selection process to take in all requirements of the product, including cost and manufacturing considerations – explain the evolution of your design with annotated sketches
- a freehand sketch of the final product identifying the key features, materials and mechanisms
- an additional detailed drawing of all custom components and an orthographic assembly drawing.

Plan

- What is the task? What am I being asked to do?
- What information will I need to collect? What tools and resources can I use to help me complete the task?
- How confident do I feel in my own abilities to complete this task? Are there any areas I think I may struggle with?

Do

- I know what I'm doing and what I am aiming to achieve in this task.
- I can identify the parts of the task that I find most challenging and devise strategies to overcome those difficulties.

Review

- I can explain what the task involved and how I approached each stage of the work.
- I can explain how I would deal with the hard elements differently next time.
- I can identify the parts of my knowledge and skills that could benefit from further development.

C3 Iterative development process

In reality, designing is not an entirely linear process. There is often a need to pause during the development process to review the suitability of the product. This might be to check that it still meets the design specifications, to see if there is scope for enhancement or to adapt the design to widen the market for the product.

A product proposal may initially have fulfilled key functional aspects of the specification but, as the design was further developed to bring in secondary functions, the cost or weight may have drifted up. The development should then be paused to see how to bring the proposal back in line with the specification.

It may also be that as the design has developed, new opportunities have opened up to allow a stronger overall design.

At various points along the design process it may be necessary to carry out tests or trials to evaluate the suitability of the product for its intended purpose. While computer modelling and other analysis techniques may help to validate some aspects of the design, it will often be necessary to build physical prototypes to confirm the suitability of a design.

If the trials have revealed any issues with the design, you may then have to go back a stage or more and repeat the development steps to address these issues before you can move forward.

Case study

Designing a battery-powered hand drill

A design team were developing a new battery-powered drill. They put together a proposal and wanted to check whether it would be suitable for a range of users, being both comfortable to grip and well balanced. The team had looked at ergonomic and anthropometric data to inform their design but wanted to carry out a validation. They made a prototype with the proposed handgrip and weighted it to simulate key components such as the battery and motor. This prototype was given to a group of potential end users to try. The users reported that the design felt nose-heavy and became tiring to use after a few minutes. The team reviewed their design in light of these findings and, by placing the motor further back, produced a more neutrally handling tool. By iterating

this process of reviewing their product and refining the design, the team ultimately generated a robust design that met both the customers' and the manufacturer's expectations for the drill.

Suppose that putting the motor further back in the body made it more balanced – but this restricted air to the motor, resulting in it generating excess heat that compromised its performance and made the drill uncomfortably warm.

Check your knowledge

- Can you generate three possible solutions to this issue?
- How might you evaluate how well each of these potential solutions addresses the issue?

D Technical justification and validation of the design solution

D1 Statistical methods

During the design you will often want to use data to help inform your understanding of the requirements for your product, improve the performance of existing or prototype designs, or validate your final design. You might refer to the following types of data.

- Ergonomic data – For example, if designing a crash helmet, you might gather data on the head circumference of potential users.

- Competitor data – You might collect data on the price and functionality of competitor products to help you decide where to place your product in the market.

- Market data – You might commission a survey of potential customers to find out what they want from a product and the relative importance they attach to factors, such as price, weight, reliability and functionality.

- Historic data – If you are planning to redesign an existing product to make it better, you can use historic data to help you understand the performance of the existing product and to target your improvements. For example, you could look at how long products tend to last, how they failed, how often consumable items were replaced, and so on.

- Experimental or trial data – You may create batches of prototypes for experimental trials or for use in gathering customer views of proposals. For example, if you are concerned about the reliability of a seal in a proposed design, running trials with a series of seals will give you information on their performance and consistency. Alternatively, if you are designing a product to be used in difficult and variable environmental conditions, you might place sensors in settings where the product is likely to be used, to collect data on temperature and humidity over a period of time.

In some cases the amount of data collected can be very large. There are various methods you can use to help you summarise and visualise large sets of data.

Graphing experimental data

Often, it helps to plot experimental data in a graph to try to identify trends or relationships.

For example, in testing a component, increasing loads are applied to a sample and the corresponding deflection of the sample is measured. In **Figure 3.21**, the deflection is plotted against the load – each cross represents a data point consisting of the load applied and the corresponding deflection. The red line is a 'line of best fit' that goes approximately through the middle of all the data points and highlights any apparent relationship between the two variables.

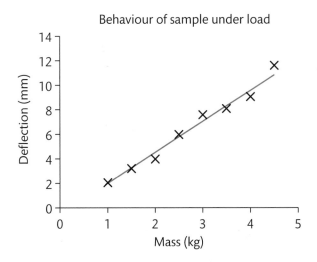

▸ **Figure 3.21** Scatter graph of deflection versus mass of load for a sample subjected to loading, together with a line of best fit

Statistical measurement

The data that you get from trials may be discrete or continuous.

▶ Discrete data can take only a fixed set of values, and typically arises from counting or classifying into categories. In the engineering design context, discrete data might come from:
  ▶ ergonomics – for example, the numbers of users involved in a trial who consider themselves left-handed, right-handed or ambidextrous
  ▶ surveys – for example, the numbers of users who claim it is very likely, likely, neither likely nor unlikely, unlikely, or very unlikely that they will buy a product
  ▶ historical information – for example, the numbers of customers who bought the small, medium or large variants of the previous product.

▶ Continuous data can take any value within a certain range, and typically arises from measuring. In the design of new products you might see continuous data in:
  ▶ ergonomics – for example, the heights of potential users of the product, which could be 1.52 m, 1.78 m or anything in between
  ▶ experiments or trials – for example, the test loads that caused various samples to fail
  ▶ market surveys – for example, the weights, lengths and heights of competitor products on the market.

Consider the data set in **Table 3.12** recording when 28 items of an existing product failed in use.

▶ **Table 3.12** Data on failure time of an existing product

Month of failure for each of 28 products that failed while in use													
24	25	26	27	28	28	29	29	30	30	30	31	31	32
32	32	32	33	33	34	34	34	35	36	36	37	38	39

If this is considered a representative sample and typical of how the product as a whole behaved in use, then you can do some analysis on the data set.

One of the first things you might want to do is find some sort of average – a single number that represents a 'central value' of how long the product lasted. The three most commonly used averages are the mean, the median and the mode.

▶ The mean is the sum of all the values in a data set divided by the number of data items. In this case we have

$$\text{mean} = \frac{24 + 25 + 26 + 26 + 28 + 28 + \ldots + 39}{28} = \frac{885}{28}$$

$$= 31.61 \text{ months}$$

▶ The median is the middle value when all the data values are arranged in order. In this case the 28 data values have already been put in ascending order, so the median is halfway between the 14th and 15th values. Since both are 32, we have

median = 32 months

▶ The mode is the most commonly occurring data value. In this case, the most frequent value (occurring 4 times) is 32, so

mode = 32 months

While an average gives you some useful information about the 'centre' of the data, it tells you nothing about how the data is spread – are all the data values quite close to the average or is there a lot of variation?

A common measure of the spread of a data set is the variance. It is defined as the mean of the squared differences between the data values and the mean of the data set. A closely related measure of spread is the standard deviation (often denoted by σ), which is the square root of the variance.

To find the variance, it is helpful to set out the calculations in a table, like **Table 3.13** for the current example.

▶ **Table 3.13** Calculation of the variance and standard deviation for the data set on product failure time

Month of failure, x	Mean month of failure for data set, $\bar{x}$	Difference, $x - \bar{x}$	Squared difference, $(x - \bar{x})^2$
24	31.61	−7.61	57.91
25	31.61	−6.61	43.69
26	31.61	−5.61	31.47
27	31.61	−4.61	21.25
28	31.61	−3.61	13.03
28	31.61	−3.61	13.03
29	31.61	−2.61	6.81
29	31.61	−2.61	6.81
…	…	…	…
…	…	…	…
38	31.61	6.39	40.83
39	31.61	7.39	54.61
Sum of squared differences			398.68
Variance (mean of squared differences)			**14.24**
Standard deviation ($\sigma = \sqrt{\text{variance}}$)			**3.77**

A small variance means that the data is closely clustered around the mean; a large variance means that the data is spread out widely. You will be able to visualise this later when you look at different ways to present data graphically.

Graphical representation of statistical data

In this section we will look at several graphical methods that allow you to visualise the distribution of data.

Histograms

Consider the data in **Table 3.12**. It should actually be continuous data, as products could fail at any point in a given month, although for practical reasons the failure times have been recorded only in whole months, turning it into a set of discrete data. In most cases where the data is genuinely continuous, or where it is discrete but there are a large number of values, it is convenient to group the data into bands (or 'classes').

In this example, there were 15 different months in which at least one product failed. You could work with the data as it is; however, as the total number of items in the sample (28) is only about double the number of values occurring, the information will be spread quite thinly – that is, the frequency of occurrence of each value is very low. To illustrate the grouping of data into classes, let us define the classes to be successive pairs of months. Then make a frequency table as in **Table 3.14** – the first column shows the different classes (nine in this case), and the second column shows the number of data values that belong to each class.

▶ **Table 3.14** Frequency table for the data set on product failure time

Month of failure	Number of failures over the two-month interval (frequency)	Mean number of failures per month over the two-month interval (frequency density)
23 or 24	1	0.5
25 or 26	2	1
27 or 28	3	1.5
29 or 30	5	2.5
31 or 32	6	3
33 or 34	5	2.5
35 or 36	3	1.5
37 or 38	2	1
39 or 40	1	0.5

From the information in **Table 3.14** you can draw a histogram to show the distribution of failure times of the product (**Figure 3.22**). The area of each bar is equal to the number of failures in that two-month period. In this case, as the intervals are all equal, the height of each bar (the frequency density) is proportional to the number of failures in that interval.

If a curve is drawn through the tops of the bars in the histogram (the red curve in **Figure 3.22**), it looks symmetrical and 'bell-shaped'. This is close to what is known as a normal distribution. For an exact normal distribution, the symmetry is perfect and the mean, median and mode will all have the same value, coinciding with the peak of the distribution.

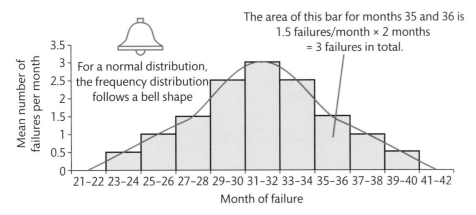

▶ **Figure 3.22** Histogram of the failure time data, which is symmetrical and seems to follow a normal distribution

If a data set follows a perfect normal distribution, you can use the standard deviation to help estimate the proportion of data lying between certain limits. This is demonstrated in **Table 3.15** for the failure time data.

▶ **Table 3.15** Proportions of data that lie within one, two or three standard deviations of the mean

Number of standard deviations	Proportion of data within these standard deviation limits	For the data in Table 3.12		
		Mean of data set (months)	Standard deviation (months)	Range of values bounded by the standard deviation limits (months)
±1	≈ 68% of data	31.61	3.77	27.8–35.4
±2	≈ 95% of data	31.61	3.77	24.1–39.2
±3	≈ 99% of data	31.61	3.77	20.3–42.9

These proportions are standard for a perfect normal distribution, and other values can be found from statistical tables.

$31.61 - 2 \times 3.77 = 24.07$
$31.61 + 2 \times 3.77 = 39.15$

If the distributon of product failure times were exactly normal, you would expect 68% of products (about 19 of the 28) to fail between 27.8 and 35.4 months. Looking back at the original data, there were indeed 19 products that failed in months 28 to 35 inclusive.

Histograms can also be useful in comparing the distributions of two or more sets of data.

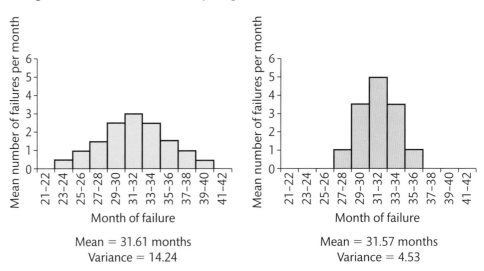

Mean = 31.61 months
Variance = 14.24

Mean = 31.57 months
Variance = 4.53

▶ **Figure 3.23** Two histograms of failure time data for 28 product items

In **Figure 3.23**, the two sets of data are both from samples of 28 items, and in both cases the mean is a little less than 32 months. There is, however, a big difference in the spread of the data (as represented by the variance or standard deviation), reflected in the much narrower shape of the second data set, in which the values generally fall much closer to the mean time.

For each of the two products in **Figure 3.23**, give a rough estimate of the month when 10%, 50% and 90% of the items will have failed.

Look at each histogram to see by what time three items (approximately 10% of the 28) had failed. Remember that each bar represents two months, so for the data from **Table 3.12** (left histogram), in months 25 and 26 an average of one product failed per month over the interval, so two products failed in these two months.

What might this mean in practice for anyone planning on using either of the products? Which situation do you think is preferable?

Skewed data

Data is not always symmetric about the mean value. It is quite possible to have data that is skewed to one side or the other, as shown in **Figure 3.24**.

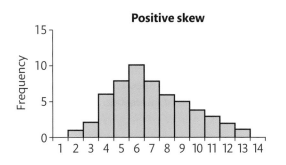

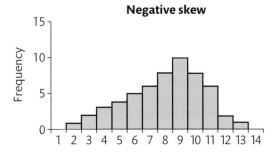

▶ **Figure 3.24** Two histograms – one with positive skew, the other with negative skew

A classic example might be if you were designing a product for which the weight of the user would be an important consideration. The mean weight of an adult male in the UK may be 80 kg, but the weight distribution has a positive skew, and there will not be the same number of men weighing 120 kg as weighing 40 kg.

In such cases the mean will no longer sit centrally, and the median may be a more representative average to use in design decisions.

Imagine that a design you are working on uses both of the products in **Figure 3.23** as components, so the two components have approximately the same mean time to failure but very different variances. Your design will fail if either of these components fails. When in use, how would you expect the time to failure of your overall product to be distributed? Normal distribution, positive skew or negative skew?

If you were to focus on trying to eliminate early failure of the overall product, which of the two components used in your design might you put most of your attention on improving?

Cumulative frequency diagrams

A cumulative frequency diagram plots a running total of the frequency up to each data value (or each class boundary in the case of grouped data). For the example of **Table 3.12**, this would be a graph of the running tally of the number of items that have failed up to a given month (see **Figure 3.25**). The highest level reached by the graph will be the total number of data values in your set. A cumulative frequency diagram is usually an S-shaped curve.

Cumulative frequency diagrams are particularly useful if you want to estimate a particular percentile of the data set. For example, suppose you want to estimate the 25th percentile of the failure time data – that is, the month in which a quarter of the items in the sample have failed but three-quarters are still working.

A quarter of the sample size is 28 ÷ 4 = 7, so on the cumulative frequency diagram you draw a horizontal line from 7 on the vertical axis across to the curve, followed by a vertical line from the curve down to the horizontal axis. This meets the horizontal axis at approximately 28, which means that by the 28th month, about a quarter of product items will have failed.

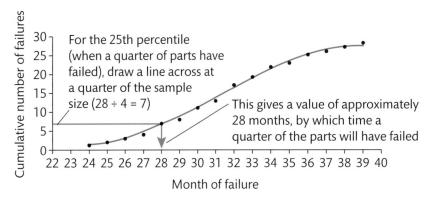

▶ **Figure 3.25** Cumulative frequency diagram for the product failure time data

Using the data collected

It is important to think carefully about how you can use the data you have gathered.

First, you need to consider the reliability of the data, any errors that might be present due to the method used to collect the data, whether you have enough data to draw meaningful conclusions, and also – crucially – what the results mean in relation to the product you are developing.

For example, suppose that the height of the user will be important for the product you are designing. Then you might want to ask:

▶ What proportion of the population do you want to target for your design? You can aim for a bigger proportion so that you have more potential sales, but this may compromise the design for the bulk of users close to the middle of the range.

▶ What is the spread of the data? Is there a big difference between the minimum and maximum values?

▶ Can you manufacture a single product, or would it need to come in different sizes?

▶ Can your product be made adjustable to cope with different heights of user? Would this have consequences, such as increased complexity, weight, cost or servicing requirements?

Sometimes you might not be so focused on data close to the mean or median and may instead be more interested in the extremities.

For example, suppose you are designing a guard for a workshop guillotine. You need to prevent fingers from getting too close to the blade, but you still need some clearance to allow sheet material to be slid into the guillotine. Would you be most concerned with the mean or median finger size or something else?

As another example, suppose you are selecting the seals for a pump intended to be used outdoors, where mean

temperatures over the whole year are 12°C but may range from –15°C in the middle of a winter night to +35°C on a summer afternoon. Should you focus your design on pump operation close to the mean temperature?

Other forms of data and graphical representation

In some situations you will need to analyse data in which the variables or factors of interest do not necessarily take numerical values.

For example, suppose you have conducted a market survey to help inform you what customers want from your products. The results are summarised in **Table 3.16**. The variable that you are collecting information on – the most important criterion customers use in choosing products – comes in distinct categories (performance, weight, price, etc.) rather than taking either discrete or continuous numerical values.

▶ **Table 3.16** Important factors in choosing a product

What is the most important factor in choosing to purchase our products?		
Criteria	**Number of respondents**	**Percentage of respondents**
Performance	16	18.8
Weight	5	5.9
Reliability	29	34.1
Product support	20	23.5
Price	15	17.6

In such cases the data can be shown in a pie chart (see **Figure 3.26**). The circle represents the whole set of data and each category (criterion) is represented by a slice whose area is proportional to the percentage accounted for by that category. A slice can be 'extracted' from the pie to draw special attention to it.

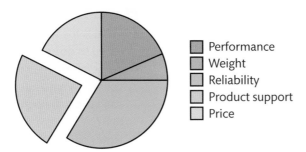

What is the most important factor in choosing to purchase our products?

☐ Performance
☐ Weight
☐ Reliability
☐ Product support
☐ Price

▶ **Figure 3.26** Pie chart displaying the criteria customers use in choosing a product

Another approach to displaying categorical data is to use a bar chart. These graphs look a little like histograms, but the variable plotted along the horizontal axis need not have numerical values or be arranged in any particular order.

For example, if several product proposals are reviewed to decide which is most suitable to take forward to the next stage, you might obtain a bar graph like the one in **Figure 3.27**.

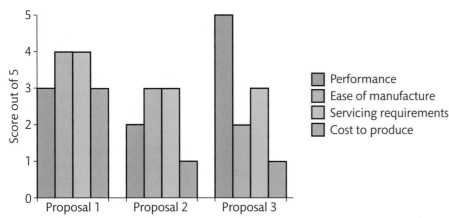

▶ **Figure 3.27** Bar chart comparing design proposals

In **Figure 3.27** the bars representing the factors 'Performance', 'Ease of manufacture' etc. are placed side by side for each proposal, but they may also be stacked on top of each other (as in **Figure 3.28**) to show the total score for each proposal – the different colours allow you to see the contribution of each criterion to the overall score.

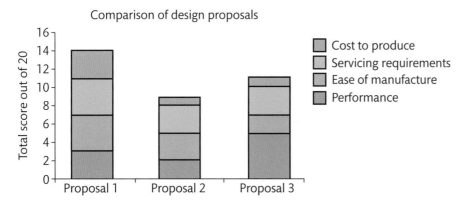

▶ **Figure 3.28** Bar chart with stacked bars

Assessment practice 3.4

You have been asked to design a pop-riveting tool (**Figure 3.29**). Your company has a reputation for designing comfortable, high-quality hand tools, and this tool is expected to maintain this tradition.

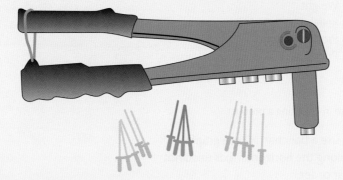

▶ **Figure 3.29** A pop riveter

Trials have been carried out on a range of potential users to determine their comfortable grip strength. The data is shown below.

Comfortable grip strength (N) of potential users									
10.4	10.4	10.4	11.1	11.1	11.2	11.2	11.2	11.2	11.3
11.3	11.8	11.9	11.9	12.0	12.0	12.0	12.0	12.1	12.1
12.2	12.6	12.6	12.7	12.7	12.8	12.8	12.8	12.8	12.9
12.9	13.0	13.0	13.4	13.4	13.4	13.5	13.5	13.6	13.6
13.6	13.6	13.7	13.7	13.8	13.8	13.8	14.0	14.0	14.1
14.1	14.2	14.2	14.2	14.3	14.3	14.4	14.4	14.4	14.4
14.5	14.5	14.6	14.6	14.6	14.6	14.7	14.7	14.9	15.0
15.0	15.0	15.1	15.1	15.2	15.2	15.2	15.2	15.3	15.3
15.4	15.4	15.4	15.8	15.8	15.8	15.9	15.9	16.0	16.0
16.0	16.0	16.1	16.1	16.2	16.6	16.6	16.7	16.7	16.8
16.8	16.8	16.8	16.9	16.9	17.5	17.5	17.6	17.6	17.6
17.6	17.7	18.4	18.4	18.4	18.4	18.5	19.2	19.2	19.2
19.2	20.0	20.0							

The pop riveter should be particularly easy to use by the 5th to 95th percentiles of your overall market.

- Calculate the mean, median and mode of the data set.
- Find the variance of the data set.
- Select a suitable interval size and plot a histogram of the data. Does it show a skew?
- Plot a cumulative frequency diagram. From this, estimate the comfortable grip forces for the 5th and 95th percentiles.
- Which of these values is more important to your design? Why?
- Investigate how a pop riveter works and, based on these results, sketch how you might optimise your design to meet the needs of possible users.

Plan
- What is the task? What am I being asked to do?
- What information will I need to collect? What tools and resources can I use to help me complete the task?
- How confident do I feel in my own abilities to complete this task? Are there any areas I think I may struggle with?

Do
- I know what I'm doing and what I am aiming to achieve in this task.
- I can identify the parts of the task that I find most challenging and devise strategies to overcome those difficulties.

Review
- I can explain what the task involved and how I approached each stage of the work.
- I can explain how I would deal with the hard elements differently next time.
- I can identify the parts of my knowledge and skills that could benefit from further development.

D2 Validating designs

As you develop and refine your design, it is important to continually look back at the product design specification to ensure that you will deliver what was intended. You are aiming to produce a product that meets customers' needs and which does so in a way that is better than competitor products. It must also satisfy any relevant regulatory constraints, and you must be able to produce it in a way that is cost-effective.

It is likely that you will make changes to your design as you move from concept towards end product, and it is important to provide rational justifications of these changes so that you can be confident that your product is as good as it can be.

Referencing against the product design specification

As your design evolves, ask yourself:

▶ Does your product still meet all the criteria in the product design specification?

▶ If not, what can be done?

▶ If the design still fulfils all requirements of the specification, can the product be improved yet further?

▶ Will improvements in one area lead to issues elsewhere?

Evaluating proposals in a weighted matrix

To evaluate possible changes and refinements to your design, you can use weighted matrices just as in evaluating ideas or concepts (see Section C1, Table 3.11). This can help you to assess whether it would be worthwhile making improvements in one area that might be detrimental in others – for example, a proposed change could make a product cheaper and more reliable but also heavier and harder to use.

Considering indirect benefits and opportunities

Can you achieve indirect benefits through your design decisions and choice of components? Could minor tweaks give extra benefits? For example, designing a casing in a particular way might also offer some degree of splash proofing.

Balancing benefits and opportunities with constraints

Always review the benefits of improving certain aspects of the product. While you are likely to have to trade off some functional aspects of the product against each other, such as strength versus weight, also keep in mind various other constraints and check that they will continue to be met.

▶ Don't get carried away with adding extra features and functionality. Check whether doing so will exceed the target price for the product, and consider whether the customer would really want those extra features.

▶ Could any changes you plan to make affect the safety of the product?

▶ Would any proposed changes cause your product to become less environmentally sound?
  ▶ Are you going to introduce materials that will be difficult to recycle or dispose of?
  ▶ Is the product going to be less energy efficient in use?
  ▶ Will it need to use more solvents or oils that require careful disposal?

Always make sure you justify any decisions made and record them in your design log.

Design for manufacturing

As you develop your product, continually review it for its ease of manufacture. Designing a product that is difficult to make will result in unnecessary expense and potentially higher levels of scrap or reject parts. This can very quickly turn a profitable product into one that will make a loss. Ask yourself:

▶ Are you using more materials or parts than you need to?

▶ Can multiple parts be combined into a single machined, cast or moulded part?

▶ Are you using more different types of parts than you need to? For example, are you using both M4 and M6 screws when M4 screws could be used for all connections?

▶ Are you using an appropriate manufacturing process for the volumes needed? For example, avoid using 3D printing for parts that need to be produced in their thousands, and avoid using injection moulding for one-off parts.

▶ Are all the parts actually designed to suit the manufacturing process that is to be used? For example, are parts that will be moulded or cast of fairly uniform thickness and designed to come cleanly out of a mould?

▶ Will parts designed to be machined be easy to make, with no internal sharp corners or deep, blind holes?

Further modifications

Always be aware that technology and markets may move on as you design your product, particularly if you are working to a long development schedule. During the design process, check regularly whether there are new materials, components, manufacturing techniques, sensors or actuators becoming available that could improve your product.

THINK ▶FUTURE

Thomas Howes
Mechanical building services engineer

I have been working as a mechanical building services engineer in a multi-disciplinary design team for the last two years. Building services engineering covers all of the mechanical, electrical and public health installations that go into a building. The majority of the multi-disciplinary design work is done with the help of 3D computer models, where the model can be used to ensure that all of the services are coordinated with one another so that there aren't any problems on site. We are often involved all the way through the design process, from concept stages to detailed design. By working with architects during the concept stages, we can influence some of the building's characteristics such as building orientation, insulation levels and shading on the building. This early input can have a large impact on the size and efficiency of the heating, ventilation and air conditioning systems, which is important if the client has any aspirations for gaining environmental certification.

In order to get into my current position, I first completed an apprenticeship, which included day release to college to complete a BTEC National in Mechanical Engineering. With my BTEC results I was able to apply for a university foundation year in engineering, which then led to completing an MEng degree in Mechanical Engineering. Having an MEng from an accredited university makes the process of becoming a chartered engineer more straightforward, but doing the BTEC was my first step to where I am now.

Focusing your skills

Becoming a confident designer

With so much underlying the design of a successful product, it can be daunting to embark on a new project. Understanding a wide range of manufacturing techniques, materials, sensors, actuators, seals, motors, bearings, human factors, environmental concerns, ergonomics, regulations and standards is a lot to take on.

The more projects you do, the more confident you will become and the stronger your underpinning knowledge will become.

Always embark on a project expecting to have to learn about a new type of material or manufacturing process or to develop a new skill. Don't expect to have all the answers. This continual development of your knowledge and skills and of the discipline itself is what makes engineering design such an exciting field.

Here are some tips to help you get started as a design engineer.

- Get used to looking at various products and trying to work out how they are made and from what materials.

- Keep a record of design features that you find interesting – a hinge on a kitchen cabinet may later serve as inspiration for a rudder control mechanism on a microlight aircraft; a sensor used in a production line might at some point in the future also be suitable for use in a piece of automotive testing equipment.

- Try to keep yourself aware of new components and materials. Use company websites and get on circulation lists to be informed about new products. Visit trade shows and exhibitions if you get the chance.

- Embrace challenge and always aim for the best – not the easiest or most obvious – product solution. That is what your competitors will do, and you want to be better than that!

Getting ready for assessment

This section has been written to help you to do your best when you take the external assessment. Read through it carefully and ask your tutor if there is anything you are not sure about.

About the assessment

This unit is assessed by a supervised task. Pearson sets and marks the task.

You will be assessed on your ability to follow a standard development process of interpreting a brief, scoping initial design ideas, preparing a design proposal and evaluating your proposal.

Make sure that you arrive in good time for each session of the assessment, and check that you have everything you need beforehand. Make a schedule for completing the task, and leave yourself enough time at the end to check through your work.

Listen to and read carefully any instructions that you are given. Marks are often lost through not reading instructions properly and misunderstanding what you are being asked to do. Ensure that you have checked all aspects of the task before starting.

At the end of the supervised assessment period, proofread your work and correct any mistakes before handing it in.

As the guidelines for assessment can change, you should refer to the official assessment guidance on the Pearson Qualifications website for the latest definitive guidance.

Preparing for the assessment

Before the assessment you will be provided with a case study related to the assessment and will be given the opportunity to carry out independent preparatory research. You can then take notes into the supervised assessment - check with your tutor how long these can be. You must work independently and should not share your work with other learners. Your tutor cannot give you feedback during the preparation period.

Make sure you plan out the preparatory work to maximise your time examining the case study. For example, questions to think about might include:

- How is the product made, or could the proposed design be realised?
- What does the customer want from the product?
- What materials are used?
- Are there any particular constraints, such as those relating to cost, reliability, weight, functionality, servicing, environmental impact or sustainability?

The following table shows the key terms typically used in the assessment.

Command or term	Definition
Client brief	Outlines the client's expectations and requirements for the product.
Design	A drawing and/or specification to communicate the form, function and/or operational workings of a product prior to it being made or maintained.
Manufacture	To make a product for commercial gain.
Project log	A document to record the progress made, key activities and decisions taken during the development of a product.

Sample assessment

Preparatory case study

Case study

You are working for a company that makes flying model aircraft. The company wants to diversify its business and move into the building of more professional drone aircraft for use by farmers, cartographers and utility companies. Market research has suggested that a well-designed aircraft should be able to sell around 1000 units per annum with a production life of 3–5 years.

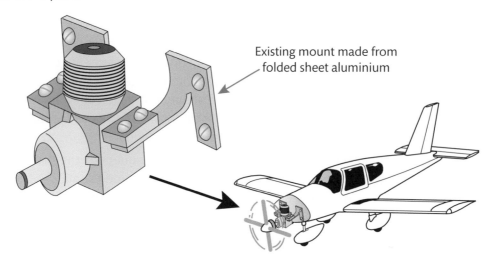

Existing mount made from folded sheet aluminium

▶ **Figure 3.30** Designing a new engine mount for drone aircraft

Your particular responsibility is the engine mount (**Figure 3.30**), which on your existing aircraft is made from folded sheet aluminium. It is expected that in their current form these mounts will not be rigid enough for the new aircraft.

This is the type of case study you may be given before the assessment. You will be given a set period of time to investigate the case study and you are allowed to compile notes, which you can later use in the supervised assessment periods - check with your tutor how many notes you are allowed to take into the assessment.

Use the case study as an exercise to practise gathering information for the design task.

To help you get started, underline specific details in the description to help you draw out the key aspects of the case study. For example:
- You are told that the company is developing a professional drone aircraft rather than a hobbyist model. Write down what this might mean for the aircraft, its reliability and quality.
- You are given some information about possible users of the product. What does this mean in terms of how the product will be used? Will the customers want to do lots of servicing? What sort of environment might the product have to work in (e.g. wet weather, possible contamination from oil or fuel)? What does this mean for the design and the materials used? Do you need to spend some time researching specific materials and manufacturing techniques?

- You are given an estimate of the likely sales volume – what does this tell you? What manufacturing techniques will suit this sort of volume? Research some of these techniques.
- You are told you will be leading the design on the engine mount. Make sure you understand what the mount does, how an engine might be connected to it and how it will connect to the aircraft.
- You are also told that rigidity is likely to be an issue. Could you improve the rigidity of the engine mount by using different materials or a different geometry? Research this and find out ways to make a structure rigid. Vibration could also be a concern – this can cause cyclic loading and so fatigue might be a worry. Research the sorts of materials and features that can make fatigue more likely. Also remember that you are designing for an aircraft, so consider what other constraints there might be on any parts.

The set task

The case study outlined a scenario in which your company wishes to move into a new market for more powerful drone aircraft. A new range of aircraft models and components must be developed to meet this need.

Your current engine mounts are made from folded sheet aluminium, but they lack rigidity and can vibrate excessively. It is recognised that you will not be able to use these existing mounts on the new drones with their more powerful engines and the high expectations of your professional users.

Your role as a design engineer is to develop an effective engine mount for the new drones, which are intended to be high-quality professional aircraft and will be expected to provide many years of reliable service.

It is expected that the engine will need servicing over time and that it ought to be possible to carry out routine servicing with the engine in place, but that engine removal should also be possible for more comprehensive overhaul.

You are expected to produce a redesigned engine mount for use in the drone aircraft.

You will need to record your design planning, design decisions, concepts, drawings, and so on in your task book, which will be taken in for assessment.

Client brief

Your company has surveyed 16 potential customers regarding how important they think various aspects of the drone are (see **Table 3.17**). For each factor they were asked to score the importance out of 5, with 5 representing high importance and 0 meaning of no importance.

▶ **Table 3.17** Customer survey scores

Factor	Customer															
	1	2	3	4	5	6	7	8	9	10	11	12	13	14	15	16
Reliability	5	3	3	5	5	3	2	5	5	5	3	4	4	4	3	5
Ease of use	2	3	4	1	5	4	3	2	5	4	3	2	3	4	3	4
Low price	1	2	3	2	1	4	3	2	1	1	2	3	2	3	2	3
Advanced functions	1	2	1	1	1	2	2	1	0	2	5	5	1	1	2	4

Here are some details about the engine in the new drone:
- four-stroke air-cooled petrol engine
- volume 30 cc
- maximum output 2.68 HP/9000 rpm
- weight 1400 g including exhaust.

Details of the locations of the mounting holes for the interface with the engine and the bulkhead can be seen in the diagrams (**Figure 3.31**).

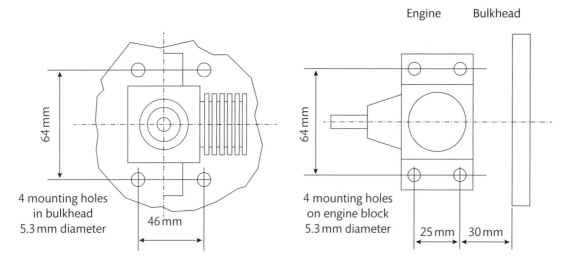

▶ **Figure 3.31** Diagrams of the engine mount, showing dimensions and the locations of the mounting holes

Activity 1

At the start of the task, create a short project time plan in your task book. During the development process you should also record:
- why changes were made to the design during each session
- action points for the next session.

Total for Activity 1: 6 marks

Activity 2

Interpret the client brief into operational requirements, which should include:
- product requirements
- opportunities and constraints
- interpretation of numerical data
- key health and safety, regulatory and sustainability factors.

Total for Activity 2: 6 marks

In this activity you demonstrate that you understand what your product needs to do. It can help to break the requirements down into key general areas and identify other more specific criteria within each area.
- Physical requirements:
  - must securely support engine
  - must utilise standard engine and bulkhead screw locations
  - must attach to the aircraft via a bulkhead
  - must transfer propellor thrust from engine through to aircraft
  - must dampen vibration from engine
  - must …

In the real assessment you will have a set period of time to complete the entire task, spread over a number of sessions. You will only be allowed a limited number of notes you prepared earlier. Try to limit yourself to using what you have on these sheets and in your head. In the supervised assessment sessions you will not be able to carry out any additional research.

Read ahead carefully to see what you will be expected to do over the entire assignment, and divide up your time accordingly. Your tutor should give you the schedule for your sessions, so use this information to organise your time.

- Assembly and servicing requirements:
  - must allow for efficient assembly
  - must allow for easy removal of engine
  - must allow for easy access to serviceable and adjustable elements of engine.
- Manufacturing requirements:
  - must be cost-effective to manufacture
  - must be easy to assemble if produced from multiple parts.

Refer back to Section B1 for other general areas to consider.

Look at the data collected from users (such as the information in **Table 3.17**). Plot the data in an appropriate graph or complete some other statistical analysis to get a feel for what your customers want from the product. For example, you may find the customers value reliability more than a low purchase price.

You also need to consider other constraints. For example:
- The engine mount is to go in a small, unmanned aircraft, so weight is likely to be a key factor.
- The mount may also be subject to fuel and oil splashes, as well as the possibility of rainwater, so you may wish to include relevant constraints.
- The mount needs to be safe – failure in use or during start-up could cause injury to users.

> On completing this activity you should have formulated a clear description of what is required of your product, and this should be referred to continually as you develop your design proposal.

Activity 3

Produce a range of initial design ideas based on the client brief, to include:
- sketches
- annotations.

Total for Activity 3: 9 marks

Try to generate as many different concepts as you can, with supporting sketches or drawings. Use morphological analysis (Section C1) to help you if you're struggling to come up with ideas, or try imagining how you might make the mount if it were to be cast, machined, moulded or assembled.

> Make sure that you add notes to each concept you have generated to draw out the key features of the proposal. Focus on getting the idea across and do not worry if your sketches lack polish.

Activity 4

Develop a modified product proposal with relevant design documentation. The proposal must consider:
- a solution
- existing products
- materials
- manufacturing processes
- sustainability
- safety
- other relevant factors.

Total for Activity 4: 30 marks

From your two or three best concepts, decide which one you will take forward as a formal proposal. Try using a weighted matrix evaluation, comparing each proposal against key design criteria. Make sure that you can also justify why your design is better than the existing product.

Write your explanations and justifications down in your task book. It should be clear to someone reviewing your work why you chose the approach you did.

Think carefully about the material you will use for making the mount – this will have functional, manufacturing, cost and sustainability implications. Your documentation should show clearly that you have considered these implications and can justify your choice of material.

The basic concept is likely to need some work to turn it into a product that can be produced commercially in the required numbers. One option that could be viable for the mount is to use casting or moulding, in which case you would need to consider draft angles and the way the mould is split.

This activity is worth half the total marks for the assessment, so make sure you do it as completely and correctly as you can. The end result should include a detail drawing (or drawings) of the part, with all key dimensions marked, a material specified and a plan for manufacture. Don't forget to consider the end-of-life options for the drone and your mount. What will happen to it? Will the part be easy to remove? Can it be recycled?

Activity 5

Your final task book entry should evaluate:
- the success and limitations of the completed solution
- the indirect benefits and opportunities
- the constraints
- opportunities for technology-led modifications.

Total for Activity 5: 9 marks

Review your proposal against your design specification. Be clear about where you think it has succeeded, but also be honest about where you think it may be more marginal or where conflicting constraints could not be resolved as well as you would have liked.

See if you can think of any indirect benefits of your design. For example, it may be that the design of the mount enables a range of engines to be accommodated, simply by placing the mounting holes in slightly different locations.

What constraints still exist? For example, in this brief you were asked to design an engine mount, but by looking at the drone as a whole it might be that a separate engine mount is not needed and it can instead be made an integral part of the aircraft fuselage.

Could there be opportunites for a more refined design if new materials become available?

Applied Commercial and Quality Principles in Engineering 4

Getting to know your unit

In this unit, you will explore how key business activities and trade considerations influence engineering organisations and are used to create a competitive advantage. You will learn why organisations need to control costs and how they make decisions by applying activity-based costing methods. You will also examine what is meant by quality and why it means different things to different people. You will investigate quality systems, including quality assurance and control. Finally, you will explore value management as a process to create value in an organisation.

How you will be assessed

This unit will be assessed by a series of internally assessed tasks set by your tutor. Throughout this unit you will find assessment practice activities to help you work towards your assessment. Completing these activities will not mean that you have achieved a particular grade, but you will have carried out useful research or preparation that will be relevant when it comes to your final assignment.

In order for you to complete the tasks in your assignment, it is important to check that you have met all of the assessment criteria. You can do this as you work your way through the assignment.

If you are hoping to gain a Merit or Distinction, you should also make sure that you present the information in your assignment in the style that is required by the relevant assessment criteria. For example, Merit criteria require you to analyse and to produce a model or complete an exercise accurately, while Distinction criteria require you to evaluate.

The assignments set by your teacher/tutor will consist of a number of tasks designed to meet the criteria in the table. They are likely to take the form of written reports but may also include activities such as the following:

▶ Investigation of local businesses, which could include visits.

▶ Research and problem-solving projects.

▶ Analysing case studies, observations or examples from real-world settings.

Assessment criteria

This table shows what you must do in order to achieve a **Pass**, **Merit** or **Distinction** grade, and where you can find activities to help you.

Pass	Merit	Distinction
Learning aim A Examine business functions and trade considerations that help engineering organisations thrive		
A.P1 Explain how key business activities influence an engineering organisation. **Assessment practice 4.1**	**A.M1** Analyse how key business activities and trade considerations influence an engineering organisation, which can create a competitive advantage. **Assessment practice 4.1**	**A.D1** Evaluate, using language that is technically correct and of a high standard, how key business activities and trade considerations combine to influence an engineering organisation, which can create a competitive advantage. **Assessment practice 4.1**
A.P2 Explain how trade considerations influence an engineering organisation. **Assessment practice 4.1**		
Learning aim B Explore activity-based costing as a method to control costs and to determine if an engineering product or service is profitable		
B.P3 Explain why an engineering organisation controls costs. **Assessment practice 4.2**	**B.M2** Produce accurately an activity-based cost model for an engineering product or service, explaining the reasons for cost controls. **Assessment practice 4.2**	**B.D2** Produce an accurate and refined activity-based costing model, during the process, for a product or service to determine the major cost areas that could impact on profitability, explaining the reasons for cost controls. **Assessment practice 4.2**
B.P4 Produce an activity-based cost model for an engineering product or service. **Assessment practice 4.2**		
Learning aim C Explore how engineering organisations use quality systems and value management to create value		
C.P5 Explain the purpose of different quality management systems and value management used by engineering organisations. **Assessment practice 4.3**	**C.M3** Analyse the purpose of different quality management systems and value management used by engineering organisations. **Assessment practice 4.3**	**C.D3** Evaluate the outcome of a value management exercise for a given engineering activity and make recommendations which include the use of quality systems to implement efficiencies in the engineering activity. **Assessment practice 4.3**
C.P6 Complete a value analysis exercise on a given engineering process. **Assessment practice 4.3**	**C.M4** Complete accurately a value analysis exercise on a given engineering process. **Assessment practice 4.3**	

Getting started

Work in groups to consider the design and manufacture of one of your mobile phones. Which company manufactured the product? What manufacturing processes do you think have been involved? What departments could there be in an organisation that manufactures a mobile phone? What are the costs of producing a mobile phone? Make notes. When you have completed this unit, think about how your knowledge and understanding of the topic has increased. Identify what new information you can add to your notes.

 A # Examine business functions and trade considerations that help engineering organisations thrive

In this section you will learn to identify how key business activities and trade considerations influence an engineering organisation and can be used to create a competitive advantage.

A1 Business functions

Engineering organisations range from small, local businesses that manufacture a product for a specific market to large-scale global businesses such as BP. Whatever the size of the business, it will have a management team that manages a range of **functions**.

Table 4.1 shows examples of key activities that take place within different business functions.

> **Key term**
>
> **Functions** – areas of activity into which a business can be organised.

Here are some functions of which you need to be aware, and the aspects of work that they cover.

▸ **Human resources (HR)** – staff issues within the organisation.

▸ **Finance** – financial dealings and costing within the organisation.

▸ **Sales** – marketing and selling the end product to customers, including advertising and promotion.

▸ **Operations** – activities involved in manufacturing the product.

▸ **Research and development (R&D)** – market research and product investigation to develop the products for manufacture; reviewing customer satisfaction.

▸ **Purchasing** – tasks such as sourcing and buying raw materials for manufacture.

▸ **Table 4.1** Key activities within different business functions

Functions	Examples of activities
Manufacturing of products and delivery of services	• Forming • Fabrication • Removal of material • Addition of material • Assembly processes • Quality checks and inspection/testing processes
Supply chain management	• Outsourcing decisions – where the organisation considers the use of an external company for a specific job • Supplier appraisal – where the organisation evaluates the use of current and alternative suppliers (e.g. of raw materials)
Marketing and sales	• Brand awareness – how the organisation wants the brand to be perceived by customers • Market research – investigating the market the product is aimed at to aid development • Sales • Customer feedback – acquiring data on customer satisfaction and responding to customer comments
Customer relations	• Meeting expectations – organisations that are successful make every attempt to ensure that their customers are satisfied • Being proactive – for example, taking the initiative to ask customers for feedback or anticipating common customer concerns

▶ **Table 4.1** – *continued*

Functions	Examples of activities
Resource management	• Sources of funding – identifying funds from internal budgets or from external streams • Resource allocation – allocating resources among different departments and projects • Stock control – ensuring that there is sufficient stock available for different departments and projects to use, and purchasing more as required
Staff recruitment	• Internal and external recruitment – appointing staff to new roles from both within and outside the organisation • Apprenticeships – setting up and managing training programmes for apprentices
Staff management	• Appraisals – evaluating and reviewing staff performance • Support and training – responding to support and training needs, and ensuring that staff have CPD (continuing professional development) opportunities
Financial	• Financial statements showing profit (the amount of money the company makes when all costs have been paid), loss (the amount of money the company loses) or break-even (where profits match loss)

> **Link**
>
> The activities associated with manufacturing of products and delivery of services are covered in Unit 2.

> **Research**
>
> Investigate an organisation through a visit or on the internet. What functions or departments does the organisation have? Make notes on what each function or department is responsible for.

A2 Trade considerations

Tendering and contracting

An organisation may choose to **contract out** certain activities. This is where a different company is used to provide specific goods or a specific service. For example, an organisation that manufactures products using injection moulding may design their own moulds but have the metal casts manufactured by a different company with specialist equipment. The companies who take up a contract to supply an organisation with goods or services are called contractors.

If an organisation decides that it wants to contract out an activity, it will put that contract 'out to tender'. Companies can then put in formal bids, stating how much they would charge to carry out the activity. Potential contractors might also provide documents such as drawings or specifications for consideration. The contracting organisation can then decide which contractor to use, often based on price and quality.

The organisation and chosen contractor will draw up a contract for the goods or service provided. The contract will contain agreed terms and conditions, with clauses to make sure that each party gets what it needs from the agreement.

These clauses are designed to protect both the organisation and the contractor (**see Figure 4.1**). For example, the organisation will have 'breach of contract' clauses, perhaps including financial penalties or rejection if it is not happy with the goods or service that the contractor provides. The contractor will have a **force majeure** clause to protect them from **liability** if something unpredictable or unavoidable affects their provision of the goods or service.

A **warranty** or **guarantee** can also be included in a contract. This is a promise made by the contractor to the organisation about the goods or services offered. An express warranty is one that is agreed and directly stated or written down; an implied warranty is not actually stated, but is a reasonable expectation that the goods or service will perform in a certain way or meet a certain standard. Implied warranties are supported by laws that govern what they cover.

An **indemnity** clause will also be a part of the contract. This is an agreement between the contractor and organisation about who is responsible if things go wrong or are not as expected. Indemnities may include **insurance** that can provide compensation for any loss or penalty.

> **Key terms**
>
> **Contract out** – to make a contract with an external supplier to carry out a specific activity.
>
> **Force majeure** – a clause that releases a company from its obligations under a contract if unpredictable, extraordinary circumstances prevent it from fulfilling the contract.
>
> **Liability** – being legally or financially responsible for something.
>
> **Warranty** or **guarantee** – a written promise to repair or replace a product if it does not work or breaks down, usually within a specified time frame.
>
> **Indemnity** – an obligation of one party to pay compensation for a particular loss suffered by another party.
>
> **Insurance** – a payment made to guard against uncertain losses.

PAUSE POINT

An engineering organisation has contracted out the manufacture of a key part of a product they sell, but the contractor misses the deadline for getting the part to the organisation, and costs are incurred. What action could the engineering organisation take? What protection could the contractors have against any financial penalties?

Hint   Consider breach of contract, force majeure and indemnities.

Extend   Research the difference between insurance and indemnity. See if you can find an example of each of these in an engineering contract.

▶ **Figure 4.1** Contracts can protect both the organisation and the contractor

Intellectual property rights

Intellectual property rights are the overall rights that an organisation has to use its own ideas and designs without another company being able to use them. Examples are patents and trademarks. If an organisation develops a new product, it can obtain a patent on the product or invention, so that only it can sell it. Trademarks can include logos and slogans that a company uses as part of its identity. Establishing these graphics or words as a trademark stops any other company using them. Registered designs are used to protect the external appearance of a product, such as its shape, colour or texture.

Theory into practice

In 1922 William Henry Hoover founded the Hoover Company. He purchased a patent in 1908 for a vacuum cleaner design that incorporated a rotating brush, and began to manufacture and market the product. His product and associated brand became so successful that nowadays a vacuum cleaner is often generically referred to as a 'Hoover', even when it is of a different brand. 'Hoovering' is now a recognised English term for vacuuming.

More recently, we use vacuum cleaners called a 'Dyson'. James Dyson invented a new way of collecting dirt when vacuuming that uses a cyclone system instead of a filtration bag. Although his vacuum cleaners cost almost twice as much as non-cyclonic vacuums when they were launched, his innovation was a success. Dyson is now a billionaire, despite selling his first upright vacuum cleaner as recently as 1993.

Think about and investigate the following questions:
1 What would have happened if William Henry Hoover had not patented his rotating brush vacuum cleaner design?
2 Why do you think customers were willing to pay twice as much for Dyson vacuum cleaners when they first came on to the market?
3 Research the designs of Dyson and other similar vacuum cleaners. Did James Dyson register his design? Why?

A3 Competitive advantage

Competitive advantage can be created by all business functions or by any combination of activity and trade considerations. Here are some ways in which an organisation can achieve competitive advantage.

▶ Innovating – An organisation must constantly develop and introduce new and exciting products to compete with other similar organisations. If it can come up with a product that is unique or which solves a problem that no other product can, the organisation may have no competition when its product is first introduced.

▶ Using new technology – Technology is developing all the time. If an organisation embraces new technology, either within its products or in the processes used to manufacture its products, this could give it the edge it needs to outperform other companies in the same market.

▶ Protecting intellectual property – To stand out from the crowd, an organisation has to be unique. By protecting its ideas, identity and brand names through patents and trademarks, a company can maintain its individuality and the uniqueness of its products.

▶ Managing costs – To make a profit, an organisation needs to manage its costs well. The costs of manufacturing a product will affect the selling price of the product. If, by better cost control, an organisation can sell a product of the same quality as those of its competitors but at a lower price, it will have a competitive advantage.

Key term

Competitive advantage – what an organisation does that allows it to outperform competitors.

Research

Think about Hoover and Dyson. How did these companies create a competitive advantage?

Case study

AgriParts Ltd

▶ An AgriParts Ltd customer fitting one of their machine parts

Jack is an employee at AgriParts Ltd, an organisation that provides machine parts for the agricultural industry. His job is to oversee the operations functions within the organisation. He has been asked to consider a new machine part, working with colleagues in finance and purchasing to decide whether AgriParts Ltd should

outsource the manufacture of the part. They must consider:

- whether AgriParts Ltd has the resources to manufacture the part in-house
- whether suppliers are available who could meet the expectations of AgriParts Ltd
- what trading terms would need to be agreed with suppliers
- whether outsourcing would result in lowered costs, giving AgriParts Ltd a competitive advantage.

Check your knowledge

1 How will Jack find out about potential suppliers?

2 What expectations would AgriParts Ltd have of suppliers?

3 What types of trading terms would need to be agreed?

4 How could reduced costs create a competitive advantage?

Assessment practice 4.1

Study a local engineering organisation and produce a written report that explains and evaluates how key business activities and trade considerations influence the organisation and how it creates a competitive advantage.

Use examples to demonstrate how a range of factors affect the organisation.

Ensure that you use correct technical terms and that your report is logically structured and easy to read.

Plan

• What is the task? What am I being asked to do?

• What information will I need to complete the task?

• How confident do I feel in my own abilities to complete this task? Are there any areas I think I may struggle with?

Do

• I know what I'm doing and what I am aiming to achieve.

• I can identify the parts of the task that I find most challenging and devise ways to overcome these difficulties.

Review

• I can explain what the task involved and how I approached its completion.

• I can explain how I would approach the hard elements differently next time.

B   Explore activity-based costing as a method to control costs and to determine if an engineering product or service is profitable

In Learning aim A you saw that costs have a major impact on an organisation's profitability and competitive advantage. An organisation must review its costs frequently in order to achieve maximum potential. Activity-based costing (ABC) is a method that an organisation can use to control costs and make judgements on its performance and profitability.

B1   Reasons for cost control

Within every function of an organisation, costs have to be evaluated and controlled to ensure that the organisation will be profitable.

Take the example of an engineering organisation that manufactures piston rods for the automotive sector. There are obvious costs that come to mind – for example, the costs of raw materials, of staff to manage and run the manufacturing plant, of machines to manufacture the product, and of packaging and shipping the product to the customer.

All of these costs have to be managed to ensure that the organisation is making money from selling the product. The income from selling the product must cover all the costs of manufacturing the product, plus an allowance to make some profit. The organisation also has to consider what other companies are charging for a similar product, and try to make the price of its own product competitive.

There needs to be a balance between profitability and competitive pricing, so it is vital that an organisation gets its costing right.

There are hidden costs that an organisation could easily miss without careful cost consideration. Looking back at the example of the piston rod manufacturer, there would be the costs of the energy needed to run the machines in the plant and of the consumables required to maintain the machines; there are also costs that would be incurred when a machine is out of operation due to maintenance. Accurate and effective costing is essential if an organisation is to be successful.

As well as carefully identifying and managing various costs, an organisation has to make decisions on cost that could concern:

▶ investment – An organisation may have to invest a considerable amount of money in the development and manufacture of a product before receiving any profit from its sale; this involves an element of risk. The organisation would have to decide how much money to invest and weigh this against how much profit could be expected from selling the product, or determine whether it is worth investing in the product at all.

▶ withdrawal – If a product is not making a profit, the organisation may decide that it is cost-effective to withdraw the product from sale. This would save on future manufacturing costs, and the organisation could invest the savings in a new product that might be more profitable.

▶ make or buy – In some cases it may cost an organisation less if it simply buys in a product (or part of a product) that is manufactured elsewhere, rather than manufacturing the product itself. This is often the reason why some companies sell products that carry their brand names but are in fact manufactured by other companies overseas.

Types of costs

Costs can be categorised into a number of types, summarised in **Table 4.2**.

▶ **Table 4.2** Types of costs

Type of cost	Description	Examples
Direct	Costs that relate directly to the manufacture of a specific product	• raw materials • **work in progress** (WIP) • direct labour costs
Indirect	Costs that are involved in but not directly related to the manufacture of a specific product	• energy • insurance • wages • consumables
Variable	Costs that can change over time	• raw materials • consumables
Semi-variable	Costs that can vary depending on demand	• overtime costs • commission • maintenance
Fixed	Costs that stay the same	• **overheads** • rent • machinery costs • **depreciation** • insurance
General/ administration	Costs relating to the administration associated with manufacturing a product	• human resources • finance • information technology

B2 Activity-based costing method

At one time, costs within engineering organisations were calculated based on the number of hours a machine would be running and the number of hours an operative would be working. However, with the introduction of new technology and the development of improved manufacturing methods, it has become more accurate to cost the manufacturing of a product by reviewing the actual costs needed to complete a job.

With activity-based costing, each cost is assigned to a category of costs – called an **activity cost pool**. Within each activity cost pool an **activity driver** is then identified. Activity drivers are the actual factors that cause the cost to increase. For example, if packaging and despatch constitutes an activity cost pool, the number of parcels being packaged and shipped would be the activity driver.

The steps of the activity-based costing method are as follows:

1 Identify the activities – the processes required to produce an output.

2 Assign resource costs to activities – including direct costs, indirect costs and general/administration costs.

3 Identify the outputs (cost objects) – the products, services or customers.

4 Assign activity costs to outputs – by using activity drivers.

The following worked example illustrates how to apply the activity-based costing method.

Key terms

Work in progress (WIP) – an incomplete product or project that is still being worked on, and which has a total cost allocated to it, largely for labour and materials.

Overheads – costs that relate to extra services involved in the manufacture of a product, such as equipment costs and heating or lighting; they are not directly related to the cost of labour or materials.

Depreciation – the reduction in an object's value over time; this could be related to wear and tear on machines or the decreasing market value of a product.

Activity cost pool – a category of costs.

Activity driver – a factor within an activity cost pool that causes costs to increase.

Worked Example

An engineering organisation manufactures jet engine assemblies (like the one shown in **Figure 4.2**) for the aerospace sector. One part of the assembly is manufactured in computer numerical control (CNC) cells. The finance director, Jessica, has been asked to produce an activity-based costing model for the CNC-machined part.

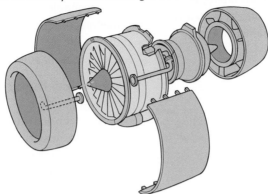

▶ **Figure 4.2** Example of a jet engine assembly

Solution

Step 1: Jessica's first step is to identify the costs associated with manufacturing the engine part.

We will refer to each engine part manufactured as a unit.

To begin with, Jessica considers how many units are sold per year and how much they are sold for.

Number of units manufactured and sold per annum	265 000
Price each unit is sold for	£5.50

These values allow her to calculate the sales revenue per year – the total amount of money made by the company from sale of the product, without considering any costs.

Sales revenue = price per unit × number of units sold

= £5.50 × 265 000 = £1 457 500

Step 2: Then, Jessica calculates the direct costs associated with manufacturing the product.

She has identified material and labour costs:

Direct material costs per unit	£0.50
Direct labour costs per unit	£0.35

Step 3: Jessica can now calculate the total direct costs of manufacturing the product:

Direct material costs = material costs per unit × number of units sold

= £0.50 × 265 000 = £132 500

Direct labour costs = labour costs per unit × number of units sold

= £0.35 × 265 000 = £92 750

Total direct costs = direct material costs + direct labour costs

= 132 500 + 92 750 = £225 250

Step 4: Next, Jessica analyses the indirect costs associated with manufacturing the product.

She considers the relevant activity cost pools and identifies the activity drivers within those activity cost pools.

Costs are then assigned per product or unit, as shown in **Table 4.3**.

▶ **Table 4.3** Using activity drivers to calculate total cost

Activity cost pool	Activity driver	Cost per unit of activity driver	Total activity per annum	Total indirect costing for the activity driver
Purchasing	Number of purchase orders received	£875	50	875 × 50 = £43750
Setting up of CNC manufacturing cells	Number of cell set-ups required	£775	100	775 × 100 = £77500
Machine testing and calibration	Number of tests that need to be carried out	£100	1000	100 × 1000 = £100000
Machine maintenance	Number of batch runs that take place	£680	200	680 × 200 = £136000
			Total indirect costs	£357250

Step 5: Once Jessica has collected information about direct costs, indirect costs and sales revenue, she can use this to calculate the organisation's profitability, as shown in **Table 4.4**.

▶ **Table 4.4** Calculating annual profit from revenue and costs

Number of units produced and sold	265000
Total direct costs	£225250
Total indirect costs	£357250
Revenue per unit	£5.50
Direct costs per unit	Total direct costs ÷ number of units = 225250 ÷ 265000 = £0.85
Indirect costs per unit	Total indirect costs ÷ number of units = 357250 ÷ 265000 = £1.35
Gross profit per unit	Revenue per unit – costs per unit = 5.50 – 0.85 – 1.35 = £3.30
Gross profit margin	Gross profit per unit ÷ revenue per unit = 3.30 ÷ 5.50 = 0.6 = **60%**

Step 6: From this analysis Jessica can conclude that manufacturing the part is profitable for the organisation.

Ⅱ PAUSE POINT Consider the activity-based costing model in the worked example. What difference would it make to the profitability of the product if the indirect cost of product packaging had been taken into account?

Hint Re-calculate the total indirect costs including the cost of product packaging (£0.30 per unit), and then substitute this value into later calculations.

Extend What other indirect costs could be included in the model?

Choose a product manufactured or service provided by an engineering company and carry out research to explore the costs of various activities associated with that product or service.

Develop an accurate activity-based costing model for the product or service. Refine your model throughout the process.

Identify the major cost areas that could affect profitability for the company.

Explain the reasons why the engineering company needs to control costs.

Plan

- What is the task? What am I being asked to do?
- How confident do I feel in my own abilities to complete this task? Are there any areas I think I may struggle with?

Do

- I know what I'm aiming to produce at the end of the task and I am clear on the steps that I need to take to achieve my goal.
- I can identify the elements of the task that I find most challenging and devise ways to overcome these difficulties.

Review

- I can explain what the task involved and how I approached its completion.
- I recognise which areas I feel less confident in and can seek ways to improve my skills in those areas.

Explore how engineering organisations use quality systems and value management to create value

In order to be successful, engineering organisations have to deliver high-quality products and services. To help ensure that outputs meet a high standard, the organisation can use quality systems and value management. In this learning aim you will study some types of quality system, why organisations use them, and the principles and processes of value analysis on a product or service.

C1 Quality systems

Engineering organisations can apply for accreditation to show that they meet international quality standards. Accreditation demonstrates to customers that the organisation has quality management systems in place – this can give the organisation a competitive advantage and customers the confidence to buy from that organisation.

The ISO series

The **International Organization for Standardization** (**ISO**) has a series of standards – the 9000 series – that provide a quality assurance system for manufacturing and service industries. These standards can be used to improve the organisation through applying quality management principles.

> **Key terms**
>
> **International Organization for Standardization** (**ISO**) – an independent organisation that develops international standards for quality, safety and efficiency.
>
> **ISO 9000** – a quality management system for manufacturing and service industries.

ISO 9000 series

To gain **ISO 9000** certification, an engineering organisation has to demonstrate that it meets 20 requirements. You need to be aware of these and keep up to date with the latest versions. The standards are updated regularly, with the most recent revisions in 2015.

1. Management responsibility – The management team must designate the responsibilities of the personnel involved in quality management. A policy for quality should be developed that is documented, understood by all, implemented, refined and monitored. A member of the management team must be designated to oversee the implementation, regular review and monitoring of the quality system.

2. Quality system – A system which demonstrates that customer requirements will be fully satisfied must be prepared and implemented.

3 Contract review – Contracts and purchase orders must be reviewed to ensure that they meet the needs of the customer. Any changes to contracts must be reviewed and agreed.

4 Design control – Projects that the organisation is involved in must be well planned, with all aspects of the project being documented and verified. Any changes must be controlled by qualified personnel.

5 Document and data control – All documents and data handled within the organisation must be controlled. This includes controlling their production, distribution, amendment and, finally, disposal.

6 Purchasing – Suppliers and contractors must be evaluated and monitored. This includes monitoring their quality assurance systems.

7 Control of supplied product – Products must be easily identified and handled and be stored correctly to ensure protection against damage or loss.

8 Product identification and traceability – Products must be able to be traced and identified throughout the processes of manufacturing, packaging and delivery to the customer.

9 Process control – All stages of product manufacture must be planned, controlled, monitored and documented.

10 Inspection – Materials supplied to the organisation, the manufacturing processes and the final testing of the product or service must all be subject to regular inspection, with accurate records kept.

11 Calibration, measuring and test equipment – The accuracy of measurements taken during product manufacture must be tested and reviewed. Calibration equipment must be regularly maintained.

12 Inspection and test status – Testing must take place as products are being manufactured and judgements made as to whether the products pass or fail those inspections. Testing of the tests must also occur.

13 Control of non-conforming product – Procedures must be in place to control products that do not pass inspection tests. Such products must not be used but should be reviewed before disposal.

14 Corrective and preventive action – Any problems identified should be reviewed and the cause of each problem corrected. The effectiveness of the corrective actions that take place must also be reviewed.

15 Handling, storage, packaging, preservation and delivery – Documents must be kept at all stages through to product delivery, with handling controls in place to prevent any product damage. Procedures must be regularly reviewed and maintained.

16 Control of quality records – All records must be collated, indexed and stored to ensure that they are easily retrievable.

17 Internal quality audits – Audits of quality procedures must be planned and regularly carried out. Results from audits should then be passed to the management team for review and to allow any issues raised to be addressed.

18 Training – Staff training needs and the provision of those needs must be addressed and recorded.

19 Servicing – Formalised procedures must be in place for the servicing of manufacturing equipment, with requirements for quality being met.

20 Statistical techniques – Statistical methods must be identified and applied in order to review the products and the processes used to manufacture them.

The ISO 9001 standard provides a planning tool for quality and to support continual improvement. Its purpose is to ensure that the quality management systems put in place by the organisation actually work to improve the business. It focuses on objectives and on the implementation of more efficient working practices. Areas of the organisation covered by ISO 9001 include equipment, facilities, people, training and services.

Research

What does the ISO 9001 mark look like? Find out by researching it on the internet.

ISO 14000 series

You also need to know about the **ISO 14000** series. This series of standards provides guidelines on how an organisation can make improvements relating to the environment, for example by reducing their **carbon footprint**. Some specific standards in the series are:

▸ ISO 14001 – relates to the reduction of waste.

▸ ISO 14006 – concerns the improvement of product quality in an environmentally positive way. It addresses the implementation, documentation, maintenance and continuous improvement of an organisation's eco-design management system.

▸ ISO 14040 – relates to life-cycle assessments, which means assessing the environmental impact of a product from the supply of raw material, through production, packaging and distribution, all the way to the end of the product's life. Guidelines for performing a life-cycle assessment are provided.

Key term

ISO 14000 – an environmental management system.

Carbon footprint – a measure of the environmental impact of a particular individual or organisation, measured in units of carbon dioxide.

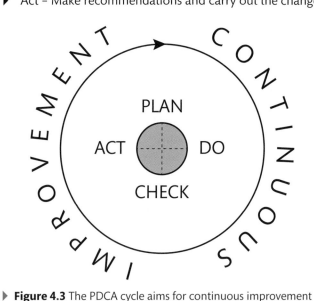

Quality assurance

Having standards is all very well, but more importantly the organisation has to attain those standards in practice. **Quality assurance** is what an organisation does to make sure that the quality requirements of its products or services are met. Quality assurance can be seen as a company-wide philosophy.

Key term

Quality assurance – the set of planned activities for ensuring that the quality requirements of a product or service are met.

Total quality management (TQM)

By committing to TQM principles, an organisation demonstrates that high quality is central to the way that all parts of the organisation work, and that the highest-quality product or service is its united goal.

In a TQM organisation:

▸ The needs of the customer are fully met and quality standards are defined by customer requirements.

▸ Clear systems and procedures for improvement are in place.

▸ A culture of continual improvement exists, involving all members of the organisation.

▸ Improvement is led by continuous review and feedback.

▸ All costs relating to quality are identified and minimised – for example, appraising production processes and consequently preventing failure of products as much as possible.

PDCA cycle

PDCA is a management model with four basic steps (plan, do, check, act) for the implementation of change. As you can see in **Figure 4.3**, PDCA is a cycle of continuous improvement – like a circle, it never ends.

The four steps of PDCA are:

▸ Plan – Identify the issue, collect data and plan the change.

▸ Do – Develop a solution and test the proposed change.

▸ Check – Verify that the proposed change will meet expectations full-scale, and compare with the original.

▸ Act – Make recommendations and carry out the change.

▸ **Figure 4.3** The PDCA cycle aims for continuous improvement

If the change has not been successful, start the cycle again with an alternative change. If the change is successful, incorporate what has been learned into a wider scope and start the cycle again.

Why implement a quality system?

Organisations can choose to implement quality systems to:

▸ benchmark against other organisations – The company compares its products and working practices with those of other businesses, particularly the companies that currently lead the field, with the aim of making its own business as good or better.

▸ ensure consistency of processes – Consistent processes must be in place to ensure that a product or service of consistent quality is the outcome.

▸ achieve conformity of the product or service to a standard – All products or services must meet a set minimum standard so that customers will feel confident about buying them.

▸ reduce unnecessary waste – Excessive purchasing, over-processing or over-production costs the organisation time and money, and should be eliminated.

▸ improve the effectiveness of the organisation – Efficiency improves cost-effectiveness.

▸ gain a competitive advantage – An organisation that has quality systems in place will have a competitive advantage over companies that don't.

▸ achieve customer satisfaction – Customer satisfaction is a measure of the overall success of a business.

▸ ensure that a product or service is fit for purpose – Any product or service that is not fit for purpose will not sell, and the organisation will be unsuccessful.

Quality control

Quality control is an essential part of any quality assurance system. Quality control methods are applied within organisations in the same way that they are applied to an object in an engineering workshop to monitor a tolerance. Inspection sampling and testing are carried out throughout a product's manufacture to ensure that quality standards are met at every stage. For example, the product's condition could be monitored through vibration or thermal analysis. Maintenance is planned and undertaken at regular intervals, and a **'right first time'** philosophy is adopted – quality control checks should be such that errors can never occur.

Theory into practice

Toyota Motor Corporation has developed a managerial approach and production system that they refer to as 'The Toyota Way'. The philosophy is centred on two considerations: continuous improvement and respect for people. In order to achieve a 'right first time' standard, any employee in the organisation's production system has the authority to stop the process at any stage if they have a quality concern.

Think about and investigate the following questions:

- What are the 14 principles behind The Toyota Way?

- What drawbacks are there in giving all employees the authority to stop the manufacturing process? Why does Toyota adopt this form of quality control despite the potential problems you have identified?

C2 The principles and processes of value management

Value management

The economic **value** of a product or service – how much it is 'worth' – is not simply equal to its price. It is a measure of how much benefit the product or service provides to the customer relative to its cost. Poor value could be due to lack of innovation, high costs incurred during certain parts of the project or inefficiencies in production processes, among other factors. In an engineering organisation, value management is used to increase the value of outcomes, which can be done by reducing waste and unnecessary costs while improving the efficiencies of activities along the process.

Value management begins with conducting a value analysis exercise for a particular product or service. To do this, you first identify the **value-added** and **non-value-added activities** associated with the product or service and then analyse all the processes involved in producing the outcome. Consider where wastes, redundancies or inefficiencies occur in the current process, and generate alternatives that could offer better value. The results of this functional analysis are used to identify improvements that could be made to the process to enhance efficiency and produce a lower-cost outcome.

Key terms

Quality control – the testing and monitoring of activities that are used to check the quality of a product or service outcome.

'Right first time' – a quality management concept which asserts that preventing defects and errors is better and more cost-effective than correcting them later.

Value – how well a product meets the needs of customers in relation to its price.

Value-added activities – activities that directly affect a product and add value to the outcome; these might include activities related to the product's design, manufacturing and innovative features.

Non-value-added activities – activities that are essential to a product and its production but do not directly add value to it; for example, product inspection, review and reporting.

Carrying out a value analysis exercise

The process of carrying out a value analysis exercise on a product or service consists of five stages.

1 Information phase
   ▸ Identify key issues.
   ▸ Identify value-added and non-value-added activities and review their costs.

2 Analysis phase
   ▸ Perform a functional analysis.
   ▸ Identify which existing processing methods are used.
   ▸ Identify key features and parts which are unnecessary.

3 Creative phase
   ▸ Generate alternatives for better-value solutions.
   ▸ Apply problem-solving tools and methods.
   ▸ Develop a least-cost solution.

4 Evaluation phase
   ▸ Assess and prioritise ideas.

5 Development and reporting phase
   ▸ Refine ideas and develop action plans.

The following worked example illustrates these steps of carrying out a value analysis exercise.

Worked Example

LightUp UK is an engineering company that manufactures torches. The company is reviewing the types of torches it manufactures that aren't selling well, including their dynamo torch.

Tom Wright works in the company and has been asked to carry out a value analysis exercise on the dynamo torch.

1 Information phase

Tom starts by writing down the cost and price of each unit of the product (**Table 4.5**).

▸ **Table 4.5** Cost and price per unit of the dynamo torch

Cost (the amount it costs the company to manufacture each torch)	£2.29
Value (the price each torch is sold at)	£4.99

Tom then gathers some information from customers about whether they think the torch is 'good value'. He finds out that there may be an issue with the dynamo mechanism inside the torch: many customers say that it breaks easily.

Next, Tom investigates the processes used to manufacture the torch. He identifies some value-added activities and non-value-added activities, as well as the costs associated with them (**Table 4.6**).

▸ **Table 4.6** Costs of value-added activities and non-value-added activities for the dynamo torch

	Activities	Cost per unit
Value-added	Machining	£0.25
	Processing	£0.60
Non-value-added	Inspection and testing	£0.30
	Transportation between different parts of the processing plant	£0.20

2 Analysis phase

Tom undertakes a functional analysis on the dynamo torch. He works with the processing team to examine current manufacturing processes, and with the product development team to discuss the product itself.

Tom identifies two key issues:

- There is a cogwheel inside the torch that is made of plastic and takes a lot of wear when the torch is in dynamo mode. This could be the cause of the customer complaints about breakage of the mechanism.
- Among the non-value-added activities, inspection and testing are essential, especially in view of the problems with the torch breaking. However, as regards transportation, two parts of the torch are currently manufactured in different locations of the plant before being brought together for assembly.

3 Creative phase

Based on what he has discovered in the information and analysis phases, Tom now generates some ideas and solutions.

First, he considers the plastic used to manufacture the cogwheel – could an alternative material be used that is more resistant to wear?

- Tom considers using a metal – could this prevent the dynamo mechanism from breaking while not adding significantly to the cost of the final product?
- Tom also considers using a different type of plastic – one that can still be injection-moulded (as this is the manufacturing process currently used by the company) but which would be more resistant to breakage. Is there a plastic that could satisfy these requirements?

Tom then considers the transport issue he identified – would it be possible to rearrange the production lines at the plant to reduce the need for transporting parts around the site?

4 Evaluation phase

After some research and discussion with various teams at the plant, Tom finds that using metal for the cogwheel would increase the weight of the product, the metal would cost more to purchase, and the part would have to be manufactured elsewhere as his company can only process plastics through injection moulding. Tom calculates that this change, despite making the dynamo mechanism less likely to break, would result in a significant increase in the cost of the torch, so it is not viable.

However, he does discover a different type of plastic that is slightly harder than the plastic used in the original design. Despite being more brittle, this plastic is less likely to break and is also suitable for injection moulding.

After discussions with the processing team, Tom concludes that it would be possible to move the second assembly-based production line closer to the injection moulding processing area. This would considerably reduce the need for transporting parts around the plant and could result in a lowering of the product manufacturing cost.

5 Development and reporting phase

Tom produces costing documentation for his managers and works with the processing team to develop an action plan to rearrange the plant layout. He will now work with the management team to help them decide whether his changes should be implemented. The change in material increases the cost of materials by £0.05 per torch, but will reduce breakage – and so increase customer satisfaction. The change in processing arrangements will reduce the cost of transportation by £0.10 per torch.

If you were one of Tom's managers, would you implement the changes he has proposed?.

PAUSE POINT   Think about the non-value-added activity of inspection and testing in the worked example. Tom did not include a full review of this activity as part of his value analysis. Should he have?

Hint   If inspection and testing were adequate, would customers have complained about a part breaking? How could the inspection and testing processes be improved?

Extend   Further inspection and testing would result in a higher cost for this activity. Recalculate the total manufacturing cost if this activity increased by £0.20 per unit. What impact might this have on the value management exercise? Would the recommendations change?

Assessment practice 4.3   C.P5   C.P6   C.M3   C.M4   C.D3

Explain and analyse the purpose of different quality management systems and value management principles used by engineering organisations.

Choose an activity or process used in a particular engineering organisation and perform a value analysis/management exercise on that activity or process. Make sure that you complete all the steps of the analysis accurately.

Evaluate the outcome of the value analysis exercise and make recommendations, including the use of quality systems to implement efficiencies to the engineering activity or process.

Determine whether further value can be created from the engineering product or service that the activity or process relates to.

Plan
- What are the components of this task? What am I being asked to do?
- What information will I need to collect in order to complete each stage of the task?
- How confident do I feel in my own abilities to complete this task? Are there any areas I think I may struggle with?

Do
- I know what I need to produce at the end of the task and I am clear on the steps that I need to take to achieve my goal.
- I can identify the elements of the task that I find most challenging and devise ways to overcome these difficulties.

Review
- I can explain what the task involved and how I approached its completion.
- I recognise which areas I feel less confident in and can seek ways to improve my skills in those areas.

THINK ▶FUTURE

Srinivas Galman
Operations manager

When I left school I completed a BTEC HNC manufacturing qualification with my current company, now known as ClipFit UK. I started with the company working on the shop floor, manufacturing casings for a variety of industrial applications that included holes ready for screws to hold the casings in place. I worked hard and took further qualifications once I'd finished my HNC, as I wanted to move from the shop floor and manufacturing into managing the business side of an engineering company.

I was appointed to the post of Operations Manager around the same time that the company moved towards a total quality management ethos, and I undertook a value analysis exercise. All aspects of the operations to manufacture the casings were reviewed to see if the company could make more profit or gain a competitive advantage. I was part of the team that looked at product development. I proposed incorporating a clip into the product that would eliminate the customer's need for screws when installing our casings. The idea was considered and developed, and a test batch was manufactured.

The customers loved it! We now manufacture casings with clips incorporated, unlike any other casing manufacturers, and have sole intellectual rights for that innovation. At first our company manufactured the clips using the same steel as the casings, but a further value analysis exercise suggested that using plastic clips manufactured by a different company could save us money. This proved to be true, so we now have an innovative yet cheaper product than before – and I got a pay rise as a result!

Focusing your skills

Referring to the ClipFit UK example above:

- What additional costs did the product change incur?
- Why did the company decide to incorporate clips into its product if this would raise costs?
- What considerations do you think went into the company's decision to have the plastic clips manufactured elsewhere?
- What types of outcome might further value analysis result in?

Getting ready for assessment

Natasha recently completed a BTEC National in Engineering and is now studying for a BTEC Higher National Diploma (HND). She wants to get a job at management level within the engineering industry and aims to specialise in quality management.

How I got started

In order to understand the way that engineering organisations operate, it was essential to do some research. I used the internet to identify some local engineering organisations and gave them a call to see if I could visit them. This wasn't possible with some of them due to health and safety issues, or simply because they said they were too busy, but two companies said they would be happy for me to visit, so I did just that. Before going along I did some research into the companies and also made sure I was aware of the content required for the assignments I had to complete. Based on this I prepared questions to ask during my visits. The information I got from the companies was invaluable in completing my assignments – it made writing them so much easier, as I was producing activity-based costing models and value analysis exercises using data from real organisations.

How I brought it all together

▸ I combined knowledge from my visits and internet research with what I had learned in class.

▸ I used a combination of written information, diagrams and images to present my evidence in a way that was easy to understand.

▸ After visiting the companies, I contacted them to ask further questions as my assignments developed (many of my peers had to rely on the internet and books from the library for this type of further research).

What I learned from the experience

When I first asked for help from companies by emailing and calling them, I got discouraged as I received very few replies. I didn't consider then that the primary purpose of these companies is to manufacture products, not help students like me. But I persevered, and that's how I eventually made some useful contacts – I shouldn't have let myself get frustrated.

When it came to producing my assignments, for my first draft I made the mistake of simply reporting what I had found out; I didn't include enough evaluation in my comments. It's really important to include analysis and evaluation in your assignments if you want to achieve the highest grade.

Think about it

▸ Are there any local engineering organisations from which you can collect information for your assignments, that you could visit to improve your understanding of the concepts, or which might be a source of real-life examples to study in your work?

▸ Have you identified the information you need to gather on existing organisations?

▸ Have you identified all of the content you need to cover from the unit specification?

Do you understand what you need to include in your assignments so that you can achieve the highest grade?

A Specialist Engineering Project 5

Getting to know your unit

Assessment

This unit will be assessed by one project, which will comprise a number of tasks to be undertaken over a 30-hour period.

Project management is fundamental to all engineering disciplines, from aerospace and computing to the manufacturing sector. The output of projects is varied, and could be a product, service, system or process relevant to a specialist area of engineering. There are also many approaches to project management. In this unit you will research an engineering problem and generate a range of solutions to the problem. You will produce a feasibility study to select the most appropriate solution given the known constraints, and make use of project-management processes over the life cycle of your project to design and develop a solution that is fit for audience and purpose. By undertaking this engineering project, you will apply, consolidate and build on the knowledge and skills you have learned across your programme.

How you will be assessed

This unit will be assessed by one project set by your tutor, which will consist of a number of tasks to be completed over 30 hours, which is the recommended time to be spent on the project tasks. Throughout this unit you will find assessment practice activities to help you work towards your assessment. Completing these activities will not mean that you have achieved a particular grade, but you will have carried out useful research or preparation that will be relevant when it comes to your final assignment.

In order for you to complete the tasks in your project, it is important to check that you have met all of the assessment criteria. You can do this as you work your way through the project.

If you are hoping to gain a Merit or Distinction, you should also make sure that you present the information in your assignment in the style that is required by the relevant assessment criteria. For example, Merit criteria require you to assess, design and perform effectively, and Distinction criteria require you to evaluate and optimise.

The project will involve the following three main activities:

▶ Investigating an engineering problem, generating ideas and scoping out at least three possible solutions, and then undertaking a feasibility study of the alternative solutions.

▶ Design a solution by applying project-management processes, and develop a technical specification with supporting documentation for the designed solution.

▶ Develop, test and review the solution, and collate a portfolio documenting the development and outcomes of the project.

You will present the evidence for your project in the form of a portfolio, which might include written reports, a logbook, planning documentation, testing documentation, printed or plotted drawings and annotated photographs of the project-development process.

Assessment criteria

This table shows what you must do in order to achieve a **Pass**, **Merit** or **Distinction** grade, and where you can find activities to help you.

Pass	Merit	Distinction
Learning aim A Investigate an engineering project in a relevant specialist area		
A.P1 Research an engineering problem based on a given theme and scope out at least three alternative solutions. **Assessment practice 5.1**	**A.M1** Assess consistently at least three solutions to an engineering problem on a given theme and recommend a preferred solution. **Assessment practice 5.1**	**A.D1** Evaluate, using language that is technically correct and of a high standard, at least three realistic solutions to an engineering problem on a given theme and justify a preferred solution. **Assessment practice 5.1**
A.P2 Outline at least three alternative solutions to an engineering problem and select a preferred solution. **Assessment practice 5.1**		
Learning aim B Develop project-management processes and a design solution for the specialist engineering project as undertaken in industry		
B.P3 Implement project-management processes, including planning, monitoring and risk and issue management. **Assessment practice 5.2**	**B.M2** Implement project-management processes, including detailed planning and monitoring and proactive risk and issue management. **Assessment practice 5.2**	**B.D2** Optimise the project-management processes and design solution, while allowing for reasonable contingency and considering constraints. **Assessment practice 5.2**
B.P4 Produce the technical specification for a solution to an engineering problem. **Assessment practice 5.2**	**B.M3** Design a coherent solution, considering alternative approaches. **Assessment practice 5.2**	
B.P5 Produce design documentation to detail the solution, including a test plan and taking account of sustainability. **Assessment practice 5.2**		
Learning aim C Undertake the solution for a specialist engineering project and present the solution as undertaken in industry		
C.P6 Produce a solution safely using project-management processes while recording progress. **Assessment practice 5.3**	**C.M4** Perform effective project-management processes while justifying refinements and demonstrating effective behaviours consistently to develop a solution that is fit for audience and purpose. **Assessment practice 5.3**	**C.D3** Optimise the project-management processes to develop a solution that is fit for audience and purpose while anticipating and resolving risks and issues and demonstrating behaviours to a professional standard. **Assessment practice 5.3**
C.P7 Perform relevant behaviours effectively while developing a solution safely. **Assessment practice 5.3**		

Getting started

In pairs, discuss projects that you have worked on before, either in class or as part of your hobbies or interests. Think about the issues and problems you had with the project. Did it all go to plan? Did you need to go back and consider other solutions? When you have completed this unit, think about how your knowledge and understanding of the topic has increased. What would you do differently next time?.

A   Investigate an engineering project in a relevant specialist area

A1 Project life cycle

If you research **project life cycle** (or **project management** life cycle), you will discover that there are many different approaches, with anything from three to ten stages.

> **Discussion**
>
> In groups, consider one of your mobile phones and discuss the different stages there might be in the project life cycle for creating such a product.
>
> (As you work through this unit, look out for boxes that follow up on this example of developing a mobile phone.)

One way to look at a project life cycle is to split the project into four stages:

▸ initiation
▸ planning and design
▸ implementation
▸ evaluation.

This is a typical industrial model, and is the one you will be following for your project.

Initiation

The first stage starts with identifying a problem clearly. Sometimes this requires research. For your project, the problem may be provided by your tutor. This may be in the form of a general theme that everyone in your class will follow, or it may be a specific idea, such as modifying a motor vehicle braking system or designing and manufacturing a part for a door lock.

The initiation stage also includes:

▸ establishing the key design features of potential solutions

▸ thinking about constraints that might apply

▸ generating ideas

▸ conducting a **feasibility study** (see Section A3).

> **Key terms**
>
> **Project life cycle** – a series of activities that are necessary to fulfil a project's objectives.
>
> **Project management** – carefully planning and guiding a project's processes from start to finish of the project.
>
> **Feasibility study** – an assessment of a proposed project to see if it can be achieved in practice.

> **Reflect**
>
> **Development of the first mobile phones: Initiation stage**
>
> Although mobile phones have made rapid advances within the past 30 years, it was in the early 1900s when the first patent was issued for wireless telephones. Initial efforts of mobile phone developers to achieve the modern goal of instant communication resulted in a product that looked like this.

▸ The first mobile phones were highly innovative – and very large!

Planning and design

The key to any successful project is planning. Good planning will save you time and resources, and help you to avoid many issues and problems.

One of the first things you should do is create a project plan, setting out the necessary steps in your project, along with approximate timings. You also need to think about the resources you will need, such as consumables, labour, materials, tooling and software.

Time planning will be very important: there is only limited time available, and other learners may want to use the same resources as you (for example, there may only be one 3D printer for additive manufacturing).

Reflect

The first mobile phones: Planning stage

The size of mobile phones has always been constrained by the technology available. When mobile phones were first developed, miniature electronic circuits did not exist. The planners and designers used technology that was well proven at the time, and created a two-way radio system linked to a single base station. The first mobile phones had limited range and were so expensive that few people could afford them.

Implementation

This stage is when the project plan is put into motion and the work of the project is done. It covers all the activities needed to develop the actual product, service, system or process. You need to keep control of the project by continuously monitoring progress against the project plan and managing any risks and issues that arise.

Reflect

The first mobile phones: Implementation stage

Motorola – the company credited with delivering the first commercial mobile phones – was on risky ground with their project but took a gamble. Twenty years later, after many **iterations**, smartphones have become capable of a wide range of functions besides making calls. Motorola showed great foresight – and careful planning and implementation were crucial to the commercial success of their mobile phones.

Key term

Iteration – repeating a step, process, version or adaptation of something, usually with the aim of improving it.

Evaluation

The final stage of the project life cycle involves looking at the lessons learned – what went well, what didn't and what can be improved. You will assess how far you have met customer expectations and whether you have achieved your goals to the correct timescale and within budget.

As you develop your project, you will use the process of iteration – repeating steps of the project or creating new and refined versions of a model – to improve your product. Make sure you keep a record of the different iterations throughout the life cycle of your project, as they will be crucial in producing your final report.

Reflect

The first mobile phones: Evaluation stage

Did the first mobile phone producers succeed? If their aim was to develop a device that people could use to communicate with each other 'on the go', then the simple answer is 'yes'.

What will come next? Smaller batteries made of graphene, quickly charging and holding charge for days? Phones shaped like a thin piece of plastic that you can fold up and keep in the bottom of your pocket? A tiny electronic device grafted to your skin?

Tip

Your own project need not be an artefact that needs manufacturing, such as a mobile phone. It could be:

- a service – for example, you could develop a service procedure for a heavy aircraft landing
- a plan – for example, create a plan for routine maintenance of a variety of machines in a workshop
- a process – for example, design a software program or package to control a system
- an interface between different objects
- a way to improve the performance of an existing product, process or service.

Hint   Close the book and put away your notes. Explain the project life cycle to a peer.

Extend   In pairs, outline to each other your initial thoughts on how you would manage and develop your project.

A2 Generating ideas and developing solutions

Research a project theme or topic

The starting point for your project may be a specific problem given to you by your tutor. Alternatively, your tutor may suggest a general theme, and you will need to identify a suitable problem under that theme to be the focus of your project.

Some typical project themes are:

- power
- energy
- motion
- forces
- security
- wind turbines
- flight
- propulsion.

Here are some possible project ideas under different engineering specialisms.

- Mechanical – design a mechanical system; manufacture a conveyor system; create a servicing and repair manual for an indexing conveyor system.
- Manufacturing – modify a manufactured product or service; create a computer numerical control (CNC) program and machine a complex product; use computer-aided design/computer-aided manufacturing (CAD/CAM) or additive machining to design, post-process, program and manufacture an assembly of various components.
- Electrical and electronic – test a system or service; design a security system and manufacture the printed circuit board for the system.
- Aeronautical – modify an aeronautical product, service or procedure; build a hydraulic rig or system; create a maintenance procedure for a hard aircraft landing.
- Computing – modify a computer service, hardware device or software product; write a program to control a robotic arm.
- Mechatronic – design and build an electronic, mechanical and computer-controlled system; design and build a system to control the temperature in a greenhouse.
- Automotive – modify an existing vehicle part or system; modify a vehicle's electronic, braking or steering system.

Remember that at this stage you should concentrate on coming up with as many ideas as possible. You are not developing a specific plan for your project yet.

Generating ideas

Once you have chosen a suitable problem, you need to generate ideas for potential solutions.

At this point, it is good to do research on the problem you will work on. Gather as much background information as you can, using all of the resources that are available to you – such as the internet, journals, libraries, databases and company websites.

Then you can start generating ideas. Here are some creativity tools you could use.

▶ Reword the problem – Try writing out the problem in a number of different ways, to help you pinpoint what is needed for its solution.

▶ Hold a group discussion or brainstorming session to come up with potential solutions – Bring in people who are familiar with the problem or theme, but invite others too; sometimes 'outsiders' have the most interesting ideas.

▶ Display the problem visually – You might want to draw a diagram or make a **mind map**.

▶ Challenge your own assumptions – Try to look at the problem from different perspectives. Just because something has traditionally been done in a certain way doesn't mean this can't be improved on.

▶ Think in reverse – Instead of thinking towards a potential solution, think back to how the problem might have arisen in the first place. This can be a good way to stimulate new ideas.

▶ Try using De Bono's Six Thinking Hats® to generate fresh ideas.

Research

Do some online research to find out more about De Bono's Six Thinking Hats® and how they can be used to generate ideas. Which of the hats is the one you 'wear' most? Try putting on the hat that appeals to you least, and see what ideas you come up with.

Developing solutions

After you have generated as many ideas as you can, try to pin things down and see what practical solutions might come out of all these ideas.

The best way to do this is by creating a specification that outlines different potential solutions. This will allow you to focus in on products, services, systems or processes that have not yet been designed, but could work for what you need.

Here are some elements your specification could include.

▶ Graphic solutions – Draw some outline sketches and diagrams, find some relevant photos, or create a **storyboard**.

▶ Process outlines – Look at the processes within each solution, and list the machine tools, systems and assemblies they might involve. A high-level flow chart or list could be helpful here.

▶ Rough costings – Think about your budget. Will these solutions come in at the right cost, accounting for items such as materials and labour? You may need to draw up a spreadsheet for this.

▶ Initial technical information – Although you may not have the full picture yet, you should already start making notes on approximate mass and volume of product parts, suggested materials and **performance parameters**.

Key terms

Mind map – a web of thoughts written out on paper or drawn on a computer, with a key word or idea in the centre, linked by lines to related ideas.

Storyboard – a graphic sequence of events used for planning operations, e.g. for 3D modelling on CAD systems or for additive manufacturing.

Performance parameters – the capabilities that a system must have in order for it to be successful or to achieve its goals.

Smartwatches

Smithsons has been producing watches for a long time, beginning with mechanical pocket watches and moving to electronic 'cheap and cheerful' digital watches as the market changed. In recent years, however, Smithsons has been losing market share and needed to develop a new product. The company management team considered the options: should Smithsons develop a smartwatch? Should it develop a luxury range and change its market focus?

Smithsons needed to act quickly as technology was advancing fast. After some market research, both smartwatches and luxury watches were discounted –

▶ The market for fitness watches has grown rapidly

they would be expensive niche items, and their markets would be difficult to break into.

During its research, Smithsons identified the boom in sports and fitness and realised that it could continue to sell its existing range of watches but supplement this with a fitness-related watch. This left the company with a simple decision – should it develop a **fitness watch** or a **tracker band** to supplement its range?

Check your knowledge

1   Is technology leading the market, or is the market pushing technology?
2   How quickly are models replaced? Why? What does 'planned obsolescence' mean?
3   Imagine you worked for Smithsons and your project was to develop a fitness watch specification. What features would you want the watch to have? What would be the price point? How would you balance the desired features against the budget?
4   Develop an outline product design specification (PDS) for a fitness watch or a tracker band.
5   Would a mobile phone app be sufficient for a customer to use instead?

Key terms

Fitness watch – a watch with advanced features to support sport and fitness activities, such as a heart monitor, workout programmes, a sleep monitor, an accelerometer and GPS; most give immediate on-screen displays of live data.

Tracker band – a thinner version of a fitness watch, usually limited to keeping track of steps, distance and calories; most trackers can synchronise with your phone, tablet or computer so that you can store and access the data on other devices.

Discussion

Discuss creativity tools that you could use and the benefits and limitations of each. Which is best for your project? Which is best for you? Remember that different people work in different ways, and will find different creativity tools helpful.

A3 Feasibility study of solutions

The purpose of a feasibility study is to determine whether a project can actually be done – whether an idea or a plan can become reality.

Feasibility studies are conducted across the engineering industry, and you will need to carry out a feasibility study as part of your project. In this study you will look at the best three potential solutions you came up with, examine the practicability of each, and then decide which one to choose as the design solution.

The following questions are important ones to ask as you consider each of the three proposals.

▶ Is the proposed solution too large or too complex?

Don't overcomplicate your project. Remember that you have only 60 hours for the project, split between a recommended 30 hours teaching time and 30 hours individual work on your project, and this includes developing your idea, various iterations of modelling and implementing the idea, writing reports of the process and preparing your portfolio.

▶ Are the benefits or improvements enough?

Imagine you were the customer or client. Would the outcome of the project meet your expectations? Is it 'fit for purpose'? Ask a friend to look over your proposed solution and listen to their feedback.

▶ Do you have the money and the skills needed?

People often take on projects that run over budget or for which they realise they don't have the necessary expertise. Do you have enough resources to solve this problem in the proposed way? If you don't have sufficient funds or the right skills, think carefully about what you can do with the resources you have. Can you ask a friend to make a part for you, while you provide some service for them in return?

▶ What are the risks?

Before embarking on a project, think about the unknown factors in your proposed solution and assess what the risks are for each. This could involve looking into unproven technology or equipment and unpredictable changes in time and costs.

▶ How sustainable is it?

You will need to consider the environmental impact of each proposed solution. Would it involve mass production? How much waste will there be? What are the environmental ramifications of using different types of materials? How might recycling fit in?

You may not be able to use environmentally friendly materials for your prototype product or service, but you can specify these for the 'real' version. Ask your tutor about the most suitable materials and their recycling potential.

▶ What are the legal constraints?

You need to think about all the legislation that might pertain to your project, such as the Data Protection Act 1998, the Health and Safety at Work etc Act 1974, and other relevant international equivalents.

Reflect

An engineering company will always investigate the feasibility of a project, checking that they have (or can acquire) all the technical skills and resources needed before they go ahead. Subcontracting can fill some of the gaps, but this needs careful management and adds to the risks. One major issue is viability – without sufficient financial resources, the project can't be realised.

Most products are modified over time. What was a viable project one day can become unviable later on due to changes in legislation, new technical developments or a product becoming obsolete. Equally, new opportunities can arise as different processes become cheaper to implement or if certain regulations are relaxed.

Analysis means studying a problem in detail to find a solution. It is a very important tool in engineering. When analysing the feasibility of a new project, the costs, the related benefits to the company and the payback time all need to be considered.

Imagine you work in an automotive plant and have been asked to investigate the possible manufacturing process for a new type of brake pedal. You need to consider whether the component should be:

- fabricated in steel
- additive machined (3D printed) as one part in a polymer
- additive machined as one part in a metal
- forged and machined
- machined from one piece of solid steel bar.

Think about and investigate the following questions (see **Figure 5.1**):

1   What are the different processes involved in these options?
2   What is the suitability of each option in terms of cost, practicality and resources?
3   Which option would you choose and why?

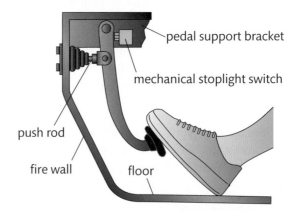

▶ **Figure 5.1** Which option would you choose for manufacturing a brake pedal?

Search online to find out more about the Data Protection Act 1998 and the Health and Safety at Work etc Act 1974.

The Health and Safety at Work etc Act 1974 is also discussed in Unit 2, Section A2.

In pairs or groups, discuss the difference between price and cost. The distinction may have a big impact on which solution you select for your project.

Imagine you see a bike for sale. At just £20 it seems a bargain – but it needs a new set of tyres, a new chain and new brakes. Does this bike have a low price but a high cost? Or does it have a high cost but a low price?

Case study

Energy sources for cars

 The first electric cars were invented in the 1880s, this one by English inventor Thomas Parker.

The development curve for electric cars has been long. Electric cars have been in operation since the late 1800s, and they have had a limited range of uses, from trolley buses to milk floats. However, recent rises in oil price have led motor vehicle manufacturers, including Toyota, Nissan and Renault, to put a greater emphasis on electric cars. Many companies have developed fully electric or **hybrid cars**, with advantages and disadvantages compared with standard petrol or diesel vehicles.

Technology is now moving towards driverless vehicles – mainly cars – which have been developed intensively in recent years (some cars can already self-park). The question is whether these products are safe, and whether the public and transport authorities will accept them.

Do some research into electric and hybrid cars. What are the advantages and disadvantages of each? Can you find out anything about the time needed for research and development, and the amount of money, manpower and resources involved?

Check your knowledge

1 What do you think were the main barriers to developing electric cars at first?

2 What stage of the project process are electric cars at now?

3 What factors could affect the project process from now on?

4 What might happen if the public accepts driverless cars?

Key term

Hybrid car – a car with a petrol engine and an electric motor.

❚❚ PAUSE POINT   Can you explain how the above case study relates to Learning aim A?

Hint   Close the book and draw a concept map of the project life cycle for an electric car.

Extend   Expand your map to explore the feasibility of all project solutions that could have been available (e.g. the use of LPG as fuel).

Reflect

Look back at your rationale for choosing the topic and solution for your project. Consider getting a second opinion, from a peer or your tutor. Ask them to work through the selection process with you.

Research an engineering problem based on a given theme. Generate at least three alternative solutions and assess the feasibility of each to select a preferred solution.

The three stages of the task are to:
- investigate an engineering problem under a given theme
- produce a specification scoping out at least three alternative technical solutions to the problem
- complete a feasibility study for each of the alternative solutions and recommend a preferred solution.

Ensure that you:
- work within the set problem or theme
- justify the feasibility of your selected solution and check with your tutor that it can be realistically implemented given your centre's resources and the time available.

Plan
- What is the task? What am I being asked to do?
- What information will I need to collect, and what resources can I access to help me complete the task?
- How confident do I feel in my own abilities to complete this task? Are there any areas I think I may struggle with?

Do
- I know what I'm doing and what I am aiming to achieve in this task.
- I can identify the parts of the task that I find most challenging and devise strategies to overcome those difficulties.

Review
- I can explain what the task involved and how I approached each stage of the work.
- I can explain how I would deal with the hard elements differently next time.
- I can identify the parts of my knowledge and understanding that require further development.

B   Develop project-management processes and a design solution for the specialist engineering project as undertaken in industry

In other units you have studied individual areas of engineering such as maths, design and electronics. In your project you will apply some of these skills as they would be applied in industry. This is your chance to be creative, apply your knowledge and bring your project to fruition.

B1 Planning and monitoring

All projects depend on good planning to be successful – but good planning won't achieve anything unless there is proper, regular monitoring as well.

When an engineering company wants to develop a new product, service, system or process or to improve its existing output, one of the first things they will do is appoint a **project manager**. The project manager will select a specialist team of personnel who have the skills required by the project, and will organise the team so that the project is completed on time, within budget and to the required specification. It is also the project manager's responsibility to make sure that resources are made available as needed and that testing procedures are followed as the project progresses.

For your project, the project manager will be you. Understanding how to plan and monitor a project will be key to turning a good idea into a great solution.

> **Key term**
>
> **Project manager** – the person who manages a project, solves problems and issues, organises and guides the project team, ensures that resources are used effectively, and carries responsibility for successful project completion. See the Think Future section at the end of this unit for a profile of a project manager.

Tools for planning

Resource plan

When we hear someone talk about resources, we often think of materials, equipment and tools. However, project resources may include many other things.

▶ Money – You will keep track of this in your budget. Putting it into your project plan will remind you that you might need to pay for some of the materials and equipment that you use.

▶ People – Your 'human resources' are probably the most valuable ones. Tutors and technicians will have a wealth of knowledge and experience that you can draw on; also, your peers can check your plans and give useful feedback. In your resource plan, make a note of the best people to call on for help or ideas.

▶ The internet – This is a rich resource for all sorts of information and support, but use it wisely and be careful not to believe all the 'facts' that you find on websites. Anyone can upload information to the internet, but it may be far from correct. See the suggested websites in this book for some sources you can trust.

▶ Books – Don't forget the library. Sometimes it may be easier to find the information you need from books and manuals on the specific area you are interested in.

▶ Materials, equipment and tools – It is worth taking time to plan your access to these carefully, so that you will have what you need when you need it, and your project can go smoothly and without delays.

Once you have a list of all the resources you will need for your project, you should set them out in a clear plan.

Theory into practice

In an engineering company, the major costs for a project tend to be labour and materials. These are the items generally classed as resources.

The more skilled or more technical the requirements are, the higher the labour costs will be. Materials costs include the cost of raw materials *and* the cost of processing, which needs to be considered carefully when planning manufacturing projects. Service and system projects tend to be very labour-intensive, and often require highly skilled staff. Process projects may have either high labour costs or high materials costs, or fall somewhere in between.

Think about and investigate the following questions:

1 What materials are the most expensive in engineering? Which cost the most to process?
2 What factors make some types of labour 'skilled'?
3 How does the information above relate to your project?

Time plan

With a recommended 30 hours for development of your project, it is important to plan your time carefully so that you can complete the work efficiently and effectively.

You may choose to write a simple time plan, perhaps using a calendar or diary. For example, you can make notes of when meetings have been arranged and when resources are available, as well as the phone numbers and email addresses of people you might call on for assistance. It doesn't matter if you choose an electronic planner or a paper one – the important thing is that you use it.

Simple time plans can already be very helpful. However, there are some special time-planning tools used in industry which you should know about.

Gantt charts and **critical path analysis** can help you set priorities and plan the order in which to do different activities.

▸ **Gantt charts** – developed by Henry Gantt in the 1910s, these are a type of bar chart that illustrates a project schedule, showing start and finish times of the different elements of the project (see **Figure 5.2**). You can create a Gantt chart in a spreadsheet and keep updating it during your project.

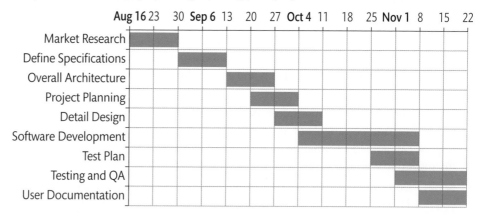

▸ **Figure 5.2** A Gantt chart for an engineering project

▸ **Critical path analysis (CPA)** – a method of scheduling the activities in a project so that the project can be completed in the most efficient way. In CPA you set out all the activities needed for a project, the time allotted for each and the relationships between the activities (e.g. whether they must be completed sequentially or can run in parallel). You then draw diagrams to organise these activities so that the project can be completed in the shortest time possible (see **Figure 5.3**). You can use CPA to reorganise a project as it progresses, to keep it on track.

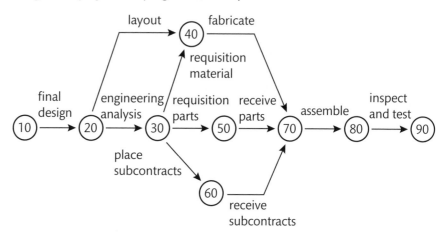

▸ **Figure 5.3** An example of critical path analysis applied to an engineering project

There is free software available for both of these time-planning tools, but you need to check that any software you download from the internet will suit your particular requirements and skills. Discuss with your tutor the best type of time plan to use for your project.

Project contingencies

You also need to plan for contingencies in your project. What will happen if a machine breaks down or equipment malfunctions? Planning for contingencies is a bit like planning a journey: you may know the quickest route, but if you hit roadworks and have to change course, you need to have an alternative plan for what to do.

Two things that people often include in a project plan are:

▶ a contingency fund – an amount in the budget that is put aside in case of needs arising from unforeseen events

▶ time buffers – a little extra time allotted at each stage, in case something takes longer than expected.

Thinking about potential problems before they happen, and having contingency plans in place, could make a huge difference to whether or not your project is completed successfully on time and within budget.

Project constraints

Earlier, you learned how legal constraints can affect some projects. You need to understand any constraints that could affect your project, and plan how to avoid, overcome or work within them. For example, if you were intending to injection-mould a copy of an automotive badge, you may discover that the design is protected by copyright. Knowing this means that you can plan to change the design so that it is not just a straight copy.

The three major constraints to all projects are often considered to be time, cost and scope. However, there are also other types of constraint that you should bear in mind.

▶ Time constraints – In most development projects, time is a major constraint. For example, motor vehicle manufacturers often give themselves a time limit of two years to develop a new model of car. Remember that in industry, time is money.

▶ Budget constraints – Few companies throw a bottomless purse at new developments. Developing products is a risky business and many proposed projects will fail – which means money wasted.

▶ Scope constraints – 'Scope' refers to the range of features that a project encompasses. If you increase the number of features of a product, you will increase the cost and probably the development time.

▶ Sustainability – What is the shelf life of a product? Is it a throwaway item? Can it be recycled? Will it biodegrade? Will any of the materials or processes involved damage the environment?

▶ Ethical constraints – In our interconnected world, there are many different ways to obtain raw materials, get processes done and find the labour needed for a project, but some of these will not be ethically acceptable, to the company or to its customers. Project planning will need to take into account the boundaries of ethical practice.

▶ Cultural constraints – Most products manufactured nowadays are intended for a worldwide market, and will be marketed, sold and used in many different countries and cultures. A company needs to check carefully that its products, supporting information and advertising will fit in with and not offend people of these cultures.

▶ Legal constraints – If a design is protected by patent or copyright, using it is an offence that could bankrupt a company. It is very easy for an independent inventor or a small business to fall foul of patent and copyright protection laws.

Tools for monitoring

Logbook

Your logbook will be the main tool you use to monitor your project. For your logbook to be useful, you need to update it often and record a wide range of information, including:

▶ the options you came up with and the reasons for deciding on your project choice

▶ notes on any key meetings, discussions and group activities you were involved in as part of your project

▶ details of the different iterations of your solution as the project progresses

▶ things that have gone well, and why you think they went well

▶ any problems you encountered, such as a lack of resources or delays due to illness

▶ any technical support you received – why you needed it and how it helped

▶ what you would do differently another time.

In an important project like this, which you have to work on in parallel with other activities, it is easy to forget things, so it's a good idea to maintain records on a weekly basis.

Unless you are given specific instructions by your tutor, you need to decide in what form you will keep your logbook. You can use a notebook, sheets of paper collected in a binder, a word processing program on a computer or a notes app on a tablet – just make sure you can print out the evidence for your portfolio.

The main purpose of your logbook is to track the progress of your project, from planning through implementation to final output. Although it is a working document, perhaps containing notes, dates, sketches and downloads, your logbook will form a substantial part of your final portfolio, so you should make sure it is complete, up to date and as neat as possible.

Progress against the plan, milestones and modifications

To successfully complete a project, you need to:

1 work out which tasks you need to do

2 plan the sequence that you will do the different tasks in

3 identify the date by which each task needs to be completed

4 monitor the tasks and their completion times, to make sure you are progressing as you should against your project plan.

If you have created a Gantt chart as part of your planning, you will be able to see your schedule clearly and know when you need to do what.

Working out the key **milestones** in your project will also be helpful. Some stages or activities will be more important than others, or will have a greater significance for the rest of the project. For example, you may need to finish the drawings by a certain time if you are going to meet your deadlines for the next stage. Showing the milestones clearly on your Gantt chart or critical path analysis diagram will help you to stay on track.

Very few projects are completed without **modifications**. Modifications are changes you make as the project progresses. These could be changes to a plan or drawing, 'tweaks' to a design or even the moving of a milestone. You should carefully record each modification you make in your logbook – these notes will become valuable evidence for your portfolio.

Here is some further advice on planning your project:

▸ A plan is no more than a plan, unless you use it – A Gantt chart can be a useful tool, but it will just be wishful thinking unless you actually use it to plan your work and monitor your progress. As you move forward, you need to adjust and update your Gantt chart, recording the reason for each modification in your logbook.

▸ 60 hours is not very long, and 30 hours is even shorter – You have a total of 60 hours to spend on this unit, and it is recommended that you spend 30 hours of that time on the practical part of your project. At first this may seem a long time, but it will go far more quickly than you think. Time is a major constraint for all projects, and many industrial projects run over their allotted time, so beware.

▸ Make sure that your timescales are achievable and realistic – Your best source of advice here is your tutor. Ask your tutor to look over your time plan and give you feedback on whether the schedule is realistic.

▸ Build in contingencies – You cannot expect every piece of apparatus to be available when you need it, or every process to run just as you would like. Think about what would happen if you fall behind at each stage. Are there ways that you could manage these risks in advance?

▸ Remember that it's not all about the end product – In this unit you will be assessed mainly on your research into possible engineering solutions and their feasibility, and your development and refinement of one chosen solution, including your planning of the project and your management of time and risks. The final product, service, system or process is a vehicle by which you demonstrate how you went through the project process.

▸ Discuss your plans with your tutor – Your tutor will have a great deal of knowledge and experience of the project process, and can identify potential bottlenecks for you.

▸ Order well in advance – If you need to order components or materials that are not available in your college, make sure that you give your tutor plenty of notice. Order systems are notoriously slow, and you don't want to be stuck waiting for a crucial part or tool. If possible, put in place alternative plans.

Tutor monitoring and peer reviews

Your tutor is your greatest resource. Your tutor will not let you proceed until they have approved your project ideas and plans and made sure they are viable, practical and safe. Your tutor will also review your project at set periods, which you may request to coincide with your milestones. Listen carefully to your tutor's advice and feedback.

Your peers – friends, classmates and colleagues – are a wonderful resource too. You can ask them to check your ideas and informally review your plans before you see your tutor. As your project progresses, your peers can give you support and encouragement – and you can return the favour.

B2 Risk and issue management

Management of **risks** and **issues** is crucial if you are to avoid a crisis. If you have to spend time and effort dealing with a crisis, it means precious time taken away from your project. By managing risks and issues well, your project will run smoothly and be more likely to succeed, and you can gain a competitive advantage.

Key terms

Milestone – an important point, stage or event in the progress or development of something.

Modification – a change or alteration, usually to make something better.

Risk – a future event that could adversely affect project processes or outcome.

Issue – an event that is currently having an effect on project processes or outcome.

Link

See Unit 4, Section A3, for more information on competitive advantage.

Risks and issues can come in many forms, and need not simply be associated with workshop activities. They can also relate to finances, health, knowledge or resources.

Risks and issues can be measured in terms of:

▶ severity – low, medium, high and extreme

▶ probability of occurrence – unlikely, likely and very likely

▶ expected impact on project – minor, moderate and major.

There is a formula for calculating the severity of a risk or issue:

$$\text{severity} = \text{probability of occurrence} \times \text{expected impact on project}$$

The resultant risk and issue severity is shown in the matrix in **Figure 5.4**.

probability of threat occurring	very likely	medium	high	extreme
	likely	low	medium	high
	unlikely	low	low	medium
		minor	moderate	major
	expected impact on the project			

▶ **Figure 5.4** A severity matrix maps the probability of a risk or issue occurring against its impact on the project

Management options

Prevention

The best option is always to stop problems and issues arising in the first place – and risk analysis can help you achieve this. You need to plan in advance. For example, if you are going to need a certain piece of hardware at a specific time, find out how many other people will want to use it. Can you work together? Will you be able to access the equipment at another time?

Reduction

If you cannot prevent a problem, you may be able to reduce the likelihood of it occurring or decrease its impact. It is important that in all your activities you reduce the risk of accident or failure as far as possible. For example, if you are going to need a certain piece of equipment at a time when it might be unavailable, what alternatives could you use?

Acceptance

This means doing nothing about a risk or issue yourself, but perhaps passing the risk on to a third party. The financial loss incurred from not completing a task or project by the deadline is often greater than the cost of paying staff overtime to get the job done on time.

Reflect

Consider a project you have completed at home, work, school or college, however small, such as mending a bike. List the risks and issues involved in this project. How did you manage these?

B3 Technical specification

After planning your project, setting up monitoring processes and considering how to manage risks and issues that might arise, you need to create the **technical specification** for your project – often referred to as the **product design specification** or **PDS**. For example, an electrical appliance will have its technical specification provided in the user manual or as a separate document. **Figure 5.5** shows a sample PDS for a computer.

	CPU	Intel Huron River Core i5 2467M / Core i3
	Operating system	Genuine Windows® 7 Professional
	LCD	11.6″ HD, 400nit, 16:9, PLS (view angle, >= 170 degrees)
System configuration	GFX Memory	Intel HD Graphics 3000 2GB / 4GB DDR3 1333 MHz
	Storage	64 / 128 GB SSD
	Webcam	Front : 2.0M Rear : 3.0M
	WWAN	HSPA+ (802.11 bgn)
	PAN	BT 3.0
	Battery	40Wh (up to 6 hours – based on Mobile Mark test)
Dimensions	Thickness/ weight	12.9mm / 860g (0.5″ / 1.89lbs)
I/O	Input	Capacitive Touch + Digitizer / AF Coating, Durable Glass
	Ports	1 x USB 2.0 (Type A), m-HDMI, Audio(in/out), micro SD slot
	Dock connector	Interface: USB 2.0, Gigabit Ethernet, DC-in, HDMI
Security	Keyboard TPM	Bluetooth keyboard (optional) Yes (optional)

▶ **Figure 5.5** Example of a product design specification

Key terms

Technical specification – detailed description of the technical requirements for a product, service, system or process. This will become the reference point for the project design.

Product design specification (**PDS**) – a document detailing the requirements in various areas that a product, service, system or process must meet.

The PDS that you draw up should:

▶ set out all parameters for an artefact

▶ provide a base for all drawings

▶ provide a base for design, manufacture, service and satisfying regulations

▶ determine what your product, service, system or process will do.

A PDS is a live document, and you will adapt and modify it many times. Expect your PDS to evolve as you work your way through your project. To get the most from your PDS, you should:

▶ get input from all parties involved with the project.

▶ start simply, with just the basic information – Your PDS will grow and evolve as your project develops.

▶ remember to answer the obvious questions – What size is the product? What is its weight?

▶ assign values and units to all the parameters – e.g. 12 mm, 4 A or 1200 rpm.

▶ be specific where you can – Don't just say the product will run off batteries; say it will run on less than 5 V DC.

▶ give an estimate if you are not yet sure – If you are still to make a final decision, provide an estimate (e.g. 2 kg approx.) to begin with; then, once you're sure, remember to put in the final values.

Tip

Use the technical specification checklist to ensure that your PDS includes all the necessary details.

A technical specification contains the requirements for:

☐ function and features

☐ interfaces (physical, software, human and electrical/electronic)

☐ materials

☐ tolerances

☐ standards

☐ security

☐ environmental impact and sustainability

☐ operational conditions

☐ process capability

☐ reliability

☐ capacity (current and future)

☐ maintenance

☐ performance.

Case study

The Airbus A380

Developing the Airbus A380 was a major design project. It took nearly 20 years to get from the initial idea to a final product. If the technical specification for the Airbus A380 covered the complete aircraft, it would be huge. Instead, it is broken down into smaller sections, with a technical specification for each of the different parts of the aircraft. While part of the information is classified, some of it is in the public domain.

Go to the A380 page on the Airbus website and look at 'Dimensions & key data'. Read through the information there.

▶ It took many years to get the Airbus A380 off the ground

Now look up the website of Rotary watches. Choose one of the watches, then find and download a product specification for it.

Check your knowledge

1 Who are the two specifications aimed at? What makes you think this?

2 Is the information in each of these specifications enough for use as a technical specification? If not, why not?

3 Could the Airbus or watch be manufactured or serviced from the data supplied? If not, what is missing?

B4 Design information

The documentation for your project also needs to include information on how you designed a solution to the problem. There are many different tools you can use to design your solution.

Design tools

▶ **Engineering drawings** – You may create these as part of a mechanical or an electrical project, for example. Engineering drawings can be produced manually or by using computer-aided design (CAD) packages. You can use CAD software to make 2D drawings, 3D models or circuit diagrams (see **Figure 5.6**).

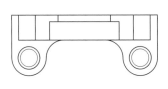
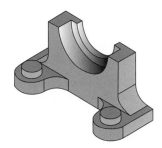
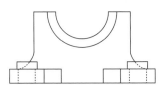

▶ **Figure 5.6** Example of engineering drawings produced using CAD software

Link

Computer-aided drawing is covered in Unit 10. See also Unit 2, Learning aim B.

▶ Simulations – You can use software to simulate pneumatic circuits, hydraulic circuits or electrical and electronic circuits.

▶ Physical modelling – You may find it helpful to create a physical **prototype** of your product. You could do this by **3D rapid prototyping** (also known as 3D printing or additive manufacturing) or by creating mock-ups in wood, cardboard or other modelling material. Some people lack good **spatial awareness**, and many of us find it hard to picture what a 2D drawing will look like in 3D – so modelling is a valuable aid. For marketing people and customers, 3D models are often better representations of a product than engineering drawings.

▶ Processes or computer programs – You can create **flow charts** and make operation sheets to help you plan your project.

▶ Other resources – These could include data tables, engineering formulae, pseudocode and outlines of key algorithms.

The best tools for you will depend on the nature of your project, as well as your own skills and preferences.

▶ Rapid prototyping can bring your design ideas to life for others

Key terms

Engineering drawing – a drawing that conveys technical information using standard layouts, projections and symbols.

Prototype – a first or preliminary version of something, from which other versions will be developed.

3D rapid prototyping (additive manufacturing) – a method of synthesising a 3D prototype, where material is added layer by layer to build an object.

Spatial awareness –the ability to visualise a flat 2D drawing as a 3D object.

Flow chart – a diagram consisting of boxes and arrows that shows a workflow or process.

Safety, sustainability and culture

An important criterion that your design must meet is that it should be safe – for yourself, for other people and for the environment.

Many different regulations may be relevant here, from national health and safety legislation to rules within specialist areas of the engineering sector. Find out which regulations are pertinent to your project, what you need to plan for and what you must monitor. Remember that you have a duty of care to yourself and anyone who will come into contact with the process or output of your project – any work that you engage in must be safe.

Link

Health and safety requirements and risk assessment are covered in Unit 2.

Tip

Make sure that the product, service, system or process you are designing is safe. Carry out a health and safety risk assessment before you do any work that presents a risk of injury or ill health.

Don't forget to take into account environmental and **sustainability** considerations relevant to your specialist area of engineering, such as recycling waste and disposing of oil and polymers.

Consider the cost and availability of materials and manufacturing processes for the potential solutions you are exploring. Some materials may be difficult to source, and some processes may be beyond your budget. Try to avoid using a material for which the supply might diminish over the life cycle of the product.

Also, think carefully about the market for your product. Is there a perceived demand for what you are doing? Who would be your likely customer? Has something similar been done before?

Remember that there are diverse cultures, attitudes and beliefs in the world. If your product or service is to be sold or operated worldwide, you will need to design variations tailored to different countries. For example, an item that works well outdoors in the UK is less likely to be successful in Dubai or Alaska. There is no need to make a different version for every country that you might market your project to (you will not have time), but consider what kinds of adaptations you would need to make.

Key terms

Sustainability – the endurance of systems and processes in the environment; the maintenance of a balance in natural systems by avoiding the depletion of resources and reducing the generation of waste.

Test plan – a document detailing the objectives, target market, personnel and processes for a specific test of a software or hardware product.

Test plans

Don't leave things to chance – you must put in place a **test plan** to monitor your project and ensure that it meets required standards. If your PDS specifies a particular standard from the British Standards or the International Organization for Standardization (ISO), your test plan must reflect this standard. For example, if your project involves welding and is safety-critical, you will need to run tests to the relevant welding standard for non-destructive testing.

Test plans may include:

▶ inspection sheets containing actual sizes, tolerances, pass/fail results and instrument used

▶ destructive and non-destructive testing, such as nick break tests

▶ material testing

▶ tensile testing

▶ maintenance checks

▶ structural checks, for example on aircraft

▶ data recorded from data loggers, such as in vibration checks.

It is only necessary to link to a recognised standard where appropriate.

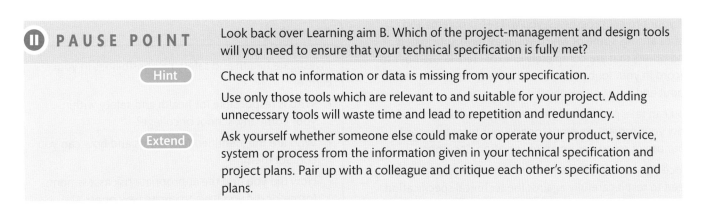

PAUSE POINT Look back over Learning aim B. Which of the project-management and design tools will you need to ensure that your technical specification is fully met?

Hint
Check that no information or data is missing from your specification.

Use only those tools which are relevant to and suitable for your project. Adding unnecessary tools will waste time and lead to repetition and redundancy.

Extend
Ask yourself whether someone else could make or operate your product, service, system or process from the information given in your technical specification and project plans. Pair up with a colleague and critique each other's specifications and plans.

Assessment practice 5.2

| B.P3 | B.P4 | B.P5 | B.M2 | B.M3 | B.D2 |

For your engineering problem and chosen solution, implement project-management processes to develop a coherent design solution.

The project-management processes that you implement should include:

- a project resource and time plan in Gantt chart format, with a critical path analysis of dependent activities, and with appropriate milestones identified
- analyses of contingencies and constraints
- documentation for tracking and monitoring progress of the project
- a project risk and issue log.

By applying these project-management processes:
- create a technical specification to meet customer requirements
- produce detailed and coherent evidence supporting the design solution for your engineering problem, including a test plan where appropriate.

Plan
- What is the task? What am I being asked to do?
- What information will I need to collect? What tools and resources can I use to help me complete the task?
- How confident do I feel in my own abilities to complete this task? Are there any areas I think I may struggle with?

Do
- I know what I'm doing and what I am aiming to achieve in this task.
- I can identify the parts of the task that I find most challenging and devise strategies to overcome those difficulties.

Review
- I can explain what the task involved and how I approached each stage of the work.
- I can explain how I would deal with the hard elements differently next time.
- I can identify the parts of my knowledge and skills that could benefit from further development.

C Undertake the solution for a specialist engineering project and present the solution as undertaken in industry

The final stage of the project is when you put all your plans into motion and finalise your portfolio. If your solution doesn't quite work as expected, don't worry – this can give you just as many (if not more) opportunities for reflection and evaluation than a successful project.

C1 Undertake and test the solution to the problem

Whether your project involves developing a piece of software, building a machine or designing a maintenance plan, your specialist skill set comes into play here.

As your project progresses, you will see problems and issues arising, some of which you may have anticipated and others that are completely unexpected. Remember to record in your logbook how each problem or issue came about and how you resolved it.

You can take photographs to provide a visual record of how your project develops, or ask your tutor to record your use of resources on an observation document.

Once you have implemented the complete solution, you need to test it carefully against the technical specification.

Use of project-management processes

The effects of all your preparatory work in planning resources, time, contingencies, monitoring and risk and issue management will become evident at the implementation stage of your project.

Status reporting and the management of risks and issues

You need to decide how frequently and in what detail and format to report on the status of your project and any risks and issues that have arisen. Are you simply going to count every single risk or issue and report the total number in your final portfolio? For example, if you report that 22 issues arose during the implementation of the project, does that actually tell anybody anything useful? It may help to think of your project as being undertaken in an industrial setting – what level of detail would the project manager want to see? They would probably want to know the severity of each risk or issue and its impact on the project development. You may choose to summarise only the major risks in a graphical form, for example, by using bar charts with supporting documents.

Safe use of resources

Always keep safety in mind while undertaking your project.

Think about the resources you have used (typically machines, workshops, tools and consumables). Have you:

▸ conducted risk assessments for each type of resource?

▸ used tools safely?

▸ operated machinery safely?

▸ asked for assistance when you were unsure of how to operate something?

If you are producing an item with a computer numerical control (CNC) machine, ask yourself – have you:

▸ checked your program?

▸ had a 'dry run' on the CNC machine?

▸ produced a service plan?

▸ followed the correct isolation procedures?

▸ checked whether work permits are required?

Research

Investigate health and safety responsibilities in your centre:

- Who is responsible for health and safety within your company, school or college?

- Who are the identified first-aiders and how can you contact them?

- How did you find the appropriate risk assessment forms, or did you have to create new ones?

- What are the evacuation procedures?

- Where can you find safety equipment, such as fire extinguishers and first-aid kits?

Troubleshooting

When problems arose in your project, how did you **troubleshoot** them?

Key term

Troubleshoot – to work out in a logical way what and where the problem is and suggest how to resolve it.

Here are some standard troubleshooting procedures that could be useful for your project too.

▸ **Half-split** – With this method you split the system or process into two halves, and test them to determine in which half the fault lies. Then you take this half and split it again, and repeat your tests on these smaller sections to see which one contains the fault. The splitting and testing steps are carried out repeatedly until you can pinpoint the location of the fault. This method is also called 'divide and conquer'.

▸ **Cause and effect analysis** – First, identify the problem and write it in a box at the end of a line (called the 'spine'). Next, consider the principal factors that could have contributed to the problem, such as issues with materials, equipment or personnel, and use these to label lines that branch off the spine. For each of these major branches, think of possible causes of the problem that are related to the factor (e.g. shortage of resources, malfunctioning equipment) and show these as shorter lines coming off the main branches. If a cause is still quite large or complex, try to break it down further. You will end up with a 'fishbone diagram', and you can then analyse it to determine or prioritise which cause(s) to deal with in order to resolve or alleviate the problem.

▶ **5 Whys** – This is a technique for discovering the root cause of a problem. After stating the problem, you use a series of questions beginning with 'Why' to dig deeper into the underlying causes of the problem. Each question should build on and be more specific than the preceding question, and the answers should enable you to focus in on the most important cause. For many problems, five questions are enough to uncover the root cause, but sometimes fewer or more iterations are needed. If you search for '5 Whys' on the internet, you will find many examples of how this technique is used to troubleshoot.

Fitness for purpose

Is the outcome of your project **fit for purpose**? The International Organization for Standardization defines (in ISO 9001) checking fitness for purpose as 'part of quality management focused on providing confidence that quality requirements will be fulfilled'. In practice, it means that your product, service, system or process should comply with the specification. If the specification is correct, the product should be fit for the purpose intended. This is an important reason why your PDS must be accurate and workable.

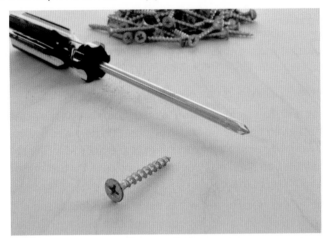

▶ Fitness for purpose does not have to be complicated

> **Key terms**
>
> **Fit for purpose** – good enough for the job it was intended to do; fulfils the specifications or requirements.
>
> **PDCA** – a four-step quality-management process where you plan, do, check and act in a continuous cycle to continually improve a product.

Testing methods and fitness for audience

How can you demonstrate that your final product, service, system or process is capable of working correctly? Is it 'fit for audience' – that is, will the customer find it user-friendly? How might you check that?

Think about what supporting documentation you will need. For example, if you have produced a physical product on a machine, you should have an inspection document to show that it functions correctly. Look back at the section on test plans in Learning aim B to remind yourself of the testing and inspection methods that can be used.

Service and maintenance plans can be checked using a **PDCA** approach. This iterative four-step management method is used in business and industry to control and continuously improve processes and products.

> **Link**
>
> The PDCA approach is discussed in Unit 4, Section C1.

You should now be approaching completion of your project. How well prepared is your portfolio? What else might you need to include?

Look back over Section C1 and make sure that you understand what evidence you need to present in your portfolio. Remember to record in your logbook when and where you used troubleshooting methods and how successful your chosen methods were.

To achieve the higher grades, your portfolio needs to concisely and logically demonstrate that the solution is fit for audience and purpose. Consider carefully what evidence must be included, given the nature of your project, and what can be simplified or omitted for conciseness.

C2 Demonstration of relevant behaviours

As your project develops, you will be assessed against various criteria. These assessment criteria concern not just the end 'outputs' – the product, service, system or process you design and the portfolio you create – but also your behaviour during the process.

'Behaviour' here doesn't just mean being 'good'! It is about showing that you can act professionally, responsibly and responsively throughout your project. So, for example, you need to show that you are able to link the different aspects of your project together, and also that you kept calm when something didn't go as smoothly as you wished.

Here are the relevant behaviours you need to show.

▶ Time planning and management – You need to complete all the activities of your project within a given time frame and in an appropriate order. Did you plan your time well? It would be unusual if everything went exactly according to your original schedule. How did you handle any delays or time conflicts that arose? What changes did you have to make?

▶ Communication and literacy – As well as being able to follow and implement instructions accurately and interpret documentation, you need to communicate effectively and professionally with others, including your peers, technical and academic staff and any outside agencies. Your portfolio should be aimed at an audience with at least the same knowledge and skills as yourself, but you still need to think about the best way to communicate for different purposes and to different groups throughout the project. Should you write, call or talk with people face to face? Would someone who does not speak English as their first language be able to understand you? When are drawings and diagrams a better way to explain? Don't forget to include the obvious here – if you did it, record it so that others will know.

▶ Commercial and customer awareness – You need to show that you understand the market for your solution, at both the commercial and the customer level. Does your product, service, system or process meet the brief you were given or had set for yourself? Is it fit for purpose? Will customers find it user-friendly?

▶ **Observable emotions** linked to successes and issues during project development – These are the impressions that others have of your attitude when you are experiencing successes or problems. Think about how you reacted, particularly when things went wrong. Did you get angry? Did you examine the issue and seek a solution? What did you find frustrating? How did you behave when things went well? Did this motivate you to move forward with your project? What are you most proud of?

▶ Individual support – It is important to recognise and acknowledge the people who helped you throughout your project, from teaching and support staff to peers. What help did you get along the way? Why was it this person that you asked? Why was their help important at that point?

Key term

Observable emotions – emotions that are displayed that others can see, rather than internal feelings.

C3 Present a solution to the problem

Evaluation of the final solution

It is important to look back and decide if the final outcome of your project meets the criteria in the initial brief or specification. In industry, a mismatch between the initial criteria and the final outcome would generate a management inquiry and – at a minimum – an action plan would need to be drawn up, outlining timescales, responsibilities and extra resources needed. In a worst-case scenario, the project might be cancelled altogether.

Ask yourself these questions:

▶ How well was the project organised?

▶ What were the issues that arose?

▶ Were resources used effectively and efficiently?

▶ Does the product, service, system or process function as it should?

▶ Does it meet its planned specification?

▶ Is it fit for purpose?

Collating a portfolio

Finally, you need to collate a project portfolio to cover all aspects of your project from start to finish. This is a crucial part of the project. Your portfolio will present your evidence, showing all the stages and iterations, the processes you have followed, the choices you have made, the problems you have encountered and the solutions you have devised along the way.

Make sure that you keep your portfolio updated, and capture evidence as soon as you can. The portfolio gives you a chance to showcase your skills and technical expertise, as well as to evaluate lessons learned and consider any improvements you could have made, particularly with reference to the success or fitness for purpose of the final product, service, process or system.

Take time and care with your presentation. Your portfolio may be sent out to an external Standards Verifier for checking, and this will be the only information they have about you and your project. Your portfolio should be a self-contained, well-structured document that you would be proud to present to a future employer.

What your portfolio should contain

Here is a list of the elements your portfolio should contain. You may wish to use this as the contents page for your portfolio, or as a checklist for the final version.

☐ Thematic title or initial idea – This was probably given to you by your tutor.

☐ Research and clarification of the problem.

☐ Possible solutions and constraints – Explain how you started to solve the problem.

☐ Initial specification of alternative technical solutions.

☐ Feasibility study – Examine each of the alternative solutions and assess whether they are workable.

☐ Technical specification – This is the PDS for your chosen solution. Have you recorded all the detailed information needed, such as actual voltage and amperage, the type of battery required (nickel cadmium, lithium ion) and so on?

☐ Project-management documents – including plans and a risk and issues log.

- ☐ Logbook of events – such as a diary, outline sketches, notes and records. It is important to maintain a logbook from the start of the project and keep it up to date; do not try to reverse engineer a logbook, as this is hard work and you are bound to miss things.
- ☐ Design documents – such as sketches, engineering drawings, simulations and flow charts. Include your initial ideas, such as sketches of different parts of the potential product, to illustrate how the project evolved.
- ☐ Artefacts for a product, service or process – such as a prototype product, computer program, pneumatic or hydraulic circuit, electronic circuit or process demonstration.
- ☐ Test documentation – such as test results, inspection reports, customer feedback, photographs and video.
- ☐ Peer reviews and tutor monitoring.
- ☐ Conclusion on the success of the solution against the project theme and initial idea – In your conclusion, look back at the initial idea or project brief and assess whether the outcome was successful or not, discuss the possible reasons why, and identify directions for improvement.

Project records

Project records are vital, for your course project and in industry. As a product is developed through multiple iterations, records enable designers to look back at the rationale for each decision made about the specification, the creation of prototypes, the types of testing carried out or the quality control and quality assurance processes that were implemented.

Product recalls can be initiated from a company's records. Inspectors can look back through product development records to assess whether a fault should have been identified and rectified at a prototype stage. This can have serious consequences for the manufacturer.

Theory into practice

Keeping clear and organised records for your portfolio is good practice for working in industry, where traceability is an important concern. Many products are identified by serial numbers, and information about customers is collected at the point of sale so that they can be contacted directly in case of product recalls. This avoids high-profile coverage of a general recall in the media, and is far more effective than placing a low-key advert in newspapers, which could easily be missed by customers?

Reflect

Think about what you have gained in doing your project – the technical knowledge you have acquired, the practical and thinking skills you have used, the new ideas you have generated and analysed, the solutions you have designed and refined, the experience in applying project-management and testing methods that you have built up.

In your future career you will make use of all of these skills and processes on a day-to-day basis. You will also need to keep learning and developing as technology advances.

Assessment practice 5.3

C.P6 | C.P7 | C.M4 | C.D3

Implement and develop the solution you designed for your engineering problem, using project-management processes to control and monitor progress and to resolve any risks and issues.

You will need to:

- Use project-management processes to develop your solution, aiming for optimal efficiency and effectiveness. This includes monitoring progress against the plans and specification, as well as anticipating and resolving risks and issues.
- Test the solution for fitness for audience and purpose, and record and analyse the results.
- Demonstrate effective and professional behaviours throughout the development of the solution.
- Collate a portfolio of evidence for the project and prepare a conclusion. Justifications should be given of any modifications and refinements made, and of any improvements to the processes and behaviours that could be applied next time.

Plan

- What is the task? What do I need to achieve?
- What tools and resources can I use to help me complete the task?
- How confident do I feel in my own abilities to complete this task? Are there any areas I think I may struggle with?

Do

- I know what I'm doing and what I am aiming to achieve in this task.
- I can identify the parts of the task that I find most challenging and devise strategies to overcome those difficulties.

Review

- I can explain what the task involved and how I approached each stage of the work.
- I can explain how I would deal with the hard elements differently next time.
- I can identify the parts of my knowledge and skills that could benefit from further development.

Further reading and resources

Books

Various textbooks cover project development, but many are quite advanced and may not always be suitable for this project. Your library may have books and magazines that focus on specific technical areas, such as graphics software, electronics, maintenance, CNC and 3D printing. Companies that supply CAD/CAM software also make available a range of learning materials such as instruction manuals and tutorials.

Bond, W.T. (1995) *Design Project Planning: A Practical Guide for Beginners*, Prentice Hall.
Explores basic project management, linking it to the design process and modelling.

Plummer, F. (2007) *Project Engineering: The Essential Toolbox for Young Engineers*, Butterworth-Heinemann.
Aimed at the young engineer who wants to enter or who has just started in project management; written with a work-based approach.

Videos

Some examples of online resources in video form:

Mechanical engineering project topics: **www.youtube.com/watch?v=ibE626hR0sk**

Mechanical engineering mini-project topics: **www.youtube.com/watch?v=jMwrkB4JQ4M**

Websites

Examples of websites that offer ideas for projects:

IEEE Real World Engineering Projects library: **www.realworldengineering.org/library_search.html**

Mechanical engineering science fair project ideas: **www.sciencebuddies.org/science-fair-projects/Intro-Mechanical-Engineering.shtml**

Electronics projects for engineering students: **www.electronicshub.org/electronics-projects-ideas/**

Nevon Projects project ideas by category: **http://nevonprojects.com/project-ideas/**

Planning maintenance projects: **www.machinerylubrication.com/Read/1330/planning-maintenance**

THINK ▶FUTURE

Reham Abdulrashid Project manager

I have been working as a project engineer for 12 years, and managing a department for three years, developing projects for the aerospace industry. I have worked on a wide variety of projects, from aerofoil design to applications of vortex generators. My specialism is airflow problems for transonic aircraft – mainly commercial aircraft, but also some smaller private jets.

My responsibilities include managing a small team of aerospace engineers, each with skills in different areas, such as design, electrical, mechanical, stress and aerodynamics, software and systems. I support them in their technical and professional activities, and make sure that all projects use the latest technologies. As manager I also have overall responsibility for delivering aerofoil design and development tasks to schedule and on time, and I'm responsible for planning and managing projects and resources, particularly financial resources.

My work involves me working closely with other professional engineers from all aeronautical disciplines, so it's really important for me to have a working knowledge of what they do, even if I don't have their expert knowledge. The ability to work effectively in teams is a key skill to develop as an engineer, which helps in every type of project. Modern projects are often so complex that no individual can have all the technical expertise that's needed. When we develop new aircraft, it can take many years, with numerous iterations – and we have to keep up with technology that is changing all the time. Think about a modern aircraft – many of its components are made of advanced composites and polymers, but not so long ago they would have been made from aluminium.

I have to keep up with changing rules too. As an aeronautical engineer, I am bound by legislation, regulations and codes of practice, because the lives of flight crew, maintenance staff and passengers are at stake.

Completing my BTEC aeronautical qualifications at college helped me understand what information I need to know and where to find it. Although I am well qualified, it is still very important to continue learning and stay up to date with changing technology, so I regularly attend continuing professional development (CPD) events. That looks good on my work profile too – it's a competitive world out there!

Focusing your skills

Using project-management processes to develop a project

Project development is a multi-skilled operation involving these essential steps:
- identifying and understanding a problem
- generating possible solutions and thinking about constraints
- investigating feasible ideas – Make sure your initial ideas are reasonable and not too outlandish; sometimes people have 'eureka' moments, but most viable ideas are developed in small steps

- resource planning and time planning – The next time you catch a flight, think about how much planning has gone into getting your luggage from the check-in desk to the aircraft
- formulating design ideas into a coherent design and creating a specification based on the customer's requirements
- developing and implementing a solution, while controlling and monitoring the progress of the project
- managing risks and issues
- evaluating the outcome of the project, reviewing whether the customer requirements were met.

Getting ready for assessment

Billy has recently completed a BTEC National in Aeronautical Engineering and is now studying for a Higher National Diploma (HND). He attended college part-time while working in an engineering company on an apprenticeship.

How I got started

I started thinking about my project really early – it was one of the things I found particularly appealing about the course in the first place, so I wanted to make the most of it. We started properly towards the end of Year 2, but by that time I knew my engineering exams were just around the corner and I had to study for them too. I was glad I got some of the thinking and planning done in advance.

How I brought it all together

I thought a lot about the project I wanted to work on. I discussed ideas with my tutor and my employer. My tutor had given us a theme of aircraft maintenance, and explained that it could cover repairs, servicing or modifications, so I took time to think about the different possibilities. It took me nearly three lessons, but eventually I decided to look at the repairs of aircraft skins. I decided to focus on a light aircraft, as my employer and the local airfield could help me with that.

My first task was research. I used the internet and the library, and talked to some of the fitters at work to find out about the various kinds of damage to the aircraft skins. I followed my tutor's advice to start a logbook at this early stage, and I took photos with my phone for it.

I then realised I'd have to look at existing repair specifications and probably come up with my own, with sketches, timelines and costings for three different solution ideas. I examined the feasibility of each idea and concluded that weight, size and strength were critical to the safety of the structural repair. My tutor showed me how to use objective testing to select the best idea.

I set aside time at home to organise and plan my project, and to keep my logbook up to date. I developed my PDS at home, and made notes on anything I didn't fully understand so I could ask about it in the next lesson.

I created sketches and CAD drawings and made a prototype. The practical part was all done in college, but my employer also allowed me to practise at work. This was very helpful as I was already used to the layout and discipline of an aircraft workshop.

I made sure that each part of the assignment was completed and my logbook included all records of my modifications, iterations, test results and risk management actions. I had plenty of photographs as evidence of development work. Finally, I drew all the documentation together to complete my portfolio. It looked OK, but my tutor had to remind me to check 'How well does it meet the criteria?' I nearly forgot to put in my reflections and behaviour evaluations before I handed in my portfolio. I knew the rules about late hand-ins, as they had been explained carefully when we started at college, so I made sure everything was submitted on time.

What I learned from the experience

I learned a lot about project management. Initially I thought the project would just be about making something, but it went a lot deeper than that. The key to a successful experience was getting ahead of the game, being organised and keeping to the timeline. Of course, it's essential to make a good choice of topic to start with – some of my classmates picked problems that were too difficult and struggled to finish!

Microcontroller Systems for Engineers 6

Getting to know your unit

The skills of designing and developing microcontroller systems are an essential part of the growing field of research and development on the 'Internet of Things' (IoT). In this unit, you will gain an understanding of the different types of microcontroller hardware and the diverse range of peripherals available that can be connected. The aim of this unit is to inspire you to design your own tailored solutions and to provide you with the skills to create them. You will learn how to code systems and realise the desired solution by producing a fully functioning prototype, together with all associated documentation.

How you will be assessed

This unit is assessed under supervised conditions. You will complete an assigned task which can be arranged over a number of sessions.

As the guidelines for assessment can change, you should refer to the official assessment guidance on the Pearson Qualifications website for the latest definitive guidance.

During the supervised assessment you will complete a practical task in which you will develop a prototype microcontroller system to solve a problem. You must complete this task using a computer. Each year a different task will be set by Pearson, for which you will need to follow a standard development process of:

▶ analysing the brief
▶ system design and program planning
▶ system assembly and coding
▶ system testing and recording the system in operation.

All external assessment tasks can be solved using a range of components given in the unit content.

To obtain a Pass you must be able to analyse client briefs, interpret operational requirements and produce plans to test the operation of your proposed system. You will need to produce a structured system which demonstrates your understanding of the relationship between the written program and the microcontroller and peripheral hardware. You will deliver a functional system that can be tested and assessed against the client brief, and you will make

judgements on how well it fits the initial brief. You will also need to follow your set development process and provide technical justifications to support your chosen solution.

For an award of Distinction you will need to show a thorough knowledge and understanding of coding principles and hardware components and demonstrate that you can apply them in both familiar and unfamiliar contexts. You will be able to analyse client briefs to interpret operational requirements and produce plans to comprehensively test the operation of the system under a variety of conditions. You will deliver an optimised system with consideration of enhanced user experience, which demonstrates thorough understanding of the relationship between the program and the microcontroller and peripheral hardware. You will carry out testing of the full functionality of the system in a structured way, taking into account unexpected events, and make critical judgements on how well what you have developed fits the brief. You will follow a structured development process for the entire project and provide technical justifications to support your final solution.

Essential preparation for the external assessment includes:

▶ attendance at all practical sessions
▶ development of programming techniques for your chosen integrated development environment (IDE)
▶ working methodically and accurately
▶ keeping accurate records and documentation.

Assessment outcomes	
AO1	Demonstrate knowledge and understanding of computer coding principles, electronic hardware components and the development process
AO2	Apply knowledge and understanding of computer coding principles, electronic hardware components and the development process to design and create a physical computer system to meet a client brief
AO3	Analyse test results and evaluate evidence to optimise the performance of a physical computer system throughout the development process
AO4	Be able to develop a physical computer system to meet a client brief with appropriate justification

In a modern day urban environment you are unlikely to be more than 20 metres away from a microcontroller system of one type or another. Take some time at home and in class to look for all the things that might include a microcontroller system within 20 metres of where you are. Make a list and add to it throughout the day when you find yourself in different places.

A Investigate typical microcontroller system hardware

With microcontrollers being integrated into devices used in every area of our lives, it is important that you understand the various types that exist and their capabilities. Programmable devices are already used in numerous everyday systems and products, such as car engine control systems, household white goods, wearable and health technology products, and environmental control and process control systems.

One of the most popular programmable devices is a microcontroller, which contains all the internal components of a computer on a single chip and runs a stored computer program to achieve an intended purpose. Microcontrollers are available in a variety of shapes and sizes, with different internal configurations to suit the required application. Some of the advantages of microcontrollers are that they are cheap and small, have low power consumption, are readily available and can be programmed to control products and systems, making them economical for developing solutions. Microcontrollers and their programs form an important part of the rapidly growing 'Internet of Things' (IoT), a network of billions of interconnected physical objects, which is bringing about the next information revolution.

Discussion

All of the systems included in the IoT will have inputs, processes and subsequent outputs. Choose one system used at home (e.g. a washing machine) and another used in class, and make a list of the separate parts of each system – inputs, the process and outputs. Discuss and review this list with your tutor as part of Assessment practice 6.1.

Different devices require different internal hardware specifications; a device such as a keyboard does not require the same specification as the microcontroller within a washing machine. Requirements that will have an effect on the size, cost and usability of any complete system include **processor speed**, memory requirements, input/output pins/ports, **bus width**, need for an internal **ADC** or **PWM**, and power requirements. Cost and availability are important considerations. It is therefore vital that you understand the internal configuration of a microcontroller so that you can select the most appropriate product to integrate into your solution to any given problem.

Key terms

Processor speed – usually refers to the maximum number of calculations per second that the microcontroller processor can perform. Microcontroller processor speeds can be variable and set by the user using a crystal depending on the application. This will be covered later as you build your own microcontroller system.

Bus width – the size of an information channel for accessing or transferring data. The larger the bus the more data can be transferred at any time. An 8-bit data bus can transmit 8 bits of data at any one time and a 16-bit data bus can transmit 16 bits of data at any one time.

ADC (analogue-to-digital converter) – hardware that converts continuous physical quantities, such as a voltage, into a digital number representing the value's amplitude. A common example in a microcontroller system is the conversion of temperature (analogue) into an 8-bit value that can be digitally displayed.

PWM (pulse-width modulation) – a modulation procedure that creates a representation of an analogue signal from a digital output. PWM adjusts the mark-to-space ratio of a square wave, which when connected via an R/C (resistor/capacitor) network will generate an analogue signal. A common application of PWM is to control the speed of a winch motor.

A1 Control hardware

Your tutor will identify a microcontroller that you will use throughout this unit. An example is the PIC16F877 microcontroller, but many others are available. In order to deliver a solution to a problem, you need to understand the basic internal architecture and configuration of the selected microcontroller and any hardware necessary.

Hardware

You need to be able to find, review and understand datasheets, not just for this unit but in every aspect of electronic engineering.

Assessment practice 6.1

Find the datasheet for the microcontroller you will be using for all practical activities throughout this unit in your centre.

Review the datasheet in terms of the microcontroller's use for the application discussed with your tutor (see the Discussion activity on the previous page).

Use the datasheet to identify aspects of the configuration that will have an effect on the cost and size of the microcontroller integrated circuit (IC).

Explain why you have selected and highlighted these aspects of the internal configuration and describe the effects they will have on the cost, size and usability of the IC in any system.

Plan
- What is the task? What am I being asked to do?
- Do I need a full datasheet or would a short one suffice?
- What are the most important aspects of the internal configuration to identify?

Do
- I have found the most detailed datasheets available and any relevant user manuals.
- I know what hardware is required and what inputs/outputs and features the microcontroller provides.
- I can summarise the main aspects of the microcontroller.

Review
- I can relate the information from the datasheet to a particular application of the microcontroller.
- I can analyse the required aspects and suggest how they could affect the cost, accessibility and ease of use of the microcontroller IC.
- I can review the minimum requirements and discuss possible trade-offs that could be considered in order to realise a complete solution.

ⅠⅠ PAUSE POINT

Can you identify the main aspects of a microcontroller and justify the inclusion of a microcontroller in a given design solution?

> **Hint**

Choose one of the microcontroller systems you identified at home (in the previous Discussion activity). List what is required to solve this particular problem and outline the solution you envisage.

> **Extend**

At this stage you will not know how to control or use many of the additional aspects of a microcontroller, such as **interrupts**, PWM and the ADC, but you should know why and when you might use them to achieve a solution. Research each of these aspects for future use and development.

> **Link**

You may revisit this datasheet investigation activity in other units of your course, such as Unit 19: Electronic Devices and Circuits, Unit 21: Analogue Electronic Circuits and Unit 22: Electronic Measurement and Testing of Circuits.

> **Key term**

Interrupt – a signal sent to a specific microcontroller input pin to mark the occurrence of an input event that requires specific immediate attention. An interrupt requests the internal processor to stop the current function and to execute a code associated with input. Interrupts can be given a priority order.

Development environment

Pearson have approved the following **integrated development environments** (IDEs) for programming the chosen microcontroller:

▶ Arduino IDE

▶ MPLAB® IDE (MPLAB® C)

▶ PICAXE® Editor

▶ GENIE® Studio or Circuit Wizard

▶ For illustrative purposes, this unit will use the PIC16F877 microcontroller and the MPLAB® IDE, with embedded C as the programming language.

Get familiar with the IDE

The IDE is the software environment that you will be using extensively throughout this unit, so it is essential that you become confident with the environment and understand the finer points of the system.

> **Key terms**
>
> **Integrated development environment (IDE)** – a software application that provides a set of tools to aid programmers in the development of software. A typical IDE consists of a source code editor, a compiler, automation and simulation tools, and a debugger.
>
> **Compiler** – a program that converts the code created into machine code (1s and 0s) that your microcontroller can interpret. Different compilers are required for different IDEs and microcontrollers.

Step by step: Creating projects in the IDE
10 Steps

1 Review the online resources. Each IDE will have supporting video tutorials to support you in using the system. For example, search for 'Microchip MPLAB®X IDE'.

▼

2 Create a new project.

▼

3 Select Mid-Range 8-bit MCUs (PIC12/16/MCP) from the Family dropdown, followed by PIC16F877.

▼

4 Select the hardware interface tool used by your centre.

▼

5 Select the **compiler** being used.

6 Name your project.
(Never use names such as Project_1, Project_2 etc. Give projects and files relevant names that link them to the output of the project. Remember that names must not include spaces.

▼

7 Create your code file, or attach a piece of code from a shared location provided by your tutor (Program_1).

▼

8 Compile and run the code.

▼

9 Explore the environment. Look at tools to:
• view graphical outputs
• step through your code
• view the memory usage
• view the port input and output values.
Spend time with the IDE and get to know where everything is. Let curiosity be your guide.

▼

10 Create another project with a different piece of code (Program_2).

Worked Example

Following the steps above, create two projects using the code from the program examples given below. The IDE will have an option to monitor the output ports, local variables and other inputs and outputs. You can view these in decimal, binary or hexadecimal form. You will learn more about these formats later, but for now you just need to build the projects.

Code for Program_1	Code for Program_2

<pre>void main (void)
{
    TRISB = 0x00;   // set port B as an output
    PORTB = 0x00;   // clear port B before start
    while(1)
    {
        PORTB = 0x55;   // output to port
    }
}</pre>
	<pre>void main (void)
	{
	    TRISB = 0x00;
	    PORTB = 0x00;
	    while(1)
	    {
	        PORTB = 0xAA;
	    }
	}</pre>

- For Program_1, the output on Port B will be either 01010101_2, 55_{16} or 85_{10}. For Program_2, the output on Port B will be either 10101010_2, AA_{16} or 170_{10}.

The subscript identifies the **base number** of the numbering system. (Numbering systems will be covered in detail later, in Section B4.)

Key terms

Base number – the number of digits used in a numbering system. You have been using base 10 (the decimal system) since you learned to count. It has 10 digits: 0, 1, 2, 3, 4, 5, 6, 7, 8, 9. Base 2 (the binary system) has 2 digits: 0 and 1.

Breadboard – a specially constructed board for physically building and testing prototype electronic circuits without the need for soldering.

Get to know your microcontroller

There are several development boards available to integrate the microcontroller with the chosen IDE (see **Figure 6.1**). Most microcontroller development boards have built-on peripherals, which remove the need to understand how the inputs and outputs interface with the microcontroller. However, in industry you do need to have an understanding of how these interface when building larger prototype systems.

Your tutor will issue you with the chosen microcontroller and your toolbox, which will include basic tools, a **breadboard** and connecting wire.

The next step is to build a basic power supply. For this you need an L7805 voltage regulator, a $220\,\Omega$ resistor and a red LED (light-emitting diode).

▶ **Figure 6.1** Two examples of development boards

> **Link**
>
> Building the basic power supply can be part of the assessment and grading for Units 19, 20, 21 and 23.

▶ Build the 5 V power supply following the circuit diagram in **Figure 6.2** and using the 5 V output to supply power to your microcontroller.

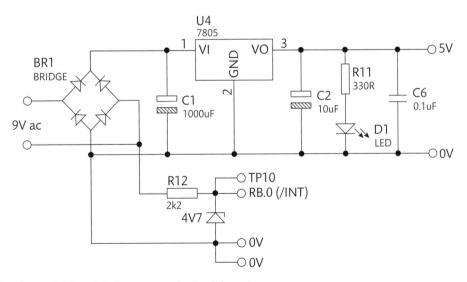

▶ **Figure 6.2** Regulated power supply circuit layout

▶ Now build the clock system, reset switch and interconnection circuitry for your microcontroller. The circuit schematic in **Figure 6.3** identifies all the necessary components.

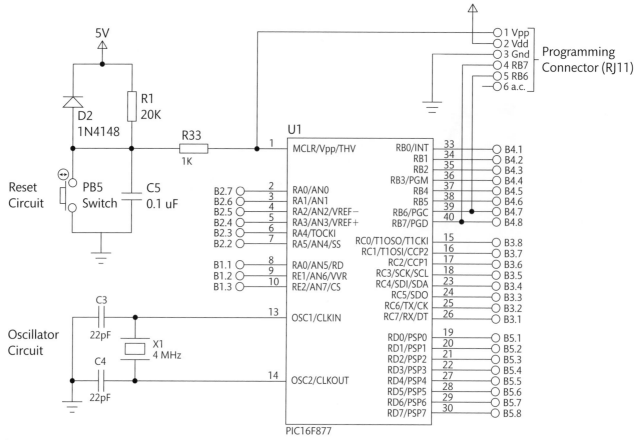

▶ **Figure 6.3** Core microcontroller circuit layout

Just as in your PC, the clock controls the timing within the processor, determining the speed at which instructions are carried out. A faster clock means a faster processor, but all have an operational limit identified on the datasheet.

You will now have a functioning system with many components and integrated circuits. It is very important that you get into the habit of downloading and collecting the datasheets for all the components that you use in any design solution. These will provide a handy reference when building circuits and testing for faults.

Your tutor will issue you with the IDE interface device to enable connection with the microcontroller so you can check that it all works.

▶ Connect up the microcontroller with the IDE, open your first project, and compile and download Program_1. Identify and resolve any building or connection errors.

Use either a **logic probe** or a multimeter to test your system. Check that the output at port B is 01010101_2.

▶ Now try Program_2. The output should be 10101010_2.

If you or other learners are getting compilation or connection errors or the incorrect output, try to help each other identify the mistakes and fix them. Peer learning will be very valuable throughout this unit.

> **Link**
>
> Common errors are described in Section B1: Programming techniques.

> **Key term**
>
> **Logic probe** – a pen-like device that indicates by means of audio and/or visual output whether a logic 1 or 0 is output on a device pin.

> **Discussion**
>
> Using a logic probe or multimeter is not always the most efficient way to identify the state of the output pins of port B. What visual methods could you use to check the outputs without the need for a multimeter or a logic probe?

226   Microcontroller Systems for Engineers

PAUSE POINT

You have now built a functioning microcontroller system, created an online project in the IDE, and developed working code which has been compiled and downloaded to the microcontroller. You will be required to do this for the external assessment, so it is important to remember this fundamental process, which will be key to a successful assessment. Now clear the breadboard and go through the whole of Section A1 again, so that the process becomes second nature.

Hint   Label and identify the different sections of the circuit, clearly identifying the component parts for ease of reference and testing.

Extend   Following on from the last Discussion activity, connect LEDs to create a visual output and/or reprogram the microcontroller to output different binary patterns. Test and confirm that the binary pattern in the code matches the visual binary pattern on the LED outputs. If not, the connections to the LEDs are incorrect.

A2 Input devices

Inputs to microcontroller devices all have their own specific characteristics and methods of operation. They can be broken down into five main areas: user inputs, temperature inputs, light inputs, movement/orientation inputs and presence inputs. You need to examine physical examples of each type and understand the operation and characteristics of the devices so that you can relate theory to practice. An example of an input device is shown in **Figure 6.4**.

▶ **Figure 6.4** Example of an input device to a microcontroller system

Reflect

For each input device that you choose to study, find, download and read the datasheet. Review the contents and create a list of questions for further discussion.

User input devices

These input devices require the user to interact physically with them before anything can be **read** by the microcontroller and a process started. These devices include simple buttons and switches.

Buttons can start, stop or interrupt a process such as a timer, and a switch can turn a process on or off. As you progress through this unit, you will build microprocessor systems and attach several user-type inputs that will affect specific processes to produce solutions to set problems.

Both buttons and switches are designed to generate either a logic 1 or a logic 0 (5 V or 0 V) input to the microcontroller. You have already built a reset switch that creates a logic 0, which is read by pin 1 and resets the microcontroller.

Discussion

Why have a logic 0 for the reset? The reset pin (pin 1) of the microcontroller has the label $\overline{\text{MCLR}}$/Vpp. If you have studied Unit 23 (Digital and Analogue Electronic Systems), you will know that a line above a letter or word represents a logic 0, which means that 0 V is required for the input to function – it is **active low**.

Another type of user input is a straightforward analogue variable resistor (potentiometer) that provides a changeable value to be interpreted by the microcontroller ADC, with separate processes being called upon depending on the analogue input.

Key terms

Read – where a microcontroller receives an input and follows a specific set of instructions depending on the input.

Active low (or active high) – identifies whether an input pin requires a logic 0 (low) or a logic 1 (high) to function.

Temperature input devices

Temperature-type inputs are integral parts of many microcontroller systems such as temperature-dependent home heating systems, refrigeration systems, washing machine systems (controlling the water temperature) and engine management systems.

The most basic type of temperature input is a **thermistor**. These devices are generally split into two types: positive temperature coefficient (PTC) and negative temperature coefficient (NTC) devices. A PTC thermistor gives an increase in resistance with an increase in temperature, while an NTC thermistor gives a decrease in resistance with an increase in temperature.

> **Key term**
>
> **Thermistor** – a thermally sensitive resistor that is manufactured to give predictable and accurate variations in resistance based on the external temperature of the device.

Thermistors can be produced to read temperatures from –500°C to over +3000°C, and they can be integrated with the microcontroller through the ADC. They are considered a very predictable and accurate means of measuring and monitoring temperature.

> **Research**
>
> In 1883 Michael Faraday discovered that the semiconductor silver sulphide had a negative temperature coefficient, paving the way for the development of thermistors. Research this and other contributions that Faraday made to engineering.

There are also specific temperature sensors for monitoring and measuring temperature. These input devices can be separate discrete components or complete integrated circuits. For example, the LM35DZ is a 0–100°C temperature sensor that comes in three versions: a metal can package, a plastic case package and a dual in-line (DIL) integrated circuit (IC).

Environmental sensors are another type of temperature input device. For example, dual packages exist for monitoring temperature and humidity that can easily be integrated within a microcontroller system. As with temperature sensors, the devices come in different packages. Some are complex ICs with built-in ADCs, internal memory and inter-integrated circuit (I^2C) interfaces, such as the SHT25 temperature and humidity sensor.

Light input devices

Light input devices are extremely useful and versatile. They can be used in applications such as street light controllers (so that the lights come on only when it becomes dark), smoke detection systems, outside security lighting, infrared receivers and night vision equipment.

The most common type of light input device is the light-dependent resistor (LDR). This is a discrete component that has a very high resistance in darkness, up to 1 million ohms (1 MΩ), but as it is exposed to light the resistance drops dramatically, allowing current to flow and enabling a voltage drop across the device to be measured by the ADC input if a variable output is required. It can also be used as a simple switch, with an input of 0 or 1 depending on whether it is dark or light.

The other main light input devices are phototransistors and photodiodes. Phototransistors work in a very similar way to normal transistors, but the base drive current is controlled by the light input to the device. They can be built into a circuit that provides a changeable analogue value or a digital input. They can be used to detect most visible and near infrared light sources, including neon, fluorescent or incandescent bulbs, LEDs and laser or even flame sources.

Movement/orientation input devices

The most common type of movement input device is a tilt switch, but this group can also include simple pressure switches. Tilt switches are non-conductive tubes that have two electrical contacts and an internal material which acts as a conductor between the contacts. Originally mercury was the most commonly used internal conductive material, but because of its toxicity its use has been reduced. There are other internal materials that can be used to trigger a tilt switch, including metal balls and electric current. Tilt switches are used in many common items such as air bags, game controllers, cameras and security devices. They are simple devices that are either ON or OFF (logic 1 or 0).

A pressure switch is operated when the pressure in the device becomes sufficient to enable an electrical connection. This can be by the connection of two physical plates or by a simple pressure-operated plunger creating a connection. Again, these devices will be either ON or OFF (logic 1 or 0).

Presence input devices

A micro-switch is a presence input device that is used widely in electronic engineering to identify the presence of objects (such as products and coins in a vending machine or paper in a printer), to count moving objects (such as on production lines) or as a tamper switch (such as in security systems). A micro-switch operates at very high speeds; only a small movement of an actuating lever is required to have a large effect on movement of the electrical contacts, achieving a reliable switched circuit.

The other most common presence devices are ultrasonic-type devices. The ultrasonic sensors emit a sound pulse, which is reflected off any object entering the field of the sound pulse. This reflection is received by the internal sensor, which produces an electrical output signal for use by the microcontroller. The electrical output can be either analogue or digital. These devices are used to calculate the distance between objects, for example in car-reversing sensors, where the output increases in frequency as the distance between car and object decreases.

Requirements for input interfacing

Signal conditioning

It is sometimes necessary to manipulate analogue input signals before any process can be carried out on that input. Signal conditioning is a catch-all phrase that describes the manipulation of, typically, an analogue signal, so it is better understood at the processing stage, for example, by removing any possible interference from the source signal.

In a system there are normally four stages:
1 Sensing stage – e.g. thermistor.
2 Signal conditioning – e.g. amplification, filtering, smoothing and level shifting of the signal.
3 Processing – e.g. ADC or PWM (this can be internal or external to the microcontroller).
4 Output – e.g. LCDs and LEDs.

Analogue-to-digital conversion

Analogue-to-digital conversion is a process that reconstructs an analogue input signal as a multi-level (digital) signal. An analogue signal such as a sine wave is continuously varying, but after being converted by an analogue-to-digital convertor (ADC) the resulting digital signal has a finite number of states. The diagram in **Figure 6.5** shows an analogue signal and its binary (4-bit) digital interpretation from the ADC. The greater the number of bits used in an ADC, the more accurate the ADC output will be. A 4-bit ADC system will allocate 16 possible discrete values to the digitised signal, whereas an 8-bit system will allocate 256 possible discrete values to the digitised signal.

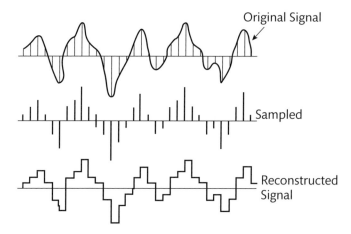

▶ **Figure 6.5** Conversion of an analogue signal to a 4-bit digital signal

The following equation can be used to find the ADC output:

$$\frac{\text{resolution of the ADC}}{\text{system voltage}} = \frac{\text{ADC reading } (x)}{\text{analogue voltage measured}}$$

For example, an 8-bit ADC with a temperature sensor that reads values between 0°C and 100°C will have 256 possible digital values for the voltage range 0–5 V (the system voltage). So if the voltage measured is 2.5 V, we can rearrange the equation to find the value of the ADC output:

$$\frac{256}{5} = \frac{x}{2.5}$$

$$\frac{256}{5} \times 2.5 = x$$

$$x = \frac{256}{5} \times 2.5 = 128$$

The ADC output will be 128, which represents 50°C.

Ⅱ PAUSE POINT	Take 20 minutes to create a table listing up to 20 input devices present in your home. Record the device or machine that the input belongs to and what type of device it is. Try to find a minimum of three inputs per device type.
Hint	List all the potential input devices that exist within a washing machine, and identify which type of input device each of them is.
Extend	Choose a type of car and identify as many input devices as you can. There are many in the engine compartment alone.

Pulse-width modulation (PWM)

PWM is the opposite of ADC – it generates an analogue signal from a digital source. This is typically achieved using computer code. From the two main elements that define the behaviour of the digital signal – the **frequency** and the **duty cycle** (which controls how long the signal stays high during each cycle) – PWM outputs a constant average voltage for each cycle (see **Figure 6.6**). It is used to control direct current (d.c.) motors, hydraulic values, pumps and many other mechanical devices.

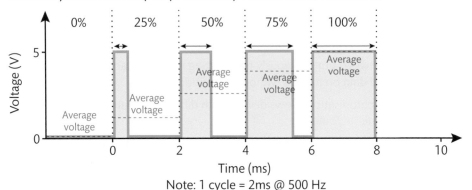

Note: 1 cycle = 2ms @ 500 Hz

▸ **Figure 6.6** PWM duty cycles and output voltages

Modular sensor board

A modular sensor board has all sensors integrated onto a single printed circuit board (PCB). Some engineers like using these boards, as they remove the need to create the physical connections between the microcontroller and the input and output devices being used. There are many types on the market, varying in size and complexity, and they can be quite cost-effective. However, by using a breadboard to develop the microcontroller, you will gain a better understanding of the interconnections, characteristics and operation of both inputs and outputs as you connect them to the microcontroller system.

Serial communications

Many PCs still use a serial port for data. The serial port transmits data one bit at a time over a single wire. Serial communication is slower than parallel communication, but fewer physical connections are required. Only three wires are needed to achieve two-way (full duplex) communications: one to send data, one to receive data, and a common ground wire. There are specific **protocols** for the transmission of data by serial communications, including the use of start/stop data bits and **parity bits** for error checking. The RS232 and USB are common standards for transmission using serial communications, and many peripheral devices are available for these types of serial port. Serial communication is an extensive subject; you do not need to study it in great detail, but you may need to use the protocols when connecting some peripherals.

Inter-integrated circuit (I²C)

I²C is a type of serial communication that enables many devices to be serially interconnected to a single microcontroller from a single serial port, using only two physical wires to control and transmit the data.

I²C communication was developed to reduce the number of connections between integrated circuits. The I²C network uses two PIC16F877 microcontroller input/output lines:

▸ Port C, line 3 (RC3) – SCL ('Serial CLock')
▸ Port C, line 4 (RC4) – SDA ('Serial DAta').

Key terms

Frequency – whether a changing current or voltage, the frequency is defined as the number of complete cycles that occur every second. The unit of measurement is Hertz (Hz).

Duty-cycle – the proportion of time that a component, system or device is in operation. This can be expressed as a percentage or as a ratio.

Protocol – a specific set of rules that are applied when two electronic devices communicate with each other to ensure effective, efficient communication.

Parity bit – a bit of data added to the end of a set of binary data to make the number of 1s in the data an even or odd number. This is used in the simplest form of error checking.

These lines are used in conjunction with the 0 V line to form the I²C bus and allow serial communication between 'bussed' devices at a data rate of up to 3.4 megabits per second in high-speed mode.

The main features of the I²C bus are:

▶ Serial, 8-bit oriented, bidirectional (two-way) data transfers between **masters** and **slaves** allow data to be transmitted at rates of up to 100 kilobits per second in the standard mode, up to 400 kilobits per second in the fast mode, and up to 3.4 megabits per second in the high-speed mode.

▶ The I²C bus was developed as a multi-master bus (no central master).

▶ **Arbitration** between simultaneously transmitting masters enables concurrent transfer of serial data on the bus without corruption.

▶ Serial clock synchronisation allows devices with different bit rates to communicate via one serial bus.

▶ Serial clock synchronisation can be used as a handshake mechanism to suspend and resume transfer.

▶ Each device connected to the bus can have a unique address controlled within the written software code, so simple master/slave relationships exist at all times. Masters can operate as master-transmitters or as master-receivers.

▶ The I²C bus can be used for testing and diagnostic purposes.

A typical I²C bus configuration is shown in **Figure 6.7**. The two communication lines, SDA and SCL, have one common pull-up resistor. The configuration is called a wired-AND structure, and the protocol allows for 1008 possible slave devices. Adding or removing I²C components on the bus will not affect the operation of any of the already connected ICs, nor will it affect the software that runs on the system.

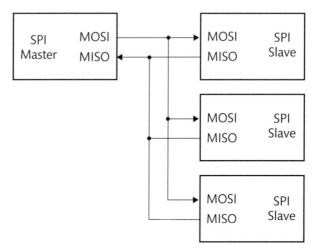

▶ **Figure 6.7** A typical I²C bus configuration

PAUSE POINT

By now, you've had the opportunity to examine many common input devices and review their datasheets. Discuss the typical operation and potential applications of the input devices you have viewed.

Hint Keep the datasheets safe and remember the potential applications – these may be vitally useful during programming tasks and will provide a starting point for the external assessment.

Extend Review and research any related manuals for the devices, including the microcontroller.

The modern motor industry

Motor vehicle manufacturers have been researching and developing electronic sensors since the early 1950s. These include water or rain sensors for convertible cars, which would raise an audible alarm at the first sign of rain, indicating that the car needs to stop and the roof needs to be closed. Nowadays most water sensors are used to control the windscreen wipers.

The process of research and development is the same regardless of the industry.

- The process starts with an idea, such as 'I would like the wipers to turn on at the first sign of rain.' (This is the 'client brief'.)
- The next step is to research available input devices that can provide an input when sensing or coming into contact with water.
- The review and analysis stage then follows, where engineers identify and develop the best method that fits with the design and ergonomics of the vehicle. Detecting water is a straightforward electronic concept, but the technology to do so would not necessarily integrate into the windscreen of a vehicle.
- In industry, the developed input device would undergo several iterations and intensive testing

before being included in a concept device (a concept car at a trade show). It would be introduced to the market in high-specification vehicles as a selling point.

The same research and development model applies to all other microcontrolled additions to modern vehicles, such as parking sensors, tracking headlights, remote locking mechanisms and tyre pressure sensors. For decades the motor industry has been developing enhanced microcontrolled monitoring systems for vehicles, but the concept has only recently spread to other industries, driving the IoT.

Check your knowledge

1 List all the factors that might have an impact on the type of input device chosen for:
   - a car parking sensor
   - a remote central heating device.

2 Can you describe the types of research needed to identify and review the device?

3 Identify at least three things in your home that could be automated, with input data used in some helpful way.

A3 Output devices

There are many output devices that can be connected to a microcontroller (see **Figure 6.8**). These can be classed into three main types: optoelectronic, electromechanical and audio. Examine physical examples of each type of output device, and refer to the many websites, supporting materials and books that provide detailed background and useful hints and tips on application. Circuit diagrams will be included with the coding examples and learner programming tasks to assist you further.

Optoelectronic output devices

LEDs

Optoelectronic devices can detect, source or control different wavelengths of light. The most basic and common optoelectronic output device is the light-emitting diode (LED). The LED is the simplest device that you can connect to – and use to obtain a visual output from – a microcontroller pin or port.

▶ **Figure 6.8** Example of an output device of a microcontroller system

There are two possible methods of connecting an external load to the microcontroller system or another type of electronic device or IC; these are current sinking and current sourcing.

▸ Current sinking – The microcontroller acts as a ground for the load circuit. Therefore, when the pin output is high, 5 V or logic 1, no current can flow in the circuit; but when the pin output is low, 0 V or logic 0, the power to the load is turned on and current flows (sinks) to the microcontroller.

▸ Current sourcing – The microcontroller supplies the load circuit with current. When the pin output is high, 5 V or logic 1, current flows in the load circuit; when the pin output is low, 0 V or logic 0, the power to the load is turned off.

PAUSE POINT

What is the point of considering current sinking and current sourcing?
Imagine you have a microcontroller that has several outputs connected to it, such as 16 LEDs. If all the LEDs happen to be on at the same time, the microcontroller will need to supply a source current of 20 mA × 16 = 320 mA, as well as any source current required by other devices. Review the datasheet and manual for the microcontroller you are using and note the current sourcing values and the current sinking values.

Hint

Do not assume that all output pins on a microcontroller have the same electronic characteristics and parameters.

Extend

Consider, from the information you have now reviewed, which method is more appropriate for the connection of devices.

Assessment practice 6.2

In this activity you will continue to gain hands-on experience of constructing and creating circuits and downloading programs to your microcontroller system.

You will need eight LEDs and eight 220 Ω resistors.

- Connect the LEDs to the chosen port.
- Create, download and run different programs to output 10101010_2, 01010101_2, 11111111_2, 11001100_2 and 00110011_2.
- Amend your code and expand the program to output values 0–255_{10} in binary.

Plan
- What components do I need?
- Which port should I use?
- What do I need to output?

Do
- I have all the necessary tools and equipment.
- I am familiar with the relevant health and safety procedures and will follow them closely throughout the task.
- I can complete the circuit construction accurately using the information provided.

Review
- Have I considered all the relevant theory that I have learned so far to build the circuit to the correct standard?
- Have I thought about the amount of code necessary to output all the values between 0 and 255 in binary?
- Have I carefully designed the code for outputting all the values required (e.g. a simple counter counting from 0 to the maximum of a 16-bit binary number)?
- Have I analysed the quality of code necessary for such a program?

Seven-segment displays

The second most common optoelectronic output device is a seven-segment display. These devices can be controlled in several ways, but there are two common methods that you will implement in the class tasks of Section B3 and which you may use to realise a product under assessment. In general, a seven-segment display has seven individual segments that have either all the anodes connected (common anode), for which a positive voltage is required, or all the cathodes connected (common cathode), which would need to be grounded for the operation of the device (see **Figure 6.9**). Therefore, eight connections are required to operate each seven-segment display, so any value above 99 would require 24 physical pin connections to a microcontroller, leaving few connections for any other inputs or outputs.

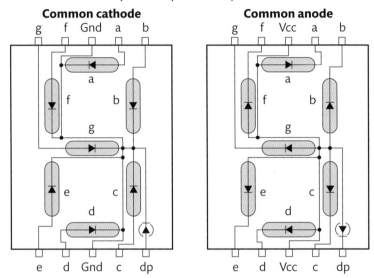

▶ **Figure 6.9** Seven-segment display with common cathode or common anode connection

Each seven-segment display requires an 8-bit pattern to be output on the connected port for each of the digits 0 to 9. The segments are individually labelled 'a' to 'g', as shown in **Figure 6.9**, and the output patterns can be calculated as shown in **Table 6.1**. For this example, we will assume a common anode display, so the output pin needs to be logic 0 to light up the segment (current sinking).

▶ **Table 6.1** Example pin connections

Digit	Common	Segment label (pin output 1 or 0)							Hex
	Px.7	a – Px.6	b – Px.5	c – Px.4	d – Px.3	e – Px.2	f – Px.1	g – Px.0	
0	1	0	0	0	0	0	0	1	0x81
1	1	1	0	0	1	1	1	1	0xCF
2	1	0	0	1	0	0	1	0	0x92
3	1	0	0	0	0	1	1	0	0x86
4	1	1	0	0	1	1	0	0	0xCC
5	1	0	1	0	0	1	0	0	0xA4
6	1	1	1	0	0	0	0	0	0xE0
7	1	0	0	0	1	1	1	1	0x8F
8	1	0	0	0	0	0	0	0	0x80
9	1	0	0	0	0	1	0	0	0x84

This is not efficient. However, the addition of some further simple circuitry allows up to eight seven-segment displays to be controlled by one port. This is achieved using a

3-8 decoder, such as the 74HCT 238, and a **BCD to seven-segment decoder**, such as the 74LS 47, which would require only seven pins to form a single port. You will test this circuit later, during the development of Class task 1 and Class task 2 in Section B3.

LCDs

The last main optoelectronic output device we will cover is the liquid crystal display (LCD). LCDs come in many variants, from just a few characters on a single-line output to multiple-line outputs that are many characters across. The main advantages of using these displays are:

▶ the range of characters that can be displayed – numbers, letters (lower and upper case) and special characters such as *, % and &

▶ the size and type of output that can be displayed – for example, the output readings from two inputs displayed simultaneously on separate lines

▶ the possibility of controlling the device in multiple modes, including 4-bit or 8-bit modes, with the 4-bit mode requiring the use of only one port of the microcontroller.

The circuit diagram for the interconnection of an LCD in 4-bit mode is shown in part of **Figure 6.19** in the 'Getting ready for assessment' section.

Electromechanical output devices

Relays

The relay is the most commonly used electromechanical output device. It provides insulation and protection between high- and low-voltage sections of a circuit. There are various types of relay, but each type usually consists of five individual parts or sections (see **Figure 6.10**):

▶ Armature – a magnetic strip through which current flows to energise the coil and produce a magnetic field. This in turn 'makes' or 'breaks' the normally open (NO) or normally closed (NC) contacts.

▶ Spring – if no voltage is applied and no current flows, the spring holds the armature in place, ensuring non-completion of the circuit.

▶ Electrical contacts – in a normally closed set of contacts, when the relay is triggered the external device will not operate; the external device is always on until the relay is triggered. In a normally open set of contacts, the opposite occurs, and the external device is off until the relay is triggered.

▶ Electromagnet – a standard iron core with coils of wire wound around it.

▶ Cover/container/package – typically transparent so that operation of the device can be observed and checked.

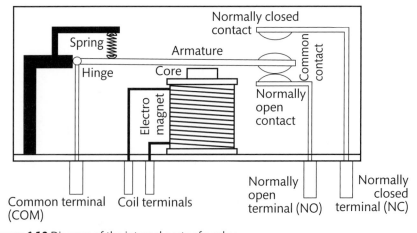

▶ **Figure 6.10** Diagram of the internal parts of a relay

Direct current motors

Direct current (d.c.) motors are standard electromechanical output devices that can be controlled directly by the microcontroller for low-voltage motors or through a relay for high-voltage motors. Small d.c. motors are typically used in small devices, such as toys and portable power tools that can be powered by batteries. Larger d.c. motors are used in lifts and hoists. The d.c. motor speed can be controlled directly from the microcontroller using PWM. (See the Pause point below.)

Servos

Even if you have not heard the term before, you will probably have used devices in which servos are an integral part – for example, a radio-controlled car or plane, a toy or a robot. Servos typically have only three wires and require a coded input signal to move a shaft into a specific angular position and keep it in that position until a new signal is received.

The servo starts at a neutral position and normally moves by approximately ±90°. Under normal operation, the duration of the pulse received by the servo control wire determines the magnitude of the angle moved. The duration and corresponding angle can vary between different servo manufacturers, but the operating principle is the same for all. For example, it could be that the central position (90°) requires a 2 millisecond pulse, and if the servo receives a 1 millisecond pulse it moves to the 0° position, whereas if it receives a 3 millisecond pulse it moves in the other direction to 180°.

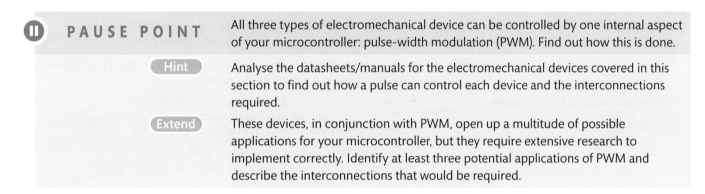

PAUSE POINT　All three types of electromechanical device can be controlled by one internal aspect of your microcontroller: pulse-width modulation (PWM). Find out how this is done.

　Hint　Analyse the datasheets/manuals for the electromechanical devices covered in this section to find out how a pulse can control each device and the interconnections required.

　Extend　These devices, in conjunction with PWM, open up a multitude of possible applications for your microcontroller, but they require extensive research to implement correctly. Identify at least three potential applications of PWM and describe the interconnections that would be required.

Audio output devices

Sound outputs are probably the most familiar of all output devices. These come in many shapes, sizes and configurations, but all have the same end result. They are simple devices that require very little connection but are very effective.

▶ Buzzer – creates different sounds or notes as a result of applying an oscillating voltage. This device can be from a single-pin output, with the frequency and duration controlled via the microcontroller program code.

▶ Siren – incorporates oscillators and amplifiers to produce a tone or various tones through speakers. This device is used in emergency vehicles such as police cars and ambulances.

▶ Speakers – require the use of amplification to boost current levels in order to drive the speaker and produce an adequate volume that can be heard by the end user.

Output interfacing requirements

The majority of output devices require some sort of additional interface, such as amplification or signal conditioning. These might include relays, PWM, I²C, signal conditioning and serial communications, which have all been covered earlier.

A4 Selecting hardware devices and system design

As part of your external assessment you will be required to analyse a given problem and devise a suitable solution. It will not be enough simply to pick an input and an output, connect the devices and produce the code. You will need to identify several potential input and output devices that could realise your chosen solution and justify your final selection using reasoned arguments. **Table 6.2** sets out the suggested format for the layout and decision-making sections of your project specification document. You can then incorporate this into your final, full project specification.

▶ **Table 6.2** Suggested format for the opening sections of your project specification

Project overview		A brief overview of the problem identified in the external assessment task.	
Outline of initial requirements	**Inputs**	• Identification of the parameter to be measured • The required accuracy and range (and why) • The electrical, mechanical and dimensional specification • Outline of the types of inputs that could realise the solution and why	
	Outputs	• The required output type or effect • The electrical, mechanical and dimensional specification • Outline of the types of output devices that could realise the solution and why	
	Microcontroller (available versus desired)	• Outline of the microcontroller that would best suit a solution, compared with what is available to you • Internal hardware	
Potential design solutions (a minimum of 3)	**Inputs**	• The specifics of the inputs to realise the solution • How these input devices will meet the requirements of the external assessment, including relevant details from the datasheet	
	Microcontroller	• The specifics of the microcontroller and why it was chosen, e.g. what internal hardware makes it suitable for this solution?	
	Outputs	• The specifics of the output devices to realise the solution • How these output devices will meet the requirements of the external assessment, including relevant details from the datasheet	
	Cost	• Materials: each device, microcontroller, packaging, circuitry • Human cost: design, production, programming and testing costs	
	Advantages and disadvantages		
Chosen design solution (an evaluation and justification of each element – referencing the datasheets and manuals)	**Inputs**	• Devices • Required interfacing • Circuit construction	• Usability of user controls: aesthetics, control and ergonomics • Cost and availability
	Microcontroller	• Type • Design • Programming: method, interface, coding input	• Usability (in chosen solution) • Power requirements
	Outputs	• Devices • Required interfacing • Circuit construction	• Usability of indication and display devices: aesthetics, visibility or audibility • Power control requirements • Cost and availability
	Circuit diagram(s)	• Ensure that these are created to the correct standards with correct labelling, headers and interconnections • Unit specification requires BS8888 and BS3939 or other relevant international equivalent	
References	• Datasheets	• Websites	• Books
Appendices	• Datasheets		

A5 Assembling and operating a microcontroller system

By now, you will already have built at least one microcontroller system, including power supply and clocking device, and connected LEDs to one of the output ports. Rather than learning the practical skills of assembling and operating a microcontroller system in isolation, you will continue to develop them as you work through the unit. Remember to follow these guidelines:

▶ Preparation – Clear your workspace, check that you have all your tools to hand, and make sure you use electrostatic protection when handling ICs.

▶ Health and safety – Ensure that the workspace is suitable, conceal any loose clothing, use all tools responsibly, and run an extractor fan when soldering.

▶ Components – Check that you have the correct **IC footprint** for each component and know how it should be connected to the breadboard.

▶ Power supply – Ensure that you identify the power requirements. To prevent damaging components, you need to use a suitable, safe power supply. Using four AA batteries as your power supply will provide you with a portable system that you can also use for research outside the formal guided learning in class.

> **Key term**
>
> **IC footprint** – identifies the locations, pattern and distance between the pins on a device.

 # B Programming techniques and coding

You may find this the most interesting part of the unit, as it is where you make everything work. You have learned how to select appropriate input and output devices, and connect them all properly, but nothing can happen until you give them a set of instructions. There are some differences between IDEs, in the structure of the code and the declarations of routines, but the basic techniques and constructs are the same regardless of which IDE you are using.

A good starting point for this unit is numbering systems, as you will be using them extensively in your programming and it is vital that you recognise and understand the outputs on ports.

B4 Numbering systems

Digital logic systems and memory devices can register only one of two possible states:

▶ a positive voltage (approximately 5 V depending on the device family) is considered a logic 1, ON, active or HIGH

▶ zero voltage (0 V) is considered a logic 0, OFF, inactive or LOW.

Each 1 or 0 is called a 'bit' of memory or a logic state.

As only the two states 0 and 1 are available in a digital device, we need a way of representing values from our common numbering system – the decimal system – using only 0s and 1s.

Decimal system

First, let us consider how the common decimal (or denary) numbering system works.

The decimal numbering system has ten distinct digits: 0, 1, 2, 3, 4, 5, 6, 7, 8 and 9. We can create any number from these digits. We write a number by ordering the digits in ascending powers of 10 from right to left. For example, **Table 6.3** describes the value that 13 658 in the decimal system represents.

10 000	1000	100	10	1
10^4 (10 to power 4)	10^3 (10 to power 3)	10^2 (10 to power 2)	10^1 (10 to power 1)	10^0 (units or 1s)
1	3	6	5	8

So the number 13 658 is made up of 10 000 + 3000 + 600 + 50 + 8.

Binary system

The binary numbering system uses a set of just two digits: 0 and 1 (the prefix 'bi' means 'two'). Using the binary system to represent a number works in exactly the same way as using the decimal system, except that we use ascending powers of 2 instead of 10. In the binary system we typically represent numbers in sets of 8 bits, called a byte; each bit can take value 0 or 1. The largest binary number that can be represented by 8 bits is 1111 1111, as shown in **Table 6.4**.

▶ **Table 6.4** The number 11111111 in the binary system

128	64	32	16	8	4	2	1 (unit)
2^7	2^6	2^5	2^4	2^3	2^2	2^1	2^0
1	1	1	1	1	1	1	1
7 – most significant bit (MSB)	6	5	4	3	2	1	0 – least significant bit (LSB)

From the table, the decimal equivalent of the binary number 1111 1111$_2$ is

128 + 64 + 32 + 16 + 8 + 4 + 2 + 1 = 255

The subscript 2 after the number identifies it as being represented in binary, so that we don't read it as eleven million, one hundred and eleven thousand, one hundred and eleven.

ⅠⅠ **PAUSE POINT** Use **Table 6.4** to help you work out the equivalent decimal value of these 8-bit binary numbers.

a) 1101 1101$_2$ b) 1010 1010$_2$ c) 0101 0101$_2$ (See next page for answers.)

You can also use **Table 6.4** to convert decimal numbers to their equivalent binary representations. Do this by subtracting each power of 2 in turn, starting with the highest power (furthest left). Write a 1 when you can subtract the power of 2 without going negative, and write a 0 when you cannot.

For example, to convert the decimal number 189 into its binary equivalent:

189 – 128 = 61 (write 1 under 128);

cannot subtract 64 (write 0 under 64);

61 – 32 = 29 (write 1 under 32);

29 – 16 = 13 (write 1 under 16);

13 – 8 = 5 (write 1 under 8);

5 – 4 = 1 (write 1 under 4);

cannot subtract 2 (write 0 under 2);

1 – 1 = 0 (write 1 under 1).

Therefore, the binary representation of 189 is 1011 1101$_2$, as shown in **Table 6.5**.

▶ **Table 6.5** Binary representation of 189_{10}

128	64	32	16	8	4	2	1 (unit)
2^7	2^6	2^5	2^4	2^3	2^2	2^1	2^0
1	0	1	1	1	1	0	1
7 – most significant bit (MSB)	6	5	4	3	2	1	0 – least significant bit (LSB)

You should now be able to see the connection between your 8-bit microcontroller and an 8-bit binary number.

An 8-bit microcontroller moves bytes of data around the processor on an 8-bit parallel data bus, and there are 256 possible combinations or 0s and 1s in a byte. When coding in C, binary numbers are represented in the following way:

X = 0b00110011;   // 0b identifies the number as binary

Larger values are represented by 16-bit numbers, 32-bit numbers or even 64-bit numbers. (Notice that the number of bits doubles each time.) In C, a 16-bit binary number can represent integer values between 0 and 65 535 for an **unsigned variable**, or between –32 768 and +32 767 for a **signed variable**.

Hexadecimal system

It is very helpful for programmers in any language to be familiar with the hexadecimal equivalents of binary numbers. For example, imagine you want to do a calculation adding 31 067 and 2 375. The binary representations of both these numbers are strings of sixteen 1s and 0s, which are tedious to write down by hand and prone to errors if you're typing them out. However, most people find 4-bit patterns easy to remember. In the hexadecimal system, each distinct 4-bit pattern is assigned a hexadecimal digit or letter. This makes it easier to represent and identify large numbers.

There are 16 distinct 4-bit patterns. The hexadecimal representation uses a set of 16 distinct digits and letters: 0, 1, 2, 3, 4, 5, 6, 7, 8, 9, A, B, C, D, E, F.

Table 6.6 shows which hexadecimal character corresponds to which 4-bit pattern.

> **Key terms**
>
> An **unsigned variable** – a variable that can take only positive values (or be zero). For example, an unsigned 8-bit number has 256 possible values, from 0 to 255.
>
> A **signed variable** – a variable that can take both negative and positive values. For example, a signed 8-bit number has 256 possible values, from –128 to +127.

▶ **Table 6.6** Working out which hexadecimal character you need

Hexadecimal character	0	1	2	3	4	5	6	7	8	9	A	B	C	D	E	F
Total value of 4-bit pattern	0	1	2	3	4	5	6	7	8	9	10	11	12	13	14	15

Table 6.7 demonstrates how to work out the equivalent hexadecimal value for the 16-bit binary number 1100 1010 0011 1110$_2$ (which is 51 774 as a decimal number).

▶ **Table 6.7** Using the hexadecimal number system

2^3	2^2	2^1	2^0	2^3	2^2	2^1	2^0	2^3	2^2	2^1	2^0	2^3	2^2	2^1	2^0
8	4	2	1	8	4	2	1	8	4	2	1	8	4	2	1
1	1	0	0	1	0	1	0	0	0	1	1	1	1	1	0
8 + 4 + 0 + 0 = 12				8 + 0 + 2 + 0 = 10				0 + 0 + 2 + 1 = 3				8 + 4 + 2 + 0 = 14			
C				A				3				E			

Answers to Pause point on previous page

a) 221_{10} **b)** 170_{10} **c)** 85_{10}

So, in C programming, you could represent the number by either

x = 0b1100101000111110;

or

x = 0xCA3E;   // 0x identifies the number as hexadecimal

Hexadecimal is often used in datasheets and IDEs to represent memory addresses or port outputs and to set up registers, so it is important to become familiar with it. Initially it may seem complicated, but learning how to use it will be worth the effort. There are many websites and apps available to help you understand the principles and practise converting between binary and hexadecimal representations.

Use **Tables 6.6** and **6.7** to help you with the calculations in the next Pause point.

PAUSE POINT

Try these conversions between binary and hexadecimal values.

1   Convert these binary numbers to hexadecimal values.

a) $1010\,0101\,1100\,0011_2$ b) $0001\,1000\,0011\,1100_2$

2   Convert these hexadecimal values to binary numbers.

a) $FA86_{16}$ b) $5A5A_{16}$

(See next page for answers.)

Parallel and serial data transfer

Figure 6.11 shows the sending of 8 bits of data along a parallel 8-bit data bus.

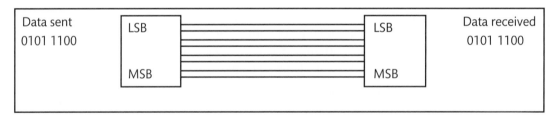

▶ **Figure 6.11** Sending 8 bits of data along a parallel 8-bit data bus

One data bit is transmitted on each individual wire in the 8-bit bus, with a byte of data being transferred simultaneously on the bus. The rate of data transfer, when the next set of 8 bits is sent, is controlled and synchronised by the processor clock. Parallel transmission requires more wires than serial transmission, and this increases the size of the internal circuitry. However, more bytes can be sent at a time and so a parallel bus has a higher data transfer rate.

Figure 6.12 shows 8 bits of data being sent along a serial bus.

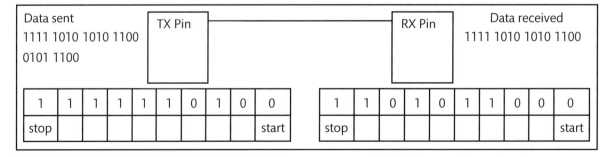

▶ **Figure 6.12** Sending 8 bits of data along a serial bus

In its simplest form, serial data transfer involves sending data sequentially (one bit at a time) on a single wire. This reduces the circuitry required but results in a lower data transfer rate. There is also the need to send additional data to indicate the start and end (stop) of each byte (**WORD**) of data and also for error checking, which has a further adverse effect on the rate of data transmission.

Serial data transmission can be either synchronous or asynchronous.

▶ Synchronous data transmission – The data is combined or grouped into **frames** or **packets** of fixed length (number of bits). The frames are sent continuously, controlled by the system clock, regardless of whether they contain data or not. A typical use is within computer networks.

▶ Asynchronous data transmission – Data is sent when it needs to be sent and not in a steady stream. Start and stop bits are used to let the receiving circuitry know when the data transmission begins and ends. A typical example is telephone line transmissions.

> **Key terms**
>
> **WORD** – a fixed number of data bits, typically 8 bits, or a byte of data.
>
> **Frame** (or **packet**) – a group of a specific number of data bits that includes a **header**, which indicates the start and the end of the data, the number of bits being transmitted and, in some instances, a form of error checking.
>
> **Header** – a collection of bits within a frame or packet that provides information about the data within it.

> **Reflect**
>
> Familiarity with numbering systems is fundamental to understanding microcontroller registers and reading code.
>
> Reflect on the numbering systems covered in this section. Consider carrying out further research on calculations using binary and hexadecimal numbers.
>
> Research available apps and online books or tutorials on working with the binary system. For example, try a search on the 'two's complement' method.

B1 Programming techniques

So far, you have:

▶ created some simple coding programs

▶ created a project on the IDE

▶ interconnected the microcontroller with the IDE interface

▶ simulated the program operation on the IDE

▶ created some simple circuitry

▶ downloaded the code

▶ run the program on the microcontroller

▶ carried out some testing.

Answers to Pause point on previous page	
1 a) $A5C3_{16}$	**b)** $183C_{16}$
2 a) $1111\ 1010\ 1000\ 0110_2$	**b)** $0101\ 1010\ 0101\ 1010_2$

Error checking

During the processes you have completed, you will have encountered several **syntax** errors when compiling the code. It is important to realise that there are structures and syntax in any programming language which *must* be followed in order to get your programs to compile and run. Everyone makes some mistakes, even the best programmers – what's important is that you be able to quickly identify and rectify the errors. In most IDEs, the compiler will stop at or highlight a line of code where there is an issue. The error may be on the line indicated, or it may be the result of an error in the line of code immediately before or after.

Below is a section of code containing some common errors that have been highlighted by the compiler. The reason for each error and how to correct the code to remove it are explained in **Table 6.8**.

> **Key term**
>
> **Syntax** – a set of rules that govern the structure, symbols and language used in specific types of computer code.

```
1./*Title: Error checking example
2.Description: program to identify the common errors on compiling a program
3.Author: A Serplus
4.Date: 01/09/xxxx
5.Version
6.*
7.
8.#include<pic16f87.h>
9.#include<stdio.h
10.
11.void Output_On(void);
12.void Output_Off(void);
13.void delay_ms(void)
14.
15.void main(void)
16.{
17.    TRISB = 0;
18.    while(1)
19.    {
20.        Otput_On();
21.        delay_ms();
22.        Output_Off();
23.        delay_ms();
24.    }
25.}
26.
27.void Output_On()
28.{
29.    PORTB = 0x55;
30.}
31.
32.void Output_Off()
33.{
34.    PORTB = 0xAA;
35.}
36.
37.void delay_ms()
38.{
39.    unsigned int chec;
40.    for(check=0;check<2000;check++)
41.        {
42.            ;
43.        {
44.}
```

▶ **Table 6.8** Errors in the example section of code and how to correct them

Line number	Error	How to correct the error
6	Multiple lines of comments are ignored by the compiler, but they must be enclosed by /* and */. Comments that are included on the same line as code can be started with // and do not need to be closed off. However, they must not run on to more than one line.	The multiple lines of comments start on line 1, which begins with /*. The comments finish on line 6 but the / is missing – it should be */
9	Compiler **header files** must always follow the format: #include<name of header with a .h extension> Here the final > is missing.	Insert the > character: #include<stdio.h>
13	The **function** delay_ms(void) has not been properly declared at the beginning of the program. The line does not end with a semicolon.	Declare the function delay_ms(void) at the beginning of the program. Insert a semicolon at the end of the line.
20	The spelling error in Otput_On means that the compiler does not recognise this function as one of those declared at the top.	Correct the spelling to 'Output_On'.
37	As a result of the error on line 13, the complier finds a corresponding error on line 37 at the function itself.	Correction of the error at line 13 and recompiling the code will remove this error.
39	The compiler does not recognise the **variable** 'chec' because there is a spelling error here.	Correct the spelling from 'chec' to 'check'.
43	The { here should be a } to match the bracket on line 41. Every opening { must have a closing }.	Change { to }

Key terms

Header file – a file that has a .h extension and includes declarations of pre-written functions, macros and global variables. In the error checking example code given, the header file allows the compiler to understand what TRISB and PORTB mean.

Function – a portion of code that has a specific function, also called a subroutine.

Variable – a memory location that will be used in a particular function (local variable) or throughout the program (global variable).

All defined constants, variables and functions should have names that are relevant to their purpose. The names are not allowed to contain spaces, but you can use underscores to separate words or use capitals to start each new word. For example, you could define

    void display_current_temperature(void);

or

    void DisplayCurrentTemperature(void);

To avoid wasting time, it is important to simulate and test your program on the IDE first, before downloading it to the microcontroller and testing again. It is much easier to identify an error if you plan and test as you go along. If you write your code, download it and connect the interface all in one go and the system does not work, where do you start to look for the error? It could be a coding error, a download error or a circuit construction error. Testing at each stage takes time, but it allows you to identify and rectify errors much more quickly and easily, especially if you are approaching the end of a time-constrained assessment where a working system is essential.

Coding practices

Each IDE and coding environment will have a different set of distinct practices that you must follow. This enables other individuals familiar with that environment to easily read the code and helps with identifying and resolving errors. It is also very important in industry that common practices are followed, so that teams can work efficiently on projects and easily pick up work from other staff when they are absent or leave the company. It is advisable to follow the structure below for all programs created using the MPLAB® IDE. This will enable you to quickly identify errors and will allow the code to be more easily assessed.

```
// Title: Standard structure and layout for all C source files
// Description: Template to use when beginning all coding
// Author: Alan Serplus
// Date: 21/01/20xx
// Version: 1a
// Microcontroller in use: PIC16F877

// List all HEADER files

// END of HEADER files

// List all definitions

// END all definitions

// List all GLOBAL/STATIC variables

// END of all GLOBAL/STATIC variables

// Declare all FUNCTIONS that will be in the program

// END of all FUNCTIONS that are in program

// Begin Main Program
void main (void)
{
    while (1)            // endless loop to ensure program continues until reset
    {

                        // function calls

    }
}

// short description of each function if necessary to explain its use and function
void nameoffunction()
```

B2 Coding constructs

Constructs for port inputs and outputs will vary depending on the coding environment. The most effective way to learn how to code for microcontrollers is by doing it. Therefore, in this section you will work with basic examples of code for specific purposes, using previously identified input and output devices. These examples will include the coding and circuitry for a rudimentary microcontroller system. Hints and tips will follow to help you to develop these programs and to understand all the elements of coding constructs.

You are not expected to become an expert in embedded coding, but you will need to gain a thorough understanding of the structure, concepts, constructs and development of embedded C. C is a very versatile language, making it suitable for multiple applications. Unfortunately, there is no fixed agreed format; many programmers will truncate (shorten) statements without commenting, making some code impossible to interpret. We will be following a sensible format that will enable any programmer to decipher your code. This will stand you in good stead for your future study at levels 4, 5 and beyond.

Registers

At this stage it is important to learn about some important **registers** within the **PIC** and how to set them up. You don't need to remember all of these, but once you use them in a program you will find them easier to understand. Refer to the operation manual for the chosen microcontroller for full details on how the registers are used to control all aspects of the microcontroller. The main registers that you will encounter when programming in this unit are listed below.

▸ TRIS bit determines if the port or pin is an input or output.

▸ PORT identifies a port.

▸ ANON bit controls the analogue-to-digital converter (ADC).

▸ CMCON is the comparator control register.

▸ INTCON is the interrupt control register.

▸ T1CON is the Timer TMR1 control register.

▸ PWM1CON is the PWM control register.

▸ PIE1 is the peripheral interrupt enable register.

▸ PCON is the power control register.

▸ ADCON1 is the ADC control register.

▸ ADRESL refers to the LOW bits of the ADC result.

▸ ADRESH refers to the HIGH bits of the ADC result.

▸ OPTION_REG is used to set the prescaler for calculating delays and other timer-driven routines.

> **Key terms**
>
> **Register** – a reserved section of memory that can be written to and read from. Registers typically control specific operations of the microcontroller.
>
> **PIC** (peripheral interface controller) – a family of specialised microcontrollers.

⏸ PAUSE POINT   The registers listed above are the most common registers used when developing simple solutions to given problems using the PIC family of microcontrollers. Identify the names and structures of the registers of the microcontroller used in your centre, and extend the list above by adding a minimum of five extra registers.

Hint   Consult the datasheet for your microcontroller and the detailed user manual to identify the registers and their functions.

Extend   Compare three different families of microcontroller. Include a comparison of the number and sizes of registers and the amount of memory reserved for the registers.

Variables

Declaring variables in C programs is straightforward – state the variable type followed by the name you wish to use. If a program contains several functions that use single letters as variables, and the same letter is used to represent different variables in different functions, a lot of confusion can result. Therefore, regardless of the language, give your variables names that are relevant and make sense within the program. Also, put comments in your code to explain what the variables mean and the values they can take. For example:

- bit 7_seg_power; // one bit
- unsigned char height; // variable called height, 8 bits in size (0 to 255)
- char temperature; // variable called temperature, 8 bits in size (–128 to 127)
- int value; // variable called value, 16 bits in size (–32 768 to 32 767), 'int' short for integer
- unsigned int test_score; // variable called test_score, 16 bits in size (0 to 65 536)
- float score; /* variable called score, 24 bits in size. In simple terms a memory location that can hold a value that has a decimal point. Not a standard C data type */
- long result; /* variable called result, a signed integer but 32 bits in size (-2 147 483 648 to +2 147 483 647) */
- double answer; /* variable called answer, 32 bits in size. In standard C this is considered a float, but has 6 digits of accuracy whereas float has only 3 */

The above examples have described the principal variable types in C: bit, char, unsigned char, int, unsigned int, float, long and double. In addition, there are:

- strings – sequences of **ASCII** characters, not typically handled by the C programming environment. As ASCII characters are treated as 8-bit numbers, C places and stores each character into an array, creating a character string. Each character is placed into the array with the first memory address used as the reference for the complete string. The unfortunate result is that strings are unnecessarily difficult to handle in C, but they are much more efficient in terms of memory allocation, as individual memory locations do not have to be declared for each value.

- arrays – a means of storing a collection of data of the same type, with the size of the array determined by the programmer. Arrays are used to store strings. The ability to access each variable by a numerical index makes arrays very efficient.

For example, in a character array, a null character (zero) is required and added by the system when the string is declared as below.

```
char LearningAim [23] = "Programming techniques";  /* a character array
(a string in C) defined with the Learning aim as its contents */
```

A character array of 23 characters has been declared because there are 21 actual a-to-z characters in 'Programming techniques', plus a space, and then a null character is added to the end of the string.

Arithmetic operators

Within C code and in most other programming languages, the standard arithmetic operations are used:

- () brackets
- / division
- - subtraction
- * multiplication
- + addition
- % modulus

You may be unfamiliar with the modulus operator. It supplies the remainder of a division operation, and is used when calculating thousands, hundreds, tens and units values for outputting to a LCD. For example:

```
int thousands;     // Memory location called thousands, 16 bits long
int hundreds;      // Memory location called hundreds, 16 bits long
int remainder;
int value;
char units;        // Memory location called units, 8 bits long
value = 1357;      // Move number 1357 into memory location value
value/1000 = thousands;       // Divide value by 1000. Answer placed into memory location thousands
                              // (thousands = 1)
value % 1000 = remainder;     // Find the modulus (remainder) and put into memory location remainder. 357
                              // placed in remainder
remainder/100 = hundreds      // Next part of calculation to find number of hundreds
```

Common logic operations in C

A vast array of coding expressions, statements, and arithmetic and logic operators are available in C – too many to cover in this unit. The most common are described here, so that you can read and follow the examples of code.

▶ = (assignment operator)

This will assign what is on the right to the memory location on the left. For example, the statement
value = 1357;
means that the number 1357 will be placed into the memory location 'value'. The statement
value = remainder;
means that the value of memory location 'remainder' will be moved into memory location 'value'.

Relational operators

The 'if statement' (see later) is used to help explain the next few operators.

▶ == (equal to)

The double equals sign means that one side equals the other.

For example, in the code
```
    if (remainder == value)
    {
            function();
    }
```
the function will only be executed (carried out) if the memory location 'remainder' equals the memory location 'value'.

▶ != (not equal to)

The exclamation mark means 'not'.

For example, in the code
```
    if (remainder != value)
    {
            function();
    }
```
the function will only be executed if the memory location 'remainder' is *not* equal to the memory location 'value'.

▶ && (Boolean operator AND)

For example, in the code
```
    if ((value != 10) && (remainder == 50))
    {
            function();
    }
```
the function will only be executed if both sides of the && symbol are true – that is, if 'value' is not equal to 10 and 'remainder' is equal to 50.

▶ ‖ (Boolean operator OR)

This is similar to the AND operator, but means that either side of the ‖ symbol is true (or both sides are true).

▶ > (greater than)

▶ For example, in the code

```
if (remainder > value)
{
        function();
}
```

The function will only be executed if 'remainder' is greater than 'value'.

The following operators work in a similar way to '>':

▶ >= (greater than or equal to)

▶ < (less than)

▶ <= (less than or equal to)

Binary bitwise operators

▶ << (left shift)

For example,

0b00011100 << 3;

will shift all the binary digits three places to the left, giving 0b11100000.

▶ >> (right shift)

For example,

0b00011100 >> 2;

will shift all the binary digits two places to the right, giving 0b00000111.

Assignment, increment and decrement operators

The main assignment operator is '=', which was introduced above. There are two other operators that assign something to the memory location on the left.

▶ ++ (increment)

For example, the statement

remainder ++;

has the same effect as

remainder = remainder +1;

It would add 1 to 'remainder' and then assign the result to memory location 'remainder'.

▶ -- (decrement)

Similarly,

remainder --;

would subtract 1 from 'remainder' and then place the result in memory location 'remainder'.

Common C statements

Conditional statements

A conditional statement is executed or not depending on whether a condition is true or false. The condition is an expression in brackets that will be evaluated. If it is true, the statement is executed. If it is false, the statement is ignored (not executed) and the program will continue on to the next instruction. In C, you can use the following types of conditional structure.

▶ if statement

For example, the following simple code fragment increments the variable 'x' by 1 if the value stored in that memory location is 100:

```
if (x == 100)
        x++;
```

If we want more than a single instruction to be executed when the condition is true, we can specify a block of instructions enclosed between braces { }.

```
if (x == 100)
{
        y = x;          // transfer the value of x to variable y, i.e. y = 100
        PORTB =y;       // output the value of y to port B
}
```

For clarity, it is probably best to include braces around each set of instructions in a conditional statement, even if it actually contains only one instruction.

▶ if/else statement

We can additionally specify what we want to happen if the condition is not fulfilled, using 'else' in conjunction with 'if' in the format

    if (condition) statement1 else statement2

For example:

```
if (x == 100)
{
        x = 0;          // reset x to 0 if currently 100
        else
        {
            ;           // do nothing
        }
}
```

Note that the use of tabbing makes it easier to see the opening and closing braces for each set of instructions.

Several if/else structures can be concatenated (combined or included end to end) to check a range of values. The following example identifies whether the present value stored in 'x' is positive, negative or neither (that is, equal to zero).

```
if (x > 0)
{
        PORTA = 2;                  // if x is greater than 0 output 2 to port A
        else if (x < 0)
        {
                PORTA = 4;          // if x is less than 0 output 4 to port A
                else
                {
                        PORTA = 0;  // if x is neither of the above output 0 to port A
                }
        }
}
```

Repetitive structures or loops

A loop repeats a set of instructions a specified number of times or while a certain condition holds.

There are several kinds of loop structures.

While loop

Its format is:

    while (condition) statement

Its function is to repeat 'statement' while 'condition' is true.

For example, the program on the right uses a while loop to count up to ten:

```
while (while_true)
{
      do_this;
}
```

```
while (i < 10)
{
      i++;
}
```

Do-while loop

Its format is:

    do statement while (condition);

Its functionality is exactly the same as the while loop, except that the condition in a do-while loop is evaluated *after* the execution of 'statement' instead of before. This means that there will be at least one execution of 'statement' even if 'condition' is never fulfilled.

```
do
{
      do_this;
}
while (condition);
```

```
do
{
      i++;
}
while (i<10);
```

For loop

Its format is:

    for (initialisation; condition; increase) statement;

Its function is to repeat 'statement' while 'condition' remains true, as with the while loop, but in addition it allows you to specify an initialisation instruction and an increment instruction. So the for loop is specially designed to perform a repetitive action with a counter. It works in the following way:

1 'initialisation' is executed – generally this sets an initial value for a counter variable. It is executed only once.
2 'condition' is checked – if it is true the loop continues; otherwise the loop finishes and 'statement' is skipped.
3 'statement' is executed – as usual, it can be either a single instruction or a block of instructions enclosed within braces { }.
4 'increase' is executed (if it is specified), and the loop goes back to step 2.

```
for (starting; while_true; do_this)
{
      statement;
}
```

```
for (i = 0; i <10; i ++)
{
      PORTA = i;      // output i to port A
}
```

The switch/case statement

When a variable can take one of several values and the statement executed or function called will be different depending on the value, you can code this using several if/else statements, but it can become untidy. An alternative is to use the switch/case statement.

```
switch (count)
{
        case 1:
                statement1;
                break;
        case 2:
                statement2;
                break;
        case 3:
                statement3:
                break;
        default:
                statement4;
                break;
}
```

```
switch (count)
{
        case 1:
                PORTA = 0x01;
                break;
        case 2:
                PORTA = 0x02;
                break;
        case 3:
                PORTA = 0x04;
                break;
        default:
                PORTA = 0x00;
                break;
}
```

B3 Structured program design

There are several tools you can use to plan the structure of your programs, including flow charts, pseudocode and decision tables.

Flow charts

Flow charts are a visual method of identifying the progress of data through a system, the functions that act on the data and the order in which these functions manipulate the data. You can use individual flow charts for each function and then link them together to explain the full program. Flow charts help you to analyse problems and write programs in high-level languages such as C and .NET. A standard set of symbols is used in flow charts. The main symbols used in software development flow charts are shown in **Table 6.9**. An example flow chart is shown in **Figure 6.13**.

▶ **Table 6.9** ANSI (American National Standards Institute) flow chart symbols

Symbol	Name	Description	Symbol	Name	Description
	Flow direction	Arrows indicating the direction of flow of data		New page or off page	Indicates that the flow chart continues on another page
	Terminal	Indicates the start or end of a process		Decision symbol	Shows that a decision is being made or that the function is branching off in a different direction
	Input or output	Indicates an input or output of data		Pre-written or pre-defined process	Identifies processes that already exist within the system, which have been tested and written into the header
	Process	A process, function or statement that has an influence on data		Delay	Another method of identifying a pre-written function, specifically a delay function
	Connector	Connects different parts of a program or different pages		Manual input	A physical input, such as an 'interrupt' button or a variable

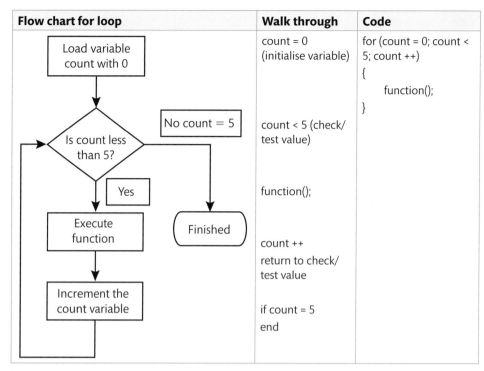

Flow chart for loop	Walk through	Code
Load variable count with 0	count = 0 (initialise variable)	for (count = 0; count < 5; count ++) { function(); }
Is count less than 5? — No count = 5	count < 5 (check/ test value)	
Yes	function();	
Execute function — Finished		
Increment the count variable	count ++ return to check/ test value	
	if count = 5 end	

▶ **Figure 6.13** A simple example flow chart describing a for loop

Decision tables

Decision tables can be used to test different combinations of software/hardware inputs and to ensure that the system operates in accordance with the set input. Decision tables are typically arranged in four quadrants:

Conditions	Condition possibilities
Actions	Action possibilities

There are many examples on the internet of how decision tables operate, most of which are related to business processes. You can investigate further, but in this unit we will concentrate on flow charts and pseudocode.

Pseudocode

Pseudocode is not a programming language itself. It is a verbal representation of a program, which describes the flow of the program and the instructions that it will be carrying out. There are no specific standards and notation, but the most common words used are (or closely resemble) the key commands in most high-level programming languages.

The main words in any pseudocode plan, which typically appear in bold and capitals, are:

BEGIN	**WHILE**	**IF THEN ELSE**
SET	**CALL**	**FOR**
INPUT	**REPEAT**, **REPEAT UNTIL**	**END**
OUTPUT	**IF**	

In Assessment practice 6.2, you output a count from 0 to 255 on an array of LEDs. Before any such task, it is good practice to make a pseudocode plan prior to writing, downloading and executing the code. The pseudocode for the program of Assessment practice 6.2 could be:

Program to count from 0 to 255 – output onto an array of 8 LEDs

BEGIN

        SET Inputs and Outputs

        REPEAT forever

                FOR a count from 0 to a maximum of 255

                        OUTPUT value to **SET** port

                        CALL a one-second delay function

                END FOR

        END REPEAT

END

Just like when you are writing real code, it is important to use indentation to make the pseudocode plan easier to read.

Each function in the program should have supporting pseudocode. Accurate pseudocode that is amended with every iteration of the code as it is being developed is essential to good programming. Pseudocode is independent of the programming language used to write the actual code, so it enables the designed solution to be coded in any language and does not require the designer and programmer to be the same person.

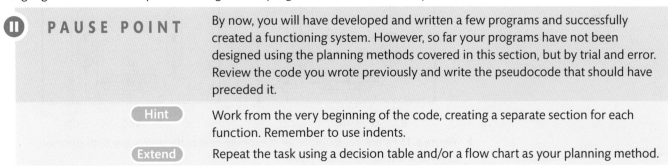

Ⅱ   **PAUSE POINT**   By now, you will have developed and written a few programs and successfully created a functioning system. However, so far your programs have not been designed using the planning methods covered in this section, but by trial and error. Review the code you wrote previously and write the pseudocode that should have preceded it.

       Hint   Work from the very beginning of the code, creating a separate section for each function. Remember to use indents.

       Extend   Repeat the task using a decision table and/or a flow chart as your planning method.

Example systems

Now it is time to leave the theory behind for a while and concentrate on more circuit construction, along with program design and execution. By working through some basic programming solutions in the following class tasks, you will gain better understanding of the interconnection between the inputs and outputs and the software code that controls the interaction.

Recall that every system has three main components, as shown in **Figure 6.14**.

INPUTS   ⟹ | Software code / PROCESS | ⟹   OUTPUTS

▶ **Figure 6.14** The three main components of any system

Make sure you understand the inputs and outputs for your particular system, how they operate and how to connect them correctly to the microcontroller. Remember that all the details you need will be on the datasheet.

Design and plan your code in stages, and test and debug as you progress through each stage of your planned code.

Class task 1: Controlled output to a seven-segment display

1   Connect the development board, using PORT D to connect to the seven-segment display.

2 Use a copy of the table below to identify how the board is connected.

Port pin	RD7	RD6	RD5	RD4	RD3	RD2	RD1	RD0
7-segment connection								

3 Create a table to identify what pattern is needed to display each number from 0 to 9 (see **Figure 6.15**). Whether your seven-segment display is the common anode or common cathode type will determine if a logic 1 or 0 is required to light each segment.

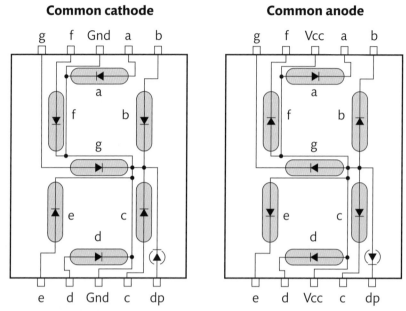

▶ **Figure 6.15** References for the letter of each segment of the display (for both common cathode and common anode types)

The three tables below give examples of the binary values required to output the digit 0, depending on how the seven-segment display is connected.

Example 1

Port pin	RD7	RD6	RD5	RD4	RD3	RD2	RD1	RD0
Seg pin	a	b	c	d	e	f	g	dp
0	1	1	1	1	1	1	0	0

Example 2

Port pin	RD7	RD6	RD5	RD4	RD3	RD2	RD1	RD0
Seg pin	dp	g	f	e	d	c	b	a
0	0	0	1	1	1	1	1	1

Example 3

Port pin	RD7	RD6	RD5	RD4	RD3	RD2	RD1	RD0
Seg pin	d	b	c	a	dp	e	f	g
0	1	1	1	1	0	1	1	0

4　Using the coding structure in your table, create programs to:

　a)　output the number 3 – save as 7seg3.c

　b)　output the number 4, followed by a delay of 1000 ms and then the number 6 – save as 7seg46.c

　c)　count from 0 to 9 with a 1000 ms delay between each digit – save as count09.c

　d)　count from 0 to 9 using a for loop and a 1D array with 11 memory – save as count09_array.c

Hints

▶　Do not forget to plan using pseudocode.

▶　Follow the template given in 'Coding practices' in Section B1.

▶　Test as you go along to avoid compounding errors.

Core code

```
void main()
{
        TRISD = 0;              // Sets port D as output
        PORTD = 0x00;           // Clear port before beginning program
        while(1)                // Endless loop
        {
                PORTD = 0xFC;   // Using connects at EG 1 this would output a 0
        }
}
```

Solution to task 1(a)

1　Obtain the seven-segment display.

2　Review the datasheet and make sure you know if your seven-segment display is common anode or common cathode. In this example it is a common anode display.

3　Create the circuit in software for simulation and to aid construction.

4　Interconnect as per the circuit diagram in **Figure 6.16**.

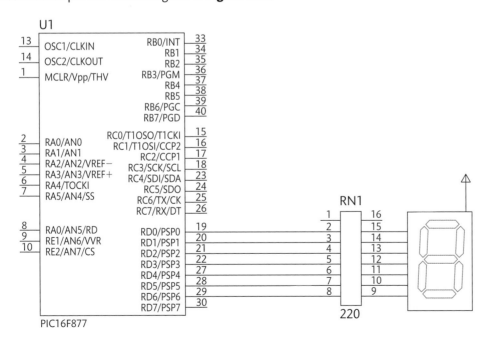

▶　**Figure 6.16** Connection for a single seven-segment display

5 Design the output table as below (n/c means no connection necessary).

Port pin	RD7	RD6	RD5	RD4	RD3	RD2	RD1	RD0	HEX
Seg pin	n/c	a	b	c	d	e	f	g	
0	0	0	0	0	0	0	0	1	0x01
1	0	1	0	0	1	1	1	1	0x4F
2	0	0	0	1	0	0	1	0	0x12
3	0	0	0	0	0	1	1	0	0x06
4	0	1	0	0	1	1	0	0	0x4C
5	0	0	1	0	0	1	0	0	0x24
6	0	0	1	0	0	0	0	0	0x20
7	0	0	0	0	1	1	1	1	0x0F
8	0	0	0	0	0	0	0	0	0x00
9	0	0	0	0	0	1	0	0	0x04

6 Design the code by making a pseudocode plan.

BEGIN

SET Port D as an output

REPEAT

OUTPUT to Port D binary pattern 0b00000110 (Hexadecimal 0x06) – digit 3 in decimal

END REPEAT

END

7 Write the code.

```c
// Title: Output on 7-segment digit 3
// Description: Proof of concept of use of IDE and circuit construction with a 7-segment display
// Author: A Serplus
// Date: 05/02/xx
// Version 1.0

// Include Header files for Micro
#include<pic16f877.h>   // these are dependent on micro used and compiler
#include<htc.h>
// end header files

// No external functions
// No global variables

// Main program starts
void main ()
{
   TRISD = 0;   // set PORT D as an output
   PORTD = 0;   // clear port pins - always good practice
   while (1)    // endless loop set up
   {
      PORTD = 0x06;    // output pattern for digit 3
   }
} // program ends
```

8 Collate the documentation, including images, circuit diagrams and analysis. We will cover this in more detail later, in Learning aim C.

Solution to task 1(b)

Stages 1–5 of the solution are the same as in task 1(a), so for this solution we will start from stage 6.

6 Pseudocode

> **BEGIN**
>> **SET** Port D as an output
>> **REPEAT**
>>> **OUTPUT** to Port D binary pattern 0b01001100 (Hexadecimal 0x4C) digit 4
>>> **CALL** a delay of 1 second (1000 milliseconds)
>>> **OUTPUT** to Port D binary pattern 0b00100000 (Hexadecimal 0x20) digit 6
>>> **CALL** a delay of 1 second. If no delay after second output, it will not be seen by the user as the first output will be output over it in a few microseconds
>> **END REPEAT**
> **END**

7 Code

```
// Title: Output on 7-segment digit 4 then 6 with a 1000ms delay
// Description: Proof of concept of use of IDE and circuit construction with a 7-segment display version 2
// Author: A Serplus
// Date: 05/02/xx
// Version 1.0

// Include Header files for Micro
#include<pic16f877.h>  // these are dependent on micro used and compiler
#include<htc.h>
// end header files
// List functions
void Delay_ms(unsigned int length);
// end function list

// No global variables

// Main program starts
void main ()
{
  TRISD = 0;          // set PORT D as an output
  PORTD = 0;          // clear port pins – always good practice
  while (1)           // endless loop set up
  {
    PORTD = 0x4C;     // output pattern for digit 4
    Delay_ms(1000);   // call the delay function – taking the value 1000 with it to use in the function
    PORTD = 0x20;     // output pattern for digit 6
    Delay_ms(1000);   // the value 1000 will be used by Delay_ms (an example of parameter passing)
  }
}    // main program ends

/******************************************************************************
* Generic delay routine using Timer 0 clocked at 250kHz i.e. a period of 4 microseconds *
* - number of loops specified by length                                       *
* - each loop waits until Timer 0 = 250 i.e. a delay of 1ms                    *
* - instruction takes 4 clock cycles, therefore with a 4MHz crystal instructions run *
* 1MHz so a prescaler /4 gives the 250kHz                                      *
******************************************************************************/
```

```
void Delay_ms(unsigned int length)
{
   unsigned int i;
   OPTION_REG = 0x01;           // prescaler = /4 i.e. TMR0 clocked at 250kHz
   for (i=0;i<length;i++)
   {
      TMR0 = 0;                 // timer used to generate a 1ms delay - set to zero
      while (TMR0<250)
      { };                      // do nothing
   }
}     // function ends
```

8 Collate the documentation, including images, circuit diagrams and analysis.

> **Key term**
>
> **Parameter passing** – a technique used to transfer a value from one function to another so that it can be used by the receiving function. For example, in the code for task 1b 'Delay_ms(1000);' takes the value 1000 and passes it to the Delay_ms function.

Solution to task 1(c)

Again, stages 1–5 of the solution are the same as in task 1(a), so for this solution we will start from stage 6.

6 Pseudocode

> **BEGIN**
> > **SET** Port D as an output
> > **REPEAT UNTIL** all 10 patterns are displayed
> > > **OUTPUT** to Port D digit pattern 0 to 7-segment display
> > > **CALL** a delay of 1 second (1000 milliseconds)
> > **END REPEAT**
> **END**

7 Code

```
// Title: Output on 7-segment display a count from 0 to 9 with a 1000ms delay
// Description: Proof of concept of use of IDE and circuit construction with a 7-segment display version 3
// Author: A Serplus
// Date: 05/02/xx
// Version 1.0


// Include Header files for Micro
#include<pic16f877.h>  // these are dependent on micro used and compiler
#include<htc.h>
// end header files
// List functions
void Delay_ms(unsigned int length);
// end functions

// Main program starts
void main(void)
{
        TRISD = 0;       // Configure port D as outputs
        PORTD = 0;       // Clear port
        while(1)
```

```
        {
                PORTD = 0x01;   // output value for displaying digit 0 to PORTD
                Delay_ms(1000);
                PORTD = 0x4F;   // output value for displaying digit 1 to PORTD
                Delay_ms(1000);
                PORTD = 0x12;   // output value for displaying digit 2 to PORTD
                Delay_ms(1000);
                PORTD = 0x06;   // output value for displaying digit 3 to PORTD
                Delay_ms(1000);
                PORTD = 0x4C;   // output value for displaying digit 4 to PORTD
                Delay_ms(1000);
                PORTD = 0x24;   // output value for displaying digit 5 to PORTD
                Delay_ms(1000);
                PORTD = 0x20;   // output value for displaying digit 6 to PORTD
                Delay_ms(1000);
                PORTD = 0x0F;   // output value for displaying digit 7 to PORTD
                Delay_ms(1000);
                PORTD = 0x00;   // output value for displaying digit 8 to PORTD
                Delay_ms(1000);
                PORTD = 0x04;   // output value for displaying digit 9 to PORTD
                Delay_ms(1000);
        }
}               // program ends

// Begin definition of function
void Delay_ms(unsigned int length)
{
        unsigned int i;
        OPTION_REG = 0x01;      // prescaler = /4 i.e. TMR0 clocked at 250kHz
        for (i=0;i<length;i++)
        {
                TMR0 = 0;        // timer used to generate a 1ms delay – set to zero
                while (TMR0<250)
                {  };            // do nothing
        }
}               // function ends
```

8 Collate the documentation, including images, circuit diagrams and analysis.

Solution to task 1(d)

Again, stages 1–5 of the solution are the same as in task 1(a), so for this solution we will start from stage 6.

6 Pseudocode

BEGIN

> **SET** Port D as an output
>
> **DECLARE** array count0_9
>
> **LOOP** through the array
>
> > **OUTPUT** to Port D array value to seven-segment display
> >
> > **CALL** a delay of 1 second (1000 milliseconds)
>
> **END LOOP**

END

7 Code

```
// Title: Output on 7-segment display a count from 0 to 9 using an array with a 500ms delay
// Description: Proof of concept of use of IDE and circuit construction with a 7-segment display version 4
// Author: A Serplus
// Date: 05/02/xx
// Version 1.0

// Include Header files for Micro
#include<pic16f877.h>     // these are dependent on micro used and compiler
#include<htc.h>
// end header files
// List functions
void Delay_ms(unsigned int length);
// end functions

// global variables
const unsigned char  BCD_to_7seg[10]  // set up an array for segment codes
                = {0x01, 0x4F, 0x12, 0x06, 0x4C, 0x24, 0x20, 0x0F, 0x00, 0x04};
// end global

// Main program starts
void main(void)
{
    unsigned char arraylocation;    // memory location 8 bits in size called arraylocation

    TRISD = 0;      // configure port D as outputs
    PORTD = 0;      // clear port
    while(1)
    {
        for (arraylocation=0; arraylocation<0x0A; arraylocation++)  // loop to start at position 0 in
                                                            // array and stop at 9

        {
            PORTD = BCD_to_7seg[ arraylocation ];   // output value of current position within array
                                                // to port D

            Delay_ms(1000);
        }
    }
}                   // program ends

// Begin definition of function
void Delay_ms(unsigned int length)
{
    unsigned int i;
    OPTION_REG = 0x01;        // prescaler = /4 i.e. TMR0 clocked at 250kHz
    for (i=0;i<length;i++)
    {
        TMR0 = 0;             // timer used to generate a 1ms delay – set to zero
        while (TMR0<250)
        { };                  // do nothing
    }
}    // function ends
```

8 Collate the documentation, including images, circuit diagrams and analysis.

Class task 2: Controlled output to multiple seven-segment displays with inputs to amend output

Now that you have successfully completed Class task 1 and feel more confident, you are ready to make it a bit more interesting. In this task you will add multiple seven-segment displays and some input buttons. You could think of this example as an input/output control for items of stock, cars in a car park or people in a building. Although the task asks you to use buttons, these could be replaced by other types of input devices, such as micro-switches, pressure switches or infrared sensors. The code will remain mostly the same – only the method of production of the input will change.

1 Identify the necessary components required to build such a system. Here you will need three seven-segment displays, a BCD to seven-segment decoder, a 3-8 decoder, a BC327 transistor to power the displays and three push buttons.

2 Review all the datasheets and ensure that you understand the necessary connections to make the IC and components operate correctly.

3 Design a circuit diagram for testing and constructing the circuit (see **Figure 6.17**).

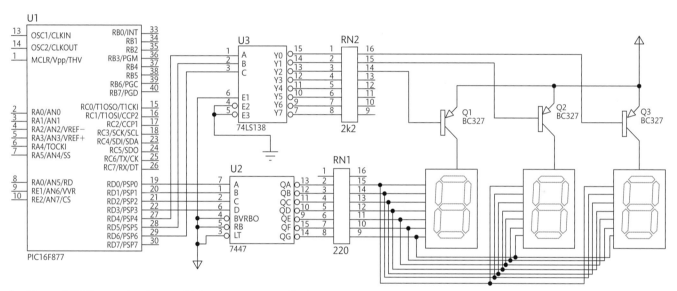

▶ **Figure 6.17** Connection for a multiple seven-segment display

4 Build the circuit. Don't forget the input switches.

5 Design the control and the output table.
The most important aspect is the identification of each of the displays and digits (the number). The digit to be illuminated is controlled by the lower nibble of the output byte on Port D (RD0, RD1, RD2 and RD3). The seven-segment display on which the digit will appear is controlled by the upper nibble (RD4, RD5 and RD6). Pin RD7 is not used and no connection is necessary. (See **Table 6.10**.)

▶ **Table 6.10** Example connection table

	Seven-segment display control				Digit control				
	RD7	RD6	RD5	RD4	RD3	RD2	RD1	RD0	
Display 0	n/c	0	0	0	1	0	0	0	Digit 8
Display 1	n/c	0	0	1	0	0	1	1	Digit 3
Display 2	n/c	0	1	0	0	1	1	0	Digit 6

This is an example of the hexadecimal representation making things much easier. Whenever you want to output to display 0, the hexadecimal will always start 0x0...; for display 1 it will start 0x1..., for display 2 0x2..., and so on. This makes it very easy to identify which display you are controlling. Display 0 should always be the furthest display to the right, counting from left to right.

6 Design the code by making a pseudocode plan. For example:

SET functions for Checking Buttons, Car In, Car Out, Reset
SET bits for the buttons RB1, RB2 and RB3
SET Global variables
BEGIN main
    SET Port D as an Output / Port B pins 1, 2 and 3 as inputs
    REPEAT
        CALL check buttons input function
        CALL refresh the display function
    END REPEAT
END

BEGIN check buttons input function
    CALL check in
    CALL check out
    CALL check reset
END check buttons

BEGIN checkin
    IF total available 299 has been reached
        Carpark full
    ELSE
        Button pressed increment carpark count
    END IF
END check in

BEGIN checkout
    IF total is 0
        Do nothing
    ELSE
        Button pressed decrement carpark count
    END IF
END check out

BEGIN reset
    IF button pressed
        Reset carpark to 0
    END IF
END check in

BEGIN refresh display to always observe illuminated digits
    OUTPUT hundreds, tens, units
END refresh

7 Write the code.

```
// Title: Output on 7-segment displays a count from a car park simulated input
// Description: Proof of concept of use of IDE and circuit construction for a car park system
// Author: A Serplus
// Date: 05/02/xx
// Version 1.0

// Include Header files for Micro
#include<pic16f877.h>      // these are dependent on micro used and compiler
#include<htc.h>
// end header files
// List functions
void Check_Buttons(void);
void Check_Button_Car_in(void);
void Check_Button_Car_out(void);
void Check_Button_Reset(void);
void Refresh_Display(void);
void delay_100us(unsigned int lengthus);
// end functions

// list define
#define Car_in   RB1;       // allocate and set name to port B pin 1
#define Car_out  RB2;
#define Reset    RB3;
// end bit selection

// global variables
int units = 0x06;
int tens = 0x19;
int hundreds = 0x22;
int full = 0x40;
// end variables

// start main program
void main(void)
{
    TRISD = 0;      // set port D as an output
    TRISB1 = 1;     // set port B pin 1 as an input
    TRISB2 = 1;
    TRISB3 = 1;
    PORTD = 0;      // clear all pins
    PORTB = 0;
    while(1)
    {
        Refresh_Display();
        Check_Buttons();
    }
}

void Check_Buttons()
{
    Check_Button_Car_in();
    Check_Button_Car_out();
    Check_Button_Reset();
}

void Check_Button_Car_in()
{
```

```
        if (hundreds==0x22 && tens == 0x19 && units == 0x09)   // has 299 been reached?
        {
            full = 0x78;                // make all displays 8
        }
        else
        if(!Car_in)                     // if button input is 0 i.e. not 1 (NOT Car_in)
        {
            units++;
            if (units>0x09)             // if the units value on display 0 is greater than 9
            {
                tens++;                 // add 1 to the tens
                units = 0x00;           // reset units to 0
                if(tens>0x19)           // if as a result tens value on display 1 is now greater than 9
                {
                    hundreds++;         // add 1 to the hundreds
                    tens = 0x10;        // reset tens to 0
                }
            }
        }
        while(!Car_in)                  // if user continues to hold down button refresh the display
        {
            Refresh_Display();
        }
}                                       // main program ends

// Begin function definitions

void Check_button_Car_out()
{
        if (hundreds==0x20 && tens == 0x10 && units == 0x00)   // check if all displays are 0
        {
            Refresh_Display();          // refresh the display as no other action needed
        }
        else
        if (!Car_out)
        {
            full = 0x70;                // reset full value
            units--;                    // reduce value
            if(units<0x00)              // if units value is less than 0 then take 1 from the tens
            {
                tens--;
                units = 0x09;           // set units value to 9
                if(tens<0x10)           // if tens value falls to less than 0 then take 1 from the hundreds
                                        // and make tens 9
                {
                    hundreds--;
                    tens = 0x19;
                }
            }
        }
        while(!Car_out)                 // if user holds down button refresh display only until released
        {
            Refresh_Display();
        }
}

void Check_Button_Reset()
{
        if(!Reset)
        {
```

```
                units = 0x00;              // reset units on display to 0
                tens = 0x10;
                hundreds = 0x20;
        }
        while(!Reset)                      // if user holds down button only refresh display is carried out
        {
                Refresh_Display();
        }
}

void Refresh_Display()
{
        PORTD = units;                     // output value to port D
        delay_100us(1);                    // minimal delay required to enable micro to carry out above
                                           // instruction before the next

        PORTD = tens;
        delay_100us(1);
        PORTD = hundreds;
        delay_100us(1);
        PORTD = full;
        delay_100us(1);
}

void delay_100us()
{
        unsigned int i;
        OPTION_REG = 0x01;                 // prescaler = /4 i.e. TMR0 clocked at 250kHz
        for (i=0;i<length;i++)
        {
                TMR0 = 0;                  // generates a delay of 0.1ms (100 microseconds)
                while (TMR0<25)    {};
        }
}
```

8   Collate the documentation including images, circuit diagrams and analysis. We will cover documentation in more detail in the next section.

PAUSE POINT   By this stage, you should be confident in applying your programming skills and be able to simulate your system on the software available or at least step through your program, observing local and global variables and/or the state of registers as the program develops. Take time to make sure your programs are annotated in a manner that both you and others can readily understand.

Hint   The ability to read code and understand what it does is an essential skill when you need to reuse pre-existing code in other applications and systems. As a programmer you should never re-invent something that already exists and which you can reuse.

Extend   Pair up with another learner and review each other's code to see how easy it is to understand. Provide mutual feedback.

C System development cycle

In the remainder of this unit you will work through the development of three more programs:

▶ an analogue-to-digital conversion example, with an output to LEDs and LCD

▶ a simple I²C program example

▶ a keypad interface.

These examples are in the 'Getting ready for assessment' section at the end of the unit.

Before you start work on them, let us outline the system development cycle, which will allow you to visualise how each stage fits within the overall portfolio of documentation required.

C1 Development processes

You are required to produce a detailed written account of your project, including justifications and analysis of the system you have developed and the methods and approaches you have chosen to realise your solution.

In Section A4, you created the opening sections of your project specification, which covered 'Outline of initial requirements', 'Potential design solutions', 'Chosen design solution', 'References' and Appendices. **Table 6.2** set out the scaffolding for your project specification. Here you will develop it further by adding extra details to enhance any submission you make as part of the external assessment.

Table 6.11 sets out what you need to add to your project specification.

▶ **Table 6.11** New sections to add to your project specification

Functional specification	• A table of essential functions of the system based on the information given in the external assessment brief. • It should also include desirable functions and potential enhancements for future development. • Typically in the format shown in **Table 6.12**.
Chosen design solution	Add the following to this section: • Detailed justifications of the inputs, outputs and microcontroller chosen for your solution and how they meet or enhance the requirements of the client brief. If only one microcontroller is available, ensure you provide details of the microcontroller that would have been the most suitable for this system if you had been able to procure it and, most importantly, explain why it would be the most suitable. • Circuit diagrams – make sure these do not become cluttered with too much circuitry and components. To avoid too much clutter, create sub-circuit diagrams drawn to the relevant British Standards.
Implementation	Position this after the 'Chosen design solution' section. Include: • A plan for implementation, with set milestones, including a logbook of actions. • A structured test plan based on the functional specification. The more detailed the functional specification the more useful the test plan will be. The easiest way to create the test plan is by adding an extra column to the functional specification called 'Achievement', with possible entries being Full/Partial/None. • A design plan for your microcontroller software code that will realise your solution and interconnect the inputs and outputs – create this using pseudocode, flow charts, decision charts, or a mixture of all of these. Remember to include any rough work along with the final presented work. • A system assembly plan – create this in stages, recording each stage as you complete it. Trying to build the whole system in one go is likely to end in failure, so break down the circuit construction and coding into specific recordable stages. This will help you to compartmentalise errors into specific stages and resolve each of them without having to review the entire solution.
Product design solution	Include: • A description of the solution and its operation, including how the solution meets the assessment brief. • A user manual describing the operation of the project solution. • A technical manual that includes all power requirements, all interconnections and all descriptions of the inputs and outputs, linking them to the sections of code that control their functionality. • A test plan of the project in operation, which demonstrates the system's functionality and identifies any errors or limits to its operation. This should also include a detailed analysis of the functional test plan (use an extended version of **Table 6.12**) and operation test plan (testing all possible inputs, outputs and processes). Your analysis should include how the solution meets the customer brief and areas/functions that could be enhanced and improved. • Fully annotated copies of all code and plans.

▶ **Table 6.12** Format of a functional specification

Functional reference	Details	Essential/Desirable/Future

C2 Documentation

Finally, you need to collate all the information neatly and concisely, taking care to include everything you have used to develop the product. This must include a portfolio of evidence produced throughout the development process:

▶ technical specification with operational requirements

▶ test plan

▶ details and justifications of input/output devices and hardware selected

▶ system connection diagrams and schematics

▶ initial design of the program structure

▶ annotated copies of all code

▶ audio-visual recording of operation with commentary

▶ test data and analysis

▶ structured project log.

Ⅱ PAUSE POINT
You should now have everything you need, but as a last check refer to the section C2 Documentation in the unit specification for a list of what needs to be in your portfolio. Do you have the items for all the programs you have created to date?

Hint
If you have not included the requisite information and paperwork, then your project cannot be graded, reviewed or considered for assessment – check and recheck your portfolio.

Extend
Review your documentation and ensure that it is specific, detailed and includes full justifications and comprehensive analysis where necessary. Good luck!

THINK ▶FUTURE

Stephen Serplus
Senior Firmware Engineer at Tyco Security Products

I develop software for security cameras, design the company's global API (application programming interface) and support the company's Open Source Software Policy. Prior to working for Tyco Security Products, I worked for 12 years with an electronics design company called Marturion. It specialised in medical electronics and software. I started as a Junior Research Engineer, progressing to Research Engineer, before being promoted to Engineering Manager. I developed a wide range of microcontroller-based systems and products, from an NIBP (non-invasive blood pressure) monitor to automated safety evacuation systems.

I initially took a BTEC National Diploma in Engineering (Electrical/Electronic/Mechanical) from Southern Regional College and then completed an Honours degree in Electronic Systems from the University of Ulster. As part of my continuing professional development (CPD), I act as a STEM Ambassador in Northern Ireland and am a Member of the Institute of Engineering Technology (IET).

Focusing your skills

Important issues to consider and skills to develop

- Make sure you understand the requirements of a project before starting the development process. A missed or misunderstood requirement can be costly and cause a project to fail.
- When designing a complex system, just deciding where to start can seem daunting. The key is to break the system down into manageable blocks.
- Selecting appropriate hardware devices for a design is crucial. Getting the selection wrong can be costly to the project in terms of both time and money.
- Circuit simulation is an important means of proving the viability of your design before you go to the expense of making a prototype PCB board.
- A vital step in the development process is to build a prototype and test its functionality – this helps to predict whether your design will work or not.

- When designing software, plan how all the main features will fit together.
- Break your code up into different layers – this not only makes for good organisation but also allows easier extension and expansion in the future.
- Follow the test-driven development approach – if it's worth building, it's worth testing!
- Be consistent in your coding style. You can download various coding styles from the internet to give you some ideas, but then choose one and stick to it.
- Use source control, such as 'bitbucket', for handling your code, and discipline yourself to update it regularly.
- Stack overflow is a great resource for finding solutions to software problems.
- Research the Agile design process and try to gain a good understanding of it, as it is used in many design companies.

Getting ready for assessment

This section has been written to help you to do your best when you take the external practical assessment. Read through it carefully and use it to help you create, develop and build the example projects. Ask your tutor if there is anything you are not sure about.

About the test

The test booklet will contain material for completion of the set task under supervised conditions. You will not receive the booklet until the assessment period.

You will complete all the work for your assessment on a computer, using appropriate hardware and software as listed in the unit content. You will use an electronic task book, which will be provided. You will not have access to the internet during the supervised periods.

All your work, a task book and one audio-visual recording must be submitted for assessment on a compact disc (CD). Your centre will provide you with access to suitable audio-visual recording equipment and software to ensure that the footage is recorded in an appropriate file format.

Make sure that your work is backed up securely and is kept until the end of the post-result service period. You must complete your work independently, and your tutor will authenticate it before it is submitted for assessment. Your centre will arrange the supervised assessment.

You must work independently throughout the supervised assessment period, and you should not share your work with other learners. Your tutor may clarify the wording that appears in this task but cannot provide any guidance on how to complete the task.

As the guidelines for assessment can change, you should refer to the official assessment guidance on the Pearson Qualifications website for the latest definitive guidance.

The assessment will be in the form of:
- a scenario
- a client brief
- planning activity
- analysis of brief
- system design
- assembly and programming
- system testing and analysis.

Preparing for the test

This unit is assessed under supervised conditions. Your preparation should include:
- understanding the format of the specification
- knowing how to locate datasheets and conduct other research
- completing the example assessment tasks below
- constructing test schedules
- analysing test schedules
- practice in circuit construction and testing
- construction and analysis of commented code
- review of previous tests, including:
  - booklet
  - task
  - marked solutions.

Hints and tips for the supervised assessment

- Make sure you fully understand the client requirements.
- Planning is vital, so review your proposed solution and plan it thoroughly. Include milestones and allow time at the end to provide the all-important analysis.
- Focus on the scenario and client brief, ensuring that you achieve the minimum requirements first. There will be time for enhancements later.

Sample solutions

We will now look at three worked examples, where solutions are developed for analogue-to-digital conversion with output to LEDs and LCD, a simple I²C and a keypad interface.

Pseudocode is provided for the first example, circuit schematics for the first two examples, and C code for all three examples. The rest of the development process is left to you. You should consider why the particular inputs and outputs have been chosen and how they meet the requirements of the project. Record your reasons and consider how you can put together a detailed justification.

Create a log of how you develop each solution, breaking it down into stages and recording all testing that you carry out. Many learners forget that every time you download a program and run it to check that it functions as expected, you are testing. It is not just the end tests that count.

Worked example A

ADC input with LED/LCD output

1 Build the circuit, referring to the diagrams in **Figures 6.18** and **6.19**.

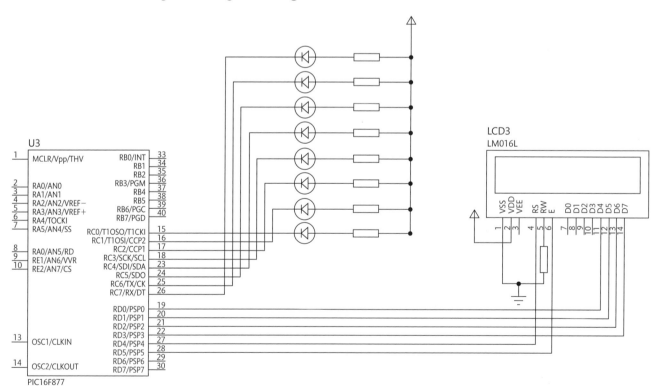

▶ **Figure 6.18** Schematic of a microcontroller with both LCD and LED outputs

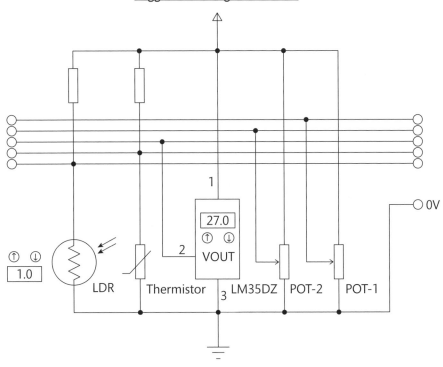

Suggested Analogue Interfaces

▶ **Figure 6.19** Schematic of analogue inputs to the microcontroller that you could use

2 Design the program using pseudocode.

SET pins and ports for LCD/LED outputs

SET constant descriptors and variables

BEGIN

    SET LCD output

    SET ADC use

    REPEAT UNTIL stopped

        CALL monitor and run ADC return value

        OUTPUT ADC value to LEDs

        OUTPUT ADC value output to LCD

    END REPEAT

END

BEGIN monitor and run ADC

    SET channel for ADC in line with Microcontroller manual

    SET mode for ADC in line with Microcontroller manual

    RETURN ADC value

END

BEGIN ADC value to LEDs

    IF value is > 100 light LSB Port connected to LEDs

    ELSE

        IF value is > 200 light bit Rx.0 and Rx.1 connected to LEDs

```
        ELSE
            IF value is > 300 light bit Rx.0, Rx.1, Rx.2 connected to LEDs
            ELSE
                IF value is > 400 light bit Rx.0, Rx.1,Rx.2,Rx.3
                ELSE
                    IF value is > 500 light bit Rx.0 to Rx.4
                    ELSE
                        IF value is > 600 light bit Rx.0 to Rx.5
                        ELSE
                            IF value is > 700 light bit Rx.0 to Rx.6
                            ELSE
                                IF value is > 800 light bit Rx.0 to Rx.7
                                ENDIF
                            ENDIF
                        ENDIF
                    ENDIF
                ENDIF
            ENDIF
        ENDIF
END


BEGIN  ADC value output to LCD
    SET LCD
    CALCULATE ADC output and convert to ASCII for output as a character
    IF ADC >= 700
        OUTPUT ADC value plus IT IS NIGHT TIME
        IF ADC >= 500
            OUTPUT ADC value plus IT IS MIDDAY
            IF ADC >= 300
                OUTPUT ADC value plus IT IS MORNING
            ENDIF
        ENDIF
    ENDIF
END
```

3 Study the following code. It is quite complex and covers many aspects. Try to step through the code and notice the links and jumps from one function to another. The code has been annotated as much as possible to help you understand its structure. Many of the functions have been created in line with the PIC16F877 user manual.

```
// Title: Analogue-to-digital conversion
// Description: ADC with an output on an array of LEDs and on an LCD
// Author: Alan Serplus / Paul H Stewart
// Date: 09/02/xxxx
// Versions 1.3a


// Include header files for Micro
#include <pic16f877.h>
#include <htc.h>
// end includes


// define constants
const unsigned char channel_0 = 0b10000001;  // set the binary value for each channel to a name - there
                                             // are 8 possible channels
const unsigned char channel_1 = 0b10001001;  // each channel is listed in the manual and include only
                                             // those needed in the program


// ADC mode select, i.e. number of channels, assign reference voltages etc.
const unsigned char mode_0  = 0b10000000;    // set the ADC mode - there are 16 in total
const unsigned char mode_1  = 0b10000001;    // include only those needed in the program
// end define constants


// defines
#define LCD_port  PORTD              // set port D and name within code as LCD_port
#define LCD_TRIS  TRISD              // rename TRISD as LCD_TRIS for ease of reference
#define    LCD_RS  RD5               // name pin RD5 as LCD_RS, i.e. LCD register select line
#define    LCD_E  RD4                // name pin RD4 as LCD_E, i.e. LCD enable signal
// end defines


// define constants
const char line1 = 0x80;            // LCD constants - when line1 is used the LCD will be set to line 1
const char line2 = 0xC0;            // LCD constants
const char line3 = 0x94;            // LCD constants
const char line4 = 0xD4;            // LCD constants
const char LCD_enable = 0x10;       // set enable line
// end constants


// declare function prototypes
void output_bar(int ADC_result);
void display_result(int ADC_result);
void config_ADC(unsigned char ADC_mode);        // set the ADC mode using constants above
int start_ADC(unsigned char channel_num);       // associated with ADC routines
void init_LCD(void);
void clock_inst(void);
void LCD_function(unsigned char function);
void output_char_string(unsigned char line_num, const char output_string[ ]);
void output_char(unsigned char character);
void clock_LCD(void);
void delay_ms(unsigned int length);             // millisecond delay routine
void delay_100us(unsigned int length);          // 100 microsecond delay routine
// end function declares

void main()
{
    int ADC_result;             // declare memory called ADC_result
    TRISC = 0;                  // set port C as an output
    PORTC = 0x00;               // clear port C
    // LCD CODE
    init_LCD();                 // initialise LCD routine
    LCD_function(0x0C);         // function to turn LCD on
    // ADC code
```

```
        config_ADC(mode_0);    // first select mode of operation
                               // mode_0 - configure RE2, RE1, RE0, RA5, RA3, RA2, RA1, RA0 as analogue inputs
        while (1)
            {
                ADC_result = start_ADC(channel_0); // call conversion routine - i/p signal on channel 0 (RA0)
                                                   // pass resulting returned int into variable ADC_result
                output_bar(ADC_result);            // call function to display ADC result sending
                                                   // parameter ADC_result
                display_result(ADC_result);
            }
}

void config_ADC(unsigned char ADC_mode)
{
    ADCON1 = ADC_mode;              // select mode of operation
}

int start_ADC(unsigned char channel_num)
{
    unsigned int ADC_result;        // declare variable that will be returned
    ADCON0 = channel_num;           // select channel number and turn ADC on
    ADON = 0x01;                    // start conversion
    while (ADON){};                 // wait for end of conversion
    ADC_result = ADRESH*256 + ADRESL;  // convert result to a 16-bit number:
                                       // ADC_result = (upper byte × 256) + lower byte
    return ADC_result;              // return 10-bit result (0 to 1023) in 16-bit format (ADC_result)
}

// BINARY OUTPUT

void output_bar(int ADC_result)    // function output_bar takes parameter ADC_result
{
    unsigned char bar;
    bar = 0b000000000;             // reset bar height to 0

    if(ADC_result>100)             // test if input light level greater than 100
    {

        bar = 0b11111110;          // YES - set bar height (or 0xFE)
    }
    if(ADC_result>200)             // test if input light level greater than 200
    {

        bar = 0b11111100;          // YES - set new bar height
    }
    if(ADC_result>300)             // test if input light level greater than 300
    {

        bar = 0b11111000;          // YES - set new bar height
    }
    if(ADC_result>400)             // test if input light level greater than 400
    {

        bar = 0b11110000;          // YES - set new bar height
    }
    if(ADC_result>500)             // test if input light level greater than 500
    {

        bar = 0b11100000;          // YES - set new bar height
    }
    if(ADC_result>600)             // test if input light level greater than 600
```

```
        {
            bar = 0b11000000;
        }
        if(ADC_result>700)                  // test if input light level greater than 700
        {
            bar = 0b100000000;              // YES - set new bar height
        }
        if(ADC_result>800)                  // test if input light level greater than 800
        {
            bar = 0b00000000;               // YES - set maximum bar height
        }
        PORTC = bar;                        // regardless of value output bar value to port B
}

/***********************************************************************
 * Function to initialise LCD. Consider further research on methods    *
 ***********************************************************************/
void init_LCD()
{
        LCD_TRIS = 0;                       // configure LCD port as an output port
        LCD_port = 0;                       // clear port lines

        delay_ms(20);                       // 20ms delay

        LCD_port = (0x03|LCD_enable);       // #1 control sequence
        clock_inst();

        delay_ms(5);                        // delay 5ms

        LCD_port = (0x03|LCD_enable);       // #2 control sequence
        clock_inst();

        delay_ms(1);                        // delay 1ms

        LCD_port = (0x03|LCD_enable);       // #3 control sequence
        clock_inst();

        LCD_port = (0x02|LCD_enable);       // #4 control sequence
        clock_inst();

        LCD_port = (0x02|LCD_enable);       // #5 control sequence
        clock_inst();

        LCD_port = (0x08|LCD_enable);       // #6 control sequence
        clock_inst();

/*************************************************
 * function set command " 0 0 0 0 1 D C B "      *
 * D = 1 LCD on                                  *
 * C = 1 cursor on                               *
 * B = 1 blink  on                               *
 *************************************************/
        LCD_port = (0x00|LCD_enable);       // #7 control sequence - switch LCD off
        clock_inst();
        LCD_port = (0x08|LCD_enable);       // #8 control sequence
        clock_inst();

/*****************************************
 * Clear display - Control word = 0x01   *
 *****************************************/
        LCD_port = (0x00|LCD_enable);       // #9 control sequence - clear display
        clock_inst();
```

```
        LCD_port = (0x01|LCD_enable);                // #10 control sequence
        clock_inst();

/***************************************************************
 * Entry mode setting                                          *
 * Entry mode command " 0 0 0 0 0 1 ID S "                     *
 * if ID = 0 no increment (during write and read operation)    *
 * if S = 0 no shift                                           *
 ***************************************************************/
        LCD_port = (0x00|LCD_enable);                // #11 control sequence
        clock_inst();
        LCD_port = (0x06|LCD_enable);                // #12 control sequence
        clock_inst();
}
        // end of initialisation

void clock_inst()
{
        LCD_E = 0;
        delay_ms(2);
        LCD_E = 1;
        delay_ms(2);
}

/***************************************************************
 * Outputs a control character to an LCD – high nibble first   *
 ***************************************************************/
void LCD_function(unsigned char function)
{
        LCD_port = ((function>>4)|LCD_enable);     // shift high nibble to lower nibble and set RS=0 and LCD_E=1
        clock_inst();                              // clock high nibble of control character

        LCD_port = ((function&0x0F)|LCD_enable);   // mask high nibble and set RS=0 and LCD_E=1
        clock_inst();                              // clock low nibble of control character
}

// LCD OUTPUT CODE
void display_result(int ADC_result)
{
        int remainder;
        unsigned char thous, hunds, tens, units;  // declare multiple variables
        thous = (ADC_result/1000) + 0x30;          // calculate number of thousands and convert to ascii (add
                                                   // 0x30)
        remainder = ADC_result%1000;               // calculate remainder
        hunds = (remainder/100) + 0x30;            // calculate number of hundreds and convert to ascii (add
                                                   // 0x30)
        remainder = ADC_result%100;                // calculate remainder
        tens = (remainder/10) + 0x30;              // calculate number of tens and convert to ascii (add 0x30)
        units = ADC_result%10 + 0x30;              // calculate number of units and convert to ascii (add 0x30)

        if(hunds == 0x37)
        {
                output_char_string(line2,"IT'S NIGHT TIME ");  // LCD will display if ADC_result is 700 to 799
        }
        if(hunds == 0x35)
        {
                output_char_string(line2,"IT'S AFTERNOON  ");  // LCD will display if ADC_result is 500 to 599
        }
        if(hunds == 0x33)
        {
                output_char_string(line2,"IT'S MORNING    ");  // LCD will display if ADC_result is 300 to 399
```

```
        }
}

/*****************************************************************
 * Outputs a character string to specified LCD line number   *
 *****************************************************************/
void output_char_string(unsigned char line_num, const char output_string[])
{
     unsigned char i, character;   // declare to character-sized memory locations named i and character

     LCD_function(line_num);       // send cursor to start of specified line number

     for(i=0;output_string[i] != 0;i++)      // output character string until string terminator
     {
         character = output_string[i];
         output_char(character);
     }
}

/*********************************************************************
 * Outputs a character to an LCD – high nibble first as in 4-bit mode *
 *********************************************************************/
void output_char(unsigned char character)
{
     LCD_port = ((character>>4)|0x30);        // get character from character string
                                              // output high nibble first with RS=1 and LCD_E=1
     clock_LCD();                             // clock high nibble of character

     LCD_port = ((character&0x0F)|0x30);      // get character from character string
                                              // output low nibble with RS=1 and LCD_E=1
     clock_LCD();                             // clock low nibble of character
}

void clock_LCD()
{

     LCD_E = 0;                               // falling edge clocks data into LCD
     delay_100us(1);                          // 0.1ms delay
}

/*********************************************************************************************
 * Generic delay routine using Timer 0 clocked at 250kHz, i.e. a period of 4 microseconds   *
 * - number of loops specified by length                                                    *
 * - each loop waits until Timer 0 = 250, i.e. a delay of 1ms                                *
 *********************************************************************************************/
void delay_ms(unsigned int length)
{
     unsigned int i;
     OPTION_REG = 0x01;                       // prescaler = /4 i.e. TMR0 clocked at 250kHz
     for (i=0;i<length;i++)
     {
         TMR0 = 0;                            // generates a 1ms delay
         while (TMR0<250) {};
     }
}

/*********************************************************************************************
 * Generic delay routine using Timer 0 clocked at 250kHz, i.e. a period of 4 microseconds   *
 * - number of loops specified by length                                                    *
 * - each loop waits until Timer 0 = 25, i.e. a delay of 0.1ms                               *
 *********************************************************************************************/
```

```
void delay_100us(unsigned int length)
{
    unsigned int i;
    OPTION_REG = 0x01;              // prescaler = /4 i.e. TMR0 clocked at 250kHz
    for (i=0;i<length;i++)
    {
        TMR0 = 0;                   // generates a 0.1ms delay
        while (TMR0<25){};
    }
}
```

4 Collate the documentation, including images, circuit diagrams and analysis.

PAUSE POINT

It may take you several sessions to get to grips with coding in C. Once you have achieved this, you should be able to see multiple applications for this type of system. Revisit the code and attempt to simplify or shorten some aspects of it, as each line of code uses up precious internal memory.

Hint

To further rationalise the code, the multiple 'if' statements can be reduced using the 'switch' decision statement.

Extend

Research header files that have pre-written common routines to reduce the complexity of the code.

Worked example B

Simple I²C

The fundamental communications procedure for all I²C operations is as follows.

- One IC that wants to talk to another must:
  - wait until it sees no activity on the I²C bus – SDA and SCL are both high, i.e. the bus is 'free'.
  - put a message on the bus that says 'it is mine' – I have **start**ed to use the bus. All other ICs then **listen** to the bus data to see whether they might be the one who will be called up (addressed).
  - provide on the **clock** (SCL) wire a clock signal – this will be used by all the ICs as the reference time at which each bit of **data** on the data wire (SDA) will be valid and can be used. The data on the data wire (SDA) must be valid at the time the clock wire (SCL) switches from 'low' to 'high' voltage.
  - put out in serial form the unique binary 'address' (name) of the IC that it wants to communicate with.
  - put a message (one bit) on the bus announcing whether it wants to **send** or **receive** data from the other chip. (There is no read/write wire in this system.)
  - ask the other IC to **acknowledge** (using one bit) that it recognised its address and is ready to communicate.
- After the other IC acknowledges that all is OK, data can be transferred.
- The first IC sends or receives as many 8-bit words of data as it wants. After every 8-bit data word, the sending IC expects the receiving IC to acknowledge that the transfer is going OK.

- When all the data has finished transferring, the first chip must free up the bus – it does this by a special message called 'STOP'. This is just one bit of information transferred by a special 'wiggling' of the SDA/SCL wires of the bus.

1 Build the circuit, referring to **Figure 6.20**.

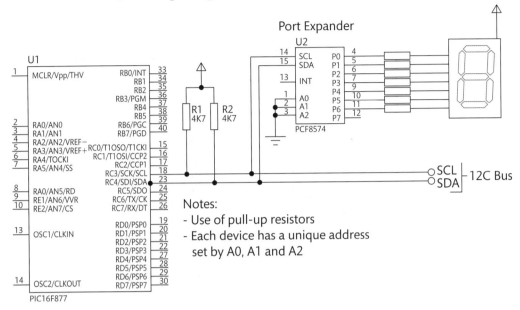

▶ **Figure 6.20** Schematic for basic communications using an I²C interface

2 Design the program using pseudocode.

3 Study the following code.

```
// Title: Simple I²C sample protocol
// Description: Example of send instruction on I²C bus
// Author: Paul H Stewart
// Date: 09/02/xxxx
// Versions 1.2b

// Include header files for Micro
#include <pic16f877.h>
#include <htc.h>
// end includes

// define constants
const unsigned char BCD_to_7seg[10]          // set up an array for segment codes as connected in circuit
                                             // diagram
    = {0x3F, 0x06, 0x5B, 0x4F, 0x66, 0x6D, 0x7D, 0x07, 0x7F, 0x6F};     // these can vary depending on
                                                                         // order of connection

// end define constants

// defines
```

```
// end defines

// define constants
// end constants

// declare function prototypes
void init_iic(void);
void i2c_write(unsigned char iic_data, unsigned char slave_address);
void delay_ms(unsigned in lengthms)
// end function declares

void main()
{
    unsigned char slave_address, iic_data, unit;

    init_iic();      // call function to initialise iic bus

    while(1)
    {
        for (unit = 0; unit<10; unit++)
        {
            slave_address = 0x40;                  // set up slave address
            iic_data = ~BCD_to_7seg[unit];
            i2c_write(iic_data, slave_address);    // call function to output data
            delay_ms(1000);
        }
    }
}

/***********************************
 * Function to initialise iic bus *
 ***********************************/
void init_iic(void)
{
            TRISC3 = 1;               // configure SCL pin as an input
            TRISC4 = 1;               // configure SDA pin as an input

            SSPADD = 0x28;            // set baud rate = 100kHz, fosc = 4MHz - see manuals for i2c and PIC
            SSPCON = 0b00101000;      // synchronous Serial Port Enable Bit SSPEN = 1
                                      // sSPM3:SSPM0 = 1000 - iic master mode
                                      // clock = Fosc/(4*(SSAPD+1)) = 100KHz
            RC3 = 1;                  // ensure SCL high
            RC4 = 1;                  // ensure SDA high
            delay_ms(10);
}

/**************************************************
 * Function to send a character over the iic bus *
 **************************************************/
void i2c_write(unsigned char iic_data, unsigned char slave_address)
{
            SEN = 1;                  // set start enable bit
            while (SSPIF == 0) {};    // wait for SSPIF to be set
            SSPIF = 0;                // clear SSPIF

            SSPBUF = slave_address;   // set up slave address
            while (SSPIF == 0){};     // wait for SSPIF to be set
            SSPIF = 0;                // clear SSPIF

            SSPBUF = iic_data;        // transmit data
            while (SSPIF == 0){};     // wait for SSPIF to be set
```

```
                SSPIF = 0;                // clear SSPIF

                PEN = 1;                  // generate a STOP condition
                                          // by setting Stop Enable Bit
                while (SSPIF == 0){};     // wait for SSPIF to be set
                SSPIF = 0;                // clear SSPIF
}

void delay_ms(unsigned in lengthms)
{
                // see previous example for code
}
```

Worked example C

Keypad interface

This system allows the integration of relays controlling magnetic locks or other security devices.

Study the circuit schematic (**Figure 6.21**) and code that have been provided, and write the pseudocode (or alternative code development plan) that should have been used to design this code. Working out the connection between the plan and the code will make the coding easier to construct for other projects.

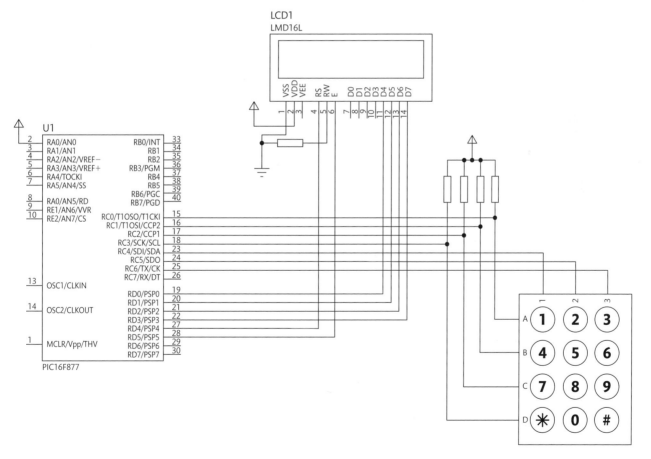

▶ **Figure 6.21** Schematic of keypad interface

Code for Worked example C:

```
// Title: Keypad interface and scanning
// Description: Example to show how to scan keypads and link to LCD or other devices
// Author: Alan Serplus
// Date: 09/02/xxxx
// Versions 1.0

// Include header files for Micro
#include <pic16f877.h>
#include <htc.h>
// end includes

// defines
#define   col_1 RC4                    // set pin 4 of port C as column 1
#define   col_2 RC5
#define   col_3 RC6
// end defines

// define constants
unsigned char keyvalue;                // set global variable keyvalue as a character (8-bit)
char   port_value;
// end constants

// declare function prototypes
void keypad_scan(void);
void init_LCD(void);
void LCD_function(unsigned char function);
void init_LCD(void);
void clock_inst(void);
void output_char_string(unsigned char line_num, const char output_string[ ]);
void output_char(unsigned char character);
void clock_LCD(void);
void delay_ms(unsigned int length);       // millisecond delay routine
void delay_100us(unsigned int length);    // 100 microsecond delay routine
// end function declares

void main()
{
    TRISC = 0x0F;            // set port C as part input and part output RC.4; RC.5, RC.6, RC.7 are
                             // outputs col 1 to 3
    keyvalue = 0;            // clear keyvalue
    init_LCD();              // function to initialise LCD
    LCD_function(0x0C);      // function to turn LCD on
    output_char_string(line1,"* Keypad Test  *");     // output character string
    output_char_string(line2,"Keyvalue is:    ");     // output character string

    /*   To output individual characters:
         First, move cursor to character position using LCD_function() function
         Second, output character using output_char() function
         NB: Cursor position automatically increments after a character has been written to the LCD. */

    while (1)                            // never-ending loop
    {
        keypad_scan();
        if (keyvalue!=0)                 // test if keyvalue has changed
        {                                // YES - output new keyvalue
            LCD_function(0xCD);          // move cursor to line 2, position 13
            output_char(keyvalue);       // output keyvalue to line 2, position 13
            keyvalue = 0;                // reset keyvalue to 0
            delay_ms(500);               // wait 0.5 seconds
        }
```

```
        }
}

void keypad_scan(void)
{
        // SCAN COLUMN 1
        col_1 = 0; col_2 = 1; col_3 = 1;          // select column 1, i.e. col 1 = 0
        port_value = PORTC;                        // variable port_value takes on port C pin values
        port_value = (~port_value & 0xFF);         // invert the values from port C
        port_value = (port_value&0x0F);            // mask off upper nibble, only monitor lower nibble
                                                   // now test if a col 1 button pressed

        switch(port_value)
        {
                case 1: keyvalue = 0x31;           // key pressed is '1' (ascii value: 0x31)
                                break;             // quit block
                case 2: keyvalue = 0x34;           // key pressed is '4' (ascii value: 0x34)
                                break;             // quit block
                case 4: keyvalue = 0x37;           // key pressed is '7' (ascii value: 0x37)
                                break;             // quit block
                case 8: keyvalue = 0x2A;           // key pressed is '*' (ascii value: 0x2A)
                                break;             // quit block
        }

        // SCAN COLUMN 2
        col_1 = 1;  col_2 = 0; col_3 = 1;          // select column 2, i.e. col 2 = 0
        port_value = PORTC;                        // variable port_value takes on port C pin values
        port_value = (~port_value & 0xFF);         // invert the values from port C
        port_value = (port_value&0x0F);            // mask off upper nibble
                                                   // now test if a col 1 button pressed

        switch(port_value)
        {
                case 1: keyvalue = 0x32;           // key pressed is '2' (ascii value: 0x32)
                                break;             // quit block
                case 2: keyvalue = 0x35;           // key pressed is '5' (ascii value: 0x35)
                                break;             // quit block
                case 4: keyvalue = 0x38;           // key pressed is '8' (ascii value: 0x38)
                                break;             // quit block
                case 8: keyvalue = 0x20;           // key pressed is '0' (ascii value: 0x20)
                                break;             // quit block
        }

        // SCAN COLUMN 3
        col_1 = 1; col_2 = 1; col_3 = 0;           // select column 3, i.e. col 3 = 0
        port_value = PORTC;                        // variable port_value takes on port C pin values
        port_value = (~port_value & 0xFF);         // invert the values from port C
        port_value = (port_value&0x0F);            // mask off upper nibble
                                                   // now test if a col 1 button pressed

        switch(port_value)
        {
                case 1: keyvalue = 0x33;           // key pressed is '3' (ascii value: 0x33)
                                break;             // quit block
                case 2: keyvalue = 0x36;           // key pressed is '6' (ascii value: 0x36)
                                break;             // quit block
                case 4: keyvalue = 0x39;           // key pressed is '9' (ascii value: 0x39)
                                break;             // quit block
                case 8: keyvalue = 0x23;           // key pressed is '#' (ascii value: 0x23)
                                break;             // quit block
        }
}
```

```
void init_LCD()
{
    // see Worked example A for code
}

void LCD_function(unsigned char function)
{
    // see Worked example A for code
}

void clock_inst()
{
    // see Worked example A for code
}

void output_char_string(unsigned char line_num, const char output_string[ ])
{
    // see Worked example A for code
}

void output_char(unsigned char character)
{
    // see Worked example A for code
}

void clock_LCD()
{
    // see Worked example A for code
}

void delay_ms(unsigned int length)
{
    // see Worked example A for code
}

void delay_100us(unsigned int length)
{
    // see Worked example A for code
}
```

Think about it

▸ You have now completed several projects, analysed the inputs and outputs, and developed coding and testing schedules. Don't stop there!

▸ Look for systems to practise your skills on in the world around you. Think of potential microcontroller systems that could be used to automate or manage the system.

▸ Try any available sample assessments and assessments from previous years to hone your skills in all aspects of this unit.

▸ You will not have access to your work outside the supervised periods, so speed and accuracy are vital during the assessment – the more you practise, the faster and more accurately you will be able to work.

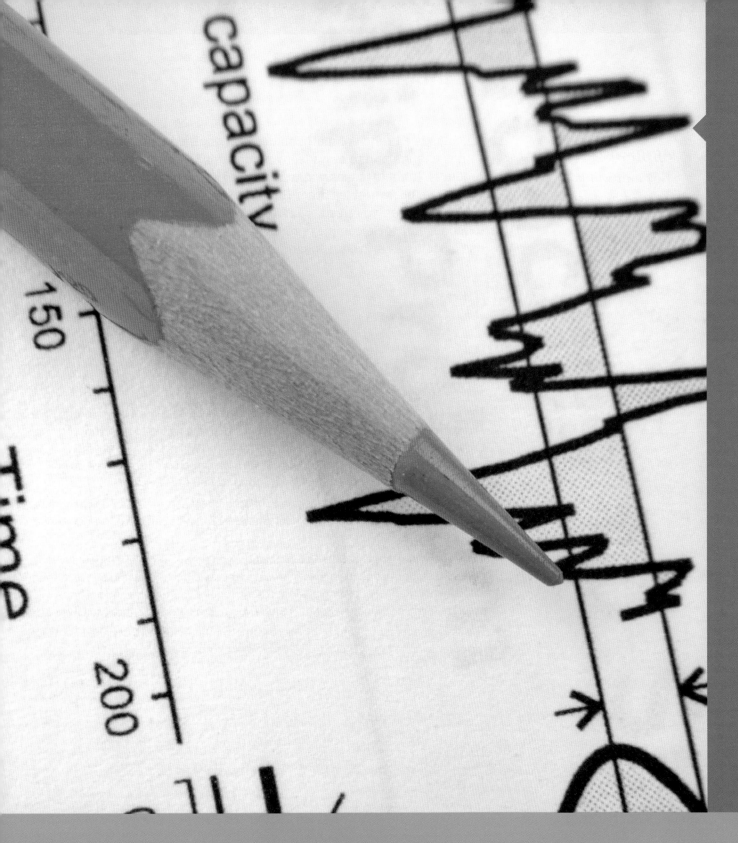

Calculus to Solve Engineering Problems 7

Getting to know your unit

Calculus is a branch of mathematics that engineers use to model how things change and to test ideas and designs before they become real products. In many cases you will want to know how a quantity changes with time, but variables other than time can also be important. As a future engineer, you need to understand and develop the skills needed to model and solve complex problems. In this unit you will learn how to use calculus as a tool for solving engineering problems.

How you will be assessed

This unit will be assessed by a series of internally assessed tasks set by your tutor. Throughout this unit you will find assessment practice activities to help you work towards your assessment. Completing these activities will not mean that you have achieved a particular grade, but you will have carried out useful research or preparation that will be relevant when it comes to your final assignment.

In order for you to achieve the tasks in your assignment, it is important to check that you have met all of the Pass assessment criteria. You can do this as you work your way through the assignments.

If you are hoping to gain a Merit or Distinction, you should also make sure that you present the information in your assignment in the style that is required by the relevant assessment criteria. For example, Merit criteria require you to solve and analyse accurately, while Distinction criteria require you to evaluate and critically analyse.

The assignments set by your tutor will consist of a number of tasks designed to meet the criteria in the table. They are likely to take the form of written reports.

Assessment criteria

This table shows what you must do in order to achieve a **Pass**, **Merit** or **Distinction** grade, and where you can find activities to help you.

Pass	Merit	Distinction

Learning aim **A** Examine how differential calculus can be used to solve engineering problems

Pass	Merit	Distinction
A.P1 Find the first and second derivatives for each type of given routine function. **Assessment practice 7.1**	**A.M1** Find accurately the graphical and analytical differential calculus solutions and, where appropriate, turning points for each type of given routine and non-routine function and compare the results. **Assessment practice 7.1**	**A.D1** Evaluate, using technically correct language and a logical structure, the correct graphical and analytical differential calculus solutions for each type of given routine and non-routine function, explaining how the variables could be optimised in at least two functions. **Assessment practice 7.1**
A.P2 Find, graphically and analytically, at least two gradients for each type of given routine function. **Assessment practice 7.1**		
A.P3 Find the turning points for given routine polynomial and trigonometric functions. **Assessment practice 7.1**		

Learning aim **B** Examine how integral calculus can be used to solve engineering problems

Pass	Merit	Distinction
B.P4 Find the indefinite integral for each type of given routine function. **Assessment practice 7.2**	**B.M2** Find accurately the integral calculus and numerical integration solutions for each type of given routine and non-routine function, and find the properties of periodic functions. **Assessment practice 7.2**	**B.D2** Evaluate, using technically correct language and a logical structure, the correct integral calculus and numerical integration solutions for each type of given routine and non-routine function, including at least two set in an engineering context. **Assessment practice 7.2**
B.P5 Find the numerical value of the definite integral for each type of given routine function. **Assessment practice 7.2**		
B.P6 Find, using numerical integration and integral calculus, the area under curves for each type of given routine definitive function. **Assessment practice 7.2**		

Learning aim **C** Investigate the application of calculus to the solution of a defined specialist engineering problem

Pass	Merit	Distinction
C.P7 Define a given engineering problem and present a proposal to solve it. **Assessment practice 7.3**	**C.M3** Analyse an engineering problem, explaining the reasons for each element of the proposed solution. **Assessment practice 7.3**	**C.D3** Critically analyse, using technically correct language and a logical structure, a complex engineering problem, synthesising and applying calculus and a mathematical model to generate an accurate solution. **Assessment practice 7.3**
C.P8 Solve, using calculus methods and a mathematical model, a given engineering problem. **Assessment practice 7.3**	**C.M4** Solve accurately, using calculus methods and a mathematical model, a given engineering problem. **Assessment practice 7.3**	

Getting started

You have already studied quite a few mathematical topics and gained skills and knowledge. Think back to what you have learned in the past and what you have found out about your best ways of learning. Try doing a few warm-up questions from your previous work. Algebra skills will be particularly useful when studying calculus. Like all skills, mathematical skills benefit from practice. When you have completed this unit, think about how your previous skills and knowledge have developed and identify some new areas to explore in the future.

 A

Examine how differential calculus can be used to solve engineering problems

A1 Functions, rate of change and gradient

Functions

You can begin to understand the relationship between measured values in an engineering problem by looking for patterns and examining how the values relate to each other. You can model the relationship between two sets of values as an equation for the **function**. A function is a relation that involves two or more variables.

If you know the equation for a function $f(x)$, you can calculate the output value of the function for different values of the input variable x. For example, if the function has equation $f(x) = x^2$, its output values when the inputs are 2 and 3 can be calculated as

$$f(2) = 2^2 = 4$$
$$f(3) = 3^2 = 9$$

This function can be written in terms of two variables as the equation $y = x^2$. You can see that the value of y depends on the value of x, so x is called the independent variable and y is called the dependent variable.

The function $y = x^2$ is an example of a **polynomial** function. It is a polynomial of **degree** 2 because the highest power of the independent variable x is 2. Polynomials such as $y = 3x^2 + 4x - 5$ are important functions for engineers to recognise.

In this unit, 'routine' functions mean functions to which differential calculus can be applied directly with no manipulation. Polynomials are examples of routine functions. 'Non-routine' functions are those that need algebraic manipulation, sometimes several steps, before they can be differentiated. Engineering problems are more likely to involve non-routine functions.

Besides polynomials, there are several other types of functions that engineers often use.

Trigonometric functions

The most common trigonometric functions are sine (sin), cosine (cos) and tangent (tan). You will also use the inverse trigonometric functions ($\sin^{-1}$, $\cos^{-1}$ and $\tan^{-1}$) on your calculator.

Exponential functions

Exponential functions take the form $f(x) = a^x$ where a is a positive constant. Note that the independent variable, x, is in the exponent. These functions have some interesting properties, such as:

▶ if $a = 1$ then $f(x) = 1^x = 1$ for all values of x

▶ if $x = 0$ then $f(0) = a^0 = 1$ for all values of a.

The second property above means that the graph of $y = a^x$ always passes through the point (0, 1), regardless of the value of a.

A particular exponential function that is very important in engineering applications is $f(x) = e^x$, where $e \approx 2.71828...$ is a constant sometimes known as Euler's number.

Logarithmic functions

Logarithmic functions are of the form $y = \log_a(x)$ where a is a positive constant, read as 'y is the **logarithm** to the base a of the number x'.

$y = \log_a(x)$ has the same meaning as $x = a^y$. For example, $100 = 10^2$, so $2 = \log_{10}(100)$; in other words, 2 is the logarithm to base 10 of the number 100.

Engineers normally use two bases for logarithms, 10 and e. Logarithms to base 10 are typically written simply as $\log(x)$, where the subscript 10 is omitted. Logarithms to base e are called 'natural logarithms' and typically written as $\ln(x)$.

Exponential and logarithmic functions are said to be inverse functions of each other. On many calculators you access 10^x as the inverse of log and e^x as the inverse of ln, or vice versa.

Key terms

Exponential function – a function of the form $f(x) = a^x$ with $a > 0$.

Logarithm – the power (or exponent) to which a base number is raised to give a particular value:
$$\text{number} = \text{base}^{\text{logarithm}}$$
Engineers commonly use logarithms to either base 10 (log) or base e (ln).

Engineers often work with functions that are more complicated combinations of the basic forms described above. For example:

▶ rational functions (ratios of polynomials) such as
$$x = \frac{s + 2}{s^2 + 3s - 6}$$

▶ more complex trigonometric functions such as
$$y = \sin^2\left(4x - \frac{\pi}{3}\right)$$

▶ more complex exponential functions such as
$$v = 12\left(1 - 0.5e^{-\frac{t}{20}}\right)$$

▶ more complex logarithmic functions such as
$$i = \ln(2t + 10).$$

Gradient and rate of change of a linear function

A linear function is a polynomial of degree 1. It can be written in the form $y = mx + c$, where m and c are constants. The graph of a linear function is a straight line, for which m is the gradient and c is the y-intercept (intersection with the y-axis).

In the worked example below, the independent variable is time and the dependent variable is distance. The gradient of the graph therefore represents how distance changes in relation to time.

Worked Example

A person standing 10 metres from you starts to walk away in straight line, moving at a constant 2 metres per second over a period of 5 seconds. Find an equation to express the distance between the person and you at any time.

Solution

Let t be the time in seconds after the person starts walking away from you, and let s be the distance between the person and you.

Using the given information on velocity (the person moves away at a rate of $2\,\text{m}\,\text{s}^{-1}$), you can make a table showing different times and their corresponding distances, like **Table 7.1**.

▶ **Table 7.1** Distances at some different times

Time, t (s)	0	1	2	3	4	5
Distance, s (m)	10	12	14	16	18	20

You can draw a graph, as in **Figure 7.1**.

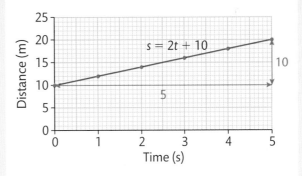

▶ **Figure 7.1** Graph of distance versus time, showing how to calculate the gradient

The graph is a straight line, with the same gradient everywhere:

$$\text{gradient} = \frac{\text{vertical change}}{\text{horizontal change}}$$
$$= \frac{\text{change in distance}}{\text{change in time}}$$
$$= \frac{20 - 10}{5 - 0} = \frac{10}{5} = 2 \, \text{m s}^{-1}$$

Because the line goes through (0, 10), the y-intercept is 10.

The equation describing the distance s at time t is $s = mt + c$ with $m = 2$ and $c = 10$; that is, $s = 2t + 10$.

The gradient of the line describes the **rate of change** of the dependent variable with respect to the independent variable. In the worked example, the gradient measures the rate at which the distance from the observer is changing with respect to time – in other words, the velocity. Velocity is a time-based function, so we can write it as $v(t)$ in function notation. Note that this is not the same as $v \times t$. In the above example, $v(t) = 2 \, \text{m s}^{-1}$, a constant value.

Gradient of a non-linear function

The graph of a non-linear function will not be a straight line, so the gradient does not remain the same throughout. To find the gradient at any point on the graph, you calculate the gradient of the **tangent** to the graph at that point.

Key terms

Rate of change – how fast the dependent variable changes as the independent variable increases. It should always be stated in the form 'the rate of change of … with respect to …'. For example, velocity is the rate of change of displacement with respect to time. The shorthand 'w.r.t.' is often used for 'with respect to'.

Tangent – a straight line that touches a curve at one point.

Worked Example

What would happen if, in the previous example, the person moving away from you had been *accelerating* from standing at $1 \, \text{m s}^{-2}$ (1 metre per second per second)? Plot the distance–time graph and find the velocity at 2 seconds and 4 seconds after the person starts moving away.

Solution

The distances at times $t = 1, 2, …, 5$ are now as shown in **Table 7.2**.

▸ **Table 7.2** Distances at some different times

Time, t (s)	0	1	2	3	4	5
Distance, s (m)	10.0	10.5	12.0	14.5	18.0	22.5

Acceleration means that the velocity is constantly changing, and therefore so is the gradient. The gradient at a given value of t can be found graphically, by drawing a tangent to the curve at the required value of t and finding the slope of that tangent, as shown in **Figure 7.2**.

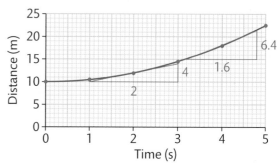

▸ **Figure 7.2** Gradient of a curve

The gradient is greater at $t = 4$ than at $t = 2$, so the rate of change of distance with respect to time (or the velocity) is greater at 4 seconds than at 2 seconds. Using the triangles drawn in **Figure 7.2**:

$$v(2) = \text{gradient at 2} = \frac{\text{vertical change}}{\text{horizontal change}}$$
$$= \frac{4}{2} = 2 \, \text{m s}^{-1}$$
$$v(4) = \text{gradient at 4} = \frac{6.4}{1.6} = 4 \, \text{m s}^{-1}$$

A2 Methods of differentiation

Graphical differentiation

As you saw above, you can find the gradient of a non-linear function at any point by drawing the tangent to the curve at that point, as in **Figure 7.3**.

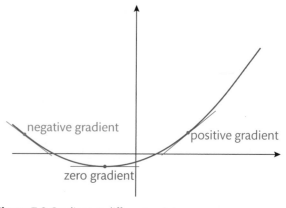

▸ **Figure 7.3** Gradient at different points on a curve

- If the curve slopes up from left to right, the gradient is positive.
- If the curve slopes down from left to right, the gradient is negative.
- If the graph is constant (horizontal), the gradient is zero.

Calculating a gradient graphically only gives an approximate value, since the position of the tangent is just an estimate. To obtain an accurate result, you need to use an analytical method.

Differentiation from first principles

Key terms

Differentiation – the process of calculating the gradient of a function at a given point, or calculating the rate of change of the dependent variable at a given value of the independent variable. The area of mathematics dealing with differentiation is called differential calculus.

Derivative – the result of applying differentiation to a function.

You can find the gradient at a point P on the graph of a function $f(x)$ by **differentiation**.

The delta method

This method gets its name from using the Greek lower-case letter delta (δ) to represent a small change in a variable. (Some texts use capital delta, Δ, but they have the same meaning.) For example, a small change in the variable x is written as δx.

On the graph of the function $f(x)$, consider a second point Q close to P. You can find the gradient of the line joining P and Q (the secant); see **Figure 7.4**. As you move Q closer to P, the gradient of the secant PQ gets closer to the gradient of the tangent to the curve at P.

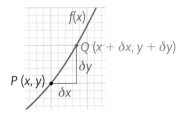

▶ **Figure 7.4** Differentiation from first principles, by using the secant to approximate the tangent

If the coordinates of P are (x, y), then the coordinates of Q are $(x + \delta x, y + \delta y)$, where δx is the small horizontal distance and δy is the small vertical distance between P and Q. Since both points lie on the graph of $f(x)$ we have

$$y = f(x)$$
$$y + \delta y = f(x + \delta x)$$

The gradient of the secant PQ is:

$$\frac{\delta y}{\delta x} = \frac{(y + \delta y) - y}{(x + \delta x) - x} = \frac{f(x + \delta x) - f(x)}{\delta x}$$

As Q gets very close to P, the gradient of the secant PQ approaches the gradient of the tangent at P. The gradient of the tangent is said to be the limit of the above expression as δx tends to zero ($\delta x \to 0$). Mathematically this is written as

$$\frac{dy}{dx} = \lim_{\delta x \to 0} \frac{\delta y}{\delta x} = \lim_{\delta x \to 0} \frac{f(x + \delta x) - f(x)}{\delta x}$$

Here $\frac{dy}{dx}$ denotes the resulting gradient of the tangent, which is called the **derivative**. It represents the rate of change of the variable y with respect to the variable x.

Using the above method of approaching a tangent with a secant to find the gradient of a function at any point is known as 'differentiation from first principles'.

Worked Example

Calculate the derivative of the function $y = x^2$ from first principles.

Solution

$$f(x) = x^2$$
$$f(x + \delta x) = (x + \delta x)^2 = x^2 + 2x\,\delta x + (\delta x)^2$$
$$\lim_{\delta x \to 0} \frac{\delta y}{\delta x} = \lim_{\delta x \to 0} \frac{(x^2 + 2x\,\delta x + (\delta x)^2) - x^2}{\delta x}$$
$$= \lim_{\delta x \to 0} \frac{2x\,\delta x + (\delta x)^2}{\delta x}$$
$$= \lim_{\delta x \to 0} (2x + \delta x) = 2x$$

This means that the gradient at any point x of the function $y = x^2$ is $2x$. This is normally written as $\frac{dy}{dx} = 2x$.

Fortunately, it is not necessary to calculate the derivatives of most common mathematical functions from first principles. This has already been done, and the results have been tabulated for most functions you will ever meet. Most engineering mathematics textbooks include a table of common derivatives.

Derivative notation

The $\frac{dy}{dx}$ form of writing a derivative is known as Leibniz notation. You can think of it as meaning $\frac{\text{difference in } y}{\text{difference in } x}$ which is the definition of gradient. Another way of writing the derivative of a function $y = f(x)$ is $f'(x)$, sometimes called Newton notation. In this book we will mainly be using Leibniz notation, but you should also be aware of the alternative notation $f'(x)$, as it is widely used.

In engineering, the variables you are working with will often be called something other than x and y. In such cases you simply replace the x and y in $\frac{dy}{dx}$ with the names of the independent and dependent variables in the particular context. In the first two worked examples of this unit, the rate of change of distance (s) with respect to time (t) can be written as $\frac{ds}{dt}$ so the velocity can be expressed as $v(t) = \frac{ds}{dt}$.

Note that the gradient $\frac{dy}{dx}$ is known as the *first* derivative of y with respect to x. Later (in Section A4) you will learn about second derivatives, and further derivatives (third, fourth etc.) can also be calculated. Both types of derivative notation can be extended to represent second and higher derivatives.

Derivative of a linear function

You have already seen that the gradient of a linear function $y = mx + c$ is m. So:

▶ if $y = mx + c$ then $\frac{dy}{dx} = m$

or, using the alternative notation,

▶ if $f(x) = mx + c$ then $f'(x) = m$

Differentiating a polynomial function

The preceding worked example gives a clue to how to differentiate a polynomial function. You saw that:

$$y = x^2$$
$$\frac{dy}{dx} = 2x$$

The general rule used for finding the derivative of a polynomial is 'multiply by the power, then subtract one from the power':

$$y = ax^n$$
$$\frac{dy}{dx} = nax^{n-1}$$

Worked Example

Differentiate the function $y = 4x^3$ with respect to x.

Solution

$$\frac{dy}{dx} = 3 \times 4x^{3-1} = 12x^2$$

Sum and difference rules

If a function is the sum (or difference) of two or more other functions of the same variable, then its derivative with respect to that variable is the sum (or difference) of the separate derivatives. This is simpler than it sounds – just differentiate each term separately and add (or subtract) the results.

You can apply this rule together with the previous rule for $y = ax^n$ to differentiate a general polynomial of the form $f(x) = a_nx^n + a_{n-1}x^{n-1} + a_{n-2}x^{n-2} + ... + a_2x^2 + a_1x + a_0$.

Worked Example

Differentiate the function $y = 4x^3 + 2x^2 - 5x + 6$ with respect to x.

Solution

Remember that $x = x^1$ and $x^0 = 1$ (so $6 = 6x^0$).

$$\frac{dy}{dx} = 3 \times 4x^{3-1} + 2 \times 2x^{2-1} - 1 \times 5x^{1-1} + 0 = 12x^2 + 4x - 5$$

Note that differentiating a constant gives zero. The derivative is the rate of change – and a constant does not change!

Differentiating fractions and surds

You can differentiate fractions such as $y = \dfrac{1}{x}$ where the independent variable is in the denominator, by using the rules of indices to rewrite the function as a power of x:

$$y = \frac{1}{x} = x^{-1}$$

$$\frac{dy}{dx} = (-1)x^{-1-1} = -x^{-2}$$

This can be simplified to: $\dfrac{dy}{dx} = -\dfrac{1}{x^2}$

You can treat surds such as $y = \sqrt{x}$ in a similar way:

$$y = \sqrt{x} = x^{\frac{1}{2}}$$

$$\frac{dy}{dx} = \frac{1}{2}x^{\frac{1}{2}-1} = \frac{1}{2}x^{-\frac{1}{2}}$$

This can be simplified to: $\dfrac{dy}{dx} = \dfrac{1}{2\sqrt{x}}$

Standard derivatives

As mentioned earlier, the derivatives of common mathematical functions – such as polynomial, trigonometric, exponential and logarithmic functions – have already been found and are given in tables in most books on engineering mathematics. **Table 7.3** is a simple table showing the derivatives of some commonly encountered functions.

▶ **Table 7.3** Table of standard derivatives

Type of function	Function $y = f(x)$	Derivative $\dfrac{dy}{dx} = f'(x)$
Constant	c	0
Polynomial	x	1
	x^2	$2x$
	ax^n	nax^{n-1}
Trigonometric (x in radians)	$\sin ax$	$a\cos ax$
	$\cos ax$	$-a\sin ax$
	$\tan ax$	$a\sec^2 ax$
Exponential	e^{ax}	ae^{ax}
Logarithm (base e)	$\ln ax$	$\dfrac{1}{x}$

Worked Example

Differentiate the function $y = 4x^5 + 2\sin 3x - 5e^{2x}$ with respect to x.

Solution

Using the derivatives of ax^n, $\sin ax$ and e^{ax} from **Table 7.3** together with the sum and difference rules:

$$\frac{dy}{dx} = 5 \times 4x^{5-1} + 2 \times 3\cos 3x - 5 \times 2e^{2x}$$

$$= 20x^4 + 6\cos 3x - 10e^{2x}$$

So far the functions we have encountered have been sums or differences of standard functions, so differentiating them has been straightforward. But what happens for functions that are put together in a more complex way?

Using the chain rule to differentiate a function of a function

A function such as $y = 2x + 3$ is said to be a simple function. A function such as $y = (2x + 3)^2$ is called a composite function, or a function of a function. In this case, $2x + 3$ is the 'inner' function, and $(\ )^2$ is the 'outer' function.

It is possible to differentiate $y = (2x + 3)^2$ by expanding brackets. This is fine as long as the outer function is a fairly low power, but could prove very time-consuming (and error-prone) if you had to multiply out something like $y = (2x + 3)^7$.

The chain rule is a method of differentiating a composite function by using a substitution for the inner function. Define the inner function to be the variable $u(x)$. Then the chain rule says that:

$$\frac{dy}{dx} = \frac{dy}{du} \times \frac{du}{dx}$$

Worked Example

Use the chain rule to differentiate the function $y = (2x + 3)^7$.

Solution

Let $u = 2x + 3$. Then the given function is $y = u^7$.

Differentiating both of these expressions:

$$\frac{du}{dx} = 2, \quad \frac{dy}{du} = 7u^6 = 7(2x + 3)^6$$

Applying the chain rule:

$$\frac{dy}{dx} = \frac{dy}{du} \times \frac{du}{dx}$$
$$= 7(2x + 3)^6 \times 2$$
$$= 14(2x + 3)^6$$

The product rule

Consider a radio signal that passes through a pre-amplifier and then a power amplifier. An input signal x is operated on by a pre-amplifier function $u(x)$, and this output is then acted upon by a power amplifier function $v(x)$ to produce a system output, as illustrated in **Figure 7.5**.

▶ **Figure 7.5** A product term generated by an amplification system

The output of this system is a **product** term, $y = u(x)v(x)$.

The product rule for differentiating a product term states:

$$\frac{dy}{dx} = v\frac{du}{dx} + u\frac{dv}{dx}$$

To apply the rule, you identify the functions $u(x)$ and $v(x)$, find the derivative of each of them, and then substitute the four expressions into the right-hand side of the rule. (You can think of the rule as a recipe and the four terms as the ingredients. You need to prepare the ingredients before you can use them in the recipe.)

Worked Example

A test signal can be modelled by the equation $y = x\sin 2x$. Use the product rule to find the derivative of y with respect to x.

Solution

Step 1: Identify $u(x)$ and $v(x)$. In most cases, take the first function as $u(x)$ and the second function as $v(x)$.
$$u(x) = x, \quad v(x) = \sin 2x$$

Step 2: Differentiate $u(x)$ and $v(x)$.

Both are standard functions, so you can use the results in **Table 7.3**.
$$\frac{du}{dx} = 1, \quad \frac{dv}{dx} = 2\cos 2x$$

Step 3: Substitute $u(x)$, $v(x)$, $\frac{du}{dx}$ and $\frac{dv}{dx}$ into the product rule formula.
$$\frac{dy}{dx} = v\frac{du}{dx} + u\frac{dv}{dx} = (\sin 2x)(1) + (x)(2\cos 2x)$$

Step 4: Simplif e result if possible.
$$\frac{dy}{dx} = \sin 2x + 2x\cos 2x$$

The quotient rule

The rule for differentiating a **quotient** looks more complicated than the rule for a product, but it contains similar ingredients. If

$$y = \frac{u(x)}{v(x)}$$

then the quotient rule states:

$$\frac{dy}{dx} = \frac{v\frac{du}{dx} - u\frac{dv}{dx}}{v^2}$$

You need to find five expressions to substitute into the quotient rule (four of them are the same as for the product rule, and there is also the v^2 in the denominator). Follow the same procedure as for the product rule.

Worked Example

A signal in a communication system can be modelled by the equation $y = \dfrac{\sin 2x}{x}$
Use the quotient rule to differentiate y with respect to x.

Solution

Step 1: Identify $u(x)$ and $v(x)$. Take the numerator to be $u(x)$ and the denominator to be $v(x)$.

$$u(x) = \sin 2x, \quad v(x) = x$$

Step 2: Differentiate $u(x)$ and $v(x)$.

$$\frac{du}{dx} = 2\cos 2x, \quad \frac{dv}{dx} = 1$$

Step 3: Square $v(x)$.

$$v^2 = x^2$$

Step 4: Substitute $u(x)$, $v(x)$, $\dfrac{du}{dx}, \dfrac{dv}{dx}$ and v^2 into the quotient rule formula.

$$\frac{dy}{dx} = \frac{v\dfrac{du}{dx} - u\dfrac{dv}{dx}}{v^2}$$

$$= \frac{(x)(2\cos 2x) - (\sin 2x)(1)}{x^2}$$

Step 5: Simplify the result if possible.

$$\frac{dy}{dx} = \frac{2x\cos 2x - \sin 2x}{x^2}$$

Ⅱ   **PAUSE POINT**

Differentiate the following functions with respect to the independent variable.

a) $y(x) = 4x^3 - 2x^2 + 5$

b) $v(t) = \sin 4t$

c) $i(t) = 5\,e^{4t}$

d) $P(T) = 2\,e^{-3T}$

e) $V = \cos 3x \sin 4x$

f) $y = \dfrac{x^2 + 2x + 1}{\sin x}$

g) $R = \sqrt{x^2 + x + 1}$

Hint   In each case, first identify the type of function (e.g. polynomial, exponential) and then select the appropriate rule to apply.

Extend   Identify which of the rules – chain rule, product rule or quotient rule – should be used to differentiate each of these functions. Then carry out the differentiation.

A3 Numerical value of a derivative

In Section A2 you looked at how to find the derivative of a simple function that is a sum, difference, product, quotient or composite of standard functions. The resulting derivative is a function itself, and you have seen that it represents the gradient of the graph or the rate of change of the dependent variable with respect to the independent variable. In engineering applications you will usually want to know the gradient of a graph at a particular point (the **instantaneous** gradient), or the rate of change at a specific value of the independent variable (instantaneous rate of change). You find this by substituting that specific numerical value of the independent variable into the derivative function.

You have seen that velocity is the derivative of displacement (distance moved in a given direction) with respect to time.

Displacement	$s = f(t)$
Velocity	$v = \dfrac{ds}{dt} = f'(t)$

The instantaneous velocity at a particular moment in time is calculated by substituting the time value into the velocity function.

Worked Example

The displacement, s metres, of a gear piston after t seconds is found by experiment to follow the equation
$$s = t^2 + 3t - 4$$
Calculate the velocity of the gear piston after 6 seconds.

Solution

Differentiate the displacement w.r.t. time to find the velocity function: $v(t) = \dfrac{ds}{dt} = 2t + 3$

To find the instantaneous velocity at 6 seconds, substitute $t = 6$ into the velocity function:
$v(6) = 2 \times 6 + 3 = 15 \text{ m s}^{-1}$

A4 Second derivative and turning points

Displacement, velocity and acceleration

Acceleration is the derivative of velocity w.r.t. time:

Acceleration	$a = \dfrac{dv}{dt}$

Since velocity is the derivative of displacement w.r.t. time, acceleration can be calculated by differentiating displacement with respect to time twice. Therefore acceleration is the *second* derivative of the displacement function $s = f(t)$. This is written as
$$a = \frac{dv}{dt} = \frac{d^2s}{dt^2} = f''(t)$$
You should read $\dfrac{d^2s}{dt^2}$ as 'd two s by d t squared'. The notation indicates that s is differentiated twice with respect to t. The operation of differentiating twice w.r.t. the variable t is represented by $\dfrac{d^2}{dt^2}$

Similar relationships hold for angular motion:

Angular displacement (radians)	$\theta = f(t)$
Angular velocity (rad s^{-1})	$\omega = \dfrac{d\theta}{dt} = f'(t)$
Angular acceleration (rad s^{-2})	$\alpha = \dfrac{d\omega}{dt} = \dfrac{d^2\theta}{dt^2} = f''(t)$

Worked Example

The angle θ radians turned by a gearwheel in t seconds is found by experiment to follow the equation $\theta = 4t + 3t^2$

a) Calculate the angular velocity of the gearwheel when $t = 5$ s.

b) Calculate the angular acceleration when $t = 8$ s.

Solution

a) First, differentiate the angular displacement w.r.t. time to find the angular velocity:
$$\omega(t) = \frac{d\theta}{dt} = 4 + 6t$$
Then substitute $t = 5$ to find the instantaneous angular velocity at that specific time:
$$\omega(5) = 4 + 6 \times 5 = 34 \text{ rad s}^{-1}$$

b) Differentiate the angular velocity w.r.t. time to find the angular acceleration:
$$\alpha = \frac{d\omega}{dt} = 6 \text{ rad s}^{-2}$$
The angular acceleration is a constant value (uniform acceleration), so when $t = 8$ s we also have $\alpha = 6$ rad s^{-2}.

Note that the differentiation must be carried out before the given values of t can be substituted.

Turning points on a function

Look at the function graphed in **Figure 7.6**.

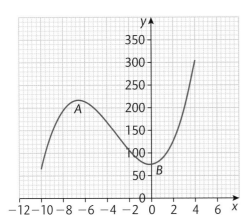

▶ **Figure 7.6** Graph of a function with two turning points

The function changes direction at the points A and B. These are called turning points.

Maximum and minimum points

In **Figure 7.6**, to the left of point A the function is increasing – the gradient is positive. Just to the right of A the function is decreasing – the gradient is negative. At A, where the function changes direction, the gradient is zero (the tangent to the curve at A is horizontal). The function is said to have a local maximum at A. Note that the y value at A is not the largest value that the function ever takes, but locally it is a 'peak'.

Just to the left of B the function is decreasing – the gradient is negative. Just after B the function is increasing – the gradient is positive. At B, where the function changes direction, the gradient is again zero. We say that the function has a local minimum at B. Again, the y value at B is not the smallest value that the function ever takes, but locally it is a 'trough'.

Local maximum and minimum points (also called maxima and minima), like A and B, occur when the gradient of the function is zero:

$$\frac{dy}{dx} = 0$$

To find all the maxima and minima of a function, you need to differentiate the function and look for the points where the derivative is zero. You can determine whether a turning point is a maximum or minimum in one of two ways.

▶ Method 1: Check gradients on either side of the turning point.

Look just to the left and just to the right of the turning point to see if the gradient is positive or negative there.

  ▶ If the gradient goes positive–zero–negative, the turning point is a maximum.

  ▶ If the gradient goes negative–zero–positive, the turning point is a minimum.

▶ Method 2: Calculate the second derivative at the turning point.

Differentiate the first derivative to get the second derivative. Then substitute the value of the turning point

to find the numerical value of the second derivative at this point.

▶ If $\frac{d^2y}{dx^2}$ is positive, the turning point is a minimum.

▶ If $\frac{d^2y}{dx^2}$ is negative, the turning point is a maximum.

Worked Example

Find and classify the turning points of the function $y = -x^2 + x + 1$.

Solution

Step 1: Differentiate the function.

$$\frac{dy}{dx} = -2x + 1$$

Step 2: Find when the derivative is zero.

$$-2x + 1 = 0$$
$$2x = 1$$
$$x = \frac{1}{2} = 0.5$$

In this case there is only one possible turning point.

Step 3: Calculate the y value of the turning point.

Substitute the x value found in Step 2 into the original function:

When $x = \frac{1}{2}$, $y = -\left(\frac{1}{2}\right)^2 + \frac{1}{2} + 1 = 1\frac{1}{4} = 1.25$

Step 4: Determine whether the turning point is a maximum or minimum.

• Method 1: Pick points just to the left and just to the right of the turning point.

When $x = 0.4$, $\frac{dy}{dx} = -2(0.4) + 1 = 0.2$

The gradient is positive – the curve is going up.

When $x = 0.6$, $\frac{dy}{dx} = -2(0.6) + 1 = -0.2$

The gradient is negative – the curve is going down.

As the curve goes up, levels out and then goes down, the turning point is a maximum.

• Method 2: Calculate the second derivative at the turning point.

$$\frac{dy}{dx} = -2x + 1 \quad \Rightarrow \quad \frac{d^2y}{dx^2} = -2$$

The second derivative is negative, so the turning point is a maximum.

To summarise, there is one turning point at the coordinates (0.5, 1.25) and it is a maximum.

You can check the result by sketching a graph of the function, as in **Figure 7.7**.

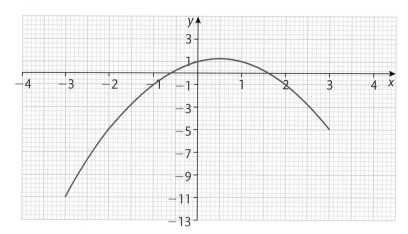

▶ **Figure 7.7** Graph of $y = -x^2 + x + 1$, showing that there is a maximum point at (0.5, 1.25)

Points of inflection

Not all turning points are maximum or minimum points. Sometimes the curve increases, levels out and then increases again, or it could decrease, level out and then decrease again. A point at which the gradient is instantaneously zero but where the curve does *not* change direction is called a point of inflection (or inflexion).

Worked Example

Find and classify the turning points of the function $y = x^3$.

Solution

Step 1: Differentiate the function.
$$\frac{dy}{dx} = 3x^2$$

Step 2: Find when the derivative is zero.
$$3x^2 = 0$$
$$x = 0$$

Step 3: Calculate the y value of the turning point.
When $x = 0$, $y = 0^3 = 0$

Step 4: Determine whether the turning point is a maximum or minimum.
When $x = -0.1$, $\frac{dy}{dx} = 3(-0.1)^2 = 0.03$. The gradient is positive – the curve is going up.

When $x = 0.1$, $\frac{dy}{dx} = 3(0.1)^2 = 0.03$. The gradient is positive – the curve is going up.
The curve does not change direction at $x = 0$, so this is a point of inflection.

If you look at the second derivative:
$$\frac{dy}{dx} = 3x^2 \quad \Rightarrow \quad \frac{d^2y}{dx^2} = 6x$$
When $x = 0$, $\frac{d^2y}{dx^2} = 6(0) = 0$

The second derivative at the turning point is neither positive nor negative.

To summarise, there is one turning point at (0, 0) and it is a point of inflection.

A sketch (**Figure 7.8**) shows what the function looks like.

Engineering applications of differentiation

Differentiation is useful when you need to optimise the values of variables. For example, you may want to maximise profits or minimise the costs of production. The following worked examples show how the techniques of differential calculus can be used to solve engineering problems.

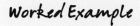

The profit for a company (in £100,000 per year) is a function of the number of components produced. The relationship between the number x of components produced (in thousands per year) and the profit P is given by

$$P = 4 + 0.03x^2 - 0.001x^3$$

Find the production level that will maximise the profit, given that the company has the capacity to produce, at most, 50 000 components per year, and calculate this maximum profit.

Solution

Step 1: Differentiate the profit function.

$$\frac{dP}{dx} = 0.06x - 0.003x^2$$

Step 2: Find any turning points by looking at where the derivative is zero.
$$0.06x - 0.003x^2 = 0$$
$$x(0.06 - 0.003x) = 0$$
$$x = 0 \quad \text{or} \quad x = \frac{0.06}{0.003} = 20$$

Step 3: Determine the nature of the turning ints.

$x = 0$ means zero production (not likely to maximise profit!).

When $x = 20$, $\dfrac{d^2P}{dx^2} = 0.06 - 0.006x = 0.06 - 0.006(20) = -0.06$

The second derivative is negative, so the turning point at $x = 20$ is a maximum.

Step 4: Check that the optimum solution found fits within the constraints of the problem.

The maximum capacity of the factory, 50 000 components per year, corresponds to $x = 50$, so the turning point at $x = 20$ fits within this capacity.

Step 5: Calculate the maximum profit.
When $x = 20$, $P = 4 + 0.03(20)^2 - 0.001(20)^3 = 8$

So the maximum profit is £800,000 per year.

A graph as in **Figure 7.9** confirms the solution.

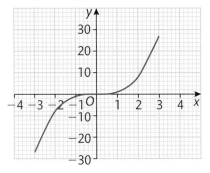

▸ **Figure 7.8** Graph of $y = x^3$, with a point of inflection at (0, 0)

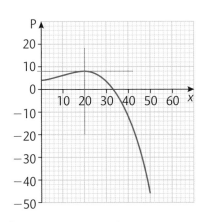

▸ **Figure 7.9** Graph of $P = 4 + 0.03x^2 - 0.001x^3$ for x between 0 and 50, showing the maximum point at (20, 8)

Worked Example

A sheet metal worker wants to make a simple box out of a rectangular sheet by cutting an identical square from each corner and folding up the sides. If the sheet measures 300 mm by 200 mm, what is the length of the cut-out square that will give a box of maximum volume? What will the maximum volume be?

Solution

Step 1: Sketch a diagram to visualise the problem.

Figure 7.10 shows the sheet and the squares to be cut out from its corners, together with the resulting box.

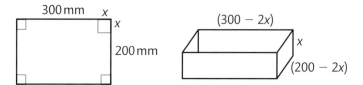

▶ **Figure 7.10** Diagram for the box problem

Step 2: Derive an equation for the volume, V, of the box.

Volume = length × width × height

$V = (300 - 2x)(200 - 2x)x$

Step 3: Differentiate V with respect to x to find any turning points.

In this case it is probably simpler to expand the brackets than to apply the product rule twice.

$$V = (60\,000 - 400x - 600x + 4x^2)x$$
$$= (60\,000 - 1000x + 4x^2)x$$
$$= 60\,000x - 1000x^2 + 4x^3$$
$$\frac{dV}{dx} = 60\,000 - 2000x + 12x^2$$

At any turning points, $12x^2 - 2000x + 60\,000 = 0$

This is a quadratic equation. It will not factorise easily, so use the quadratic formula to solve it.

First, simplify by dividing through by the common factor 4.

$$3x^2 - 500x + 15\,000 = 0$$
$$x = \frac{-b \pm \sqrt{b^2 - 4ac}}{2a} \quad \text{with } a = 3,\ b = -500,\ c = 15\,000$$
$$x = \frac{-(-500) \pm \sqrt{(-500)^2 - 4(3)(15\,000)}}{2(3)}$$
$$x = 127.4 \quad \text{or} \quad x = 39.2$$

The answer cannot be $x = 127.4$ as then the width of the box, $(200 - 2x)$, would be negative.

Therefore the only relevant turning point is $x = 39.2$.

Step 4: Check that the turning point found is a maximum.

When $x = 39.2$, $\dfrac{d^2V}{dx^2} = 24x - 2000 = 24(39.2) - 2000 = -1059.2$

The second derivative is negative, so this turning point is a maximum.

Step 5: Calculate the maximum volume.

When $x = 39.2$, $V = (300 - 2 \times 39.2)(200 - 2 \times 39.2)(39.2) = 1\,056\,305$

So the maximum volume of the box is $1.056 \times 10^6\,mm^3$, obtained when the length of each cut-out square is $39.2\,mm$.

Research

Investigate how the product rule can be extended to functions that take the form of three factors multiplied together. Try using the three-factor product rule to differentiate the volume function for the box in the worked example.

Assessment practice 7.1

A.P1   A.P2   A.P3   A.M1   A.D1

Use differential calculus to answer the following questions. It is important that you explain and show all the steps in your calculations clearly.

1   Differentiate the following functions with respect to the independent variable.

a) $y = 4x^3 - 2x^2 + 5$ b) $y = \sin 4x$

c) $y = 3\cos 2x$ d) $y = \ln 3t$ e) $T = e^{-2t}$

2   Use the chain rule (function of a function rule) to differentiate these functions.

a) $y = (x + 2)^3$ b) $v = (\sin 3t)^2$

c) $y = \sqrt{1 + x}$ d) $R = \sqrt[3]{x^2 + x + 1}$

3   Use the product rule to differentiate these functions.

a) $y = e^t \ln t$ b) $v = \sin 3x \cos 4x$ c) $s = t^4 e^{-3t}$

4   Use the quotient rule to differentiate these functions.

a) $y = \dfrac{x^2 + 2x + 1}{\sin 2x}$ b) $r = \dfrac{6\cos\theta}{\theta^3 + 2}$

5   For the following functions, find any turning points and determine whether they are maxima, minima or points of inflection.

a) $y = x^3 - 6x^2 + 4$ b) $a = 2t^3 + 3t^2 - 7$

6   Find the dimensions of a closed cylinder of volume $1000\,cm^3$ that is made from a minimum amount of sheet metal (ignoring any overlaps for seams).

For each of the questions in this assessment practice activity, use the following stages to guide your progress through the task.

Plan
- What techniques will be involved in answering the question?
- How confident do I feel in my own abilities to answer this question accurately? Are there any areas I think I may struggle with?
- Will a sketch help me to visualise the problem or check my solution to it?

Do
- I can organise my work logically and systematically, annotating the steps of my solution to explain my approach.
- I understand my thought process and can explain why I have decided to approach the question in a particular way.
- I can identify where I have gone wrong and adjust my thinking to get myself back on course.

Review
- I can explain to others how best to approach similar problems.
- I can effectively review and check mathematical calculations to prevent or correct errors.
- I can identify the type and style of questions that I find most challenging and devise strategies, such as additional purposeful practice, that will help me to overcome any difficulties.

Examine how integral calculus can be used to solve engineering problems

Key terms

Integration – the inverse process of differentiation, or the process of summing small increments to find the whole (see Section B2). The area of mathematics dealing with integration is called integral calculus.

Inverse of a function or operation – carrying out the opposite actions in the reverse order.

Integral – the result of applying integration to a function.

B1 Integration as the reverse/inverse of differentiation

Integration is the reverse mathematical process to differentiation, or the **inverse** operation of differentiation.

If you differentiate the function $y = 3x^2 + 4x + 5$ you get the derivative $\frac{dy}{dx} = 6x + 4$. So we say that an **integral** of $6x + 4$ is $y = 3x^2 + 4x + 5$.

Integral notation

The integral sign $\int$ is an elongated letter 's' (you will see in Section B2 why 's' is used). The function being integrated (the integrand) is written after the integral sign, followed by 'd' and the variable with respect to which the integration is carried out. For example, $\int (6x + 4)\, dx$ means integrate the function $6x + 4$ with respect to the variable x. Note that everything between $\int$ and dx is integrated, so the brackets are not strictly necessary.

Constant of integration

You saw above that because the derivative of $y = 3x^2 + 4x + 5$ is $\frac{dy}{dx} = 6x + 4$, the integral of $6x + 4$ is $y = 3x^2 + 4x + 5$, which is written as $y = \int (6x + 4)\, dx = 3x^2 + 4x + 5$.

The relationship between $6x$ and $3x^2$ is clear, and so is the relationship between 4 and $4x$. However, there is not enough information in the expression $\frac{dy}{dx} = 6x + 4$ to give us the constant value 5. In fact, if you change the 5 in $y = 3x^2 + 4x + 5$ to any other constant, the derivative will still be $6x + 4$ because the derivative of any constant is zero. So all you can say for sure is that:

▸   if $\frac{dy}{dx} = 6x + 4$ then $y = \int (6x + 4)\, dx = 3x^2 + 4x + c$

where c represents an arbitrary constant, called the constant of integration. An integral that includes an arbitrary constant is called an *indefinite* integral.

In applications, if you want to know the actual value of c, you will need to substitute specific values of x and y before you can calculate c.

Ⅱ PAUSE POINT

Differentiate:

a) $y = 3x^2 + 4x + 1$

b) $y = 3x^2 + 4x + 2$

c) $y = 3x^2 + 4x + 3$

What do you notice about the results?

Hint   What rule(s) should you use to differentiate these functions? How are the functions similar or different?

Extend   What implications do your results have for the integral $\int (6x + 4)\, dx$?

In the following worked example, you will look at how the process of differentiation is reversed to find integrals of simple functions. Remember to add the constant of integration.

Worked Example

In each of the following cases, note the sequence of operations used to obtain the derivative and then reverse them to find the integral.

a) If $y = x^2$ then $\dfrac{dy}{dx} = 2x$. Find $\int 2x\,dx$.

b) If $y = 2\sin(3x)$ then $\dfrac{dy}{dx} = 6\cos(3x)$. Find $\int 6\cos 3x\,dx$.

Solution

a) From $y = x^2$ to $\dfrac{dy}{dx} = 2x$

the sequence of operations is: Multiply by the power 2.
                                   Subtract one from the power.

To find the inverse: Add one to the power.
                                   Divide the coefficient by the new power.

$$\int 2x\,dx = \int 2x^1\,dx = \frac{2x^{1+1}}{(1+1)} = x^2 + c$$

b) From $y = 2\sin(3x)$ to $\dfrac{dy}{dx} = 6\cos(3x)$

the sequence of operations is: Multiply by the coefficient of x (3).
                                   Change sin to cos.

To find the inverse: Change cos to sin.
                                   Divide by the coefficient of x (3).

$$y = \int 6\cos(3x)\,dx = \frac{6\sin(3x)}{3} = 2\sin(3x) + c$$

Standard integrals

Fortunately, you do not have to work out the integrals of standard functions by reversing differentiation step by step. Similar to derivatives, the integrals of standard functions are provided in tables (like **Table 7.4**) in most books on engineering mathematics. It is important, however, to learn how to use the tables correctly.

▶ **Table 7.4** Table of standard integrals

Type of function	Function $y = f(x)$	Integral $\int f(x)\,dx$
Zero	0	c
Constant	k	$kx + c$
Polynomial	ax^n	$\dfrac{ax^{n+1}}{n+1} + c \quad (n \neq -1)$
Trigonometric (x in radians)	$\sin ax$	$\dfrac{-\cos ax}{a} + c$
	$\cos ax$	$\dfrac{\sin ax}{a} + c$
Exponential	e^{ax}	$\dfrac{e^{ax}}{a} + c$
Reciprocal	$\dfrac{1}{x}$	$\ln x + c$

Worked Example

Find the following integrals.

a) $\int x^4\,dx$ b) $\int \sin(3x)\,dx$ c) $\int (x^2 + 9x)\,dx$ d) $\int e^{-3t}\,dt$

Solution

a) The integrand is a polynomial term x^n with $n = 4$.

$$\int x^4\,dx = \frac{x^{4+1}}{4+1} + c = \frac{x^5}{5} + c$$

b) The integrand is of the form $\sin(ax)$ with $a = 3$.

$$\int \sin(3x)\,dx = \frac{-\cos(3x)}{3} + c$$

c) The integrand is a polynomial with two terms. Integrate each term separately and add the results.

$$\int (x^2 + 9x)\,dx = \int x^2\,dx + \int 9x\,dx = \frac{x^3}{3} + \frac{9x^2}{2} + c$$

d) In this case the independent variable is t, and integration is w.r.t. this variable t rather than x.

The integrand is of the form e^{ax} with $a = -3$.

$$\int e^{-3t}\,dt = \frac{e^{-3t}}{-3} + c = -\frac{e^{-3t}}{3} + c$$

There are two points of interest in part d) of the worked example.

▶ The coefficient 9 multiplies the result of integrating the standard function $x = x^1$:

$$\int 9x\,dx = 9\int x^1\,dx = 9 \times \frac{x^2}{2}$$

▶ Although the two separate integrals should each have an arbitrary constant, only one constant of integration is needed for the overall integral.

All of the above examples involve the indefinite integral – they include the constant c. In the next worked example you will see how the value of c can be determined from additional information.

Worked Example

The velocity v of a particle at time t is given by the equation $v = 10t + 6t^2$.

Find the displacement as a function of time if the initial displacement is 5 m.

Solution

As velocity is the derivative of displacement w.r.t. time, displacement is the integral of velocity w.r.t. time.

Step 1: Integrate the velocity function w.r.t. t to find the general formula for displacement.

$$s = \int (10t + 6t^2)\,dt = \frac{10t^2}{2} + \frac{6t^3}{3} + c$$

$$s = 5t^2 + 2t^3 + c$$

Step 2: Calculate the constant c by substituting the initial condition into the general solution obtained in Step 1.

In this case the given initial condition is $s = 5$ when $t = 0$.

$$5 = 5(0)^2 + 2(0)^3 + c$$

$$c = 5$$

Step 3: Write down the formula for displacement with the particular value of c you found.

$$s = 5t^2 + 2t^3 + 5$$

Definite integrals

As you have seen, an indefinite integral is a result of integration that contains the independent variable and a constant c whose value is undetermined. In an engineering problem the indefinite integral gives a *general solution* to the problem, such as $s = 5t^2 + 2t^3 + c$ in the worked example. If the value of c can be worked out from additional information (such as initial conditions), then you get a *particular solution* to the problem, like $s = 5t^2 + 2t^3 + 5$ in the worked example. Note that a particular solution is still a function involving the independent variable.

When solving engineering problems, you will often need to substitute actual values of the independent variable into the solution, obtaining a definite numerical value. This can be done directly within the integration process, and the result is called a *definite* integral. For instance, if in the worked example above you were asked to find the displacement between the times $t = 1$ and $t = 3$, you could write this as the definite integral

$$\int_1^3 (10t + 6t^2)\, dt.$$

The numbers 3 and 1 at the top and bottom of the $\int$ sign are called the *limits* of integration – these are the values of the independent variable between which the solution is evaluated. The general form of a definite integral is

$$\int_a^b f(x)\, dx,$$ where a is the lower limit of integration, b is the upper limit of integration, $f(x)$ is the function being integrated and dx shows that you are integrating with respect to the variable x. You will see in Section B2 that the definite integral is actually the area between the graph of the function, the horizontal axis and the limits of integration. For example,

$$\int_1^3 (10t + 6t^2)\, dt$$ is the area bounded by the graph of $y = 10t + 6t^2$ and the x-axis between $t = 1$ and $t = 3$.

Calculating a definite integral

The steps of working out the numerical value of a definite integral are:

1. Find the indefinite integral (but you can omit the arbitrary constant c).
2. Substitute the value of the upper limit into the integral.
3. Substitute the value for the lower limit into the integral.
4. Subtract your answer for the lower limit from your answer for the upper limit.

You must carry out the integration to find the indefinite integral *before* you substitute the limits.

You do not need to include the constant c because it will be the same in steps 2 and 3 and will therefore cancel out when you subtract the results in step 4.

Note the different types of brackets used in the following worked example. Try to use the same system in your own work to make it easier to track the stages in your solution.

Worked Example

Calculate these definite integrals.

a) $\displaystyle\int_2^5 e^{-3t}\, dt$ b) $\displaystyle\int_2^4 \sqrt{x}\, dx$

Solution

a) $\displaystyle\int_2^5 e^{-3t}\, dt = \left[\frac{e^{-3t}}{-3}\right]_2^5$

Note that the indefinite integral (without c) is enclosed in square brackets, with the limits of integration at the top and bottom of the right bracket.

Now substitute the limits into the integral and subtract:

$$\left(\frac{e^{-3(5)}}{-3}\right) - \left(\frac{e^{-3(2)}}{-3}\right) = \left(\frac{e^{-15}}{-3}\right) - \left(\frac{e^{-6}}{-3}\right)$$

$$= \frac{e^{-15} - e^{-6}}{-3} = \frac{e^{-6} - e^{-15}}{3}$$

$$\approx 8.26 \times 10^{-4}$$

b) To integrate $\sqrt{x}$, rewrite as a power and then use the formula for integrating x^n.

$$\int_2^4 \sqrt{x}\, dx = \int_2^4 x^{\frac{1}{2}}\, dx$$

$$= \left[\frac{x^{\frac{3}{2}}}{3/2}\right]_2^4 = \left[\frac{2x^{\frac{3}{2}}}{3}\right]_2^4$$

$$= \left\{\frac{2(4)^{\frac{3}{2}}}{3}\right\} - \left\{\frac{2(2)^{\frac{3}{2}}}{3}\right\}$$

$$= \frac{2(8) - 2(2.828)}{3} \approx 3.448$$

Integration by substitution

You can integrate a composite function (function of a function) by using a substitution. This is essentially reversing the chain rule, and is most easily demonstrated through examples.

Worked Example

Calculate the following integrals by substitution.

a) $y = \int (2x + 3)^4\, dx$ b) $y = \int \cos(3x + 4)\, dx$

Solution

a) The integrand $(2x + 3)^4$ is a composite function, where the 'inner' function is $2x + 3$ and the 'outer' function is $(\ )^4$.

Substitute $u = (2x + 3)$. Then $(2x + 3)^4 = u^4$, which looks simpler to integrate.

However, the integration variable must also be converted from x to u; that is, you need to replace dx with an equivalent expression involving du.

$$u = (2x + 3) \Rightarrow \frac{du}{dx} = 2 \Rightarrow dx = \frac{du}{2}$$

$$y = \int (2x + 3)^4 \, dx = \int u^4 \frac{du}{2} = \frac{1}{2} \int u^4 \, du$$

$$= \frac{1}{2}\left(\frac{u^5}{5}\right) + c = \frac{u^5}{10} + c = \frac{(2x + 3)^5}{10} + c$$

b) The integrand $\cos(3x + 4)$ is a composite function, where the 'inner' function is $3x + 4$ and the 'outer' function is cos.

Substitute $u = 3x + 4$, which gives $\frac{du}{dx} = 3$ and so $dx = \frac{du}{3}$

$$y = \int \cos(3x + 4) \, dx = \frac{1}{3} \int \cos(u) \, du$$

$$= \frac{1}{3} \sin(u) + c = \frac{1}{3} \sin(3x + 4) + c$$

Integrating products

This method is sometimes known as 'integration by parts'. The rule can be written in two ways – the second way is more common and will be used in the examples that follow. In both forms, u and v are functions of x, although in the formulae they are written simply as u and v rather than $u(x)$ and $v(x)$ for clarity.

$$\int u \left(\frac{dv}{dx}\right) dx = uv - \int v \left(\frac{du}{dx}\right) dx$$

$$\int u \, dv = uv - \int v \, du$$

When using this method, the most important decision is choosing which function in the product to use as u and which as dv. Looking at the rule, the function you choose as u must be relatively easy to differentiate, because you need du in the second term on the right-hand side; and whatever you choose as dv needs to be easy to integrate since you also need v in the right-hand side of the formula.

Worked Example

Use integration by parts to find the following integrals.

a) $y = \int x \sin(x) \, dx$ b) $\int_0^2 x^2 e^x \, dx$

Solution

a) The integrand is a product of two terms, x and $\sin(x)$.

Let $u = x$ and $dv = \sin(x) \, dx$. Find the terms v and du appearing in the right-hand side of the formula:

$u = x$	$dv = \sin(x)\,dx$
$\dfrac{du}{dx} = 1$	$v = \int dv$
$du = dx$	$= \int \sin(x)\,dx$
	$= -\cos(x)$

Substitute into the formula:

$$\int u \, dv = uv - \int v \, du$$

$$y = \int x \sin(x) \, dx = (x)(-\cos(x)) - \int (-\cos(x)) \, dx$$

$$= -x \cos(x) + \sin(x) + c$$

b) $\displaystyle\int_0^2 x^2 e^x \, dx$ is a definite integral. You need to find the indefinite integral first, and then evaluate at the limits.

When one of the terms in the product is a power of x, it is usually a good idea to let that term be u.

In this case let $u = x^2$ and $dv = e^x \, dx$. Now find v and du:

$u = x^2$	$dv = e^x\,dx$
$\dfrac{du}{dx} = 2x$	$v = \int dv$
$du = 2x\,dx$	$= \int e^x\,dx = e^x$

Substitute into the formula:

$$\int u \, dv = uv - \int v \, du$$

$$\int x^2 e^x \, dv = (x^2)(e^x) - \int (e^x) \, 2x \, dx$$

$$= x^2 e^x - \int 2x \, e^x \, dv$$

This does not seem to have helped a great deal, as the final term is not a standard integral but still the integral of a product. However, this term itself can be integrated by parts.

To integrate $\int 2x \, e^x \, dv$ by parts, let $u = 2x$ and $dv = e^x \, dx$.

$u = 2x$	$dv = e^x\,dx$
$\dfrac{du}{dx} = 2$	$v = \int dv$
$du = 2\,dx$	$= \int e^x\,dx = e^x$

Then, substituting into $\int u \, dv = uv - \int v \, du$ gives

$$\int 2x \, e^x \, dx = (2x)(e^x) - \int e^x \, 2 \, dx$$

$$= 2x \, e^x - 2 \int e^x \, dx$$

$$= 2x \, e^x - 2e^x$$

Combine the results from the two integrations by parts. You can omit the constants of integration as you are going to calculate the definite integral.

$$\int x^2 e^x \, dx = x^2 e^x - \int 2x \, e^x \, dv$$

$$= x^2 e^x - (2x \, e^x - 2e^x) = e^x(x^2 - 2x + 2)$$

$$\int_0^2 x^2 e^x \, dx = [e^x(x^2 - 2x + 2)]_0^2$$
$$= \{e^2((2)^2 - 2(2) + 2)\} - \{e^0((0)^2 - 2(0) + 2)\}$$
$$= 2e^2 - 2 \approx 12.78$$

B2 Integration as a summating tool

Area under a curve

In engineering and science, there are many situations in which an area beneath a graph represents the magnitude (size) of a physical quantity. For example, the work done (W joules) by an object moving against a constant force of F newtons through a distance of s metres in the direction of the force is given by $W = Fs$.

A work diagram is a graph of force plotted against distance moved. The area under the graph in the work diagram is equal to the work done.

Worked Examples

1 Calculate the work done by a machine applying a constant force of 10 N to raise an object through a height of 5 m.

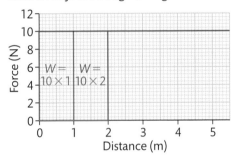

▶ **Figure 7.11** Area under a constant (horizontal) force–distance graph

Solution

For a constant force, the force–distance graph is a horizontal line (see **Figure 7.11**). The work done in moving the object a certain distance is the area between the line of the function and the x-axis up to the x value corresponding to the distance moved – that is, the area of the rectangle with base x and height 10.

The work done in moving the object 5 m is $W = 10 \times 5 = 50\,$J.

2 Calculate the work done by a force of 8 N in stretching a spring by 2 mm (see **Figure 7.12**). The spring constant is 4 N mm⁻¹.

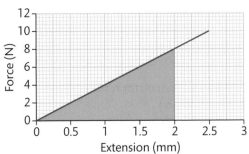

▶ **Figure 7.12** Area under a linear force–distance graph

The work done is equal to the area of the shaded region. This can be calculated simply as the area of the triangle:

$$W = \frac{1}{2}\text{base} \times \text{height} = \frac{1}{2} \times 0.002 \times 8 = 8 \text{ mJ}$$

Integration from first principles

The areas in the two examples above are easy to calculate as they are simple geometric shapes. How can you calculate the area if the function is non-linear? The summation method is based on splitting up the area under a graph into thin strips of equal width. You find the area of each strip and then sum all these areas to give the total area.

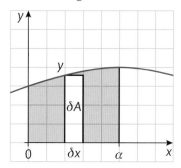

▶ **Figure 7.13** The area under the curve can be approximated by summing the areas of rectangles

Integration extends this summation idea to make the strips so narrow that the top of each strip is almost flat – so each strip can be considered a rectangle. The area of each rectangle is $\delta A = y\,\delta x$, where y is the height and δx is the width of each thin rectangular strip (see **Figure 7.13**). The total area under the graph is approximated by the sum of the areas of all the strips in the region. As δx gets smaller, you use more strips to cover the region, and the calculated area will become a closer approximation to the actual area. In other words, the actual area is the limit approached by the calculated area as $\delta x \to 0$ (δx tends to zero). Following the notation used in Section A2 for differentiation and using Σ to represent 'sum', the actual area A under a graph between $x = a$ and $x = b$ can be written as

$$A = \lim_{\delta x \to 0} \sum_{x=a}^{x=b} y(x)\,\delta x$$

This is defined to be the definite integral

$$A = \int_a^b y(x)\,dx$$

Mean and root mean square values of a function

Physical quantities such as distance, velocity, acceleration, current and voltage often vary with time. As an engineer, you may need to find the mean value of such a function over a certain period of time.

In **Figure 7.14**, the area of the shaded region under the function between the limits a and b is $\int_a^b f(t)\,dt$.

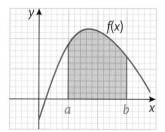 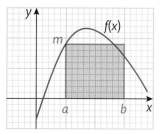

▶ **Figure 7.14** Graphical interpretation of the mean value – the area under the curve in the left diagram is equal to the area of the rectangle in the right diagram

If m is the mean value of the function over the same range (from $t = a$ to $t = b$), then by definition of the mean, the area of the shaded rectangle of height m in the right diagram should be the same as the shaded area under the curve in the left diagram:

$$m(b - a) = \int_a^b f(t)\,dt$$

Therefore, the mean value of the function $f(t)$ in the interval $a \leqslant t \leqslant b$ is $m = \dfrac{1}{(b - a)} \int_a^b f(t)\,dt$

Worked Example

Calculate the mean value of the current

$i(t) = 4 - \cos\left(\dfrac{t}{3}\right)$ over the time interval $0 \leqslant t \leqslant 10$.

Solution

$$\text{Mean current } \bar{i} = \frac{1}{(b - a)} \int_a^b f(t)\,dt$$

$$= \frac{1}{(10 - 0)} \int_0^{10} 4 - \cos\left(\frac{t}{3}\right)\,dt$$

Evaluate the definite integral:

$$\int_0^{10} \left(4 - \cos\left(\frac{t}{3}\right)\right)\,dt$$

$$= \left[4t - \frac{1}{\frac{1}{3}}\sin\left(\frac{t}{3}\right)\right]_0^{10}$$

$$= \left[4t - 3\sin\left(\frac{t}{3}\right)\right]_0^{10}$$

$$= \left\{4(10) - 3\sin\left(\frac{10}{3}\right)\right\} - \left\{4(0) - 3\sin\left(\frac{0}{3}\right)\right\}$$

$$= 40 - 3\sin\left(\frac{10}{3}\right) \approx 40.57$$

(Make sure your calculator is set to radian mode for the last calculation.)

The mean current is

$$\bar{i} = \frac{1}{10}\int_0^{10}\left(4 - \cos\left(\frac{t}{3}\right)\right)\,dt = \frac{40.57}{10} = 4.057 \text{ A}$$

The area defined by an integral is the area between the graph of the function and the x-axis. If the graph is below the x-axis, the integral will have a negative value. This means that for an alternating function such as a sine wave, the integral over a complete cycle will be zero, as the function is symmetrical above and below the x-axis so that the positive and negative areas cancel each other out.

If you need to calculate the mean magnitude of an alternating periodic function over a cycle, you could calculate the total integral by breaking it down into separate areas above and below the x-axis, or you could calculate the integral over a positive half-cycle and double the result because of the symmetry. With each method you need to take great care in choosing the limits for the integrals that you calculate. A sketch of the function will help you see which parts of the graph are above the x-axis and which are below.

Alternatively, to get around the difficulty with negative areas, you can calculate the root mean square (RMS) value of the function (see **Figure 7.15**). This involves squaring the function so that all values become non-negative. The steps are as follows.

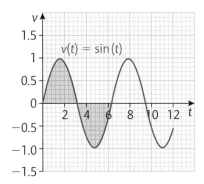

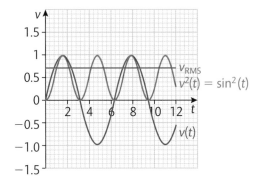

▸ **Figure 7.15** RMS value of a sine wave

1 Square the function to make all the values positive: $(f(t))^2$

2 Find the mean of the squared function: $\dfrac{1}{(b-a)}\int_a^b (f(t))^2\,dt$

3 Take the square root of the mean of the squared function: $\sqrt{\dfrac{1}{(b-a)}\int_a^b (f(t))^2\,dt}$

The RMS value of the function $f(t)$ in the interval $a \leqslant t \leqslant b$

is $f_{RMS} = \sqrt{\dfrac{1}{(b-a)}\int_a^b f(t)^2\,dt}$

Worked Example

Calculate the RMS value of the voltage $v = 240 \sin(100\pi t)$ over the time interval $0 \leqslant t \leqslant 20$ ms.

Solution

$$f_{\text{RMS}} = \sqrt{\frac{1}{(b-a)} \int_a^b f(t)^2 \, dt}$$

$$v_{\text{RMS}} = \sqrt{\frac{1}{(20 \times 10^{-3} - 0)} \int_0^{20 \times 10^{-3}} (240 \sin(100\pi t))^2 \, dt}$$

$$= \sqrt{\frac{240^2}{20 \times 10^{-3}} \int_0^{20 \times 10^{-3}} \sin^2(100\pi t) \, dt}$$

First, evaluate the definite integral. $\sin^2(x)$ is not one of the standard integrals. Use a trigonometric identity to change it into a form involving standard integrals: $\sin^2(x) = \dfrac{1 - \cos(2x)}{2}$

$$\int_0^{20 \times 10^{-3}} \sin^2(100\pi t) \, dt$$

$$= \int_0^{20 \times 10^{-3}} \left(\frac{1 - \cos(200\pi t)}{2} \right) dt$$

$$= \int_0^{20 \times 10^{-3}} \frac{1}{2} - \frac{\cos(200\pi t)}{2} \, dt$$

$$= \left[\frac{t}{2} - \frac{\sin(200\pi t)}{2(200\pi)} \right]_0^{20 \times 10^{-3}}$$

$$= \left\{ \frac{20 \times 10^{-3}}{2} - \frac{\sin(200\pi \times 20 \times 10^{-3})}{2(200\pi)} \right\} - \left\{ \frac{0}{2} - \frac{\sin(200\pi \times 0)}{2(200\pi)} \right\}$$

$$= 10^{-2} - \frac{\sin(4\pi)}{400\pi} = 0.01$$

Therefore

$$v_{\text{RMS}} = \sqrt{\frac{240^2}{20 \times 10^{-3}} \times 0.01} = 169.7 \, \text{V}$$

B3 Numerical integration

In engineering you will sometimes need to calculate the integrals of functions that cannot be expressed in terms of the standard functions in **Table 7.4** or cannot be integrated using standard techniques such as substitution or integration by parts. You may also have a graph of the function plotted from a collection of data points but not be able to write it as a mathematical equation. In such cases you could use the summation method to estimate the area represented by the integral. This method – based on splitting up the area under the graph into thin strips – generates several numerical integration rules that give different degrees of accuracy, depending on the shape of the function.

Trapezium (trapezoidal) rule

If you divide the area into narrow strips, where the top edge of each strip is a line joining two points on the function's graph (as shown in **Figure 7.16**), then each strip is a trapezium.

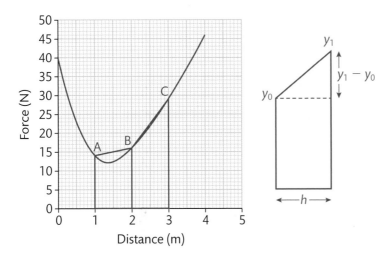

▶ **Figure 7.16** Dividing the area under a curve into trapezia

Suppose that you are using n strips of equal width h, and the y values of the end points of the top edges of the trapezia are $y_0, y_1, y_2, ..., y_{n-1}, y_n$ from left to right.

A trapezium is made up of a rectangle and a triangle.

So the area of the first trapezium is $A_1 = y_0 h + \frac{1}{2}h(y_1 - y_0)$ which simplifies to $A_1 = \frac{1}{2}h(y_0 + y_1)$.

The area of the second trapezium is $A_2 = \frac{1}{2}h(y_1 + y_2)$

If you were using only two strips as in **Figure 7.16**, then the approximate area under the curve obtained by adding together the two strips is
$A = \frac{1}{2}h(y_0 + y_1) + \frac{1}{2}h(y_1 + y_2) = \frac{1}{2}h(y_0 + 2y_1 + y_2)$

Notice that the 'end values' (y_0 and y_2) are only used once, but 'middle values' (y_1 in this case) appear twice (as the end of one strip and the start of the next).

In general, for n strips of width h, the trapezium rule estimates the area as
$A = \frac{1}{2}h(y_0 + 2y_1 + 2y_2 + 2y_3 + 2y_{n-1} + ... + y_n) = \frac{1}{2}h(y_0 + 2(y_1 + y_2 + ... + y_{n-1}) + y_n)$

You will get a more accurate estimate of the area if you make the width of the strips smaller and use more strips (that is, decrease h and increase n). The choice of strip width should strike a compromise between being small enough to give an accurate result but not so small as to have too many values to calculate.

Worked Example

Use the trapezium rule to estimate the area between the graph of $f(x) = e^{x^2}$ and the x-axis for values of x between 0.5 and 1.3, using a strip width of $h = 0.1$.

Solution

Step 1: Calculate values of y for x values from 0.5 to 1.3 at intervals of 0.1. These are shown in **Table 7.5**.

▶ **Table 7.5** Values of $f(x) = e^{x^2}$ for x between 0.5 and 1.3

x	y	
0.5	1.284 025	$= y_0$
0.6	1.433 329	$= y_1$
0.7	1.632 316	$= y_2$
0.8	1.896 481	$= y_3$
0.9	2.247 908	$= y_4$
1.0	2.718 282	$= y_5$
1.1	3.353 485	$= y_6$
1.2	4.220 696	$= y_7$
1.3	5.419 481	$= y_8$

Step 2: Substitute the values into the trapezium rule, rounding to a reasonable degree of accuracy.

$$A = \tfrac{1}{2}h(y_0 + 2(y_1 + y_2 + \ldots + y_7) + y_8)$$
$$= \tfrac{1}{2}(0.1)(1.2840 + 2(1.4333 + 1.6323 + \ldots + 4.2207) + 5.4195)$$
$$= 2.085$$

Remember that the result obtained by the trapezium rule is only an approximation, as the trapezia do not match the strips under the graph precisely – some are a little high, some a little low.

You could simplify the calculations by using a spreadsheet, since you need only write the equation for the function once. The screenshots in **Figure 7.17** show the equations used in Microsoft Excel® 2013.

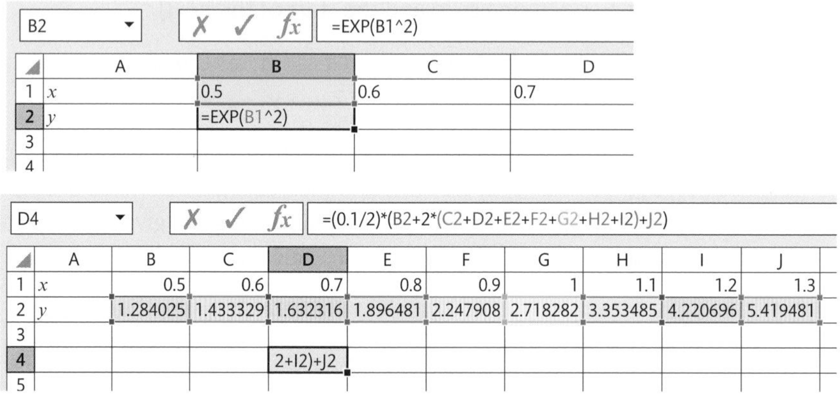

▶ **Figure 7.17** Calculating an area using the trapezium rule in a Microsoft Excel® workbook

Simpson's rule

Another summation method for calculating areas, known as Simpson's rule, uses a quadratic curve as the top edge of each strip. This gives a more accurate approximation to the actual area under the function's graph. Simpson's rule requires an *even* number of strips – and therefore an *odd* number of y values. The values are used in a similar way to those in the trapezium rule. Simpson's rule states:

$$\text{Area} \approx \frac{h}{3}\{y_0 + 4(y_1 + y_3 + y_5 + \ldots) + 2(y_2 + y_4 + y_6 + \ldots) + y_n\}$$

Using Simpson's rule on the worked example above gives

$A = \frac{0.1}{3}\{1.2840 + 4(1.4333 + 1.8965 + 2.7183 + 4.2207)$

$+ 2(1.6323 + 2.2479 + 3.3535) + 5.4195\}$

$= 2.075$

Note that there is always one fewer value in the 'even-numbered y values' group than in the 'odd-numbered y values' group.

Mid-ordinate rule

The mid-ordinate rule is a summation method that splits the area into rectangular strips. But rather than using the end points, the height of each strip is the y value at the *middle* of the strip. The first value is labelled $y_{0.5}$, the second $y_{1.5}$ and so on, as shown in **Figure 7.18**.

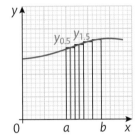

▶ **Figure 7.18** Dividing the area into rectangular strips for the mid-ordinate rule

In general, for n strips of width h, the mid-ordinate rule estimates the area as $A = h(y_{0.5} + y_{1.5} + y_{2.5} + \ldots + y_{n-0.5})$

These summation methods for estimating the numerical value of an integral are fine when actual values of the functions are known – or can be calculated – at given intervals, but remember that they provide only an approximate answer and do not give a general mathematical way of integrating any function.

① PAUSE POINT

Water flows at a constant rate through a venturi (constricted section of pipe). You measure the pressure at various points along the venturi, using a series of pitot tubes set at equal distances. You note the height of water in each tube, which represents the pressure at that point. The data is shown in **Table 7.6**. You plot a graph of the height against the horizontal position of each tube. Use the readings to calculate the area under the graph.

▶ **Table 7.6** Height of water in tubes at a series of horizontal distances along the venturi

Distance (mm)	0	20	40	60	80	100
Height (mm)	800	780	760	760	780	800

Hint   Use the heights as the y values and the spacing between consecutive tubes as h in the trapezium rule.

Is it possible to use Simpson's rule? If it is, then compare the answers. If not, then explain why.

Extend   Set up a spreadsheet to calculate the area.

Use differential calculus to answer the following questions. It is important that you explain and show all the steps in your calculations clearly.

1 Integrate the following functions with respect to the independent variable.

a) $3x + \cos 4x$ **b)** $4e^t + \cos\left(\dfrac{t}{2}\right)$

c) $\dfrac{1}{2y}$ **d)** $\dfrac{1}{3x} + 7$

2 Find each of these integrals using a suitable substitution.

a) $\int \cos(x - 3)\,dx$ **b)** $\int e^{9t-7}\,dt$

c) $\int x\sqrt{4x - 4}\,dx$ **d)** $\int_5^8 \dfrac{1}{(v - 2)^4}\,dv$

3 Calculate the following integrals using integration by parts.

a) $\int xe^{-x}\,dx$ **b)** $\int_0^1 xe^x\,dx$

4 The velocity of a decelerating particle at time t seconds is given by $v = e^{-(t^3+1)}$. Use a numerical method to calculate the distance travelled by the particle in the time interval 0 to 2.5 seconds.

5 Calculate the area between the curve $y = x(x - 3)$, the x-axis and the limits $x = 0$ and $x = 5$. (Hint: Sketch the curve first to see where the area lies.)

6 Calculate the RMS value of the voltage waveform $v = 12 \sin(4t + 2)$ in the interval $0 \leqslant t \leqslant 2$.

For each of the questions in this assessment practice activity, use the following stages to guide your progress through the task.

Plan

- What techniques will be involved in answering the question?
- How confident do I feel in my own abilities to answer this question accurately? Are there any areas I think I may struggle with?
- Will a sketch help me to visualise the problem or check my solution to it?

Do

- I can organise my work logically and systematically, annotating the steps of my solution to explain my approach.
- I understand my thought process and can explain why I have decided to approach the question in a particular way.
- I can identify where I have gone wrong and adjust my thinking to get myself back on course.

Review

- I can explain to others how best to approach similar problems.
- I can effectively review and check mathematical calculations to prevent or correct errors.
- I can identify the type and style of questions that I find most challenging and devise strategies, such as additional purposeful practice, that will help me to overcome any difficulties.

C   Investigate the application of calculus to the solution of a defined specialist engineering problem

C1   Thinking methods

Reductionism

Do you find solving problems difficult? This may be because you try to tackle the whole problem in one go. A better approach is to break the problem down (reduce it) to a fundamental level to make it easier to solve. This approach is known as **reductionism**. In other words, if you want to solve a complex problem, first try to simplify it as much as possible. In order to do this, you may have to learn some new skills and techniques.

Key terms

Reductionism – a way of thinking that breaks down a problem into simpler parts.

Synectics – a way of applying creativity in problem-solving.

Synectics

Synectics is a type of creative problem-solving (CPS) technique that brings the creative process into activities such as 'brainstorming' and 'lateral thinking'. You are encouraged initially to 'suspend disbelief', to set aside concepts of what is a 'good' or 'correct' idea and to think of alternative ways of looking at the problem. This leads to 'springboards' for ideas. In this approach, 'absurd' ideas are positively encouraged as a way of broadening your understanding of the problem. The 'springboards' generated are intended to be new starting points for exploration of the problem, rather than finished new solutions. You may identify new approaches that are not yet practical and then develop them into feasible courses of action. Alternatively, you may decide to abandon a particular approach and follow another 'springboard'. The aim is to come up with a way of tackling the problem that you feel confident in because you have investigated it in depth.

Pólya's problem-solving method

George Pólya was a Hungarian mathematician whose book *How to solve it* was published in 1945. In his book Pólya identified four stages in problem-solving.

1 Understand the problem – This sounds obvious, but it is exactly where you may be blocked. Make sure you understand the words used to describe the problem. Do you know what you have to do? Can you put the problem into your own words? Would a sketch help you visualise what is going on? Do you have enough information to start generating a solution?

2 Devise a plan – This is the stage where you can use thinking methods such as reductionism and synectics. Have you met this type of problem before? Can you break it down into simpler parts? What are the possible paths of solution to be followed? What are the processes involved? What are the relationships within the problem? Can you eliminate some possibilities? Which path and process promises the greatest likelihood of success? Decide which mathematical tool best suits the problem. What equations might be useful in solving the problem?

3 Carry out the plan – Work carefully, following the plan you made in the previous stage. Use paper and pencil, calculator and computer tools where appropriate. Constantly check your solutions. Are they realistic? Do they make sense? Is your answer the right order of magnitude? If it is not, then think about any changes that you can make. Do you get the same answers if you use alternative methods? Be persistent.

4 Look back – Reflect on what you have done. What went well? What did not go so well? What can you use to help solve future problems?

C2 Mathematical modelling of engineering problems

In engineering design it is useful to be able to predict how a product will perform in service. Investing resources in setting up manufacturing machinery and supply chains is expensive, so if engineers can use mathematical or computer methods to simulate the characteristics of a product or to conduct cost–benefit analyses, this will save time and money as well as aid decision-making. Using mathematical tools in this way is called mathematical modelling. It allows engineers to determine optimal (or near optimal) solutions before making important decisions and investing significant resources.

As part of this unit's assessment, you will need to model and solve a complex engineering problem from your specialist area of study.

First, you need to investigate a suitable problem that will require the application of calculus methods in its solution. This could be a case study from a local engineering organisation, or a problem that you have researched using books or the internet.

The problem-solving process begins with defining the problem, setting out all the relevant information gathered on it, and producing a proposal for how to solve it. You can use the thinking methods from Section C1 to break the problem down into a series of manageable steps to obtain a solution. In each step you should be able to apply appropriate calculus methods that you have learned. These may include both analytical and numerical techniques.

In the next two sections (C3 and C4), we will use a single worked example to illustrate the process you should follow for your chosen engineering problem.

C3 Problem specification and proposed solution

Here are a number of questions to ask yourself when specifying the engineering problem that you intend to solve.

▸ What is the problem? In your own words, say *who* needs *what* and *why*.

▸ Can you find more than one source of information to help you?

▸ Can you find an existing method to solve the problem? Can you think of a better way?

▸ How will you know if your solution is reasonable?

▸ How will you determine what went well and what could be improved?

▸ What will be in your final report?

Worked Example

Statement of the problem

Water flows from a river into a reservoir. The amount of water in the river depends on the season.

The flow of water in the river, w, has been modelled by the equation

$$w(t) = 2 + \sin\left(\frac{\pi}{180}t\right)$$

where t is measured in days, with $t = 0$ corresponding to 1 January, and w is measured in thousands of cubic metres ($10^3\,m^3$) per day.

Water is released from the reservoir at a constant rate of $2000\,m^3$ per day.

At the beginning of the year there is $200 \times 10^3\,m^3$ of water in the reservoir.

a) Write a function $f(t)$ for the amount of water in the reservoir on any day.

b) Determine the amount of water in the reservoir on 1 May ($t = 120$).

c) Find the rate of change of the volume of water in the reservoir on 1 September ($t = 240$).

d) Determine at what time there will be a maximum volume of water in the reservoir.

e) Calculate the maximum volume of water in the reservoir.

f) Find the average amount of water in the reservoir in the first three months of the year.

Ideas to solve the problem

Worked Example (continued)

What is the problem?

Calculate how much water is in the reservoir on any day. This can be broken down into:

- Find a way to work out the difference between what enters and leaves the reservoir each day.
- Use that to work out how much water there will be at the end of the day.

What do you need to know?

There are three variables that affect the amount of water in the reservoir:

- what you start with
- how much enters
- how much leaves.

How can you calculate the three variables?

- You are given that $200 \times 10^3\,m^3$ of water is in the reservoir at the start ($t = 0$).
- How much water enters the reservoir each day depends on the flow of water from the river that day. You are told that the water in the river flows at a rate of $w(t) = 2 + \sin\left(\frac{\pi}{180}t\right)$ thousands of m^3 per day.
- You know that water leaves the reservoir at a constant rate of $2000\,m^3$ per day.

How can you find the amount on any day?

The amount on any day is the sum of the changes for all the days up to that point.

The 'sum of' tells you to integrate – you need to integrate a function to get the total.

Can you make any approximations or simplifications?

365 days in a year is an awkward number. You could approximate the length of a year to 360 days, since 360 is a multiple of 180. That would simplify the $\sin\left(\frac{\pi}{180}t\right)$ term. Remember to set your calculator to radian mode.

Keep numbers simple by taking all volumes in $10^3 \, m^3$ rather than m^3.

How can you find the maximum volume?

The clue is in the question. You find a maximum by differentiating and calculating when the derivative equals zero. You can use the second derivative to check whether you have a maximum or minimum.

How can you calculate the mean volume?

To find the mean volume of water in the reservoir, calculate the total amount of water that has been in the reservoir over the period and divide by the number of days. Use the formula $m = \frac{1}{(b-a)} \int_a^b f(t) \, dt$.

Make a plan for the solution

- Write an equation for how the volume of water in the reservoir changes each day.
- Use the answer to write a function for how much water is in the reservoir on any day.
- Substitute $t = 120$ into the function to calculate the volume in the reservoir that day.
- Substitute $t = 240$ into the equation for the rate of change of water per day.
- Differentiate the equation for the amount of water in the reservoir with respect to time.
- Calculate when the rate of change (first derivative) equals zero.
- Find the second derivative and check whether this point gives a maximum or minimum.
- Calculate the maximum volume.
- Use the formula for the mean value in the first three months (t between 0 and 90).
- Check whether the answers are sensible using another method (e.g. graphical).

C4 Solution implementation

Worked Example (continued)

Solution

a) Net rate of change of water in reservoir = water entering per day – water leaving per day

In units of $10^3 \, m^3$ per day:

$$r(t) = w(t) - 2$$
$$= \left(2 + \sin\left(\frac{\pi}{180}t\right)\right) - 2$$
$$r(t) = \sin\left(\frac{\pi}{180}t\right)$$

The total change in the amount of water in the reservoir up to the Tth day after the start of the year is the integral of $r(t)$ with respect to t between the limits 0 and T: $\int_0^T \sin\left(\frac{\pi}{180}t\right) dt$

So the amount of water in the reservoir on any day T is the original volume + the change up to day T:

$$f(T) = 200 + \int_0^T \sin\left(\frac{\pi}{180}t\right) dt$$

The definite integral is a standard integral. Evaluating it gives

$$f(T) = 200 + \left[-\frac{180}{\pi}\cos\left(\frac{\pi}{180}t\right)\right]_0^T$$

$$= 200 + \left\{\left(-\frac{180}{\pi}\cos\left(\frac{\pi}{180}(T)\right)\right) - \left(-\frac{180}{\pi}\cos\left(\frac{\pi}{180}(0)\right)\right)\right\}$$

$$= 200 + \left(-\frac{180}{\pi}\cos\left(\frac{\pi T}{180}\right) + \frac{180}{\pi}\right)$$

$$f(T) = 200 + \frac{180}{\pi}\left(1 - \cos\left(\frac{\pi T}{180}\right)\right)$$

b) When $T = 120$:

$$f(120) = 200 + \frac{180}{\pi}\left(1 - \cos\left(\frac{\pi \times 120}{180}\right)\right)$$

$$= 200 + \frac{180}{\pi}\left(1 - \cos\left(\frac{2\pi}{3}\right)\right) \approx 285.94$$

On 1 May there will be $285.94 \times 10^3 \, \text{m}^3$ of water in the reservoir.

c) The rate of change of volume of water in the reservoir was found in part a) to be

$$r(t) = \sin\left(\frac{\pi}{180}t\right) \times 10^3 \, \text{m}^3/\text{day}$$

When $T = 240$:

$$r(240) = \sin\left(\frac{\pi}{180}(240)\right) = \sin\frac{4\pi}{3}$$

$$= -0.866 \times 10^3 \, \text{m}^3 \text{ per day}$$

This is a negative value, so on 1 September the volume is falling by $866 \, \text{m}^3$ per day.

d) To find the maximum of the volume function $f(t)$, look for where its derivative is zero.

But the derivative is the rate of change of the volume, which we know is $r(t)$, so

$$f'(t) = r(t) = \sin\left(\frac{\pi}{180}t\right) \times 10^3 \, \text{m}^3 \text{ per day}$$

The derivative is zero when $\sin\left(\frac{\pi}{180}t\right) = 0$, which happens when t is a multiple of 180.

Within one year the possibilities are day 0, 180 or 360.

Common sense says it will be mid-year (day 180), but you need to verify this.

Check the second derivative:

$$f''(t) = r'(t) = \frac{d}{dt}\left(\sin\left(\frac{\pi}{180}t\right)\right) = \frac{\pi}{180}\cos\left(\frac{\pi}{180}t\right)$$

At $t = 180$, $f'(180) = \frac{\pi}{180}\cos\left(\frac{\pi}{180}(180)\right) = \frac{\pi}{180}\cos(\pi) = -\frac{\pi}{180}$

This is negative, so $t = 180$ gives a maximum.

There will be a maximum volume of water in the reservoir on day 180.

e) $f(180) = 200 + \left(\dfrac{180}{\pi} - \dfrac{180}{\pi} \cos\left(\dfrac{\pi}{180}(180) \right) \right) = 200 + \left(\dfrac{180}{\pi} + \dfrac{180}{\pi} \right) = 314.6$

The maximum volume of water in the reservoir is $314.6 \times 10^3\,\text{m}^3$.

f) Over the first 3 months (t between 0 and 90):

$$m = \frac{1}{(b-a)} \int_a^b f(t)\,dt$$

$$= \frac{1}{(90-0)} \int_0^{90} 200 + \frac{180}{\pi}\left(1 - \cos\left(\frac{\pi}{180}t \right) \right) dt$$

$$= \frac{1}{90} \int_0^{90} 200 + \frac{180}{\pi} - \frac{180}{\pi}\cos\left(\frac{\pi}{180}t \right) dt$$

$$= \frac{1}{90} \left[200t + \frac{180}{\pi}t - \left(\frac{180}{\pi} \right)^2 \sin\left(\frac{\pi}{180}t \right) \right]_0^{90}$$

$$= \frac{1}{90} \left\{ \left(200(90) + \frac{180}{\pi}(90) - \left(\frac{180}{\pi} \right)^2 \sin\left(\frac{\pi}{180}(90) \right) \right) \right.$$
$$\left. - \left(200(0) + \frac{180}{\pi}(0) - \left(\frac{180}{\pi} \right)^2 \sin\left(\frac{\pi}{180}(0) \right) \right) \right\}$$

$$= 200 + \frac{180}{\pi} - \frac{360}{\pi^2}$$

$$= 220.8$$

The average volume of water in the reservoir in the first 3 months is $220.8 \times 10^3\,\text{m}^3$

Do the answers seem reasonable?

All the answers are of the same order of magnitude as the values given in the problem. You could check the results by drawing graphs of $r(t)$ and $f(t)$ over the year and comparing the values. **Figure 7.19** shows the graph of $f(t)$, with estimation of the gradient at $t = 240$.

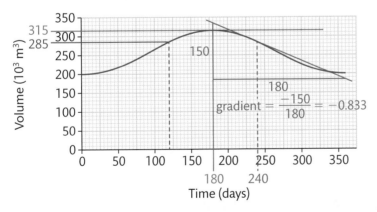

$$\text{gradient} = \frac{-150}{180} = -0.833$$

▶ **Figure 7.19** Check the results for the reservoir question by plotting the graph of the volume function

You could check the mean value by using a numerical method to calculate the area under the curve, and then divide by the number of days.

The values in **Table 7.7** were calculated in Microsoft Excel®. The area was estimated using the trapezium rule, and then the mean was obtained by dividing the area by the number of days (90).

Table 7.7 Values of $f(t) = 200 + \dfrac{180}{\pi}\left(1 - \cos\left(\dfrac{\pi t}{180}\right)\right)$ calculated at equally spaced t values

Day	0	10	20	30	40	50	60	70	80	90
Volume	200.0	200.87	203.46	207.68	213.40	220.47	228.65	237.70	247.35	257.30

Area = 19 882

Mean = 220.9

You should give reasoned explanations for any differences between the values – remember that you are critiquing the methods. For example, the trapezium rule and graphical methods only give approximate values. How accurate would you expect them to be? Are they in the same ball-park as the values you calculated using analytical approaches? What are the percentage differences? Are these differences acceptable?

Assessment practice 7.3

C.P7 C.P8 C.M3 C.M4 C.D3

Your boss has provided you with a power supply of fixed voltage output and output resistance and set you the task of finding the value of load resistance that will provide the maximum power transfer to the load. The simple circuit is shown in **Figure 7.20**.

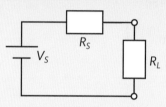

▶ **Figure 7.20** Diagram of circuit

You have heard that the load should match the output resistance, but need to prove this to your boss mathematically.

After some thought you decide to carry out the following steps.

1 Write equations for:
   - the current in the circuit, I, in terms of V_S, R_S and R_L
   - the power P_L dissipated in the load in terms of I_L and R_L.

2 Show that the maximum power transfer with respect to R_L occurs when $R_L = R_S$.

3 Compare the result with a graph of P_L against R_L for $V_S = 24$ V and $R_S = 1200\,\Omega$ over the region $0\,\Omega \leqslant R_L \leqslant 1800\,\Omega$.

4 Calculate the power dissipated in the matching load in a time of 2 minutes.

Write a short reflection on the activity, identifying what went well and not so well. What would you do differently if faced with a similar problem in the future?

Plan
- What is the task? What am I being asked to do?
- What information will I need? What tools and methods can I use to help me complete the task?
- How confident do I feel in my own abilities to complete this task? Are there any areas I think I may struggle with?

Do
- I know what I'm doing and what I am aiming to achieve in this task.
- I can identify the parts of the task that I find most challenging and devise strategies to overcome those difficulties.

Review
- I can explain what the task involved and how I approached its completion.
- I can explain how I would deal with the hard elements differently next time.
- I can identify the parts of my knowledge and skills that could benefit from further development.

Further reading and resources

Websites

Maths is Fun: **www.mathsisfun.com**

A lot of useful skills and knowledge organised into categories that are easy to find.

mathcentre: **www.mathcentre.ac.uk/students/topics**

A set of resources split into topics designed to support self-study.

Book

Croft, T. and Davison, R. (2015) *Mathematics for Engineers: A Modern Interactive Approach*, Pearson.

THINK ▶FUTURE

Gary Capewell Aerospace engineer

I have worked in the aerospace industry for over 30 years, initially within the airframes precision sheet metal sector. I was part of a team tasked with building one-off projects, ranging from components for use in space programmes to the latest Euro fighter, always taking a 'hands-on' approach. The work was varied and exciting, and calculations were at the heart of everything I did – from calculating material requirements to CNC programming to pattern development – so each day would add further experience.

When I moved from an airframes business to Rolls Royce, the leading aero-engine manufacturer in the world, my role changed. I was looked on as a company expert, making critical decisions on a daily basis. I had to ensure that standards were adhered to, but at the same time I was always looking to challenge by using process innovation.

I could only move to the next level of my career through further education, by refreshing and updating my knowledge and skills. Engineering is constantly changing, and a business has to evolve in order to stay competitive. Successfully completing a part-time degree course changed my career almost overnight. I now manage a team of people in planning and control, guiding the company on production-rate readiness for all global projects – 'can we deliver, on time, to the customer'. Since moving to Rolls Royce, I have been learning a new set of skills: the theory behind what helps a company survive, cash flow, lead time, takt time, value stream mapping and lean management principles, as well as applied science, which taught me how the machines I had used over many years actually worked. Acquiring these new skills involved learning new maths concepts and techniques. Calculus made me want to learn more – I could predict an outcome by using data to calculate rates of change and the cumulative effect of small changes. I could hold conversations at a higher level with greater confidence and got more respect from my peers..

Focusing your skills

Using mathematical skills to solve engineering problems

It is important to be able to apply mathematical methods to solve engineering problems. Here are some simple tips to help you do this.

- Gather all the information you can to understand the nature of the problem – for example, inspection and quality assurance, calculation of central tendencies and dispersion, forecasting, reliability estimates for components and systems, customer behaviour, condition monitoring and product performance.
- Use critical thinking when extracting information from the description of the problem, and identify appropriate mathematical methods to solve the problem.

- Become confident in the mathematical tools you need to use – such as arithmetic, algebraic, statistical, estimation and approximation skills. Practise the techniques to build confidence.
- Develop a plan to solve the problem.
- Critically analyse the solution obtained.
- Reflect on the problem-solving process and the solution obtained and make refinements if necessary.
- Repeat the process until you are happy with the outcome.
- Present and explain the solution to the problem in a clear and logical way.

Getting ready for assessment

Kuldeep is working towards a BTEC National in Engineering. He was preparing to carry out the assignment for Learning aim C. He knew that in the assignment he would have to use different calculus techniques to solve an engineering problem. Kuldeep shares his experience below.

How I got started

First, I looked through all my notes on differentiation and integration. I paid particular attention to the worked examples in the notes. I used the examples to identify the main skills, such as the rules for differentiating and integrating standard functions.

How I brought it all together

▶ I went through the worked examples carefully, writing out the solutions myself.

▶ I then tried answering the same questions myself, without looking at the example.

▶ I then changed the numbers slightly and tried again until I was confident in solving the problems.

▶ Once I was confident with the basic rules, I attempted questions that involved more complicated rules such as the product rule.

▶ When I was happy with the examples in my notes, I found other questions in textbooks and on the internet so I could practise using calculus to solve engineering problems.

What I learned from the experience

I was pretty sure that I knew how to answer the basic differentiation and integration questions, but I was worried that I would not understand when and how to use the different techniques in real problems. I think practising on questions taken from different books and websites helped because they did not all use the same words. I got used to having to work out what to do. I found the questions in textbooks useful because they gave the answers at the back, but they tended to be more maths-based and not so much related to engineering. Some of the problems I found on the internet were more complicated than anything I was likely to do, so I had to choose carefully.

Once I worked out what the problem in the assignment meant, and how calculus fitted in, I was able to plan a series of steps to solve it. Getting lots of practice with the basic skills really helped.

Think about it

▶ Have you identified the key methods that you could use for differentiating and integrating routine and non-routine functions?

▶ Do you fully understand the problem you are investigating?

▶ Do you know where to look for help?

▶ Have you planned your time?

▶ Are you keeping track of your progress? Do you need to find any more information to help you complete your solution?

▶ Does your solution seem reasonable?

▶ Have you reflected on what you have done? What went well? What did not go so well? What can you use to help solve future problems?

Further Engineering Mathematics

8

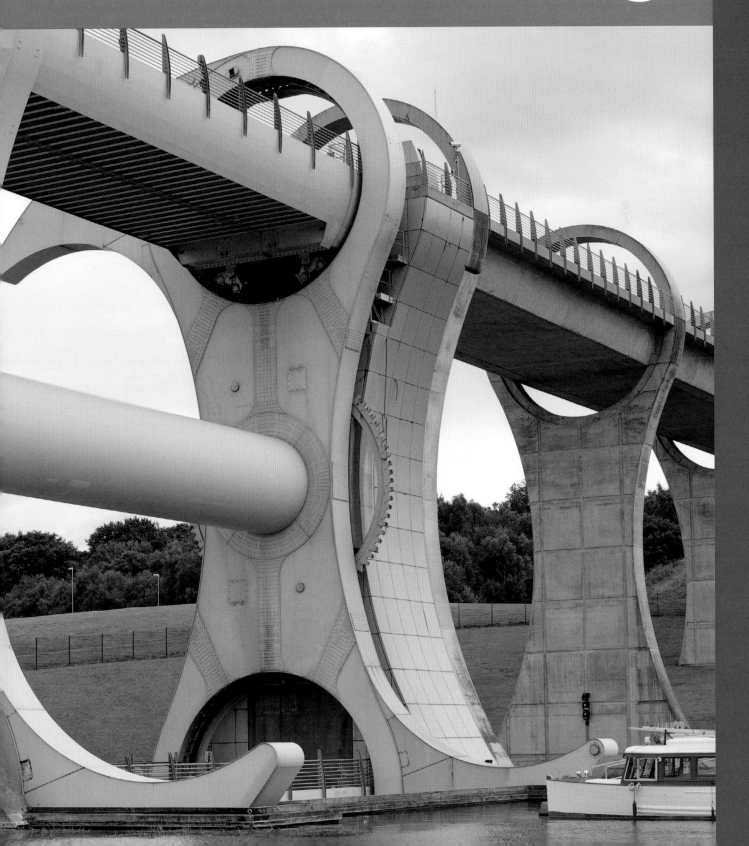

Getting to know your unit

Mathematics is a way in which engineers model the world around them and test their ideas and designs before these become real products. Statistics might be used during manufacturing processes as part of the quality control (QC) system and to determine the in-service reliability of a product. As an engineer you need to understand and develop the skills required to solve complex problems and analyse data.

How you will be assessed

This unit will be assessed by a series of internally assessed tasks set by your tutor. Throughout this unit you will find assessment activities to help you work towards your assessment. Completing these activities will mean that you carried out useful research or preparation that will be relevant when it comes to your final assignment.

In order for you to achieve the tasks in your assignment, it is important to check that you have met all of the assessment criteria. You can do this as you work your way through the assignment.

The assignments set by your tutor will consist of a number of tasks designed to meet the criteria in the table.

Assessment criteria

This table shows what you must do in order to achieve a **Pass**, **Merit** or **Distinction** grade, and where you can find activities to help you.

Pass	Merit	Distinction

Learning aim **A** Examine how sequences and series can be used to solve engineering problems

Pass	Merit	Distinction
A.P1 Solve given problems using routine arithmetic and geometric progression operations. **Assessment practice 8.1**	**A.M1** Solve given problems accurately using routine and non-routine arithmetic and geometric progression operations. **Assessment practice 8.1**	**A.D1** Evaluate, using technically correct language and a logical structure, engineering problems using non-routine sequence and series operations, while solving all the given problems accurately using routine and non-routine operations. **Assessment practice 8.1**
A.P2 Solve given problems using routine power series operations. **Assessment practice 8.1**	**A.M2** Solve given problems accurately using routine and non-routine power series operations. **Assessment practice 8.1**	

Learning aim **B** Examine how matrices and determinants can be used to solve engineering problems

Pass	Merit	Distinction
B.P3 Solve given problems using routine matrices and determinant operations. **Assessment practice 8.2**	**B.M3** Solve given problems accurately using routine and non-routine matrices and determinant operations. **Assessment practice 8.2**	**BC.D2** Evaluate, using technically correct language and a logical structure, engineering problems using non-routine matrices, determinant and complex operations, while solving all the given problems accurately using routine and non-routine operations. **Assessment practice 8.2**

Learning aim **C** Examine how complex numbers can be used to solve engineering problems

Pass	Merit	Distinction
C.P4 Solve given problems using routine complex number operations. **Assessment practice 8.2**	**C.M4** Solve given problems accurately using routine and non-routine complex number operations. **Assessment practice 8.2**	

Learning aim **D** Investigate how statistical and probability techniques can be used to solve engineering problems

Pass	Merit	Distinction
D.P5 Solve an engineering problem using routine central tendency, dispersion and probability distribution operations. **Assessment practice 8.3**	**D.M5** Solve an engineering problem accurately using routine and non-routine central tendency, dispersion and probability distribution operations, providing an explanation of the process. **Assessment practice 8.3**	**D.D3** Evaluate the correct synthesis and application of statistics and probability to solve engineering problems accurately involving routine and non-routine operations. **Assessment practice 8.3**
D.P6 Solve an engineering problem using routine linear regression operations. **Assessment practice 8.3**	**D.M6** Solve engineering problems accurately using routine and non-routine regression operations, providing an explanation of the process. **Assessment practice 8.3**	

Getting started

You have already studied quite a few mathematical topics, and gained skills and knowledge. Think back to what you have learned in the past. Try doing a few warm-up questions from your previous work. List your mathematical skills and make a truthful estimate of your ability with them. Like all skills, mathematical skills benefit from practice. When you have completed this unit, see how your previous skills and knowledge have developed and identify some new ones to add to your list.

 A # Examine how sequences and series can be used to solve engineering problems

A1 Arithmetic and geometric progressions

Sequences

One way that you can begin to understand the relationship between measured values in an engineering problem is to look for patterns by putting the numbers in sequence. This makes it easier for you to examine how the values relate to each other. You can express the relationship as a rule to identify when values fit into a sequence. In some cases, the rule can be written as an equation.

Look at the following sequences:

▸ {1, 2, 3, 4, 5} is the sequence of the first five positive integers. It could also be called the first five counting numbers or simply the numbers 1 to 5. This is a **finite sequence**. (Note that more than one rule can describe the same sequence.)

▸ {1, 2, 3, 4, 5, ...} is a similar sequence, but the three dots '...' indicate that the sequence continues indefinitely. This is an **infinite sequence**.

▸ A sequence does not have to increase. For example, {5, 4, 3, 2, 1} is the sequence counting down from 5 to 1. This is a finite sequence.

▸ A sequence can contain the same number many times. For example, {1, 0, 1, 0, 1, ...} is an alternating sequence of 1s and 0s. This is an infinite sequence.

A sequence is a set, but not all sets are sequences. The examples above highlight two differences:

▸ The elements of a sequence are in order.

▸ A number can be repeated in a sequence, but in a set each number can appear only once.

One famous sequence was described by an Italian known today as Fibonacci in his *Book of Calculations*.

Research

Investigate the Fibonacci sequence. How is it formed? What is the link between numbers in the sequence (called Fibonacci numbers)?

Who was Fibonacci? When was he alive?

What important system did he introduce to Europe?

Progressions

The Fibonacci sequence is actually a progression because it meets the criterion of each element being related to previous elements by the same rule. Each element is the sum of the two previous elements:

$$x_n = x_{n-1} + x_{n-2}$$

The Fibonacci sequence occurs naturally in such examples as the number of seeds in a seed head and the way a snail shell grows. Look at the patterns made by the petals of a flower. Do they follow a Fibonacci **progression**?

Key terms

Sequence – a set of numbers written in order according to some rule. Each value in the sequence is called an element, a term or a member.

Progression – a sequence in which each term is related to the previous term or terms by a uniform law.

All progressions are sequences, but not all sequences are progressions.

▶ Fibonacci numbers in nature

Series

The sum of the n-term sequence $\{a_1, a_2, a_3, ..., a_n\}$ is

$$S_n = a_1 + a_2 + a_3 + ... + a_n$$

For example, the sum of the first four terms in the Fibonacci sequence
$\{1, 1, 2, 3, 5, 8, 13, ...\}$ is

$$S_4 = 1 + 1 + 2 + 3 = 7$$

⏸ PAUSE POINT

1   Write the following sequences:
    a) the first five odd numbers
    b) the numbers counting down from 50 in fives
    c) the number of days in each month starting from January
    d) the square of integer values.

2   Identify which of the above are finite sequences.

3   Identify which are infinite sequences.

4   Identify which are progressions.

Hint   Check that you understand the similarities and differences between a sequence and a progression.

Extend   What is the difference between a finite sequence and an infinite sequence?

1   What is the 20th term of the Fibonacci sequence? (Try to calculate it without writing out all the values.)

2   Calculate S_5 for the sequence $\{2, 4, 6, 8, 10, 12, 14, ...\}$.

Arithmetic progressions

The sequence in the second 'Extend' question of the Pause Point above is an example of an **arithmetic progression**. The **common difference** between terms is 2. **Table 8.1** shows how the numbers in this progression are formed.

Key terms

Series – the sum of the terms in a sequence.

Arithmetic progression – a sequence of numbers where the difference between any two successive terms is a constant. This constant is called the **common difference** (d).

▶ **Table 8.1** How the numbers in the progression $\{2, 4, 6, 8, 10, 12, 14, ...\}$ are generated

Position in sequence	$n = 1$	$n = 2$	$n = 3$	$n = 4$	$n = 5$	$n = 6$	$n = 7$	...
Value	2	4	6	8	10	12	14	...
How formed	2	2 + 2	2 + (2 × 2)	2 + (2 × 3)	2 + (2 × 4)	2 + (2 × 5)	2 + (2 × 6)	...

▶ The 1st term is $a_1 = 2 + (2 \times (0))$
▶ The 2nd term is $a_2 = 2 + (2 \times (1))$
▶ The nth term is $a_n = 2 + (2 \times (n - 1))$

For any arithmetic progression, with initial value a and common difference d, the terms are

$\{a, (a + d), (a + 2d), (a + 3d), ..., (a + (n − 1)d), ...\}$

The last term of a finite arithmetic progression is

$l = a + (n − 1)d$

Research

Investigate how you can show that the three equations for finding the sum of an arithmetic progression are actually the same. You can find the proof in many textbooks or on the internet.

Worked Example

An oil rig drills to a depth of 1000 metres. Estimate the cost of drilling if the first metre costs £200, with an increase of £22 for each further metre drilled.

Solution

This is an arithmetic progression with $a = 200$, $d = 22$, $n = 1000$.

The cost of drilling to a depth of 1000 m is the 1000th term in the progression:

$$a_{1000} = 200 + (22 × (1000 − 1))$$
$$= 200 + 21\,978 = 22\,178$$

The cost is £22 178.

Arithmetic series

The sum of the first n terms in an arithmetic progression is

$S_n = a + (a + d) + (a + 2d) + (a + 3d) + ... + (a + (n − 1)d)$

This can also be written in two slightly different ways:

$S_n = \frac{1}{2}n(a + l)$

where l is the last term in the sum, or

$S_n = \frac{1}{2}n(2a + (n − 1)d)$

Key term

Arithmetic series – the sum of the terms in an arithmetic progression.

Worked Example

Find the sum of the first 15 odd numbers.

Solution

Step 1: Identify the first term, the common difference and the number of terms in the sum:

$a = 1, d = 2, n = 15$.

Step 2: Substitute the values into a formula for S_n.

Using $S_n = \frac{1}{2}n(2a + (n − 1)d)$ directly:

$$S_n = \frac{1}{2}(15)(2 × 1 + (15 − 1) × 2) = 225$$

Alternatively, first calculate the last term l and then use $S_n = \frac{1}{2}n(a + l)$:

$$l = 1 + (15 − 1)(2) = 29$$
$$S_n = \frac{1}{2}(15)(1 + 29) = 225.$$

The **arithmetic mean** is the sum of the arithmetic series divided by the number of terms:

$$\text{arithmetic mean} = \frac{S_n}{n}$$

For example, the arithmetic mean of the first 15 odd numbers is:

$$\frac{S_{15}}{15} = \frac{225}{15} = 15$$

PAUSE POINT

Calculate:

a) the 9th and 17th terms of the arithmetic progression with first term 10 and common difference 4

b) the 12th and 20th terms of the arithmetic progression with first term 60 and common difference −6

c) the 4th and 10th terms of the arithmetic progression with first term 2 and common difference $\frac{1}{4}$.

Hint   Follow the steps in the worked examples carefully.

Extend   Calculate:

a) the time taken to make 200 components if the first one takes 5 minutes and each additional one takes 2 minutes

b) the sum of the first 20 even numbers

c) the arithmetic mean of the first 20 even numbers.

Geometric progressions

The **geometric progression** {1, 3, 9, 27, 81, 243, ...} has first term $a = 1$ and **common ratio** $r = 3$.

In **Table 8.2** we look at the numbers in this progression more closely.

▶ **Table 8.2** How the numbers in the progression {1, 3, 9, 27, 81, 243, ...} are formed

Position in sequence	$n = 1$	$n = 2$	$n = 3$	$n = 4$	$n = 5$	$n = 6$
Value	1	3	9	27	81	243
How formed	1	1×3^1	1×3^2	1×3^3	1×3^4	1×3^5

The nth term is $1 \times 3^{n-1}$

In general, you can write a geometric progression with first term a and common ratio r as $\{a, ar^1, ar^2, ar^3, ..., ar^{n-1}\}$.

You can also have a negative and/or fractional common ratio.

> **Key term**
>
> **Geometric progression** – a sequence of numbers where each element is multiplied by a constant value to form the next element in the sequence. This constant multiplier is called the **common ratio** (r).

Worked Example – Motor speeds

The motor powering an oil drill has five speeds. These vary from 10 rpm to 160 rpm in a geometric progression. Find the common ratio, and make a table showing the five required speeds.

Solution

Extract the data given in the question:

$a_1 = 10$, $a_5 = 160$, $n = 5$

The nth term is given by $a_n = a_1 r^{(n-1)}$, so

$a_5 = 160 = 10r^{(5-1)}$

$16 = r^4$

$r = \sqrt[4]{16} = 2$

Now you can make a table showing the five speeds, as in **Table 8.3**.

▶ **Table 8.3** The five speeds of the motor

$n = 1$	$a_1 = 10 \times 2^0 = 10.0$ rpm
$n = 2$	$a_2 = 10 \times 2^1 = 20.0$ rpm
$n = 3$	$a_3 = 10 \times 2^2 = 40.0$ rpm
$n = 4$	$a_4 = 10 \times 2^3 = 80.0$ rpm
$n = 5$	$a_5 = 10 \times 2^4 = 160.0$ rpm

Worked Example – Depreciation costs

An engineering company installs a new machine for £75 000. The machine's value depreciates at a rate d of 15% per annum (a year).

a) What will the machine's value be after seven years?

b) How long will it take for the value to be less than £37 500 (half the initial value)?

Solution

a) The value of the machine in successive years forms a geometric progression.

The initial value is taken to be at year 0: $a_0 = 75\,000$.

The common ratio is $r = 1 - d = 0.85$.

Note that because we start at year 0, the equation for the nth term of the geometric progression (representing the value after n years) will be $a_n = a_0 r^n$.

Therefore the value after seven years ($n = 7$) will be

$a_7 = 75\,000 \times 0.85^7 = £24\,043.28$ or £24 043 (to the nearest pound).

b) To find how long it will take for the machine's value to equal £37 500, solve the equation:

$37\,500 = 75\,000\,(0.85)^n$

Rearrange to make n the subject:

$\dfrac{37\,500}{75\,000} = (0.85)^n$ \qquad Divide both sides by 75 000.

$\frac{1}{2} = (0.85)^n$

$\log\left(\frac{1}{2}\right) = n\log(0.85)$ \qquad Take logs of both sides and use the rules of logarithms.

$-0.30103 = n(-0.07058)$

$n = \dfrac{-0.30103}{-0.07058} = 4.265$

It will take 4.3 years for the machine to depreciate to less than £37 500.

Geometric series

The sum of a **geometric series** is given by

$$S_n = a + ar + ar^2 + ar^3 + \ldots + ar^n$$

This can also be written as

$$S_n = \frac{a(1 - r^n)}{(1 - r)} \text{ provided that } r \neq 1$$

Research

Investigate how you can show that the two equations for the geometric series are equivalent.

Worked Example

Find the sum of the first five terms of the geometric series $1 + 3 + 9 + \ldots$

Solution

Identify the parameters:

$a = 1$, $r = 3$, $n = 5$

Substitute these values into the formula:

$$S_n = \frac{a(1 - r^n)}{(1 - r)}$$

$$S_5 = \frac{1 \times (1 - 3^5)}{(1 - 3)} = \frac{1 - 243}{1 - 3} = \frac{-242}{-2} = 121$$

Convergence of a geometric series

If the common ratio of a geometric progression is a proper fraction (either positive or negative, but smaller than 1 in size), then each term will get smaller as n gets bigger. In other words, the terms become less significant the further along the series you go.

For example, if $a = 4$ and $r = \frac{1}{4}$, the geometric series is

$4 + 1 + \frac{1}{4} + \frac{1}{16} + \ldots$

You reach a point where adding more terms does not make any noticeable difference to the total – the series converges to a final value. The **sum to infinity** of a convergent infinite geometric series is

$$S_\infty = \frac{a}{(1 - r)} \text{ provided that } -1 < r < 1$$

Key terms

Geometric series – the sum of the terms in a geometric progression.

Convergent series – a series for which S_n heads towards a specific value as n (the number of terms in the sum) increases.

Divergent series – a series for which S_n does not settle down to a definite value as n increases.

Ⅱ PAUSE POINT

Calculate

a) the sum of the first six terms of the geometric series $2 + 8 + 32 + ...$

b) the sum to infinity of the geometric series with first term $a = 10$ and common ratio $r = \frac{1}{2}$.

Hint   Use your research into the effect of the common ratio on convergence.

Research

Investigate how the value of the common ratio affects whether a geometric series is *convergent* or *divergent*.

Worked Example

Find the sum to infinity of the geometric series $4 + 1 + \frac{1}{4} + \frac{1}{16} + ...$

Solution

Here $a = 4$ and $r = \frac{1}{4}$, so $-1 < r < 1$ is true.

The sum to infinity of the geometric series is

$$S_\infty = \frac{4}{\left(1 - \frac{1}{4}\right)} = \frac{4}{\frac{3}{4}} = \frac{16}{3} = 5\frac{1}{3}$$

The sum never quite reaches this value, but gets close enough for all practical purposes.

A2 Binomial expansion

A **binomial** is an algebraic expression with two terms connected by a plus or minus sign, such as $(a + b)$, $(x - 3)$ or $(2x + 3)^2$.

A general binomial expression can be written as $(a + b)^n$.

You probably already know how to expand a squared expression such as $(2x + 3)^2$ using FOIL (First–Outside–Inside–Last), but what about an expression like $(2x + 3)^{22}$?

One interesting way to tackle the problem uses a pattern that was known to the ancient Chinese but is more often referred to as Pascal's triangle. (Blaise Pascal was a French mathematician and physicist of the 17th century.)

Research

Investigate Pascal's triangle and the patterns that can be found in it.

The numbers in Pascal's triangle relate to the terms in the expansion of $(a + b)^n$, called the binomial expansion. The terms of the binomial expansion are written in sequence. The first term is a raised to the power n. In the second term, the power of a reduces by 1 while the power of b increases by 1. This pattern continues until the power of a is reduced to zero and you are left with just b raised to the power n.

Pascal's triangle gives the **coefficients** of the terms involving a and b in the binomial expansion.

The top line of Pascal's triangle relates to $(a + b)^0$. (Remember that anything raised to the power zero is 1.)

		1			$(a + b)^0$	1
	1		1		$(a + b)^1$	$a + b$
1		2		1	$(a + b)^2$	$a^2 + 2ab + b^2$
1	3		3	1	$(a + b)^3$	$a^3 + 3a^2b + 3ab^2 + b^3$

The triangle can be continued indefinitely following the same pattern – each number is obtained by adding together the two nearest numbers in the line above it.

Note that the coefficients are symmetrical as long as you write the terms of the binomial expansion strictly in sequence, with the power of a decreasing and the power of b increasing.

This method of expanding $(a + b)^n$ works, but you would need a very big sheet of paper to find the coefficients in the expansion of $(2x + 3)^{22}$. Luckily there is another way to determine the expansion, by using the formula

$$(a + b)^n = a^n + na^{n-1}b^1 + \frac{n(n - 1)}{2!}a^{n-2}b^2$$
$$+ \frac{n(n - 1)(n - 2)}{3!}a^{n-3}b^3 + ... + b^n$$

This equation uses 'factorial' notation: $2! = 2 \times 1$, $3! = 3 \times 2 \times 1$ and so on. Most scientific calculators have a factorial function (!).

Key terms

Binomial – a polynomial with two terms.

Coefficient – a number which multiplies a variable.

n **factorial** ($n!$) – n multiplied by $(n - 1)$ multiplied by $(n - 2)$ and so on, until multiplied by 1.

The coefficients of a binomial expansion are actually given by a function called 'n choose r', which is the number of ways of choosing r objects from a set of n objects.

There are several ways of writing 'n choose r', such as $\binom{n}{r}$ or $_nC_r$. It is given in terms of factorials as

$$_nC_r = \frac{n!}{r!(n-r)!}$$

Most scientific calculators have a $_nC_r$ function.

Worked Example

Find the coefficients of the terms in the expansion of $(a + b)^4$.

Solution

The first term corresponds to $r = 0$:

$$\binom{4}{0} = \frac{4!}{0!(4-0)!} \quad \text{(Note that, by convention, } 0! = 1\text{)}$$

$$= \frac{4 \times 3 \times 2 \times 1}{(1)(4 \times 3 \times 2 \times 1)} = 1$$

The second term corresponds to $r = 1$:

$$\binom{4}{1} = \frac{4!}{1!(4-1)!}$$

$$= \frac{4 \times 3 \times 2 \times 1}{(1)(3 \times 2 \times 1)} = 4$$

The third term corresponds to $r = 2$:

$$\binom{4}{2} = \frac{4!}{2!(4-2)!}$$

$$= \frac{4 \times 3 \times 2 \times 1}{(2 \times 1)(2 \times 1)} = 6$$

By symmetry, you know that the coefficient of the fourth term is the same as the coefficient of the second term (4), and the coefficient of the fifth term is the same as the coefficient of the first term (1).

So $(a + b)^4 = a^4 + 4a^3b^1 + 6a^2b^2 + 4a^1b^3 + b^4$.

A binomial expansion can be written in 'shorthand' using 'sigma' notation, where the Greek letter Σ stands for 'sum':

$$(a + b)^n = \sum_{r=0}^{r=n} \binom{n}{r} a^{n-r} b^r$$

Summary – binomial expansions

- A binomial $(a + b)^n$, where n is a positive integer, expands to give terms in a^n, $a^{n-1}b^1$, $a^{n-2}b^2$, $a^{n-3}b^3$ etc. The final term is b^n.
- The sum of the powers in each term equals n.
- The coefficient of the first and last terms is 1.
- The coefficient of the second and penultimate terms is n.
- The coefficients can also be obtained using the 'n choose r' function.
- The first term in the expansion has $r = 0$.
- The expansion of $(a + b)^n$ can be written as

$$(a + b)^n = a^n + na^{n-1}b^1 + \frac{n(n-1)}{2!} a^{n-2}b^2$$
$$+ \frac{n(n-1)(n-2)}{3!} a^{n-3}b^3 + \dots + b^n$$

  or $(a + b)^n = \sum_{r=0}^{r=n} \binom{n}{r} a^{n-r} b^r$.

  where $_nC_r = \frac{n!}{r!(n-r)!}$.

If the first term in the binomial expression is 1, then we get the simpler formula

$$(1 + x)^n = 1 + nx + \frac{n(n-1)}{2!}x^2 + \frac{n(n-1)(n-2)}{3!}x^3 + \dots + x^n$$

This is valid for positive integer values of n, but what about other values? The binomial expansion for $(1 + x)^n$ can still be carried out when n is not a positive integer, but only if the value of x lies between -1 and $+1$. The expansion gives an infinite series. In applications, you have to decide at what point further terms in the sum become so small as to be insignificant.

ⅠⅠ PAUSE POINT Check the conditions for the binomial expansion of $(1 + x)^{-1}$ to be a convergent infinite series.

Hint Try simple numbers in the binomial expansion.

Extend For the following binomial expressions, calculate the values obtained using the binomial expansion formula and check the answers against values given by a calculator:

$$\frac{1}{(1 + 0.2)} = (1 + 0.2)^{-1}$$

$$\frac{1}{(1 + 2)} = (1 + 2)^{-1}$$

Worked Example

You have just built an electronic circuit that gives an oscillating signal. The periodic time of the oscillation is given by the formula $T = 2\pi\sqrt{LC}$, where L and C are the values of the inductor and capacitor in the circuit. When you measure the values used, you find that the inductor value is 2% above the nominal value, and the capacitor value is 1% below its nominal value. Use the binomial expansion to calculate the expected percentage error in T.

Solution

Step 1: Write the equation for T using the actual values of the inductor and capacitor, which will be $1.02L$ (2% above nominal value) and $0.99C$ (1% below nominal value).

The measured value of T is $T = 2\pi\sqrt{(1.02L)(0.99C)}$

Step 2: Rewrite T with the coefficients as binomial terms:

$$T = 2\pi\sqrt{((1 + 0.02)L)((1 - 0.01)C)}$$
$$= 2\pi\sqrt{LC(1 + 0.02)(1 - 0.01)}$$
$$= 2\pi\sqrt{LC}\sqrt{(1 + 0.02)}\sqrt{(1 - 0.01)}$$
$$= 2\pi\sqrt{LC}(1 + 0.02)^{\frac{1}{2}}(1 - 0.01)^{\frac{1}{2}}$$

Step 3: Apply the binomial expansion formula to $(1 + 0.02)^{\frac{1}{2}}$ and $(1 - 0.01)^{\frac{1}{2}}$ separately.

$(1 + 0.02)^{\frac{1}{2}}$ has $x = 0.02$ and $n = \frac{1}{2}$

$$(1 + 0.02)^{\frac{1}{2}} = 1 + \left(\frac{1}{2}\right)(0.02) + \frac{\left(\frac{1}{2}\right)\left(\frac{1}{2} - 1\right)}{2!}(0.02)^2 + \frac{\frac{1}{2}\left(\frac{1}{2} - 1\right)\left(\frac{1}{2} - 2\right)}{3!}(0.02)^3 + \dots$$
$$= 1 + 0.01 - 0.00005 + 0.0000005 + \dots \approx 1.00995$$

$(1 - 0.01)^{\frac{1}{2}}$ has $x = -0.01$ and $n = \frac{1}{2}$

$$(1 - 0.01)^{\frac{1}{2}} = 1 + \left(\frac{1}{2}\right)(-0.01) + \frac{\left(\frac{1}{2}\right)\left(\frac{1}{2} - 1\right)}{2!}(-0.01)^2 + \frac{\frac{1}{2}\left(\frac{1}{2} - 1\right)\left(\frac{1}{2} - 2\right)}{3!}(-0.01)^3 + \dots$$
$$= 1 - 0.005 - 0.0000125 - 0.0000000625 + \dots \approx 0.99499$$

Step 4: Calculate the expected error in T:

$$(1 + 0.02)^{\frac{1}{2}}(1 - 0.01)^{\frac{1}{2}} \approx 1.00995 \times 0.99499 = 1.00489$$

so the expected percentage error in T is $0.00489 = 0.489\%$.

Summary – binomial expansion of $(1 + x)^n$

The binomial expansion of $(1 + x)^n$ for all positive integer values of n:

$$(1 + x)^n = 1 + nx + \frac{n(n - 1)}{2!}x^2 + \frac{n(n - 1)(n - 2)}{3!}x^3 + \dots + x^n$$

The expansion can also be used for values of n that are not positive integers, provided that x lies between -1 and $+1$.

A3 Power series

A **power series** is a way of writing a function as an infinite series of power terms:

$$f(x) = a_0 + a_1x + a_2x^2 + a_3x^3 + \ldots$$

You can use the series to estimate the value of a function.

Research

Revise the derivatives of basic functions.

Taylor series

You can also write a function as a **Taylor series** about the point a:

$$f(x) = a_0 + a_1(x - a) + a_2(x - a)^2 + a_3(x - a)^3 + \ldots$$

To calculate the values of the coefficients a_0, a_1, a_2 etc., you need to know how to differentiate the function:

$$f(x) = f(a) + \frac{f'(a)}{1!}(x - a) + \frac{f''(a)}{2!}(x - a)^2 + \frac{f'''(a)}{3!}(x - a)^3 + \ldots$$

where f', f'', f''' are the first, second and third derivatives of the function $f(x)$.

Worked Example

Express $f(x) = \sin x$ as a power series.

You need to know that
- the derivative of $\sin x$ is $\cos x$
- the derivative of $\cos x$ is $-\sin x$.

Solution

Use the Taylor series formula:

$$f(x) = f(a) + \frac{f'(a)}{1!}(x - a) + \frac{f''(a)}{2!}(x - a)^2 + \frac{f'''(a)}{3!}(x - a)^3 + \ldots$$

Choose $a = 0$ to get simple powers of x:

$$\sin x = f(0) + \frac{f'(0)}{1!}(x) + \frac{f''(0)}{2!}(x)^2 + \frac{f'''(0)}{3!}(x)^3 + \ldots$$

The coefficients are:

$$a_0 = f(0) = \sin 0 = 0$$

$$a_1 = \frac{f'(0)}{1!} = \frac{\cos 0}{1!} = \frac{1}{1} = 1$$

$$a_2 = \frac{f''(0)}{2!} = \frac{-\sin 0}{2!} = -\frac{0}{2} = 0$$

$$a_3 = \frac{f'''(0)}{3!} = \frac{-\cos 0}{3!} = \frac{-1}{6} \text{ etc. So}$$

$$\sin x = 0 + (x) + 0(x)^2 + \frac{-1}{6}(x)^3 + \ldots = x - \frac{1}{6}x^3 + \ldots$$

Note that the sine function does not contain any even powers.

Key term

Maclaurin series – a Taylor series with $a = 0$.

The worked example uses a special case of the Taylor series where $a = 0$. This is known as a **Maclaurin series**:

$$f(x) = f(0) + \frac{f'(0)}{1!}x + \frac{f''(0)}{2!}x^2 + \frac{f'''(0)}{3!}x^3 + \ldots$$

❙❙ PAUSE POINT

Practise using power series expansions.

Use the Maclaurin series to write the first four terms of the expansion of $f(x) = e^x$.

Hint   Follow the worked example carefully, then try again without using the example.

Extend   Calculate the sum of the first four terms of the expansion of $f(x) = e^x$, for $x = 0$.

Compare your result with the value for e^0 obtained from your calculator.

Did you predict the answer?

Assessment practice 8.1

Solve the following problems involving sequences and series. It is important that you show all the steps in your calculations clearly. You also need to give a brief explanation of each step.

1 Identify the type of progression in each case:
   a) {3, 5, 7, 9, ...}
   b {5, 10, 20, ...}.

2 Calculate:
   a) the fifth term for each progression in question 1.
   b) the sum of the first five terms for each progression in question 1, using the appropriate formulae.
   c) the sum to infinity of the geometric progression which has $a = 4$ and $r = 0.5$.

3 Calculate the time taken to make 500 components if the first one takes 4 minutes and each additional one takes 3 minutes.

4 A vehicle has a five-speed gearbox. The gears vary from $3.00:1$ to $0.72:1$ in a geometric progression. Find the common ratio, and make a table showing the gears for the five speeds.

5 An engineering company installs a new computer system for £50 000. If value of the system depreciates at a rate of 20% per annum, what will its value be after three years?

6 Draw Pascal's triangle and use the values to expand $(x + 2)^5$.

7 Expand $(x + 2)^5$ using the binomial theorem and compare the answer with the one from question 6.

8 Expand $(x + 2)^{0.5}$ using the binomial theorem.

9 Expand $(x + 2)^{-0.5}$ using the binomial theorem.

10 Write the first four terms of the Maclaurin expansion for $f(x) = e^x$.

11 a) Calculate the value of e^1 using the first four terms of the Maclaurin expansion.
   b) Compare the value obtained in part a) with the value obtained using a calculator.

Plan
- What techniques will I need to use to complete the task? How am I going to approach each problem and keep track of my progress?
- How confident do I feel in my abilities to complete this task?
- Are there any areas I think I may struggle with?

Do
- I know what strategies I could employ for the task. I can determine whether these are right for the task.
- I can assess whether my strategies are working and, if not, what changes I need to make.
- I can identify when I've gone wrong and adjust my thinking/approach to get myself back on course.

Review
- I can evaluate whether I have succeeded in solving the problem.
- I can identify the elements in the task that are difficult for me and explain how I would approach them differently next time.

B Examine how matrices and determinants can be used to solve engineering problems

B1 Matrices

You have already used two simultaneous equations to solve problems involving two variables. Many engineering problems contain multiple variables. In order to solve them, you need as many equations as there are variables. **Matrix** arithmetic provides ways to write data in a compact form and to solve multiple simultaneous equations.

Types of matrix

A matrix is a rectangular array made up of a number of rows and columns. The number of rows and the number of columns describe the matrix size or order (rows × columns). For example:

> **Key term**
>
> **Matrix** – a set of data arranged in rows and columns.

$\begin{pmatrix} a & b & c \\ d & e & f \end{pmatrix}$ is a 2 × 3 matrix

$\begin{pmatrix} a & b \\ c & d \\ e & f \end{pmatrix}$ is a 3 × 2 matrix

$(a \quad b \quad c \quad d)$ is a 1 × 4 matrix

$\begin{pmatrix} a & b & c & d & e \\ f & g & h & i & j \end{pmatrix}$ is a 2 × 5 matrix

Some matrices have special names:

▸ An $n \times 1$ matrix is called a 'column' matrix.
▸ A $1 \times n$ matrix is called a 'row' matrix.
▸ A matrix with an equal number of rows and columns (for example, 2 × 2 or 3 × 3) is said to be **square**.

The entries in the matrix (the a, b, c etc. in the examples above) are the **elements** of the matrix. These are sometimes expressed by their coordinates within the matrix, for example $\begin{pmatrix} a_{1,1} & a_{1,2} \\ a_{2,1} & a_{2,2} \\ a_{3,1} & a_{3,2} \end{pmatrix}$, where the first subscript is the row number and the second subscript is the column number.

The diagonal from top left to bottom right of a square matrix is called the **leading diagonal**.

A square matrix that has values of 1 on the leading diagonal and all other elements 0 is called an **identity matrix** (or **unit matrix**). For example, $\begin{pmatrix} 1 & 0 & 0 \\ 0 & 1 & 0 \\ 0 & 0 & 1 \end{pmatrix}$ is the 3 × 3 identity matrix.

A **null matrix** (or **zero matrix**) is a matrix in which all the elements are zero.

A square matrix where all the elements either below or above the diagonal are zero is called a **triangular matrix**. For example, $\begin{pmatrix} 1 & 2 \\ 0 & 3 \end{pmatrix}$ has zero below the diagonal, so it is an upper triangular matrix; $\begin{pmatrix} 4 & 0 & 0 \\ 7 & -1 & 0 \\ -9 & 11 & 6 \end{pmatrix}$ has only zeros above the diagonal, so it is a lower triangular matrix.

A matrix is **transposed** when its rows are written as columns and its columns written as rows, for example $\begin{pmatrix} a & b \\ c & d \\ e & f \end{pmatrix}^{\mathrm{T}} = \begin{pmatrix} a & c & e \\ b & d & f \end{pmatrix}$, where the superscript T denotes 'transpose'.

Matrix calculations

You need to follow a number of rules to carry out calculations using matrices. Some of them might appear odd at first, but they are sensible rules once you understand the meaning of a matrix.

Matrix addition and subtraction

Addition and subtraction can only be carried out when the matrices involved are of the same order (that is, they have the same size and shape). This is because corresponding elements in each matrix are combined.

For example, if $A = (1 \quad 2 \quad 3)$ and $B = (4 \quad 5 \quad 6)$, then $A + B = (1 + 4 \quad 2 + 5 \quad 3 + 6) = (5 \quad 7 \quad 9)$.

It follows that $A + B$ and $B + A$ give the same result. In other words, $A + B = B + A$.

Note that the resulting matrix is of the same order as the matrices being added.

In general:

If $A = \begin{pmatrix} a & b & c \\ d & e & f \end{pmatrix}$ and $B = \begin{pmatrix} g & h & i \\ j & k & l \end{pmatrix}$,

then $A \pm B = \begin{pmatrix} a \pm g & b \pm h & c \pm i \\ d \pm j & e \pm k & f \pm l \end{pmatrix}$

Scalar multiplication of a matrix

Imagine calculating $3A$ by adding $A + A + A$:

$$3A = 3\begin{pmatrix} a & b & c \\ d & e & f \end{pmatrix} = \begin{pmatrix} a & b & c \\ d & e & f \end{pmatrix} + \begin{pmatrix} a & b & c \\ d & e & f \end{pmatrix} + \begin{pmatrix} a & b & c \\ d & e & f \end{pmatrix}$$

$$= \begin{pmatrix} 3a & 3b & 3c \\ 3d & 3e & 3f \end{pmatrix}$$

This tells us that to multiply a matrix by a constant value (a scalar), simply multiply each element by that value. The resulting matrix will be of the same order as the original matrix. This also works for negative and fractional scalars.

Matrix multiplication

Multiplying two matrices A and B can only be carried out if the number of columns in matrix A equals the number of rows in matrix B – see **Figure 8.1**.

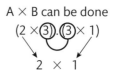

A × B can be done

(2 × ③).(③ × 1)

2 × 1

The resulting matrix is a (2 × 1)

▸ **Figure 8.1** Testing whether matrix multiplication is possible

If $A = \begin{pmatrix} a & b & c \\ d & e & f \end{pmatrix}$ and $B = \begin{pmatrix} g \\ h \\ i \end{pmatrix}$, then the product

$A \times B = \begin{pmatrix} a & b & c \\ d & e & f \end{pmatrix}\begin{pmatrix} g \\ h \\ i \end{pmatrix}$ is calculated by

combining each row of A with each column of B (see **Figure 8.2**).

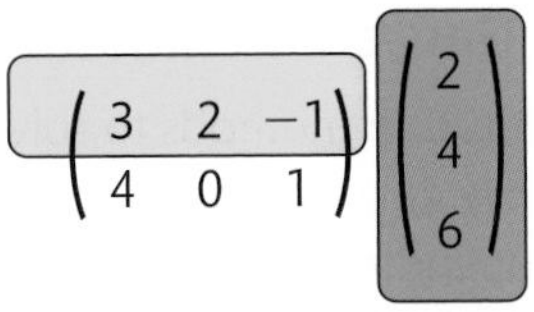

$$\begin{pmatrix} 3 & 2 & -1 \\ 4 & 0 & 1 \end{pmatrix} \begin{pmatrix} 2 \\ 4 \\ 6 \end{pmatrix}$$

▶ **Figure 8.2** To multiply two matrices together, combine each row of the left matrix with a column of the right matrix.

Row 1 of A and column 1 of B combine to give $(a \times g) + (b \times h) + (c \times i) = ag + bh + ci$.

Row 2 of A and column 1 of B combine to give $(d \times g) + (e \times h) + (f \times i) = dg + eh + fi$.

So the resulting 2×1 matrix product is $\begin{pmatrix} ag + bh + ci \\ dg + eh + fi \end{pmatrix}$.

In this case you cannot find $B \times A$ (see **Figure 8.3**).

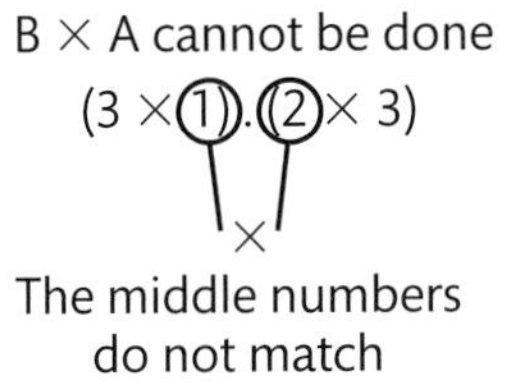

B × A cannot be done

$(3 \times \textcircled{1}) . (\textcircled{2} \times 3)$

×

The middle numbers
do not match

▶ **Figure 8.3** Why it isn't possible to calculate B × A

Larger matrices can be multiplied by repeating this process for all combinations of a row from the first matrix and a column from the second. For example:

if $A = \begin{pmatrix} a & b & c \\ d & e & f \end{pmatrix}$ and $B = \begin{pmatrix} g & h \\ i & j \\ k & l \end{pmatrix}$

then the test shows it is possible to multiply A by B (see **Figure 8.4**).

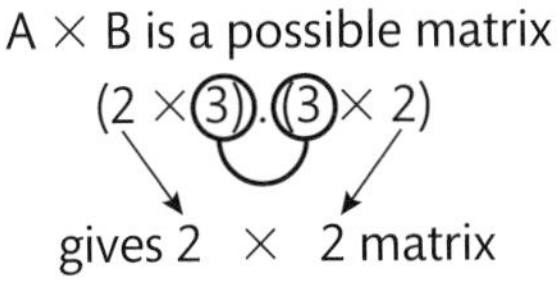

A × B is a possible matrix

$(2 \times \textcircled{3}) . (\textcircled{3} \times 2)$

gives 2 × 2 matrix

▶ **Figure 8.4** The product A × B will be a 2 × 2 matrix

For $A \times B$ the element in:

▶ row 1 column 1 results from combining row 1 of the first matrix with column 1 of the second matrix

▶ row 2 column 1 results from combining row 2 of the first matrix with column 1 of the second matrix

▶ row 1 column 2 results from combining row 1 of the first matrix with column 2 of the second matrix

▶ row 2 column 2 results from combining row 2 of the first matrix with column 2 of the second matrix.

The product is

$$AB = \begin{pmatrix} ag + bi + ck & ah + bj + cl \\ dg + ei + fk & dh + ej + fl \end{pmatrix}$$

Writing equations in matrix form

Suppose that, in analysing an electrical circuit, you had this pair of simultaneous equations:

$$2I_1 + 3I_2 = 12 \qquad (1)$$
$$-3I_1 + 4I_2 = 8 \qquad (2)$$

You could write them in matrix form as

$$\begin{pmatrix} 2 & 3 \\ -3 & 4 \end{pmatrix} \begin{pmatrix} I_1 \\ I_2 \end{pmatrix} = \begin{pmatrix} 12 \\ 8 \end{pmatrix}$$

The general matrix form of a set of simultaneous linear equations is $AX = Y$

In this example:

▶ $A = \begin{pmatrix} 2 & 3 \\ -3 & 4 \end{pmatrix}$ – the matrix of coefficients of the variables

▶ $X = \begin{pmatrix} I_1 \\ I_2 \end{pmatrix}$ – the column matrix of variables

▶ $Y = \begin{pmatrix} 12 \\ 8 \end{pmatrix}$ – the column matrix of values on the right-hand side of the equal sign.

Ⅱ PAUSE POINT Revise the rules for combining matrices, then answer these questions without looking at the rules.

1 What size does a matrix need to be if you want to add it to a 2 × 4 matrix?

2 What size will the resulting matrix in question 1 be?

3 How would you multiply a matrix by 5?

4 How would you divide a matrix by 5?

5 a) Can you find the product AB if A is a 2 × 3 matrix and B is a 3 × 5 matrix?

b) If you can, what will be the size of the resulting matrix?

c) In this case can you find the product BA? If not, explain why.

Solving simultaneous equations by Gaussian elimination

You have probably used elimination and substitution methods to solve simultaneous equations. One of the difficulties you might have encountered is that each pair of equations may be solved in a different way, depending on the values in the equations. **Gaussian elimination** provides you with a standard procedure that can be applied to any set of simultaneous linear equations. The procedure is as follows.

Step 1: Create an augmented matrix of numbers from A and Y.

Step 2: Create a triangular matrix by successively eliminating variables so that only one variable remains in the bottom row, one more variable in the row above, and so on up to the top row where all the variables remain.

Step 3: Read off the value of the variable in the bottom row.

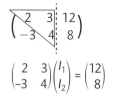

$$\begin{pmatrix} 2 & 3 & \vdots & 12 \\ -3 & 4 & \vdots & 8 \end{pmatrix}$$

$$\begin{pmatrix} 2 & 3 \\ -3 & 4 \end{pmatrix}\begin{pmatrix} I_1 \\ I_2 \end{pmatrix} = \begin{pmatrix} 12 \\ 8 \end{pmatrix}$$

▶ **Figure 8.5** Augmented matrix for applying Gaussian elimination to the system

Step 4: Substitute its value into the row above, which then allows you to calculate the value of the second variable. Continue substituting the values of the variables you have found into the row above, until you have reached the top row and found them all.

In the two-equation example above, the –3 in the lower left corner has to be converted to a zero first, which eliminates its variable (I_1) in that row (see **Figure 8.5**).

At this stage you only need to know how to solve problems with two variables.

Worked Example

Solve these two simultaneous equations using Gaussian elimination.

$3x + 2y = -3$ (1)

$5x + 3y = -4$ (2)

Solution

Step 1: Write the set of equations in matrix form $AX = Y$, so $\begin{pmatrix} 3 & 2 \\ 5 & 3 \end{pmatrix}\begin{pmatrix} x \\ y \end{pmatrix} = \begin{pmatrix} -3 \\ -4 \end{pmatrix}$

Step 2: Merge A and Y into an augmented matrix: $\begin{pmatrix} 3 & 2 & | & -3 \\ 5 & 3 & | & -4 \end{pmatrix}$

Step 3: The aim is to get a zero in place of the element in the lower left corner, 5.

To do this, first multiply row 1 by 5 and row 2 by 3:

$$\begin{pmatrix} 3 \times 5 & 2 \times 5 & | & -3 \times 5 \\ 5 \times 3 & 3 \times 3 & | & -4 \times 3 \end{pmatrix} = \begin{pmatrix} 15 & 10 & | & -15 \\ 15 & 9 & | & -12 \end{pmatrix}$$

Then subtract row 2 from row 1 to get $(0 \quad 1 | -3)$

Replace the second row in the augmented matrix with these new values: $\begin{pmatrix} 3 & 2 & -3 \\ 0 & 1 & -3 \end{pmatrix}$.

Step 4: Convert this augmented equation back to simultaneous equations:

$3x + 2y = -3$ (1)

$0x + 1y = -3$ (3)

Step 5: Equation (3) has no x variable and gives $y = -3$.

Step 6: Substitute this value of y into equation (1):

$$3x + 2(-3) = -3$$
$$3x - 6 = -3$$
$$3x = 3$$
$$x = 1$$

The solution is $x = 1$ and $y = -3$.

Check the answer by substituting these values back into the original equations:

$$3(1) + 2(-3) = -3$$
$$5(1) + 3(-3) = -4$$

So the solution is correct.

⏸ **PAUSE POINT**

Write this pair of simultaneous equations in matrix form:
$$3x + 2y = 2$$
$$4x + y = 4$$

Hint　Follow the worked example carefully.

Extend　Solve the simultaneous equations by Gaussian elimination.

B2 Determinants

Determinant of a 2 × 2 matrix

The **determinant** of a matrix A can be written in a number of ways. The most common forms are $\det(A)$ or $|A|$.

You can calculate the determinant of a 2 × 2 matrix using a simple formula:

$$\det\begin{pmatrix} a & b \\ c & d \end{pmatrix} = \begin{vmatrix} a & b \\ c & d \end{vmatrix} = ad - bc$$

Worked Example

Find the determinant of $\begin{pmatrix} 1 & 2 \\ 3 & 4 \end{pmatrix}$

Solution

$$\det\begin{pmatrix} 1 & 2 \\ 3 & 4 \end{pmatrix} = \begin{vmatrix} 1 & 2 \\ 3 & 4 \end{vmatrix} = (1 \times 4) - (2 \times 3) = 4 - 6 = -2$$

Inverse of a 2 × 2 matrix

You can use the determinant of a **square matrix** to find the **inverse** of the matrix.

For a 2 × 2 matrix $A = \begin{pmatrix} a & b \\ c & d \end{pmatrix}$, the inverse is given by

$$A^{-1} = \frac{1}{|A|}\begin{pmatrix} d & -b \\ -c & a \end{pmatrix}$$

Because all elements of the matrix need to be divided by $|A|$, if the determinant is zero then the inverse is undefined.

A square matrix that does not have an inverse is called a **singular matrix**. The determinant of a singular matrix is 0.

Key terms

Determinant – a useful value calculated from the elements of a **square matrix**.

Inverse of a matrix – a matrix which when multiplied by the original matrix gives the identity (**unity**) **matrix**.

Unity matrix – a matrix with a 1 in each position in the leading diagonal and 0 in every other position.

Worked Example

Find the inverse of $A = \begin{pmatrix} 1 & 2 \\ 3 & 4 \end{pmatrix}$.

$$\det(A) = \begin{vmatrix} 1 & 2 \\ 3 & 4 \end{vmatrix} = (1 \times 4) - (2 \times 3) = -2$$

$$\text{so } A^{-1} = \frac{1}{-2}\begin{pmatrix} 4 & -2 \\ -3 & 1 \end{pmatrix} = \begin{pmatrix} -2 & 1 \\ \frac{3}{2} & -\frac{1}{2} \end{pmatrix}.$$

Solving two simultaneous equations using the inverse matrix

You have already seen that the matrix form of a set of simultaneous equations is $AX = Y$, where X is the column matrix of variables. Therefore the solution to the problem should be of the form $X = \ldots$, which means that we need to rewrite the equation $AX = Y$ to make X the subject. To do this, we would have to 'divide both sides by A', but there's a slight problem – division by a matrix is not yet defined. We get around this by multiplying both sides of the equation by the inverse of A:

$$A^{-1}AX = A^{-1}Y$$

The inverse A^{-1} 'cancels out' the matrix A, so we get

$$X = A^{-1}Y$$

Worked Example

Use an inverse matrix to find the solution to these simultaneous equations considered in an earlier worked example:

$$3x + 2y = -3 \qquad (1)$$
$$5x + 3y = -4 \qquad (2)$$

Solution

Step 1: Write the equations in matrix form as

$$\begin{pmatrix} 3 & 2 \\ 5 & 3 \end{pmatrix}\begin{pmatrix} x \\ y \end{pmatrix} = \begin{pmatrix} -3 \\ -4 \end{pmatrix}$$

Step 2: Calculate the determinant of $A = \begin{pmatrix} 3 & 2 \\ 5 & 3 \end{pmatrix}$:

$$\begin{vmatrix} 3 & 2 \\ 5 & 3 \end{vmatrix} = (3 \times 3) - (2 \times 5) = 9 - 10 = -1$$

Step 3: Use the formula for A^{-1}:

$$A^{-1} = \frac{1}{-1}\begin{pmatrix} 3 & -2 \\ -5 & 3 \end{pmatrix} = \begin{pmatrix} -3 & 2 \\ 5 & -3 \end{pmatrix}$$

Step 4: Calculate $X = A^{-1}Y$:

$$X = \begin{pmatrix} -3 & 2 \\ 5 & -3 \end{pmatrix}\begin{pmatrix} -3 \\ -4 \end{pmatrix}$$

First, check if the multiplication is possible. The matrices being multiplied are 2×2 and 2×1. The middle numbers agree, so you can multiply.

Combine each row of A with the column matrix Y:

$$X = \begin{pmatrix} (-3) \times (-3) + 2 \times (-4) \\ 5 \times (-3) + (-3) \times (-4) \end{pmatrix} = \begin{pmatrix} 1 \\ -3 \end{pmatrix}$$

So the solution to the simultaneous equations is $x = 1$ and $y = -3$.

This agrees with the answer obtained using Gaussian elimination.

Determinant of a 3 ×3 matrix

You can find the determinant of a 3×3 matrix by using the values in the first row and their minors. The **minor** of a matrix element is the determinant of the smaller matrix left when you cover up the row and column of that element (see **Figure 8.6**).

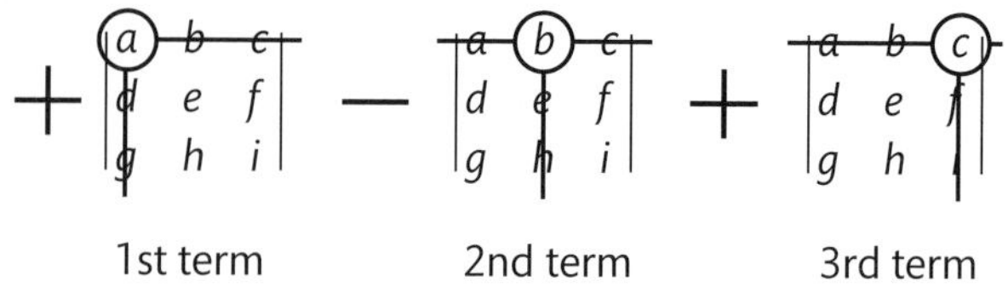

1st term   2nd term   3rd term

▶ **Figure 8.6** Using minors of a 3×3 matrix to find its determinant

For $A = \begin{pmatrix} a & b & c \\ d & e & f \\ g & h & i \end{pmatrix}$, $|A| = a\begin{vmatrix} e & f \\ h & i \end{vmatrix} - b\begin{vmatrix} d & f \\ g & i \end{vmatrix} + c\begin{vmatrix} d & e \\ g & h \end{vmatrix}$.

Worked Example

A system of ropes and pulleys has three ropes under tension. The tension in each rope satisfies these equations:

$$2T_1 + T_2 - 4T_3 = 2 \qquad (1)$$
$$3T_1 + T_2 = 6 \qquad (2)$$
$$-4T_2 + 6T_3 = 1 \qquad (3)$$

Find the determinant of the coefficient matrix.

Solution

In matrix format the system of equations is

$$\begin{pmatrix} 2 & 1 & -4 \\ 3 & 1 & 0 \\ 0 & -4 & 6 \end{pmatrix}\begin{pmatrix} T_1 \\ T_2 \\ T_3 \end{pmatrix} = \begin{pmatrix} 2 \\ 6 \\ 1 \end{pmatrix}.$$

The coefficient matrix is $A = \begin{pmatrix} 2 & 1 & -4 \\ 3 & 1 & 0 \\ 0 & -4 & 6 \end{pmatrix}$.

Then

$$|A| = 2\begin{vmatrix} 1 & 0 \\ -4 & 6 \end{vmatrix} - 1\begin{vmatrix} 3 & 0 \\ 0 & 6 \end{vmatrix} + (-4)\begin{vmatrix} 3 & 1 \\ 0 & -4 \end{vmatrix}$$

$$= 2(1 \times 6 - 0 \times (-4)) - 1(3 \times 6 - 0 \times 0)$$
$$+ (-4)(3 \times (-4) - 1 \times 0)$$

$$= 12 - 18 + 48 = 42$$

Cramer's rule

Cramer's rule is a way of solving multiple simultaneous equations using determinants. Consider this pair of simultaneous equations:

$$a_1 x + b_1 y = c_1$$
$$a_2 x + b_2 y = c_2$$

First write the equations in matrix form:

$$\begin{pmatrix} a_1 & b_1 \\ a_2 & b_2 \end{pmatrix} \begin{pmatrix} x \\ y \end{pmatrix} = \begin{pmatrix} c_1 \\ c_2 \end{pmatrix}$$

Make three determinants:

$$|D| = \begin{vmatrix} a_1 & b_1 \\ a_2 & b_2 \end{vmatrix} \qquad |D_1| = \begin{vmatrix} c_1 & b_1 \\ c_2 & b_2 \end{vmatrix} \qquad |D_2| = \begin{vmatrix} a_1 & c_1 \\ a_2 & c_2 \end{vmatrix}$$

Then the solution is given by $x = \dfrac{|D_1|}{|D|}$ and $y = \dfrac{|D_2|}{|D|}$ provided that $|D|$ is not equal to zero.

Worked Example

Find the currents, i_1 and i_2 in the circuit shown in **Figure 8.7**.

Solution

Using Kirchhoff's current rule, the circuit gives the equations

$$300i_1 - 200\,i_2 = 10$$
$$-200i_1 + 320\,i_2 = -20$$

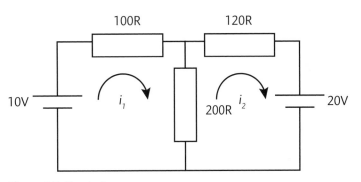

▶ **Figure 8.7** Circuit diagram

Step 1: Write the equations in matrix form

$$\begin{pmatrix} 300 & -200 \\ -200 & 320 \end{pmatrix} \begin{pmatrix} i_1 \\ i_2 \end{pmatrix} = \begin{pmatrix} 10 \\ -20 \end{pmatrix}$$

Step 2: To apply Cramer's rule, form and calculate three determinants

$$|D| = \begin{vmatrix} 300 & -200 \\ -200 & 320 \end{vmatrix} = 300 \times 320 - (-200) \times (-200) = 96\,000 - 40\,000 = 56\,000$$

$$|D_1| = \begin{vmatrix} 10 & -200 \\ -20 & 320 \end{vmatrix} = 10 \times 320 - (-200) \times (-20) = 3200 - 4000 = -800$$

$$|D_2| = \begin{vmatrix} 300 & 10 \\ -200 & 20 \end{vmatrix} = 300 \times (-20) - 10 \times (-200) = -6000 + 2000 = -4000$$

Step 3: Find i_1 and i_2.

$$i_1 = \frac{|D_1|}{|D|} = \frac{-800}{56\,000} = -0.014 \text{ A (to 3 d.p.) and } i_2 = \frac{|D_2|}{|D|} = \frac{-4000}{56\,000} = -0.071 \text{ A (to 3 d.p.).}$$

The negative answers mean that the current flows in the opposite direction to the one chosen (that is, the current flows anti-clockwise).

a) Write this pair of simultaneous equations in matrix form $AX = Y$.

$$2T_1 + 3T_2 = 11$$
$$T_1 - 2T_2 = 2$$

b) Calculate the determinant of matrix A, $|A|$ and its inverse, A^{-1}.

Hint   Use the worked examples to help you answer the questions. Try to repeat the steps of the solutions without looking at the examples.

Extend   Solve the simultaneous equations using the inverse matrix and using Cramer's rule. Compare the two methods.

Assessment practice 8.2   B.P3   B.M3   BC.D2 (part)

Use these matrices to answer questions **1–3** below.

$$A = \begin{pmatrix} 1 & 2 \\ 3 & -1 \end{pmatrix}$$

$$B = \begin{pmatrix} 6 & -1 \\ 5 & 3 \end{pmatrix}$$

$$C = \begin{pmatrix} -1 \\ 1 \end{pmatrix}$$

$$D = \begin{pmatrix} 5 & 6 & 4 \\ -3 & 2 & 1 \\ -1 & 0 & 0 \end{pmatrix}$$

$$E = \begin{pmatrix} 1 & 3 & 2 \end{pmatrix}$$

1   Find the value of each matrix expression. Give a reason if the operation is not possible.
   a) $A + B$ b) $B + A$ c) $2B$
   d) AC e) DE f) ED

2   Calculate the determinant of:
   a) A b) C c) D

3   Calculate the inverse of:
   a) B b) D c) E

4   The following equations were obtained by applying Kirchhoff's current rule to a circuit:

$$7i_1 + 9i_2 = 3$$
$$5i_1 + 7i_2 = 1$$

   where i_1 and i_2 represent the currents in separate branches.
   Find the values of i_1 and i_2 using
   a) Gaussian elimination
   b) the inverse matrix method
   c) Cramer's rule.

5   Use matrix and determinant methods to solve these simultaneous equations describing a mechanical system.

$$2T_1 + 3T_2 - T_3 = 53$$
$$T_1 - 2T_2 + 3T_3 = 16$$
$$T_2 + T_3 = 27$$

   Compare the answers and the methods.

Plan
- What techniques will I need to use to complete the task? How am I going to approach each problem and keep track of my progress?
- How confident do I feel in my abilities to complete this task?
- Are there any areas I think I may struggle with?

Do
- I know what strategies I could employ for the task. I can determine whether these are right for the task.
- I can assess whether my approach is working and, if not, what changes I need to make.
- I can identify when I've gone wrong and adjust my thinking/approach to get myself back on course.

Review
- I can evaluate whether I have solved each problem correctly.
- I have compared the different methods and can assess which method is more suitable for which type of problem.
- I can identify the elements in the task that are difficult for me and explain how I would approach them differently next time.

Examine how complex numbers can be used to solve engineering problems

C1 Complex numbers

Complex numbers extend the range of numbers from positive and negative values on the number line to two dimensions. Any complex number can be written as $a \pm jb$, where a is the horizontal component along the familiar number line (called the **real part**) and b is the vertical component along a perpendicular direction (called the **imaginary part**). The symbol j indicates the 'imaginary' component. (Note: mathematicians use 'i' for the imaginary unit, but engineers use 'j' to avoid confusion with the symbol for electric current.)

Complex numbers make it convenient to perform mathematical operations with **phasor** quantities. A phasor is a rotating vector. The instantaneous value of a sine wave at any point is equal to the vertical distance from the tip of the phasor to the horizontal axis (see **Figure 8.8**).

Key term

Complex number – a number that has the format $a \pm jb$. The 'complex operator' or 'imaginary unit' j satisfies the equation $j^2 = -1$.

Link

Phasors are discussed in **Unit 1**, in the sections '**A2 Trigonometric methods**' and '**G Single-phase alternating current**'.

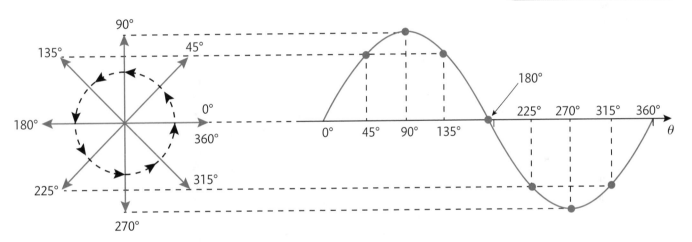

▶ **Figure 8.8** Relationship between a phasor (rotating vector) and a sine wave

A phasor that starts at the 3 o'clock position and rotates anti-clockwise represents the sine wave $y = R \sin(\omega t)$, where R is the length of the phasor and $\omega = 2\pi f (\text{rad s}^{-1})$ is the angular speed (here f is the frequency of the sine wave oscillation in units of Hz).

A phasor that starts at the 12 o'clock position represents the function $y = R \cos(\omega t)$, where R and ω have the same meanings as above.

Because 90° is $\frac{\pi}{2}$ radians, you can see that $R \cos(\omega t) = R \sin\left(\omega t + \frac{\pi}{2}\right)$.

Look at the two waves A and B and their phasor representations in **Figure 8.9**.

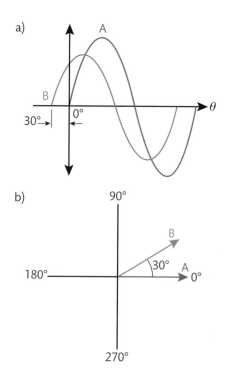

In particular, a phasor with rectangular form $a + jb$ has

▸ modulus $|a + jb| = \sqrt{a^2 + b^2}$

▸ argument $\arg(a + jb) = \tan^{-1}\left(\dfrac{b}{a}\right)$.

The complex number system lets you add, subtract, multiply and divide phasor quantities that have both magnitude and angle (see the section on complex number arithmetic below).

Numbers on the complex plane

If you plot complex numbers on a set of perpendicular axes (the complex plane), you get an **Argand diagram** (see **Figure 8.11**).

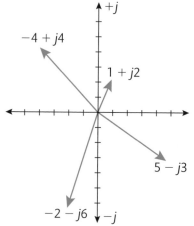

Rectanglar notation

Polar notation

▸ **Figure 8.11** Two examples of Argand diagrams

▸ **Figure 8.9** (a) Wave B is a phase-shifted version of the sine wave A. (b) The corresponding phasor B makes an angle of 30° with phasor A

In **Figure 8.9** the phasor B leads the phasor A by 30°, or $\dfrac{\pi}{6}$ radians. We know the equation for A is $y = R\sin(\omega t)$. The equation for B is $y = R\sin\left(\omega t + \dfrac{\pi}{6}\right)$.

A phasor can be written in a number of ways. The most common forms are the following:

▸ **polar** notation – where the **magnitude** (also called the **modulus**) R and the **phase angle** (also called the **argument**) θ are together expressed as $R\angle\theta$

▸ **rectangular** (or **Cartesian**) notation – where the phasor is resolved into horizontal (a) and vertical (b) components to give $a + jb$.

Figure 8.10 shows how to convert between these two forms of notation.

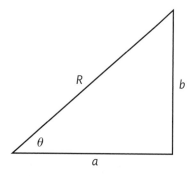

▸ **Figure 8.10** Conversion between polar and rectangular forms of representing a phasor

The j operator is associated with the vertical component of the phasor. It represents an anti-clockwise phase shift of 90° or $\dfrac{\pi}{2}$ rad. A j shift followed by another j shift gives j^2, which is equivalent to a 180° phase shift. This corresponds to the wave being inverted, or multiplied by –1. Therefore j satisfies the equation $j^2 = -1$.

Complex number arithmetic

Addition and subtraction

You add (or subtract) the 'real' and 'imaginary' parts of an imaginary number separately. For example:

$$(2 + j3) + (4 + j5) = (2 + 4) + (j3 + j5) = 6 + j8$$

$$(1 - j2) - (3 + j4) = (1 - 3) + (-j2 - j4) = -2 - j6$$

Multiplication

You can multiply in rectangular notation the same way as multiplying two binomial terms, using FOIL. For example:

$$(2 + j3)(4 + j5) = 8 + j10 + j12 + j^2 15 = 8 + j22 + j^2 15$$

But you know that $j^2 = -1$, so the last term becomes -15. Therefore

$$(2 + j3)(4 + j5) = 8 + j22 + j^2 15 = 8 + j22 - 15 = -7 + j22$$

You can also multiply two complex numbers in polar notation. To do this, multiply the magnitudes and add the phase angles. For example:

$$2 + j3 = \sqrt{13} \angle \left(\tan^{-1}\frac{3}{2}\right) \text{ rad}$$

$$4 + j5 = \sqrt{41} \angle \left(\tan^{-1}\frac{5}{4}\right) \text{ rad}$$

So $(2 + j3)(4 + j5) = \sqrt{13}\sqrt{41} \angle \left(\tan^{-1}\frac{3}{2} + \tan^{-1}\frac{5}{4}\right)$

$$= 23.09 \angle 1.8788 \text{ rad}$$

Converting this back to rectangular notation (using unrounded values) gives

$$23.09\cos(1.8788) + j\,23.09\sin(1.8788) = -7 + j22$$

Division

To divide two complex numbers in rectangular notation, you write the division as a fraction and then multiply both numerator and denominator by the **complex conjugate** of the denominator.

A complex number and its complex conjugate have the same real and imaginary parts, except that the j terms have opposite signs. In other words, $a + jb$ and $a - jb$ are complex conjugates of each other. Multiplying a pair of complex conjugates gives a real number: $(a + jb)(a - jb) = a^2 - jab + jba - j^2 b^2 = a^2 + b^2$.

For example, to divide $(2 + j3)$ by $(4 + j5)$ using complex conjugates:

$$\frac{(2 + j3)}{(4 + j5)} = \frac{(2 + j3)(4 - j5)}{(4 + j5)(4 - j5)} = \frac{8 - j10 + j12 - j^2 15}{16 - j20 + j20 - j^2 25}$$

$$= \frac{8 + j2 + 15}{16 + 25} = \frac{23 + j2}{41} = \frac{23}{41} + j\frac{2}{41}$$

You can also divide in polar notation: simply divide the magnitudes and subtract the phase angles.

So $\dfrac{(2 + j3)}{(4 + j5)} = \dfrac{\sqrt{13}\angle\left(\tan^{-1}\frac{3}{2}\right)}{\sqrt{41}\angle\left(\tan^{-1}\frac{5}{4}\right)} = \dfrac{\sqrt{13}}{\sqrt{41}} \angle\left(\tan^{-1}\frac{3}{2} - \tan^{-1}\frac{5}{4}\right)$

$$= 0.563\angle 0.0867 \text{ rad}$$

Converting this back to rectangular notation (using unrounded values) gives

$$\frac{(2 + j3)}{(4 + j5)} = 0.563\cos(0.0867) + j\sin(0.0867)$$

$$= 0.56 + j0.05 \text{ (to 2 d.p.)}$$

> ### Research
>
> Most scientific calculators have polar-to-rectangular and rectangular-to-polar conversion functions. Find out how to carry out these functions on your calculator, and practise using them.

Raising to a power

Raising a complex number to a power n is easy to do in polar notation using De Moivre's theorem:

$$(R\angle\theta)^n = R^n\angle n\theta$$

In other words, raise the magnitude to the power n and multiply the phase angle by n.

Worked Example

An electrical circuit has an inductive reactance of $(3 + j4)\,\Omega$ connected in series with a capacitive reactance of $(0 - j2)\,\Omega$ and a resistor of $(15 + j0)\,\Omega$. Calculate the total impedance and the current flowing in the circuit if the voltage is $V = 12\angle\frac{\pi}{3}$ rad.

Solution

The total impedance is $Z = (3 + j4) + (0 - j2) + (15 + j0)$

$= (18 + j2)\,\Omega$. The current in the circuit is $i = \dfrac{V}{Z}$.

To do the division, you can convert Z to polar notation:

$$18 + j2 = \sqrt{18^2 + 2^2} \angle\left(\tan^{-1}\frac{2}{18}\right)$$

$$= 18.11\angle 0.1107 \text{ rad}$$

so $i = \dfrac{12\angle\frac{\pi}{3}}{18.11\angle 0.1107} = \dfrac{12}{18.11}\angle\left(\frac{\pi}{3} - 0.1107\right)$

$$= 0.663\angle 0.9365 \text{ rad.}$$

The current is 0.663 A at a phase angle of 0.9365 rad. This can also be written as $i = 0.663\sin(\omega t + 0.9365)$.

Answer the following questions involving complex numbers. It is important that you show all the steps in your calculations clearly. You also need to give a brief explanation of each step.

1 You are given the complex numbers

$z_1 = 3 + j$ and $z_2 = 4 - j$

a) Plot z_1 and z_2 on an Argand diagram.

b) Calculate the following:

    i) the sum $z_1 + z_2$

    ii) the conjugate of z_1

    iii) the conjugate of z_2

    iv) the product $z_1 z_2$ in rectangular form

    v) z_1 and z_2 in polar form

    vi) the product $z_1 z_2$ in polar form

    vii) $(z_2)^4$ in polar form using De Moivre's theorem

    viii) $(z_1)^3$ in polar form using De Moivre's theorem

    ix) the quotient $\frac{z_1}{z_2}$ in rectangular form

    x) the quotient $\frac{z_1}{z_2}$ in polar form.

2 Two circuits are connected in series. The impedances of the circuits are $z_1 = 50 + j3$ and $z_2 = 20 - j2$, and the supply voltage is $v = 100 + 5j$.

a) Calculate:

    i) the total impedance $z_T = z_1 + z_2$

    ii) the current flowing in the circuit, $i = \frac{v}{z_T}$

b) What is the amplitude of the current?

c) What is the phase angle of the current?

Plan

- What techniques will I need to use to complete the task? How am I going to approach each problem and keep track of my progress?
- How confident do I feel in my abilities to complete this task?
- Are there any areas I think I may struggle with?

Do

- I know what strategies I could employ for the task. I can determine whether these are right for the task.
- I can assess whether my approach is working and, if not, what changes I need to make.
- I can identify when I've gone wrong and adjust my thinking/approach to get myself back on course.

Review

- I can evaluate whether I have solved each problem correctly.
- I have compared the different methods and can assess which method is more suitable for which type of problem.
- I can identify the elements in the task that are difficult for me and explain how I would approach them differently next time.

D Investigate how statistical and probability techniques can be used to solve engineering problems

Key term

Probability distribution – The distribution of values of a variable can be displayed graphically, e.g. as a histogram or pie-chart. A probability distribution plots the probability of a given value against the range of values instead of the actual number. A very common distribution is known as the normal distribution. A normal distribution gives a bell-shaped curve, symmetrical about the mean.

D1 Statistical techniques

Data can be gathered from many sources. It can be classified as **discrete** or **continuous**.

▸ Data is said to be **discrete** if it can take on values from a specific set of numbers. For example:
- The number of people living in a house can be 0, 1, 2, 3, ... but not, say, 2.4 or 3.5.
- The number of cars produced in a factory in a week must be a positive whole number or zero.

▸ Data is said to be **continuous** if it can take on any value within a range. Continuous data usually comes from measurements. Examples include:
- the diameter of a spindle turned on a lathe
- the volume of liquid in a container.

Measures of central tendency (averages)

When you analyse experimental data, it is useful to look for a single number that typifies the data – the **average** value of the data. There are three important averages that are often used:

▶ **arithmetic mean** – defined as $\dfrac{\text{sum of all the data values}}{\text{number of data items in the sample}}$. This can be expressed mathematically as $\bar{x} = \dfrac{\sum_{i=1}^{i=n} x_i}{n}$

▶ **median** – the middle value of the data when all the values are placed in order from lowest to highest. To find the middle position, add 1 to the number of data items in the sample and divide by 2.

▶ **mode** – the data value that occurs most often. If more than one value occurs the same number of times, the data is said to be multi-modal.

Raw data is **ungrouped**. For example, if the number of components produced per hour is measured over ten hours and recorded as 10, 11, 10, 9, 12, 11, 10, 10, 9, 11, this is ungrouped data.

This small amount of data can be analysed quite simply:

$$\text{mean} = \frac{10 + 11 + 10 + 9 + 12 + 11 + 10 + 10 + 9 + 11}{10} = 10.3$$

To find the median, first write the numbers in order:

9, 9, 10, 10, 10, 10, 11, 11, 11, 12

In this case the median is halfway between the fifth and sixth values. These numbers are both 10, so the median is 10.

To find the mode, tally the values as in **Table 8.4**.

▶ **Table 8.4** Tally chart to make a frequency table

Value	Tally	Frequency
9	//	2
10	////	4
11	///	3
12	/	1

The value 10 occurs most frequently, so the mode is 10.

Tallying the data when it is collected can make analysis simpler, especially for large sets of data. The data in the tally chart is already ordered, which also helps in identifying the median.

Graphical representation of data

Tables of data contain numerical values, but patterns and relationships are not always easy to find. Graphical and other visual methods allow you to see potential patterns without the need for calculations.

A **pie chart** (**Figure 8.12a**) is a useful way of comparing relative sizes of data in specific categories, for example, the number of different types of industry in an area.

A **bar chart** (**Figure 8.12b**) is an alternative way of displaying values of similar types of data to pie charts.

A **histogram** looks similar to a bar chart, but is used for displaying the frequency of grouped data. It displays the spread of values, for example, the number of components within a range of measured values (see the worked example below).

a)

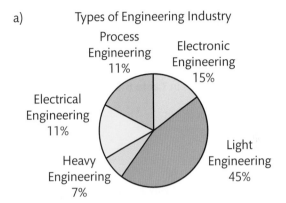

Types of Engineering Industry

Process Engineering 11%
Electronic Engineering 15%
Electrical Engineering 11%
Light Engineering 45%
Heavy Engineering 7%

b)

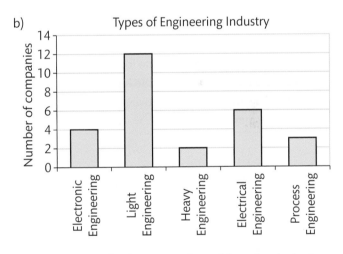

Types of Engineering Industry

▶ **Figure 8.12** Graphical representations of data: a) a pie chart, and b) a bar chart

> **Research**
>
> Revise the different ways of representing data as
> - pie charts
> - bar charts
> - histograms
> - cumulative frequency diagrams.
>
> Make sure you understand the difference between bar charts and histograms.

Worked Example

Consider the data presented above:

10, 11, 10, 9, 12, 11, 10, 10, 9, 11

Calculate the mean of the data using a frequency table.

Solution

The mean is $\bar{x} = \dfrac{\sum xf}{\sum f}$, where

- f is the frequency of each distinct value taken by the data
- $\sum f = N$ is the number of data items in the sample
- xf is the contribution of each value x to the total
- $\sum xf$ is the sum of all the contributions, that is, the total of all the data values.

Tabulate these values as in **Table 8.5**, and then use the formula to calculate the mean.

▶ **Table 8.5** Frequency table

x	f	xf
9	2	18
10	4	40
11	3	33
12	1	12
	$\sum f = 10$	$\sum xf = 103$
Mean, $\bar{x}$	$= \dfrac{\sum xf}{\sum f}$	$= 10.3$

Grouped data

Larger sets of data are more easily analysed if they are grouped.

Key terms

Classes – the groups into which the data is organised.
Lower class boundary (lcb) – the lowest value of the class.
Upper class boundary (ucb) – the highest value of the class.
Class width – the difference between ucb and lcb.
Midpoint – median of the class.

Worked Example

A machine is set to produce bolts with nominal diameter 25.00 mm. A random sample of 60 bolts is measured and provides the frequency distribution in **Table 8.6**.

a) Calculate the mean diameter.

b) Draw a histogram of the data and mark the **upper** and **lower class boundaries** for the second class.

▶ **Table 8.6** Frequency distribution of bolt diameter

Diameter (x)	Frequency (f)
23.3–23.7	2
23.8–24.2	4
24.3–24.7	10
24.8–25.2	17
25.3–25.7	16
25.8–26.2	8
26.3–26.7	3

Solution

a) The data is continuous. Expand the frequency table to show the lcb, ucb and **midpoint** of each class (see **Table 8.7**). Note that the ucb of the first class is the lcb of the second class, and so on. Averaging the lcb and ucb for each class gives the midpoint of that class.

The midpoint will be used to represent each class in the calculation of the mean.

▶ **Table 8.7** Calculation of the mean using the midpoint values

Diameter, x	Frequency, f	lcb	ucb	Midpoint, x_{mid}	$x_{mid}f$
23.3–23.7	2	23.25	23.75	23.50	47
23.8–24.2	4	23.75	24.25	24.00	96
24.3–24.7	10	24.25	24.75	24.50	245
24.8–25.2	17	24.75	25.25	25.00	425
25.3–25.7	16	25.25	25.75	25.50	408
25.8–26.2	8	25.75	26.25	26.00	208
26.3–26.7	3	26.25	26.75	26.50	79.5
	$\Sigma f = 60$				$\Sigma xf = 1508.5$

	Mean, $\bar{x}$	$= \dfrac{\Sigma xf}{\Sigma f}$	$= 25.14$

The mean diameter of the bolts is 25.14 mm.

b) Using the information in **Table 8.7**, plot a histogram as in **Figure 8.13**.

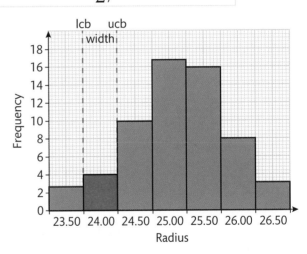

▶ **Figure 8.13** Histogram of the bolt diameter data

A **cumulative frequency graph** shows the cumulative total frequency up to each of the class boundaries. For example, **Table 8.8** shows grouped data on the marks gained in a mathematics test.

▶ **Table 8.8** Frequency table for marks gained in a mathematics test

Marks	Number of learners (frequency)	Cumulative frequency	
11–20	2	2	2
21–30	11	2 + 11	13
31–40	19	13 + 19	32
41–50	36	32 + 36	68
51–60	42	68 + 42	110
61–70	31	110 + 31	141
71–80	13	141 + 13	154
81–90	6	154 + 6	160

The cumulative frequency graph is shown in **Figure 8.14**. At each ucb value along the horizontal axis, the cumulative frequency up to that point is plotted on the vertical axis.

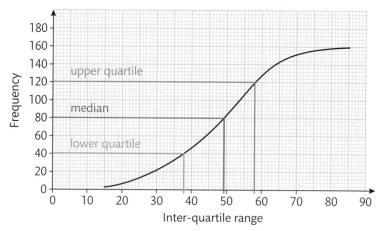

▶ **Figure 8.14** Cumulative frequency graph of the mathematics test marks

Because the final cumulative frequency value is 160, the total number of learners who took the test was 160. So if all the marks were ordered from lowest to highest, the median mark would be around the 80th value. To obtain the median mark from the cumulative frequency graph, find 80 on the vertical axis, draw a horizontal line across to the curve followed by a vertical line down to the horizontal axis, and then read off the value on the horizontal axis.

The **lower quartile** is one-quarter of the way up the ordered list of marks (around the 40th), and the **upper quartile** is three-quarters of the way up (around the 120th). The **inter-quartile range** is the difference between the upper and lower quartiles and gives a measure of the **spread** of data values.

The frequency scale can be split up in other ways than into quartiles. For example, the range can be divided into 100 equal steps called percentiles. These can be used to calculate the inter-percentile range. The range between the 75th and 100th percentiles is equivalent to the upper quartile.

Measures of dispersion (spread)

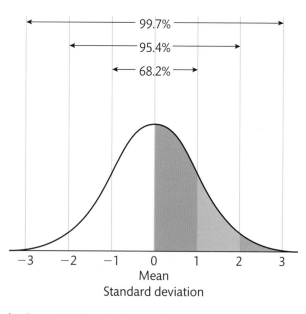

▶ **Figure 8.15** Standard normal deviation

If you were manufacturing components, ideally you would want them all to be the same. You know, however, that there will always be some variation. The variation is likely to follow a 'normal' or 'bell-curve' distribution about the mean (see **Figure 8.15**). Many things follow a **normal distribution**, for example, heights of people.

The average variation from the mean will be zero, because in general you expect there to be as many values above the mean as there are below it. You can calculate a useful measure of the variation by first squaring all the differences from the mean to make the values positive, and then averaging them. The resulting value is called the **variance**:

$$\text{variance} = \frac{\sum_{i=1}^{j=n} (\overline{x} - x_i)^2}{n}$$

The square root of the variance is called the **standard deviation**. It is usually written as σ (lower-case Greek letter sigma).

$$\text{standard deviation} = \sigma = \sqrt{\text{variance}} = \sqrt{\frac{\sum_{i=1}^{j=n} (\overline{x} - x_i)^2}{n}}$$

Worked Example

Calculate the variance and standard deviation of the bolt diameters in the previous worked example.

Solution

Step 1: Calculate the mean. In the previous worked example you found this to be 25.14 mm.

Step 2: For each data value (the midpoint of each class in this case), calculate its deviation from the mean, $(\bar{x} - x_i)$. Then square each answer to get $(\bar{x} - x_i)^2$.

Step 3: To find the average of these values $(\bar{x} - x_i)^2$, multiply each one by the corresponding frequency. The calculation results from steps 2 and 3 are summarised in **Table 8.9**.

▶ **Table 8.9** Expanded frequency table for calculating the variance

Diameter, x	Frequency, f	$x_{mid} = x_i$	$(\bar{x} - x_i)$	$(\bar{x} - x_i)^2$	$(\bar{x} - x_i)^2 f$
23.3–23.7	2	23.50	1.64	2.6896	5.3792
23.8–24.2	4	24.00	1.14	1.2996	5.1984
24.3–24.7	10	24.50	0.64	0.4096	4.096
24.8–25.2	17	25.00	0.14	0.0196	0.3332
25.3–25.7	16	25.50	−0.36	0.1296	2.0736
25.8–26.2	8	26.00	−0.86	0.7396	5.9168
26.3–26.7	3	26.50	−1.36	1.8496	5.5488
	$\Sigma f = 60$				$\Sigma(\bar{x} - x_i)^2 f = 28.546$

Step 4: Calculate the variance using the formula

$$\frac{\sum_{i=1}^{i=n}(\bar{x} - x_i)^2}{n} = \frac{28.546}{60} = 0.4758$$

Step 5: Take the square root to find the standard deviation:

$$\sigma = \sqrt{\text{variance}} = 0.6898$$

Scatter diagrams

A scatter diagram (XY plot) is used to visualise the relationship between two sets of data. The independent variable is plotted on the horizontal (X) axis, and the dependent variable on the vertical (Y) axis (see **Figure 8.16**).

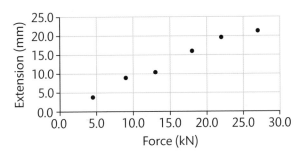

▶ **Figure 8.16** Scatter diagram

Research

Investigate probability distributions, and in particular the 'normal' distribution. Find out how the standard normal distribution curve can be used to predict the percentage of population in a given range.

In this case the plotted points appear to show that the extension increases as the applied force increases. However, you can see that the points do not fit exactly on a straight line, but the values are scattered closely about one, with some further away from the straight line than others.

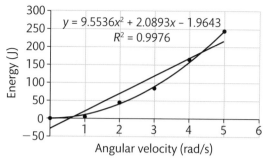

Linear regression (line of best fit)

When you take measurements, there is always a degree of tolerance. This means that in a practical situation involving measurements of two variables (one independent and one dependent), even when the dependent variable is known to be directly proportional to the independent variable, it is unlikely that the data points will lie exactly on a straight line. If you expect a linear relationship between the variables, you could plot the data points and try to draw a line 'by eye' that best describes the relationship – for example, you could draw a line that has as many data points below as above it.

An alternative to such a 'hit-or-miss' method of fitting data is to use a statistical analysis tool. The **linear regression** model generates an equation of the form $y = mx + c$ to fit the data. The values of m and c are given by

$$m = \frac{N \sum xy - \sum x \sum y}{N \sum x^2 - (\sum x)^2} \text{ and } c = \overline{y} - m\overline{x}$$

where N is the number of pairs of measurements, $\overline{y}$ is the mean of the data values for the dependent variable and $\overline{x}$ is the mean of the data values for the independent variable.

A statistical quantity that measures how well the regression line fits the data is the (Pearson's) **correlation coefficient**. This is given by the formula

$$r = \frac{N \sum xy - \sum x \sum y}{\sqrt{N \sum x^2 - (\sum x)^2} \sqrt{N \sum y^2 - (\sum y)^2}}$$

and produces a value between –1 and +1 that indicates the degree of correlation between the variables: +1 means total positive correlation, 0 means no correlation and –1 means total negative correlation.

Linear regression is a very useful tool to determine a relationship between the variables. However, not all relationships are linear. You can use a spreadsheet, such as Microsoft® Excel to investigate other potential relationships that fit the plotted points, such as exponential, logarithmic etc. (see **Figure 8.17**).

▶ **Figure 8.17** Using a spreadsheet to plot regression lines

The data in Figure 8.17 represents an experiment to relate the rotational energy in a flywheel to its angular velocity. Theory says that the relationship is $E = \frac{1}{2}I\omega^2$. A linear trendline clearly does not match the data as you might expect. The polynomial trendline gives a much better fit. The spreadsheet displays the equation of the regression line and the R-squared value that gives a measure of how well the data fits the mathematical model (equation). A value of 1 is a perfect fit. The regression line is close, but does not quite match the theoretical value. The benefit of using a software application is that other types of trendline can easily be tried to find a line of best fit.

Most graphical calculators and spreadsheet programs can calculate the linear regression line and the correlation coefficient. **Figure 8.18** shows the regression line calculated for the worked example data using Microsoft® Excel.

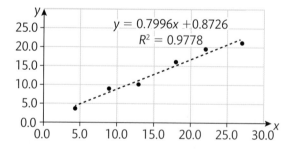

▶ **Figure 8.18** Calculated regression line for sample data

Research

Investigate how to use your graphical calculator or a spreadsheet such as Microsoft® Excel to find the regression line and correlation coefficient for a set of data.

Worked Example

The data in **Table 8.10** was obtained during a tensile test of a steel specimen. The values are plotted as a scatter diagram in Figure 8.16. Assuming a linear relationship, determine the regression line of extension on force and calculate the regression coefficient, explaining what the value means.

▶ **Table 8.10** Measurements of tension and extension in a tensile test

Tension (kN)	Extension (mm)
4.5	3.8
9.0	8.9
13.0	10.4
18.0	16.0
22.0	19.6
27.0	21.3

Solution

Make a table (like **Table 8.11**) to calculate each of the terms in the regression line formulae.

▶ **Table 8.11** Table for calculating the regression line

Tension (kN)	Extension (mm)			
x	y	x^2	y^2	xy
4.5	3.8	20.25	14.44	17.1
9.0	8.9	81.00	79.21	80.1
13.0	10.4	169.00	108.16	135.2
18.0	16.0	324.00	256.00	288
22.0	19.6	484.00	384.16	431.2
27.0	21.3	729.00	453.69	575.1
$\Sigma x = 93.5$	$\Sigma y = 80$	$\Sigma x^2 = 1807.25$	$\Sigma y^2 = 1295.66$	$\Sigma xy = 1526.7$

Now apply the formulae for the regression line:

$$m = \frac{N \sum xy - \sum x \sum y}{N \sum x^2 - (\sum x)^2} = \frac{6(1526.7) - (93.5)(80.0)}{6(1807.25) - (93.5)^2}$$

$$= 0.7996$$

$$\bar{x} = \frac{\sum x}{N} = \frac{93.5}{6} = 15.583$$

$$\bar{y} = \frac{\sum y}{N} = \frac{80.0}{6} = 13.333$$

$$c = \bar{y} - m\bar{x} = 13.333 - (0.7996)(15.583) = 0.873$$

So the equation of the regression line is $y = 0.8x + 0.87$. The regression coefficient is

$$r = \frac{N \sum xy - \sum x \sum y}{\sqrt{N \sum x^2 - (\sum x)^2} \; \sqrt{N \sum y^2 - (\sum y)^2}}$$

$$= \frac{6(1526.7) - (93.5)(80.0)}{\sqrt{6(1807.3) - (93.5)^2} \; \sqrt{6(1295.7) - (80.0)^2}}$$

$$= 0.989$$

This is a positive value close to 1, so there is strong positive correlation between the two variables and the linear regression equation is a good fit to the data.

Statistical investigation

Inspection and quality assurance, calculation of central tendencies and dispersion, forecasting, reliability estimates for components and systems, customer behaviour, condition monitoring and product performance are all examples of engineering applications that rely upon the application of statistical techniques. The best way to understand them is to carry out a statistical investigation of your own.

A common approach to carrying out a statistical investigation follows the Problem–Plan–Data–Analysis–Conclusion (PPDAC) cycle.

Problem: In the problem stage you need to formulate the question to be answered, identifying what data to collect, why it is important, where and how to collect it.

Plan: In the plan stage you identify details such as the sample size, how the data is to be gathered and recorded.

Data: The data stage is concerned with how you organise and manage the data. Tables and tally sheets are typical tools for clearly recording data so it can be manipulated.

Analysis: In the analysis stage you explore the data, looking for relationships in grouped/ungrouped data, using graphical and analytical methods to identify central tendencies, dispersion (spread) and rogue values.

Conclusions: The conclusions should relate back to the initial problem. You need to include statistical vocabulary where appropriate, for example, scatter plot, histogram, mean and standard deviation.

Assessment practice 8.4

Solve the following problems involving statistical and probability techniques. It is important that you show all the steps in your calculations clearly. You also need to give a brief explanation of each step.

1 You measured the masses of 50 metal castings in kilograms and noted the results:

4.6	4.7	4.5	4.6	4.7	4.4	4.8	4.3	4.2	4.8
4.7	4.5	4.7	4.4	4.5	4.5	4.6	4.4	4.6	4.6
4.8	4.3	4.8	4.5	4.5	4.6	4.6	4.7	4.6	4.7
4.4	4.6	4.5	4.4	4.3	4.7	4.7	4.6	4.6	4.8
4.9	4.4	4.5	4.7	4.4	4.5	4.9	4.7	4.5	4.6

a) Arrange the data in eight equal classes between 4.2 and 4.9 kilograms.

b) Determine the frequency distribution.

c) Draw the frequency histogram.

d) Is the data normally distributed?

e) Calculate the mean and standard deviation of the masses.

f) Calculate the upper and lower quartile values.

g) Calculate the inter-quartile range.

2 The data in **Table 8.12** results from an experiment you carried out to determine the relationship between frequency and the inductive reactance for an electrical circuit.

▶ **Table 8.12** Data on frequency and inductive reactance

Frequency, f (kHz)	5	10	15	20	25	30	35	40	45
Inductive reactance, X_L (Ω)	30	65	90	130	150	190	200	255	270

a) Draw a scatter graph for this data.

b) What does the scatter graph tell you?

c) Estimate the inductive reactance for a frequency of 32 kHz from the graph.

d) Calculate the line of regression of inductive reactance on frequency (y on x).

e) Calculate the regression coefficient.

f) What does the regression coefficient indicate?

g) Use the equation for the line of regression to predict the inductive reactance for a frequency of 32 kHz.

h) Compare the calculated value with the value that you read from the graph.

Plan

- What techniques will I need to use to complete the task? How am I going to approach each problem and keep track of my progress?
- How confident do I feel in my abilities to complete this task?
- Are there any areas I think I may struggle with?

Do

- I know what strategies I could employ for the task. I can determine whether these are right for the task.
- I can assess whether my approach is working and, if not, what changes I need to make.
- I can identify when I've gone wrong and adjust my thinking/approach to get myself back on course.

Review

- I can evaluate whether I have solved each problem correctly.
- I have compared the different methods and can assess which method is more suitable for which type of problem.
- I can identify the elements in the task that are difficult for me and explain how I would approach them differently next time.

Further reading and resources

Websites

Maths is Fun: **www.mathsisfun.com**
A lot of useful skills and knowledge organised into categories that are easy to find.

mathcentre: **www.mathcentre.ac.uk/students/topics**
A set of resources split into topics designed to support self-study.

THINK ▶FUTURE

Shaun Spilsbury

Design engineer

I've been working in the engineering industry for 29 years. Throughout this time the industry has never stood still and it continues to develop new technologies across all engineering sectors. I believe that design is at the heart of all engineering.

A design engineer realises ideas and brings to life new products or processes, from customer requirements, through concept, prototype, final design and tests, to manufacture, commission and disposal. Mathematical modelling and simulation are important ways of testing out new ideas before constructing expensive projects, so the design engineer has to have well-developed mathematical skills. The design engineer plays a pivotal role in an extremely diverse range of industries. You have many different and interesting sectors to choose from – including mechanical, electrical and electronic, aerospace, marine, oil and gas, nuclear and renewables (such as wind turbines, solar power) – and new areas are developing all the time.

One thing is for sure – it's important to look for a company and position where you will continue to learn and be challenged in an environment that supports your development. In engineering, every day is a school day and rarely are two days the same. Working in a challenging environment will not only keep you interested while developing your knowledge and competencies, but also be extremely rewarding both professionally and financially. Engineers are sought after the world over, and if you have a desire to work in different parts of the world, then working in this industry provides plenty of opportunities. When working in engineering design, you will be participating in an industry that is shaping the world we live in.

Focusing your skills

Using mathematical skills to solve engineering problems

It is important to be able to apply mathematical methods to solve engineering problems. Here are some simple tips to help you do this.

- Gather all the information you can to understand the nature of the problem – for example, inspection and quality assurance, calculation of central tendencies and dispersion, forecasting, reliability estimates for components and systems, customer behaviour, condition monitoring and product performance.
- Use critical thinking when extracting information from the text of the problem.
- Identify appropriate mathematical methods to solve the given engineering problem.
- Be confident in the mathematical skills you need to use, such as arithmetic, algebraic, statistical, estimation and approximation techniques.
- Practise the skills to build confidence.
- Develop a plan to solve a problem.
- Critically analyse the solution obtained.
- Reflect on the problem-solving process and the solution obtained, and make refinements if necessary.
- Repeat the process until you are happy with the outcome.
- Present and explain the solution to the given engineering problem.

Getting ready for assessment

Abbie is working towards a BTEC National in Engineering. She was revising for a time-constrained assignment for Learning aim A. She knew that in the assignment she would have to solve engineering problems using the techniques specified in the Learning aim. This was her experience.

How I got started

First I collected all my notes on this topic and put them into a folder. I decided to divide my work into sub-topics because I felt it would be easier to revise them individually. I used textbooks and the internet to identify worked examples of typical questions to help revise the information. I started with problems that used the main skills, such as identifying arithmetic and geometric progressions. Then I worked on more complex ones about solving engineering problems to try to achieve all the criteria.

How I brought it all together

▸ I went through the worked examples carefully, writing out the solutions myself.

▸ I then tried answering the same questions myself without looking at the example.

▸ I then changed the numbers slightly and tried again until I was confident in solving the problems.

▸ When I was more confident I used questions from other textbooks and the internet.

What I learned from the experience

I was pretty sure that I knew how to answer the questions, but I was worried that I would not understand them. I think it helped to use questions from different places for practice because they did not all use the same words, so I got used to having to work out what to do. I found the questions in textbooks useful because they gave the answers at the back.

I read through the questions carefully before starting to answer them. I worked out how long to spend on each question, so that I would have time to check through for any mistakes. I had planned to answer the questions in a different order from the way they were given, but in the end I did not need to. Doing lots of practice really helped.

Think about it

▸ Have you identified what topics you need to revise?

▸ Do you know which problems are routine and which are non-routine?

▸ Have you planned your time? Don't spend too long on any one problem.

Computer-Aided Design in Engineering

10

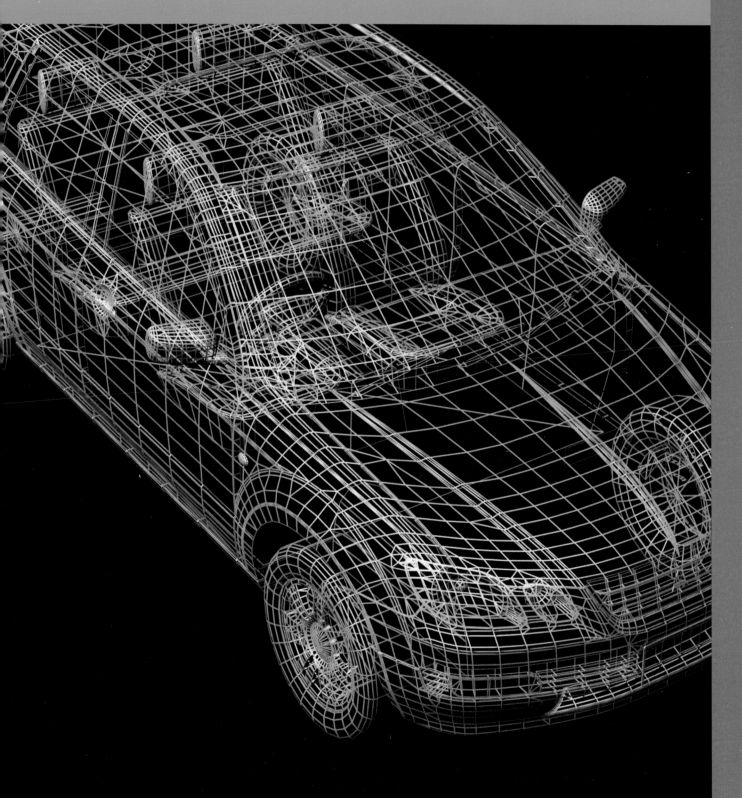

Getting to know your unit

As a future engineer it is important to be able to produce and interpret engineering drawings that help individuals and organisations to communicate ideas, design and manufacture products and improve product performance. Computer-aided design (CAD) is used within the engineering industry alongside other processes to develop, improve and maintain cutting-edge products and systems. In this unit you will acquire the skills to develop two-dimensional (2D) detailed drawings and three-dimensional (3D) models using a CAD system.

How you will be assessed

This unit will be assessed by a series of internally assessed tasks set by your tutor. Throughout this unit you will find assessment practice activities to help you work towards your assessment. Completing these activities will mean that you have carried out useful preparation that will be relevant when it comes to your final assignment.

In order for you to complete the tasks in your assignments, it is important to check that you have met all of the Pass assessment criteria. You can do this as you work your way through each assignment.

If you are hoping to gain a Merit or Distinction, you must be able to use more complex drawing features and ensure that the CAD drawings you produce are accurate and precise.

All assignment evidence will be in the form of a portfolio of drawings.

The CAD software you use to complete this unit will depend on what your centre has available. You will be learning to use the software by viewing demonstrations from your tutor, referring to specialist books, accessing online videos and tutorials, and working through tutorials within the software or provided by the software company.

It is important to remember that a CAD system is simply a tool for producing documents and models that describe a product in a more efficient way than using traditional drafting and production techniques. The learning aims in this unit cover many common requirements, which are usually not repeated in each learning aim. However, these shared requirements are the key building blocks that you should understand in order to meet this unit's goals of developing real-life drawings and models. As with any engineering tool, you need hands-on experience to gain proficiency, and it is important that you do not consider any single learning aim in isolation.

Link

Unit 2, Learning aim B ('Develop 2D computer-aided drawings that can be used in engineering processes'), covers many of the techniques and concepts required by this unit. Unit 2 also includes additional background material that you will find helpful when you practise using CAD software to produce your own drawings.

A range of CAD packages is commercially available. Here are some that you might have encountered before or may use in the future:

▶ Autodesk is one of the best-known software suites for producing 2D and 3D models using, most commonly, AutoCAD software. It is used across the majority of engineering sectors.

▶ CATIA (Computer-Aided Three-dimensional Interactive Application) is CAD software that can be used for design and modelling in 2D and 3D, and also for large-scale virtual reality simulations.

▶ SolidWorks is computer-aided design software that can often be found in schools, colleges and industry. It is used for creating 2D and 3D models, including testing and simulation.

▶ PTC Creo (formerly called Pro/Engineer) enables 2D and 3D models to be produced and integrated with computer-aided manufacturing (CAM).

▶ 2D Design by TechSoft is a basic design package for producing 2D drawings and integrating with CAM. It is widely used in schools as an introduction to CAD/CAM.

Assessment criteria

This table shows what you must do in order to achieve a **Pass**, **Merit** or **Distinction** grade, and where you can find activities to help you.

Pass	Merit	Distinction

Learning aim **A** Develop a three-dimensional computer-aided model of an engineered product that can be used as part of other engineering processes

Pass	Merit	Distinction
A.P1 Create models and drawings of at least five 3D components from an assembled product, and apply a material to two or more components. **Assessment practice 10.1**	**A.M1** Produce accurate models and drawings that mainly meet an international standard of an assembled 3D product containing at least seven well-orientated components and apply a material to all components. **Assessment practice 10.1**	**A.D1** Refine models and drawings to an international standard of an accurate and correctly orientated 3D assembled product that is fit for purpose, applying appropriate materials to all components and create a drawing template. **Assessment practice 10.1**
A.P2 Create a model and drawings of an assembled product containing at least five components with two or more components well orientated. **Assessment practice 10.1**		

Learning aim **B** Develop two-dimensional detailed computer-aided drawings of an engineered product that can be used as part of other engineering processes

Pass	Merit	Distinction
B.P3 Create, using layers, drawings of at least six 2D components from an assembled product. **Assessment practice 10.2**	**B.M2** Produce, using accurate layers, drawings that mainly meet an international standard of an assembled product containing at least ten accurate and well-orientated components. **Assessment practice 10.2**	**B.D2** Refine, using accurate layers from a master layer, drawings to an international standard of an accurate and correctly orientated 2D assembled product that is fit for purpose. **Assessment practice 10.2**
B.P4 Create a 2D assembly drawing containing at least six components, with at least two components well orientated. **Assessment practice 10.2**		

Learning aim **C** Develop a three-dimensional computer-aided model for a thin-walled product and a fabricated product that can be used as part of other engineering processes

Pass	Merit	Distinction
C.P5 Create partially rendered models and drawings of at least two 3D components from a thin-walled assembled product. **Assessment practice 10.3**	**C.M3** Produce an accurate model and drawings that mainly meet an international standard of at least two well-orientated and fully rendered 3D components from a thin-walled assembled product. **Assessment practice 10.3**	**C.D3** Refine drawings to an international standard of two accurate and correctly orientated 3D models with realistic rendering that are both fit for purpose. **Assessment practice 10.3**
C.P6 Create partially rendered models and drawings of at least four 3D components from a fabricated assembled product. **Assessment practice 10.3**	**C.M4** Produce accurate model drawings that mainly meet an international standard of at least four well-orientated and fully rendered 3D components from a fabricated assembled product. **Assessment practice 10.3**	

Getting started

Work in groups to find out more about the different types of CAD software on the market at the moment.

If you have used any type of CAD software before, share your experiences with your group.

Are different types of software better for 2D and for 3D drawings? Why?

Are different types of software used within different sectors of engineering?

A   Develop a three-dimensional computer-aided model of an engineered product that can be used as part of other engineering processes

Key terms

Orthogonal drawing – a drawing that represents a 3D object in several 2D views, typically plan, front and side views in a specific layout.

Bill of materials (BOM) – a table containing information about the parts in a drawing. CAD software can automatically generate this for you.

Detail view – an enlarged, more detailed view of part of a drawing.

This learning aim involves a practical drawing activity where you will produce a 3D model of a product and demonstrate that you can determine the material properties of its components. You will need to produce a printed or plotted portfolio of drawings supplemented by screenshots that demonstrate the processes you have used and the editing that has occurred as your drawings progressed. Your portfolio of drawings will need to include **orthogonal drawings**, 3D shaded or solid models, a parts list/**bill of materials** and **detail views**.

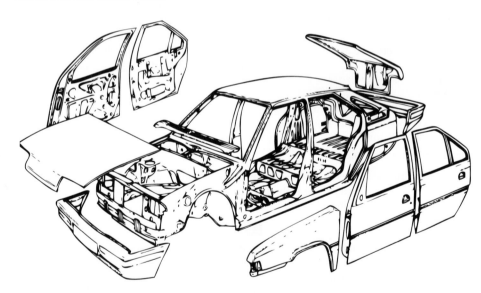

▶ **Figure 10.1** Example of an orthogonal drawing

A1 3D parametric modelling

Computer-aided design systems have revolutionised the way that engineers work. Today, you will find very few offices where drawing boards and other drawing tools are still used. Instead, draftspersons produce their drawings on computers, which require few people to operate. Other advantages include:

- automated tracking of multiple drawings for a complex design
- consistency of style
- automated error checking
- automatically generated parts lists
- the possibility of manufacturing products straight from a drawing
- easily edited and adapted drawings
- high-accuracy drawings.

The 2D drawings can be converted into 3D models for review on-screen, or they can be output to 3D printers or laser cutters for prototype manufacture. 3D models can be tested on-screen to assess their suitability for various applications before manufacture is planned.

Parametric modelling is the production of a drawing where parameters such as dimensions, features and relationships are used to define a model and allow us to control its shape and size. This enables the CAD operator to easily develop a product, capture its behaviour and make changes.

The first stage of parametric modelling is to configure the settings. To set up your parametric modeller, you can define the origin position, the units you will use, the size of the on-screen grid and whether and how to **snap** to the grid. You should also set the file type for saving and the planes that you are working to (e.g. XY, XZ and YZ).

> **Key term**
>
> **Snap** – a function that is used to make the cross-hairs/cursor on the screen jump to grid points. A grid can be set to various sizes and used with the snap function to enable accurate and speedy drawing production. Snap can also be set to jump to the ends of lines, midpoints of lines or centres of circles.

Sketching commands

You will use a variety of sketching commands to draw directly on the computer screen. These commands define entities such as straight and curved lines, circles and polygons. They also let you edit the entities, such as by adding chamfers or fillets.

- Line – To draw a line you either indicate its start and end positions or specify its starting point and length by coordinate entry. There are many types of line available, including:
  - centre lines
  - hidden lines
  - dashed lines
  - construction lines.

Figure 10.2 shows some examples of frequently used line types.

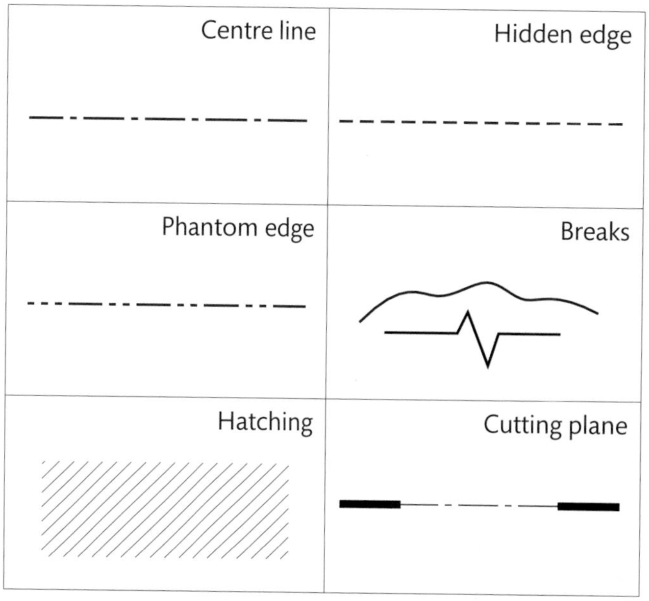

> **Figure 10.2** Examples of line types often used in engineering drawings

> **Link**
>
> Unit 2, Table 2.4, also shows some standard line types used in engineering drawings.

> **Research**
>
> Explore how the different lines identified in **Figure 10.2** and Unit 2, Table 2.4, are used in engineering drawings.

- Arc – You can use the arc command to draw a curve between two points. An arc is part of the circumference of a circle, and usually you will need to input the circle's centre point and radius and the end point of the arc. Some CAD packages provide more accurate ways of defining an arc's characteristics. Alternatively, you may be able to draw an arc and then set constraints to achieve an exact size.

- Circle – This command can be used to define a circle of a specified radius or diameter about a specified centre point. Some CAD software will draw a circle that passes through two or three specified circumference points (**Figure 10.3a**) or a circle that is tangential to another object (**Figure 10.3b**).

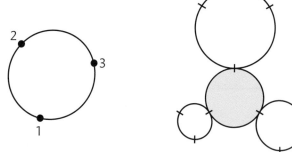

a   b

▸ **Figure 10.3** a) A circle defined by three points on the circumference; b) tangential circles

▸ Fillet – This refers to the rounding of an edge or corner. A fillet can also be applied to an edge on a solid or surface.

▸ Dimension – Dimensions allow you to label entities on your drawing. You will have drawn the entities to a certain scale, but labelling dimensions means that engineers can read them without having to measure the feature. Measuring a feature on a drawing – perhaps using a ruler for lengths or a protractor for angles – will introduce inaccuracies and may not even be possible if the drawing is on-screen. Most CAD software allows you to customise how dimensions are displayed; for example, to indicate the extent of a dimension, dimension lines could end in arrow heads or small marks at right angles to the dimension line.

Display commands

You use display commands to alter how you view the drawing you are working on. You might want to do this because the drawing is large, isn't legible or cannot be worked on accurately when displayed at its normal size.

▸ Pan – Panning allows you to move an area of a drawing in any direction so that a particular entity is displayed in a convenient part of the screen for you to inspect or work on.

▸ Zoom – Zooming allows you to view an area of a drawing in more detail (by zooming in) or to view a larger area of a drawing (by zooming out). Zooming is often used in conjunction with panning; for example:

1 Zoom out to see more of a drawing and locate a particular entity you'd like to work on.
2 Pan the drawing so that the entity you want to work on is positioned in the centre of the computer screen.
3 Zoom in so that you can work on the detail of the entity.

▸ Orbit – This function allows you to view an object from various angles, which could be useful if you want to see how an entity sits in relation to other entities of the design. Note that 'orbit' is a three-dimensional *display* command, unlike the *editing* command 'rotate' (see next page).

Editing commands

You use editing commands to make changes to your drawing. For example, once you have drawn (sketched) an entity you might want to move it to a new position, change its orientation or even delete it completely if you don't like it. Most CAD software allows you to undo the previous change you requested – this is very useful because you can then try out an alteration or correct mistakes.

> **Tip**
>
> Find out how to 'undo' a command in the CAD software you will be using.

▸ Erase – Use this command to remove an object, line or feature from a drawing. This is useful because designs often change as a project progresses.

▸ Extend – This command lets you extend an object to a pre-set boundary. You may want to extend an existing line to join with a particular point elsewhere on the drawing.

▸ Trim – Trimming (sometimes called cutting) is used to remove unwanted parts of lines. It is often simpler to draw a single line and then remove sections of it than to draw several individual lines. For example, to produce the semicircle in **Figure 10.4c** you may find it easier to draw **Figure 10.4a** of a line and a circle, and then trim off one half of the circle as shown in **Figure 10.4b**, than to draw a semicircle in the first place.

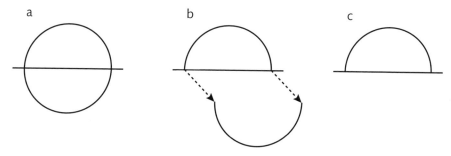

▸ **Figure 10.4** Using trimming to draw a semicircle

> **Tip**
>
> Before starting to draw, always look carefully to see if there is an easier and quicker way of producing an object. Commands such as 'trim' and 'rotate' often provide efficient methods of sketching entities. With practice, you'll find your own ways of using the commands to your advantage.

▸ Rotate – You can turn an entity or a group of entities about a given point by a specified angle. Unlike the display command 'orbit', 'rotate' is a two-dimensional editing command. 'Rotate' provides an easy way to draw lines at specified angles. You may also find it useful to draw a complex object in an orientation that is easy to work with and then, when the object is complete, group all the component parts and rotate and move the group to its final position on a drawing.

Construction commands

You now need to develop the skills to draw 3D components that, when combined, can be used to produce a complete 3D model. Fundamentally, there are three stages to this:

1 Draw a 2D sketch.

2 Convert the 2D sketch to a 3D sketch. There are a variety of techniques to do this, some of which are discussed below.

3 Add a material to the sketch to make the drawing look more realistic.

The detailed steps for each of these stages will depend on the particular CAD software you are using, but the principles are basically the same.

Construction commands can be grouped into categories, including:

▶ 3D primitives – These are pre-defined basic shapes, such as cubes and cylinders, that you can use as a basis for producing more complex objects.

▶ 3D creation – This is the process of drawing a solid 3D object from an initial 2D sketch. Two important techniques for achieving this are extrusion and revolving. **Figure 10.5a** shows the extrusion of a 2D rectangle (in this case the extrusion is also tapered). **Figure10.5b** shows the same rectangle that has been rotationally extruded (revolved) about a point so that the 3D object formed has curved edges. You can construct complex 3D objects by using these techniques in whole or in part.

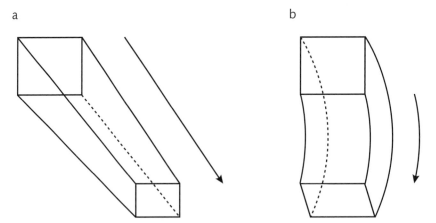

a   b

▶ **Figure 10.5** 3D creation using a) the extrude technique, where the extrusion is also tapered, and b) the revolve technique

▶ 3D modify – Once you have created a simple or complex shape you can modify it, keeping its 3D features. For example, you can insert holes through the object or change sharp edges to chamfered (angled) edges. You can even move an entire face of the object.

▶ 3D Boolean – Boolean algebra is type of maths that allows you to perform logic calculations. The Venn diagrams in **Figure 10.6** show how two sets can be combined using Boolean operations. CAD software provides these operations to allow you to combine multiple parts or remove parts from a solid to create more complex 3D shapes. This is illustrated with a block and cylinder in **Figure 10.7**, using commands for addition, subtraction and intersection.

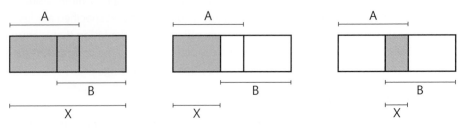

▶ **Figure 10.6** Simple Venn diagrams illustrating the Boolean operations of a) union, b) difference, and c) intersection

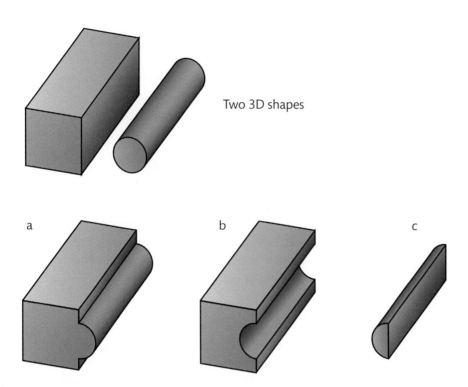

Two 3D shapes

a   b   c

▶ **Figure 10.7** Results of applying to a block and cylinder the 3D Boolean operations of
a) addition, b) subtraction, and c) intersection

▶ 3D assembly – The assembly operations allow you to assemble existing objects
using commands such as 'place' and 'constrain'. Very often the position of an entity
in a larger assembly is pre-determined by other features. For example, if you are
drawing a new gearbox as part of a larger series of drawings of a piece of machinery,
then the positions of the input and output drive shafts are likely to have been
given to you, thereby constraining where the features of your drawing will be.
Constraining a feature could be as simple as fixing the end point of a line. See also
the discussion on assembly constraints in Section A3.

▶ 3D analysis – A powerful feature of many CAD packages is the ability to analyse
a design while it is still in the drawing stage. CAD modelling simulates a design,
allowing you to analyse a 2D region or 3D solid to produce text reports on
properties such as stress and mass. As with any modelling process, you can try out
designs without having to build the finished product. This saves money and reduces
the risk of failure.

Ⅱ PAUSE POINT   Close the book and list as many drawing commands and features as you can.
Describe the purpose of each one.

Hint   Begin by listing those you have used before.

Extend   Pair up and compare your list with your partner's. Explain to your partner how to use
any commands and features that you have listed but they have not.

A2 Develop 3D components

As summarised in the previous section, you use the 2D and 3D commands to create
and modify a 3D model. Some engineering components are fairly common to many
engineering drawings, so you must demonstrate that you can produce these, including:

As previously mentioned, although some commands and features are common to all the different CAD software packages available, how you actually perform these functions will be specific to the particular package you are using.

All the techniques covered in the preceding text will be useful to you. Practise them and the following processes on drawings of your own in the CAD software you are familiar with.

▶ Male and female threads – Male threads are those found on screws and bolts, and are often called external threads. Female threads are those found on nuts, and are often called internal threads.

▶ Plain and drilled holes – A plain hole is any portion of a solid where the material has been removed. If the material has not been removed all the way through to the other side, the hole is called a blind hole. A drilled hole is similar to a plain hole except that it has a circular cross-section.

▶ Countersunk and counterbored holes – These are, respectively, conical and straight-edged holes that lead into a smaller hole, thus allowing the fixing (such as the head of a screw that goes into the smaller hole) to lie flush with the material's surface.

▶ Fillets and chamfers – These are, respectively, curved and angled edges on a material surface.

▶ Combining solid objects, including the use of Boolean operations.

▶ 2D sketching onto 3D faces – use this for modifying your 3D entities.

▶ Modifying the 3D model, including addition of features to existing geometry (e.g. cut-outs, projected geometry, new holes and extrusions).

▶ Modelling, including:
   ▶ applying materials to a mass, surface area or volume
   ▶ calculating the **density** of a material
   ▶ calculating the **principal moments** and **moments of inertia** of a body about its principal axis to see how differences in material affect a model.

Key terms

Density – the mass per unit volume of a material, calculated by dividing the mass by the volume.

Principal moment – the tendency of a force to rotate an object about a turning point, also known as torque; it is calculated by multiplying the force by the perpendicular distance of the force from the turning point.

Moment of inertia – an object's resistance to a change in its rotational direction; this is dependent on the distribution of mass within the object with respect to the axis of rotation.

A3 Develop a 3D model

After you have gained enough practice using the CAD commands described in Section A1, you can combine or assemble the 3D components to form a complete 3D model. This could be a simple assembly, such as a screw and nut, or it might be a more complex model containing many components, such as a vehicle.

Tip

Don't worry if you can't get your drawings correct the first time. You learn to use software by getting things wrong and then working out how to put them right. As long as you save your work regularly, making mistakes will help you progress. Try to first work out by yourself what you did wrong, and only ask your tutor or peers if you're really stuck.

Placement of 3D components

Components placed within an assembly drawing are positioned by 'degrees of freedom'. These determine how freely you can place a component within your workspace. The first component placed is set as 'grounded', or fixed, within the space, and at this point it has no degrees of freedom. Further components can then be placed along the three translational (or linear) degrees of freedom in the directions of the X-axis, Y-axis and Z-axis, or along the three rotational degrees of freedom about the X-axis, Y-axis and Z-axis. **Figure 10.8** shows the degrees of freedom available.

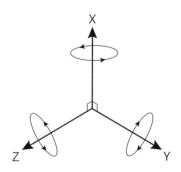

▶ **Figure 10.8** Translational and rotational degrees of freedom with reference to the X-, Y- and Z-axes of a Cartesian coordinate system

Assembly constraints and relationships between components

You can apply assembly constraints to edges, faces or any other parts of an object's geometry. Applying an assembly constraint to an assembly drawing removes degrees of freedom. You need to be able to use the following assembly constraints:

▶ Mate – This constraint places components together face-to-face (i.e. flush against one another), for example when positioning two cubes side by side. Edges can also be mated in this way.

▶ Angle – Edges or planar faces on two components can be positioned at a specified angle to each other; for example, you can position a face of one cube at 45° to a face of another cube.

▶ Insert – This combines a mate constraint between planar faces and a mate constraint between the axes of two components, for example to position a bolt within its hole.

▶ Tangent – This causes two components to make contact at one point, called the point of tangency; for example, you can use it to position a circle halfway along the edge of a rectangle.

Assembly constraints will limit how a drawing can be modified. This means that the software won't allow you to make changes that would break the constraint.

Storyboarding

When you have learned to use a range of drawing commands and features, you need to consider carefully how to apply these to produce more complex drawings. It is essential that you plan the creation of your drawings. It can be frustrating to work for many hours only to find that a constraint that you set up at the start was incorrect and can only be corrected by redrawing or repositioning many objects.

Step by step: Storyboarding the production of a hollow cylinder   `5 Steps`

1 Sketch the outer circle.

▼

2 Extrude the circle to make a solid cylinder.

▼

3 Change the view to show the top face of the cylinder.

▼

4 Sketch an inner circle onto the top face.

▼

5 Project the inner circle through the outer circle and subtract material to produce a hollow cylinder. These steps are illustrated in **Figure 10.9**.

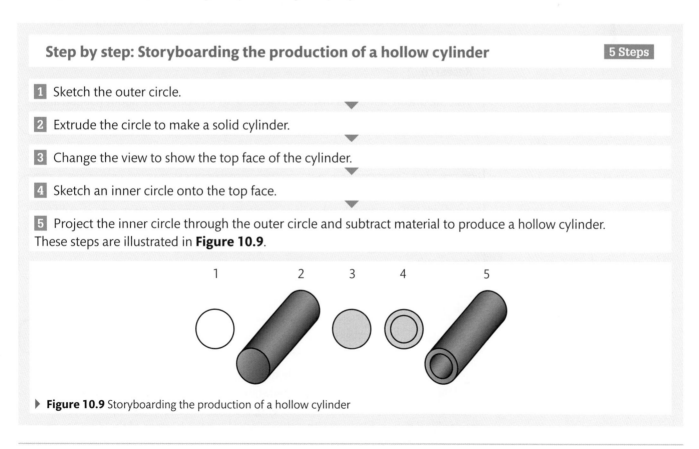

▶ **Figure 10.9** Storyboarding the production of a hollow cylinder

A4 Output of drawings from a model

After you have created drawings in your CAD package, you will need to output them in an appropriate form. First, you need to decide on a suitable format for the drawing and create an engineering template. This template must meet recognised standards, such as BS8888 (see below and Unit 2).

Your drawing must include a title block. Companies often have their own style of title block, but although the designs may differ between companies, they will share some common features, such as:

▶ the name of the person who produced the drawing

▶ the date on which the drawing was produced

▶ a projection symbol to indicate whether the drawing is in first- or third-angle projection

▶ a scale to indicate at what magnification the drawing is printed

▶ a title to indicate what it is that the drawing is showing

▶ a drawing number

▶ the company logo

▶ revisions that have been made to the drawing.

Drawing standards

The term 'engineering drawing' refers specifically to a drawing that has been produced to scale and that conforms to recognised drawing standards. This type of drawing communicates technical information in graphical form for the primary purpose of manufacturing an end product. A technical drawing must conform to standards concerning symbols, lines, terminology, scale and other features.

British Standards relating to engineering drawings, such as BS8888, are updated regularly to reflect technological advances and the needs of modern manufacturing. The main function of BS8888 is to bring together all the international standards to which engineers and designers need to prepare technical documents.

Research

Visit the British Standards Institution website to find out more about BS8888 and other drawing standards. To access this website go to www.pearsonhotlinks.co.uk.

Discussion

The drawings that you produce will have to meet international standards. These standards tell you exactly what lines, symbols, layouts and formats you should use on a drawing. Think about why it is important that drawings are produced to set standards. How might it affect manufacturers if drawings weren't produced to set standards?

Link

Unit 2 also discusses BS8888.

Your CAD software is likely to provide drawing templates that you could use. These templates may conform to international standards other than BS8888, but again they will share many common features. **Figure 10.10** shows an example of a simple standardised drawing layout with a title block along the bottom of the drawing space.

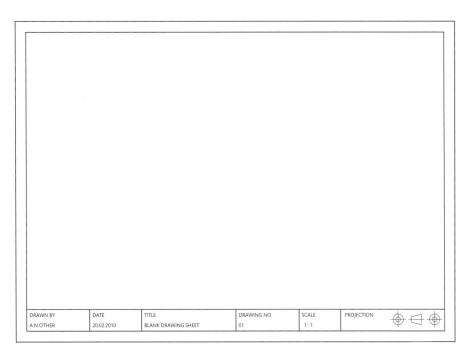

DRAWN BY	DATE	TITLE	DRAWING NO	SCALE	PROJECTION	
A N OTHER	20.02.2010	BLANK DRAWING SHEET	01	1 : 1		

▶ **Figure 10.10** Example of a standardised drawing layout provided by a CAD package

You also need to choose a template that will suit the size and orientation of the paper on which you'll print your drawing.

Assembly and component drawings

You will be required to create both assembly and component drawings as part of a project.

Assembly drawing

An assembly drawing is used to demonstrate how a product should be assembled (see **Figure 10.11**). It shows the relative positions of the components of a product or part of a product. The drawing might show a complete product or a small part of a more complex product. For example, for an electric toothbrush, an assembly drawing might show the items that make up the oscillating brush head and how the various components fit together.

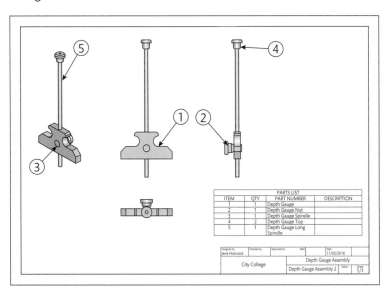

▶ **Figure 10.11** Example of a CAD-produced assembly drawing

Component drawing

A component drawing is a depiction of a single component of a product. It often comprises a set of detailed views, called detail drawings (see **Figure 10.12**), which describe the component to be manufactured. These views might consist of one or more drawings on a single sheet to make important design details easier to see. Detail drawings can provide information about dimensions and tolerances, as well as materials and surface finishes.

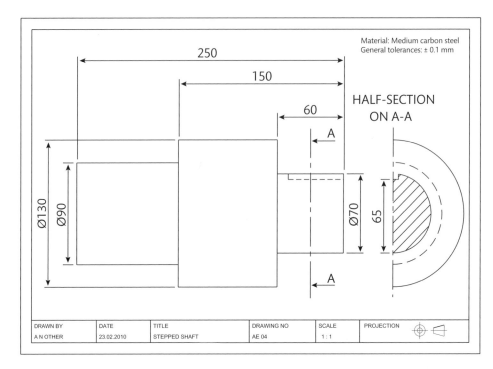

▶ **Figure 10.12** Example of a detail drawing

Detail drawings often use section views to show features of a component that are hidden by the usual orthographic views, such as the inside of an object. A section view depicts a component as if it had been cut along an imaginary plane. **Figure 10.13** shows an example where a cylinder has been sectioned.

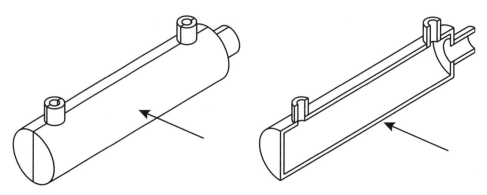

▶ **Figure 10.13** Section view of a cylinder

Here are some terms that you are likely to come across in relation to assembly and component drawings.

▶ Drawing views – You can portray a three-dimensional object on a two-dimensional drawing by showing the object from different points of view, called elevations. Together they form an orthographic projection of the object.

▶ Production – The drawings that you produce will go a long way towards enabling your product to be manufactured, but you also need to document other information, such as a list of parts and quantities. The documentation usually includes the following:

    ▶ diagram – indicates a component's function and how it connects to other components

    ▶ parts list – an inventory of all the items making up an assembly

    ▶ part drawing – a drawing of a single component of an assembly

    ▶ production drawing – a drawing that shows all the information necessary for an item to be produced.

▶ Scale – You must consider scale both when creating and when printing a drawing.

> **Link**
>
> Orthographic projection, Drawing views and Scale are discussed in Unit 2, Section B1 (Principles of engineering drawing).

> **Tip**
>
> Here are some health and safety tips for working on a computer:
> - Always adopt good posture when using a computer – make sure you are sitting or standing comfortably, and position your monitor so that your eyes are in line with the top of the screen with approximately 500 mm between the monitor and your face.
> - Avoid sitting in the same position for long periods.
> - Avoid eye strain by adjusting monitor brightness and contrast to settings that are comfortable for you. Look away from the screen regularly and focus on objects at various distances.
>
> Remember that laptops are generally unsuitable for prolonged working, primarily because they promote casual posture.

Assessment practice 10.1   A.P1 A.P2 A.M1 A.D1

A design company asks you to develop a presentation model and associated drawings of a hard disk drive to show to a customer.

- Create a drawing template to use with the drawings you produce. Ensure that your template contains all of the information required for a drawing produced to the British Standards specification.
- Create models and drawings of at least seven 3D components that together will form one assembled product. Add materials to the components.
- Create a model and associated drawings of the assembled product containing at least seven well-orientated components.

Ensure that your drawings are produced to an international standard and are fit for purpose.

Plan
- What is the task? What am I being asked to do?
- What techniques and features of the CAD software will I need in order to complete the task?
- How confident do I feel in my own abilities in using the tools? Are there any areas I think I may struggle with?

Do
- I know what I'm doing and what I am aiming to achieve.
- I can identify the parts of the task that I find most difficult and devise strategies to overcome those difficulties.
- I can identify when I've gone wrong and adjust my thinking/approach to get myself back on course.

Review
- I can explain what the task involved and how I approached the work.
- I can identify the parts of my knowledge and skills that require further development.

B Develop two-dimensional detailed computer-aided drawings of an engineered product that can be used as part of other engineering processes

This learning aim involves a practical drawing activity where you will create a 2D model of a product using **layers**. You will need to produce a printed or plotted portfolio of drawings supplemented by screenshots that demonstrate the processes you have used and any editing that has occurred as your drawings developed. You will also need to make a drawing template for the output of the professional drawings, which should include a general arrangement, component drawings and a section view produced to international standards.

As discussed in Learning aim A, making 2D drawings is part of the process of developing 3D models, but you will also need to produce 2D drawings for their own sake. To become proficient in doing this, it is important that you familiarise yourself with the following basic commands and terms.

B1 2D drawing commands

Configuration of a 2D CAD system

Your CAD system will operate using pre-set configuration values for many parameters. You can use the default values set by the manufacturer, or you can customise the parameter values to suit your own requirements. Different CAD systems will allow you to configure different parameters, but a basic set of configurable parameters might include:

▸ Coordinate system – This defines how you specify and display locations on-screen. The following three coordinate systems are available for use in most CAD packages:
  ▸ **Absolute coordinates** – based on an 'origin', which is the intersection of the X-axis and Y-axis (i.e. the point (0, 0)). If you know the exact X and Y values of a point, then you can use this type of coordinates.
  ▸ **Relative coordinates** – based on the last point that you entered. It is convenient to use relative coordinates when you know the location of a point in relation to a previously entered point.
  ▸ **Polar coordinates** – these use a distance and an angle from a given point to specify a location. An absolute polar coordinate is measured from (0, 0). A relative polar coordinate is measured from the last point entered.

▸ Drawing limits – These define the extent of your drawing, sometimes called the boundary.
▸ File paths – You can specify the route to the folder where your drawings will be stored.
▸ Grid – Set whether or not a grid is displayed to aid layout; define the dimensions of the grid.
▸ Layers – Choose the colours in which different layers are displayed.
▸ Snap – Specify whether or not placed entities will be 'snapped' to the grid.
▸ Units – Define the units that dimensions are displayed in.

Tip

Find the customisable parameters in your CAD software. Familiarise yourself with how to change these to suit your project.

Discussion

Figure 10.14 shows a line drawn from point (–3, 1) to point (4, 5). Which coordinate system has been used?

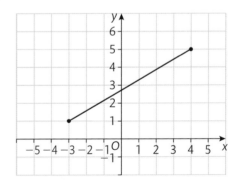

▸ **Figure 10.14** Line drawn between two points

Use of 2D drawing commands

You have already met many of the 2D drawing commands available in a CAD system, such as line, arc and circle. Another one that you will find useful is:

▸ Polyline – a sequence of line segments connected to create a single object, which is easier to edit and move than a series of individual lines.

2D display commands

The most commonly used 2D display commands, pan and zoom, were discussed in Section A1.

2D modify commands

With experience you will learn that, rather than drawing new objects from scratch, it is often quicker to modify existing objects. Sometimes you will also need to alter objects that you have drawn to meet requirements. You have already met some modify commands, such as erase, extend, trim and rotate, in Section A1 under 'Editing commands'. Some other useful commands are:

▶ Array – creates a pattern from multiple copies of an object or sketch. Arrays can be rectangular or circular (polar); for example, the magnets in a linear motor might be arranged in a linear array as in **Figure 10.15a**, and the coils in a motor might be presented in a circular array as in **Figure 10.15b**.

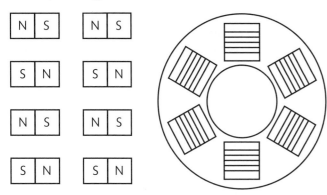

▶ **Figure 10.15** Examples of arrangements using a) a linear array and b) a polar array (Artwork courtesy of Autodesk)

▶ Copy – creates an exact copy of an object or sketch.

▶ Mirror – creates an image of a selected object reflected about a line, which could be at any angle. When the mirror line passes vertically or horizontally through the centre of the object, this operation is sometimes referred to as 'flipping' the object. See **Figure 10.16** for examples of mirroring.

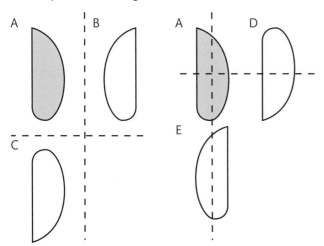

▶ **Figure 10.16** Images B and C are mirrored images of object A; Images D and E are flipped images of object A

▶ Move – relocates a selected object or sketch. You can specify the exact distance and direction of the move.

▶ Stretch – changes the dimensions of an object by a certain multiplicative factor. To use this command, draw a selection window that encloses a sketch or an object either completely or partially. The portion of the object that is partially enclosed will be stretched, while parts that are completely enclosed will be moved.

▶ Undo – reverses the previous command.

B2 Development of 2D engineering drawings

A range of commands and tools can be used to develop your engineering drawings, such as drawing and modify commands, layers, blocks, symbols and dimensions. You have already met and used drawing commands, modify commands and dimensions elsewhere in this unit. There are two other key features that can help you develop your drawings: layers and blocks/symbols.

> **Tip**
>
> Make sure that you practise using a variety of commands – only through practice will you become proficient in the techniques required to produce professional engineering drawings.

Layers

A simple way to understand layers is to think of a paper drawing that has a series of acetates or overlays on top of it. The paper drawing shows the main outline of a product, and each of the acetates shows a type of additional information, such as hidden detail, dimensions and annotations. When viewed with all the acetates in place, the drawing appears very complex, but viewing it with a single individual layer clearly shows one particular aspect of the drawing.

The advantages of a CAD drawing that contains layers are that you can easily switch between the layers, transfer information between them and switch them on or off at any time to make editing and interpretation easier. The number of layers that can be created within a CAD drawing is unlimited, as is the number of objects that can be drawn onto them. Using layers is a great way of organising a complex drawing.

Editing one layer can affect what happens with features across all layers. However, it is possible to freeze layers so that they cannot be accidentally selected and drawn on.

Tip

Give the layers clear names for easy reference, interpretation and retrieval. If you name your layers 'Main Outline', 'Dimensions' and 'Section View', it will be much easier to understand what each one contains when you come back to your drawing later than if you had used default names such as 'Layer 1', 'Layer 2' and 'Layer 3'.

A 'frozen' layer will be invisible until you deliberately 'thaw it out'. You can also use the 'lock' command to prevent changes to a layer – locked layers are visible but faded out.

Symbols and blocks

CAD systems contain a wide range of pre-constructed symbols and drawing objects, which can be retrieved from a symbols library. You can add to these libraries by installing software add-ons for specific applications within an engineering sector, or you can design your own symbols, which is useful if you frequently require a complex (or even simple) specialised object in your drawings. Using pre-drawn symbols and objects can speed up drawing production considerably.

The use of blocks can also save a lot of time. A block consists of one or more objects collected together and given a name. Blocks can be inserted from, for example, a block library or tool palette. Drawing files can also be inserted as blocks, and a block reference within a current drawing can be re-inserted into the same drawing. You can insert the same block into a drawing again and again, and a change to just one of these repeated blocks will be replicated across all of them. A block uses less memory space than the individual objects it is composed of, so the equivalent drawing file will be much smaller.

Dimensioning

Dimensions largely consist of projection lines leading away from an object edge-headed line with the size of the space between the projection lines being shown on an arrow between the projection lines. Linear dimensions are represented in this way; radii, diameters and arcs are represented differently. The CAD system you are using will have its own method of adding dimensions to a drawing, this often consists of selecting the dimension button within the software and then selecting the object, or outer edges of an object, where the dimension needs to be applied. You then left click and drag away from the object, releasing the mouse button when you hare happy with positioning. Your CAD software will automatically show a size for the dimension, normally in mm, based on the original parameters set up at the start of the drawing. This dimension can be customised, however, often by double-clicking on it directly and then inputting the dimension you want displayed.

Dimensions have to be represented on a drawing to British Standards which your CAD software will automatically apply for you. Some rules for you to consider when adding dimensions to your drawing:

▶ Dimensions must be positioned on the outside of objects.

▶ Position dimensions around an object the same distance away from the edges of that object.

▶ Position dimensions from each other the same distance as from the edge of the object.

▶ Longer dimensions should be positioned outside shorter dimensions.

▶ Do not allow projection lines to overlap.

▶ All parts of the drawing must be dimensioned – there should be enough information to allow the product to be manufactured from the drawing.

▶ Think carefully about the dimensions you need and where they are best positioned – do not repeat dimensions.

B3 Output of 2D drawings

The issues to consider when outputting your drawing to a printer or plotter are discussed in Section A4. Information relating to the creation of assembly and component drawings can also be found in Section A4 Assembly and Component drawings.

⏸ PAUSE POINT   Close the book and list as many requirements of outputting drawings as you can. Describe the purpose of each requirement.

Hint   Begin by listing the requirements that you have already encountered in previous work.

Extend   Pair up and compare your list with your partner's. Explain to your partner those requirements that you have listed but they have not.

Reflect

Ajay is a new employee of a small engineering firm that develops and manufactures specialist part prototypes for the aerospace industry. To date, the company has made minimal use of CAD and has employed Ajay to introduce the use of CAD software to enable easy sharing of technical drawings internationally and to test, virtually, the performance of different materials on their products.

- What hardware will Ajay need to produce and output drawings?
- What functions will the software that Ajay chooses need to have?
- The design team have had minimal training in the use of CAD. How should Ajay introduce and develop the required skills?

Assessment practice 10.2   `B.P3` `B.P4` `B.M2` `B.D2`

A design company asks you to develop a 2D model and associated drawings of a gearbox to show to a customer.

- Create a drawing template to use with the drawings you produce. Ensure that your template contains all of the information required for a drawing produced to the British Standards specification.
- Create, using accurate layers from a master layer, drawings of at least ten accurate and well-orientated components. Combine these components to produce a 2D assembly drawing.

Ensure that your drawings are produced to an international standard and are fit for purpose.

Plan
- What is the task? What am I being asked to do?
- What functions of the CAD software will I need to use in order to complete the task?
- How confident do I feel in my own abilities in using the tools? Are there any areas I think I may struggle with?

Do
- I know what I'm doing and what I am aiming to achieve.
- I can identify the parts of the task that I find most difficult and think of ways to overcome those difficulties.
- I can identify when I've gone wrong and adjust my approach to get myself back on course.

Review
- I can explain what the task involved and how I approached the work.
- I can identify the parts of my knowledge and skills that require further development.

C Develop a three-dimensional computer-aided model for a thin-walled product and a fabricated product that can be used as part of other engineering processes

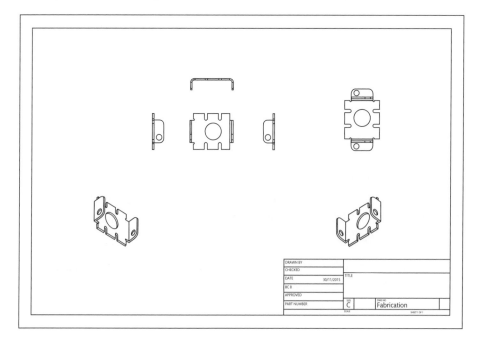

Tip

Make sure that the product you choose meets the qualification requirements; but you also don't want to pick something too complex, which might be difficult to complete satisfactorily and on time. Discuss the suitability of your choice of product with your tutor.

This learning aim involves a practical drawing activity in which you will create partially rendered 3D models and drawings of a fabricated assembled product constructed from thin-walled components. For example, the product could be a small hairdryer or a computer mouse, but the final choice is yours. You will need to produce a printed or plotted portfolio of drawings supplemented by screenshots that demonstrate the processes you have used and any editing that has occurred as your drawings have progressed. You will also need to create a drawing template for the output of your drawings, which should include orthogonal drawings, a 3D shaded/solid model, a section view of the thin-walled components and a detail view of the fabrication.

This learning aim allows you to put into practice many of the techniques discussed earlier in this unit for producing 2D drawings and converting them into 3D models.

C1 3D modelling commands

CAD packages include specific modelling commands that let you convert a 2D sketch into a 3D model:

▶ Rotate about an axis – Use this command to move an object around a 3D axis.

▶ Revolve – This command creates a solid by rotating a 2D shape around an axis.

▶ Extrude – This command allows you to add or remove material along a linear path.

Sheet metal parameters and fabrication commands

CAD systems can be used to create sheet metal designs and models, such as for car panels or mounting components, quickly and effectively. As you can see from **Figure 10.17**, the sheet metal needs to be manipulated, either manually or by machine, into different shapes.

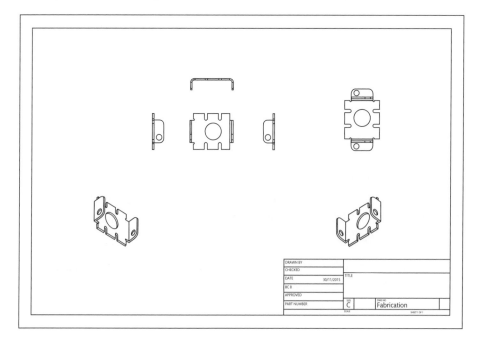

▶ **Figure 10.17** Example of sheet metal fabrication in CAD

▸ Folding and bending – When sheet metal is bent or folded the material is stretched and squashed. CAD software contains sets of parameters that specify how a sheet metal part is modelled. The bend relief transition type, radius value and shape and size are all defined. Default rules exist within the software and can be used directly, but these rules can also be edited to suit your particular requirements.

When a fold is created within the software, it forms around a straight bend line sketched onto a face. The fold will end at the face edges. No material is added to the object to create a fold.

A bend can be created by entering radii, which the software then applies to the object. The bend can be deleted, edited or added to. Parts can also be created from a template file. Material is added to the object to create the bend.

▸ Corner reliefs – These can be two-bend or three-bend, with a range of shapes and sizes available. Again, there are pre-defined rules within the software, but it is possible to edit these and specify your own.

You can also use the sheet metal commands to set a material's thickness and to create flanges, holes and slots.

Construction thin-walled commands

A number of CAD commands are available specifically for drawing the features often required by thin-walled components:

▸ Loft – creates a 3D solid by 'lifting' one 2D cross-section through to a second 2D cross-section.

▸ Shell – enables you to hollow out a part or object.

▸ Work plane – a face or plane on or within an object that can be selected, sketched onto and modified.

▸ Emboss – raises a feature from the original surface, creating a bumpy effect.

C2 Development of 3D components

The preceding sections have given you many of the techniques needed to transform your 2D drawings into 3D models of product components.

▸ Creating 2D sketches of components – basic sketching, dimensioning and modification commands are covered in Sections A1, B1 and B2.

▸ Converting 2D sketches into 3D components – construction and fabrication commands are discussed in Sections A1, A2 and C1.

▸ Producing 3D features of the components – relevant techniques are described in Sections A2 and C1.

C3 Development of a 3D model

The placing of 3D components and the assembly of them into a 3D model of the product are discussed in Sections A1 (under 'Construction commands') and A3.

To make your models more realistic, you can add surface rendering and light effects.

Effective rendering on a drawn 3D object is essential when you are presenting a model to a client or design team. It enables you and others to visualise how the finished product will look in the real world. **Figure 10.18** shows an example.

> **Tip**
>
> Explore the construction features in your CAD package and learn how to use them. Practise the commands as you develop your 3D model so that you improve your efficiency in using them. This way, when you need to model a design feature you will be armed with the necessary techniques.

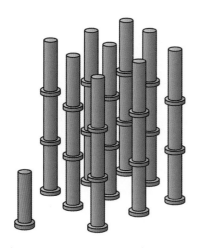

▶ **Figure 10.18** Example of a 3D model created from a single object using a rectangular array. Autodesk Inc. No modifications allowed

Rendering involves adding light effects, materials and textures to a 3D model or wireframe. You can even apply environmental effects, such as fog. Lighting effects include reflections and shadows. CAD software has pre-set automatic virtual lighting settings that can be used for a basic render effect, but these cannot be adjusted.

Light effects also include ray tracing, where paths of light rays are used to create reflections and refractions, resulting in more realistic images. **Figure 10.19** shows two drawings for comparison: one has been ray traced, showing reflection and refraction, and the other has not been ray traced.

▶ **Figure 10.19** Application of ray tracing to the left drawing results in the drawing on the right, which shows reflection and refraction effects

C4 Output of product drawings

Drawings associated with 3D models must be produced to BS8888 or other equivalent standards. The expectations for drawing outputs are covered in detail in Section A4.

Assessment practice 10.3

| C.P5 | C.P6 | C.M3 | C.M4 | C.D3 |

The design company in Assessment practice 10.2 requires 3D models and associated drawings of a computer keyboard.

- Create a drawing template to use with the drawings you produce. Ensure that your template contains all of the information required for a drawing produced to the British Standards specifications.
- Create rendered models and drawings of at least two 3D thin-walled components within the assembled product.
- Create rendered models and drawings of at least four 3D components from the fabricated assembled product.

Ensure that your drawings are produced to an international standard, include realistic rendering and are fit for purpose.

Plan
- What is the task? What am I being asked to do?
- What functions of the CAD software will I need to use in order to complete the task?
- How confident do I feel in my own abilities in using the techniques? Are there any areas I think I may struggle with?

Do
- I know what I'm doing and what I am aiming to achieve.
- I can identify the parts of the task that I find most difficult and think of ways to overcome those difficulties.
- I can identify when I've gone wrong and adjust my approach to get myself back on course.

Review
- I can explain what the task involved and how I approached the work.
- I can identify the parts of my knowledge and skills that require further development.

Further reading and resources

Websites

The following websites provide user support and tutorials for several CAD software packages.

Autodesk: **https://knowledge.autodesk.com**

AutoCAD: **www.autocadtutorials.net**

CADTutor (AutoCAD): **www.cadtutor.net**

Dassault Systemes SOLIDWORKS: **www.solidworks.com**

ProE Tutorials: **www.proetutorials.com**

THINK ▶FUTURE

Charlotte Woods
CAD manager in a design company serving the motor sport industry

When I was at school I had no idea that I would be working within the engineering industry. I discovered an interest in computer-aided design while at college, which, along with my interest in motor sport and car design, spurred me on to taking higher qualifications. I now have an Honours degree in computer-aided engineering that has enabled me to achieve the career I have now.

I manage a design team that produces 2D and 3D CAD drawings for the motor sport industry, liaising with designers to ensure that they get what they want in terms of the design drawings and 3D models we produce. Together with my team, I get to see the latest design developments within our sector and contribute to them, which I think is amazing.

Focusing your skills

Producing 2D and 3D drawings

- What types of 2D drawing can be produced using CAD?
- How is this beneficial to a design company?
- How is this beneficial to the manufacturing industry?
- What types of 3D drawing can be produced using CAD?
- How is this beneficial to industry?
- How can CAD be used to test components?

Getting ready for assessment

Jack is working towards a BTEC National in Engineering. He attends college part-time as part of his apprenticeship with an engineering company, specialising in computer-aided design and manufacture. Jack had limited experience of using CAD systems before starting his apprenticeship and has had to teach himself many of the skills required alongside his college course. Jack shares his experience of producing the required portfolio of drawing evidence to support Unit 10 completion.

How I got started

First, I reviewed the CAD skills I needed for each drawing type required by the course specification and wrote a list of all the commands and skills needed that were new to me. I practised using each of these commands and features to produce drawings of simple objects at first. Then I added extrusions to them, set parameters and modified the shapes. When I found understanding how to use some of the commands and features difficult, I used the Help menu in the software to guide me and accessed as many online tutorials as I could. My college tutor also gave me support when I was struggling with a concept.

It is a good idea to ask your school/college if there are student versions of the software available so that you can practise at home. I found it very helpful to make drawings of physical objects in my house because I could pick up and measure the actual object I was creating a virtual image of. I had fun applying different materials, textures and effects onto these objects in the CAD program to see what they would look like if made from alternative materials.

How I brought it all together

▶ Once I had tried out all the skills I needed, I applied these to produce the drawings required by the assignments.

▶ I read through the unit content and assessment criteria carefully to make sure I understood what was needed.

▶ I produced a drawing template and checked that it met international standards.

▶ I produced a storyboard for every drawing.

▶ I recorded my progress regularly by using 'print screen' – I pasted these images into a separate document and made brief supporting notes to explain which skills each one demonstrated.

▶ I kept an electronic and physical portfolio of all the drawings I produced.

What I learned from the experience

I learned a lot through trial and error – making mistakes helped me remember what I'd done wrong and not to make the same mistake again. I learned very quickly that it's really important to save my work regularly because I nearly completed a drawing during one session and then the whole college network crashed; I had to start the drawing again from scratch. I also learned how valuable storyboarding is – it saves you a lot of time when you are producing a complex drawing with many layers and features if you plan its production carefully first. Finally, I underestimated the amount of time it would take to produce each drawing. Even if you know how to use all of the skills, make sure you allow plenty of time to actually complete the work.

Think about it

▶ Have you practised all of the skills you need to produce the drawings required?

▶ Have you carefully planned your drawings by producing a storyboard for each?

▶ Do you understand what producing a drawing to an international standard means?

▶ Have you allocated sufficient time to produce the drawings needed?

Electronic Devices and Circuits 19

Getting to know your unit

In this unit you will investigate analogue electronic circuits based on diodes and transistors and combinational and sequential logic digital circuits. As part of the unit you will use software to simulate circuits, construct them safely and use typical bench instruments to test them. Electronic circuit designers make frequent use of software to simulate design ideas before building prototype circuits and testing them in the process of developing final products.

How you will be assessed

This unit will be assessed by a series of internally assessed tasks set by your tutor. Throughout this unit you will find assessment activities to help you work towards your assessment. Completing these activities will mean that you have carried out useful research or preparation that will be relevant when it comes to your final assignment.

In order for you to achieve the tasks in your assignment, it is important to check that you have met all of the assessment criteria. You can do this as you work your way through the assignment.

The assignments set by your tutor will consist of a number of tasks designed to meet the criteria in the table.

Assessment criteria

This table shows what you must do in order to achieve a **Pass**, **Merit** or **Distinction** grade, and where you can find activities to help you.

Pass	Merit	Distinction
Learning aim A Explore the safe operation and applications of analogue devices and circuits that form the building blocks of commercial circuits.		
A.P1 Simulate, using captured schematics, the correct operation of at least one diode, transistor and operational amplifier circuit. Assessment practice 19.1	**A.M1** Simulate, using accurately captured schematics, the correct operation of at least one diode, transistor and operational amplifier circuit. Assessment practice 19.1	**A.D1** Evaluate, using language that is technically correct and of a high standard, the operation of at least one diode, transistor and operational amplifier circuit, comparing the results from safely and accurately conducted simulations and tests. Assessment practice 19.1
A.P2 Build at least one diode, transistor and operational amplifier circuit safely and test the characteristics of each one. Assessment practice 19.1	**A.M2** Build at least one diode, transistor and operational amplifier circuit safely and test the characteristics of each one accurately. Assessment practice 19.1	
A.P3 Explain, using the simulation and test results, the operation of at least one diode, transistor and operational amplifier circuit. Assessment practice 19.1	**A.M3** Analyse, using the simulation and test results, the operation of at least one diode, transistor and operational amplifier circuit. Assessment practice 19.1	

Pass	Merit	Distinction

Learning aim B Explore the safe operation and applications of digital logic devices and circuits that form the building blocks of commercial circuits.

B.P4

Simulate, using captured schematics, the correct operation of at least one combinational logic circuit and two sequential logic circuits.

Assessment practice 19.2

B.M4

Simulate, using accurately captured schematics, the correct operation of at least one combinational logic circuit minimising the gates and at least two sequential bidirectional logic circuits.

Assessment practice 19.2

B.D2

Evaluate the operation of at least one combinational logic circuit minimising the gates and two sequential bidirectional logic circuits, comparing the results from safely and accurately conducted simulations and tests.

Assessment practice 19.2

B.P5

Build at least one combinational logic circuit and two sequential logic circuits safely and test the characteristics of each one.

Assessment practice 19.2

B.M5

Build at least one combinational logic circuit minimising the gates and at least two sequential bidirectional logic circuits and test the characteristics of each one accurately.

Assessment practice 19.2

B.P6

Explain, using the simulation and test results, the operation of at least three logic circuits.

Assessment practice 19.2

B.M6

Analyse, using the simulation and test results, the operation of at least three logic circuits.

Assessment practice 19.2

Learning aim C Review the development of analogue and digital electronic circuits and reflect on your own performance.

C.P7

Explain how health and safety and electronic and general engineering skills were effectively applied during the development of the circuits.

Assessment practice 19.3

C.M7

Recommend improvements to the development of the electronic circuits and to the relevant behaviours applied.

Assessment practice 19.3

C.D3

Demonstrate consistently good technical understanding and analysis of the electronic circuits, including the application of relevant behaviours and general engineering skills to a professional standard.

Assessment practice 19.3

C.P8

Explain how relevant behaviours were applied effectively during the development of the circuits.

Assessment practice 19.3

Getting started

Do you understand the difference between analogue and digital circuits? What electronic applications have you already studied? Have you constructed any electronic circuits? What methods have you used? Look through the unit content and make a list of what you think you already know something about. As part of the unit you need to compare your initial skills and knowledge with your final ones and see how they have developed. Be honest!

Explore the safe operation and applications of analogue devices and circuits that form the building blocks of commercial circuits

A1 Safe electronic working practices

Electronics safety

Working with electronic circuits can be dangerous if you do not take sensible precautions. Here are some safety guidelines to keep you safe as you work.

▶ Only work with low-voltage supplies and never work on a circuit while power is applied.

▶ Check your work carefully before connecting power to a circuit.

▶ Keep your work area clean, dry and tidy.

▶ Be very careful when you solder because a hot soldering iron can easily burn you.

▶ Always work in a well-ventilated space, as soldering can give off potentially harmful fumes.

▶ Wear safety glasses when constructing circuits.

▶ Always follow the guidelines in appropriate risk assessments.

▶ Be able to identify people with responsibility for health and safety and always follow their instructions.

As well as protecting yourself and others, you should also take care to avoid damage to components. Some components are sensitive to static electricity, and you should take precautions to avoid damage by electrostatic discharge (ESD). A good start is to keep your work area clean and tidy so that there are no materials, such as plastic bags, that might build up a static charge. Handle circuit boards by their edges where possible. Keep sensitive components in their packaging until you need to use them. You might carry out construction in a designated area with static dissipation work surfaces and floor and proper grounding methods, such as a wrist strap.

Emergencies

You must know what to do, and what **not** to do, in case of fire, electrocution or other emergency.

> **Research**
>
> For your workplace and work practices:
> - Find out who is responsible for health and safety.
> - Find out who the identified first-aiders are and how to contact them.
> - Find the appropriate risk assessments.
> - Familiarise yourself with the evacuation procedures.
> - Find out where you can find safety equipment such as fire extinguishers and first-aid kits.

Schematic capture and simulation

One of the important skills that you need to develop is the ability to understand and create schematic circuit diagrams. There are many software packages that allow you to do this – see **Figure 19.1** for an example. They range from free downloads from the internet to expensive professional packages used in industry. They all have certain things in common. A key factor is the number of components stored in their libraries and how easy it is to use them. You want to be able to find the components you need quickly so that you can drag and drop them into your drawing window. Each component you include in your design is given a unique identifier, for example R1. You then need to be able to connect your components. This is usually done just by dragging from point to point.

The libraries also contain a variety of signal sources, sensors and actuators. You need to practise using the software applications so that you can work with them confidently and efficiently.

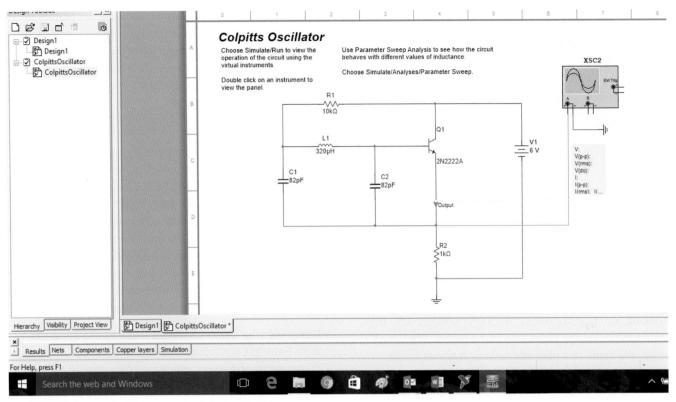

▶ **Figure 19.1** Typical schematic capture and simulation software. (Photo Courtesy of National Instruments)

The clever part of the software lies in the way that the libraries do not just contain the circuit symbols – they also store details of the characteristics of the devices, such as the breakdown voltage and power limit of a Zener diode. You can select the values of common components, such as resistors and capacitors, and change them easily to modify your designs. The software uses this information to enable you to simulate the operation of the circuit. You can include a whole range of virtual instruments, such as voltmeters, ammeters, oscilloscopes and some more complex ones to measure signals at any point in the circuit.

The majority of packages use a program called SPICE to carry out the simulation. This is embedded in the software, so you do not really need to know a lot about it at this stage.

Your software package also needs to allow you to design and simulate digital circuits. The digital libraries contain the basic gates and standard logic families. Some packages allow you to build what are called hierarchical designs, where you design a unit, then use several of the units to build a bigger one and so on.

Digital circuits can have timing problems, which can sometimes be difficult to solve. Logic analysers exist that can compare signals at multiple points in a circuit to check that changes take place at the correct time, but they can

be very expensive. Simulation allows you to check that signals have enough time to stabilise before they can affect another part of the circuit.

Circuit construction techniques

Prototyping methods

After the initial design and simulation, the next step is to build a prototype circuit to test that your design works in practice. There are many prototyping methods available. Plug-in systems are useful in the early stages because you can easily move components about and try different values. Soldered systems are more robust, but are not so easy to modify if you need to make changes.

The most common type of plug-in system is called a **breadboard** (see **Figure 19.2**). This is a plastic block with holes arranged in groups connected by spring clips. The holes are typically arranged at 0.1 inch spacing. The spacing between blocks across the centre-line is 0.3 inches, which is the same as the space between pins on a dual in-line package (DIP) used by many integrated circuits. You insert components such as resistors, capacitors and connecting wire links into the holes to complete the circuit. You may find breadboards frustrating if you try to use them for large complicated circuits, because you may have problems with connections and keeping components apart. You may also find that some component leads will not fit the holes.

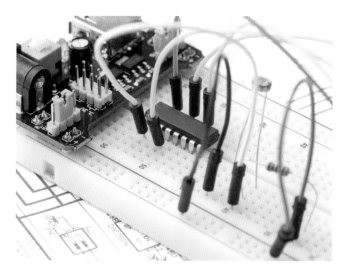

▶ **Figure 19.2** Breadboard prototype circuit

Veroboard (see **Figure 19.3**) has a grid of holes at 0.1 inch spacing. You fit components through the holes, with the components on the blank side of the board. The board is usually made of a phenolic resin or is sometimes a fibreglass-reinforced board. The reverse side of the board has parallel copper tracks running the length of the board. Component leads and connecting links are soldered to the tracks. Dual in-line packages fit the matrix. The layout needs to be planned carefully. You have to cut the tracks in places to separate components that are not connected. It is best to plan the layout on a grid before starting construction.

You can use an alternative to Veroboard, a strip board, to build a more permanent circuit, and the connections will be more reliable. You can buy several variants of strip board.

▶ **Figure 19.3** Veroboard

An alternative, perforated grid board just has plated-through holes to mount the components. You solder the components to the board and join components point-to-point, either by soldering or (less commonly) by wire wrapping. This method is useful for small, simple circuits with only a few components.

Permanent construction methods

Once you have simulated your design and tested that it works by using a prototype method, you will probably want to make a more permanent circuit that can be economically reproduced. Your design software may have a facility to convert your schematic design into a printed circuit board (PCB) layout. Some software will automatically route the connecting tracks for you. More complex circuits may require double-sided PCBs, which have tracks on both sides of the board. A big advantage of PCBs is that they can be mass produced.

Designing your own PCB means that you can allow for different-sized component leads, chips, surface mount components and so on, which might have been difficult to allow for on the prototype.

Testing electronic circuits

It is important that you know what the expected voltages, currents and waveforms should be at key points in the circuit you are testing. This is so you can compare them with your measured values. The main instruments that you need to be able to use to test analogue circuits are a stabilised bench supply, a multimeter, an oscilloscope and a function generator.

Make sure that you know how to connect and use a **multimeter** correctly to measure voltages and currents. You can also check resistor values and test diodes and, on most multimeters, transistors provided that the circuit is not powered.

A **function generator** is a source of signals of different frequencies. The functions available are normally sine, triangular and square wave. You can vary the frequency and amplitude of the signal and may also be able to adjust the duty cycle of the waveform. Most function generators you are likely to use will be audio-frequency generators with a range of about 0.03 Hz to 3 MHz. They may have 'low' and 'high' output impedances to allow for matching to different loads – for example, low impedance to connect directly to a loudspeaker or high impedance to connect to an oscilloscope.

An **oscilloscope** plots voltage on the vertical axis against time on the horizontal axis. You can alter the sensitivity of the voltage axis by adjusting the **gain**

(V cm⁻¹). Dual-trace oscilloscopes allow you to compare two signals simultaneously. This is really useful when you are measuring the gain of an amplifier, because you can see the input and output signals on the same screen at the same time. You can control the gain of each trace separately. The **time base** (s cm⁻¹) alters the speed at which the spot travels across the screen. Normally you would adjust the time base to view just over two complete cycles of the signal(s) you are measuring. It is important to make sure that the gain controls are in the 'calibrate' position if you want to take accurate measurements. You may also have to adjust the 'stability control' to make sure that each trace across the screen starts at the same point so that you will see a steady trace.

Figure 19.4 shows some basic equipment for electronic testing. More complex instruments that you might use occasionally are a Bode plotter, a spectrum analyser and a frequency counter.

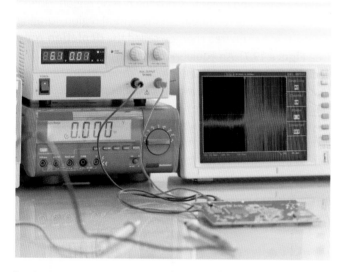

▶ **Figure 19.4** Basic electronic test equipment

A multimeter and an oscilloscope can also be used to test digital circuits, though you may need to use some additional instruments.

A **logic probe** gives a visual indication of whether a signal is at logic level 0 or logic level 1. It may be switchable to use with different logic families. A logic probe can be used with a logic pulser to check that a logic level changes as expected. Many digital circuits use multiple logic inputs and outputs. You can use a logic clip to connect onto an integrated circuit (IC) to monitor all the pins at the same time. A **logic analyser** is a more complex test instrument that captures many signals simultaneously and displays them on-screen. You can set up the analyser to trigger on a given event, collect samples and display them. The analyser can check for such things as signals that never change their logic level ('stuck at' faults) or signals that always change together (potential short circuit). You can also check for timing problems, which could cause glitches. You can use a word generator to generate bit patterns.

Follow these steps to prepare for the work you will be doing in this unit:

1 Get to know the measuring instruments that you need to use.

2 Examine the type of multimeter that you will be using. Practise measuring low-voltage direct current (d.c.) and alternating current (a.c.) voltages and currents in simple circuits.

3 Use the multimeter to measure resistor values and note what happens when you measure values in circuits. (Make sure the power is not connected!)

4 Set up a function generator to give a sinusoidal output of frequency 1 kHz. Connect the signal to an oscilloscope and adjust the amplitude to 2 V. Set the oscilloscope voltage gain so that the full trace is visible and it is as high as possible. Adjust the oscilloscope time base to give approximately two complete cycles on the screen.

5 See what happens if you change the function to output a triangular waveform, then a square waveform.

A2 Diode devices and diode-based circuits

Semiconductors

The most commonly used semiconductor material is silicon. Silicon has the atomic number 14 and is in Group IV of the periodic table. A silicon atom has four valence electrons, which form covalent bonds with four other atoms to form a crystal structure (see **Figure 19.5**). Covalent bonds are very strong, so at low temperatures there are very few electrons 'free' to move around. Therefore, **intrinsic** silicon behaves more like an **insulator**. As the temperature is increased, the atoms gain energy and more covalent bonds can be broken. The more you raise the temperature, the greater the number of bonds broken and hence the number of charges free to conduct. Whenever an electron gains sufficient energy to break a bond, the resulting covalent link is no longer electrically balanced. In effect, the 'hole' that remains has a positive charge that is equal in magnitude to the charge on an electron.

Key term

Conductor – a material that has very low resistance and easily passes an electric current. Most good conductors are metals, such as copper, aluminium and silver.

Insulator – a material that has very high resistance. It is generally a non-metal. Examples of good insulators include plastics and rubber.

Semiconductor – a material such as silicon (Si), germanium (Ge) or gallium arsenide (GaAs) that has electrical properties somewhere between those of a conductor and an insulator.

- Pure semiconductor materials are called **intrinsic** semiconductors.
- **Extrinsic** semiconductors contain very small amounts of impurity chosen to increase the conductivity. The introduction of measured amounts of impurity is called **doping**.

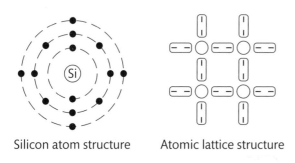

Silicon atom structure   Atomic lattice structure

▶ **Figure 19.5** Silicon atom and atomic lattice structure

You can increase the conductivity of an intrinsic semiconductor by introducing small amounts of impurity. This is known as **doping**. **Figure 19.6** illustrates doping of a silicon lattice.

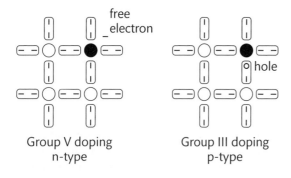

Group V doping   Group III doping
n-type   p-type

▶ **Figure 19.6** Effect of doping on a silicon lattice

For an element from Group V of the periodic table, such as antimony (Sb), each atom has five valence electrons and is said to be **pentavalent**. If you replace a silicon atom with an antimony atom, each antimony atom provides one more electron than is needed to complete the covalent bonds in the crystal lattice. Each of these electrons then contributes to conduction. A Group V impurity is called

a **donor**, and the doped material is referred to as **n-type** because electrons have a negative charge.

For an element from Group III of the periodic table, such as boron (B), each atom has three valence electrons and is said to be **trivalent**. If you replace a silicon atom with a boron atom, then each boron atom provides one fewer electron than is needed to complete the covalent bonds in the crystal lattice. This leaves a 'hole', and each of these holes contributes to conduction. A Group III impurity is called an **acceptor**, and the doped material is referred to as **p-type** because holes have a positive charge.

The PN-junction diode

If you dope a single crystal of silicon with acceptors on one side and donors on the other, then a **PN-junction** is formed. Initially there is a concentration gradient of charge across the crystal, so electrons will move towards the p-type region and holes towards the n-type region. Holes and electrons will recombine in the central region, forming a zone that has no free charges, called the **depletion zone** (see **Figure 19.7**). An equilibrium state will be reached such that the width of this depletion zone is more or less constant. The actual width will depend on several parameters, including the relative doping concentrations of p-type and n-type, as well as the temperature. There is a potential barrier to the movement of free charge carriers across the boundary.

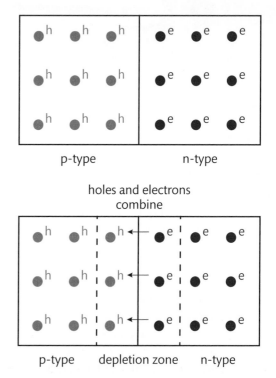

p-type   n-type

holes and electrons
combine

p-type   depletion zone   n-type

▶ **Figure 19.7** Formation of a depletion zone in a PN-junction

Anode – the connection to the p-type material.

Cathode – the connection to the n-type material.

Forward bias – describes a diode where the anode is more positive than the cathode.

Reverse bias – describes a diode where the anode is more negative than the cathode.

Barrier potential – the minimum forward bias voltage needed for conduction to start.

Majority carriers – the charge carriers due to doping, which are holes in p-type material and electrons in n-type material.

Minority carriers – the holes in n-type material and electrons in p-type material due to thermally generated hole–electron pairs in the silicon lattice.

Forward-biased PN-junction

If you connect the positive terminal of a suitable voltage supply to the **anode** of a PN-junction (left diagram in **Figure 19.8**), the voltage effectively narrows the depletion zone. When the applied potential is sufficient to overcome the **barrier potential**, electrons and holes can move freely across the junction, both contributing to the overall current.

Reverse-biased PN-junction

Reversing the voltage supply effectively widens the depletion zone, establishing a greater barrier to the movement of **majority carriers** (right diagram in **Figure 19.8**).

The n-type region contains not only majority free electrons from the donor atoms but also minority holes that have been created by thermal agitation of the silicon lattice. Similarly, there are minority electrons in the p-type region. These will find the bias conditions such that the electric field (potential gradient) across the junction 'pulls' them across. Although this reverse current is normally very small, it is extremely important. If the reverse voltage is high, then the **minority carriers** are accelerated and 'avalanche breakdown' can occur, which is usually non-recoverable.

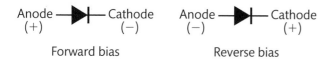

Forward bias   Reverse bias

▶ **Figure 19.8** PN-junction diode biasing

The type of diode that you are most likely to use has a standard PN-junction as described above. Such diodes can be grouped into small-signal types, used for low-current applications, and rectifier diodes, used for high-current and high-voltage applications.

The **light emitting diode** (**LED**) is another common type of diode. If you forward-bias an LED, current flows and light is produced. There are a variety of colours available. Developments in LED technology are fast finding their way into lighting and display applications, such as the backlighting of television screens. **Laser diodes** also produce light, but a major difference is that the light produced by laser diodes is coherent. They are used in DVD and CD players and laser pointers.

Photodiodes are used for detecting light. There are several types, and they are typically used in reverse bias. They are widely employed in communications applications.

Zener diodes are used in reverse bias to provide a stable reference voltage. They typically have a narrow depletion zone due to the doping concentrations of the p-type and n-type regions. They break down at a known voltage (called the Zener voltage). You need to limit the current with a series resistor to avoid permanent damage.

Investigate the types of diodes mentioned above and their typical applications.

Here are some additional diode types that you could investigate:
- Schottky diode
- tunnel diode
- varactor (varicap) diode.

Rectification

The ability of a diode to conduct current in one direction and block it in the other is useful in a number of applications. The most common is **rectification** – converting an alternating current (a.c.) signal into a direct current (d.c.) signal.

In the rest of this section we explore circuits used to develop a low-voltage d.c. regulated power supply. For each type of circuit:

▶ Use a computer-aided design (CAD) package to draw the schematic.

▶ Simulate the output using virtual meters and an oscilloscope.

▶ Connect the circuit using a prototype system and measure the signals.

▶ Compare the results you obtained from measurement and from simulation.

Half-wave rectifier

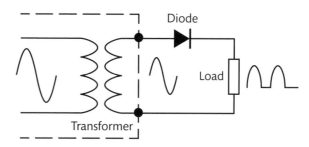

▶ **Figure 19.9** Half-wave rectifier

Figure 19.9 shows a half-wave rectifier circuit. The diode is forward biased on the positive half-cycle, so it conducts. The current through the load will be the same shape as the voltage supply producing it. The diode is reverse biased on the negative half-cycle, so no current will flow through the load on the negative half-cycle. The diode has to be chosen so that it can withstand the **peak inverse voltage** (**PIV**).

Full-wave rectifier

We consider two types of full-wave rectifier circuit.

▶ **Centre-tapped transformer**

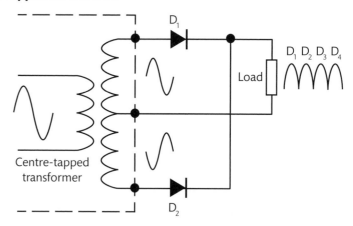

▶ **Figure 19.10** Full-wave centre-tapped transformer rectifier

Figure 19.10 shows one circuit for full-wave rectification, the **bi-phase rectifier**. The voltages in each half of the transformer secondary are in opposite phases to each other. The centre tap is taken as the reference point. Diode D_1 is forward biased on the positive half-cycle and diode D_2 is reverse biased. Current flows through the load (from top to bottom as drawn) via D_1. On the negative half-cycle D_1 is reverse biased and D_2 forward biased. Current flows through the load (from top to bottom as drawn) via D_2. The result is that current flows on both half-cycles through the diode that is forward biased. Current cannot flow in the other direction. The output stays at the same polarity all the time, but it is no longer alternating – it is pulsed d.c. The frequency is twice the supply frequency. Note that each diode needs to be able to withstand twice the PIV of the half-wave rectifier.

A multimeter connected across the load would not read the peak value of the waveform; it would read an average value over a cycle, which can be found from calculating the area under the curve and dividing by the time period. It can be shown that this average value is given by

$$V_{avg} = \frac{V_{peak}}{\pi}$$

▶ Full-wave bridge rectifier

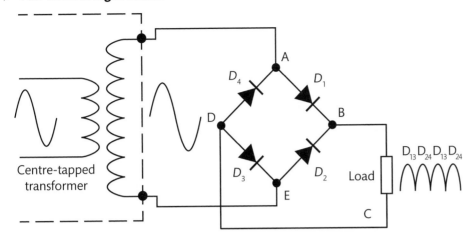

▶ **Figure 19.11** Full-wave bridge rectifier

The full-wave bridge rectifier (**Figure 19.11**) uses four diodes, D_1, D_2, D_3 and D_4, connected in such a way that opposite pairs conduct together. The current through the load is steered in the same direction (see **Figure 19.12**).

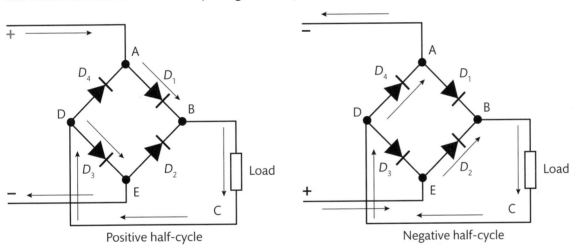

Positive half-cycle   Negative half-cycle

▶ **Figure 19.12** Current flow in full-wave bridge rectifier

Practical supplies often use bridge rectifier blocks, which have the four diodes built as a single circuit element.

Smoothing (filter)

The simplest way to smooth the pulse output of a rectifier is to use a reservoir capacitor (see **Figure 19.13**). This capacitor charges on the rising edge of the waveform and discharges on the trailing edge. The capacitor value has to be chosen so as to reduce the ripple voltage.

Zener diode regulator

You have already seen that Zener diodes are used in reverse bias. They are designed to give a fixed reverse breakdown voltage. For example, an NXP BZX79-C5V1, 113 Zener diode, 5.1 V 5% 0.5 W, has a Zener voltage of 5.1 V ± 5% and a

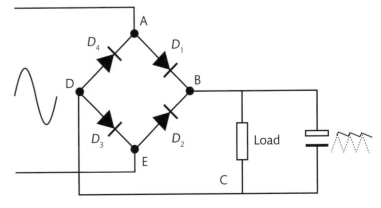

▶ **Figure 19.13** Full-wave bridge rectifier with capacitor smoothing

maximum power dissipation of 0.5 W. Note that the Zener diode must be able to carry all of the output current if the load is disconnected. The series resistor protects the Zener diode by limiting the current that can flow (see **Figure 19.14**).

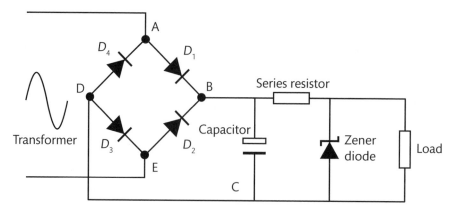

▶ **Figure 19.14** Full-wave bridge rectifier with capacitor smoothing and Zener diode regulation

Worked Example

Calculate the minimum preferred value of series resistor for a 5.1 V Zener diode regulated power supply if the input voltage to the regulator is 12 V and the maximum power rating of the Zener diode is 0.5 W.

Solution

The maximum current that can flow through the Zener diode is $I_{max} = \dfrac{P}{V_z} = \dfrac{0.5}{5.1} = 0.09804$ A or 98.04 mA.

The minimum value of the series resistor is $R_s = \dfrac{V_s - V_z}{I_{max}} = \dfrac{12 - 5.1}{98.04 \times 10^{-3}} = 70.4\ \Omega$.

Select the nearest preferred value of resistor: $R_s = 82\ \Omega$.

Ⅱ PAUSE POINT

1 Explain the difference between conductors, insulators and semiconductors.
2 What is doping and how does it affect the properties of a semiconductor material?
3 Identify different types of diode from their circuit diagram symbols. Explain how they work and what they are used for.

Hint
Use the internet to find out more details about the construction and operation of diodes. Use a CAD package to draw schematic circuit diagrams and simulate their operation. Build prototype circuits and practise taking measurements.

Extend
Compare the outcomes from the simulations and measurements you have taken. Do they match what you expect?

A3 Transistor devices and transistor-based circuits

There are two main classes of transistor: the **bipolar junction transistor (BJT)** and the **field-effect transistor (FET)**. The transistor principle was first reported in 1948 by Bardeen and Brattain. The term 'transistor' was coined as a contraction of 'transfer resistor', indicating that a large current is controlled by a smaller one. Further work by Shockley led to the development of the bipolar structure and the subsequent evolution of a range of electronic amplifying and switching elements. Shockley proposed the structure of field-effect transistors in 1948, but their development was restricted until the benefits of their use in integrated circuits were recognised.

Bipolar transistor

The bipolar junction transistor is a three-terminal device. Each terminal is connected to one doped region of either an NPN or a PNP 'sandwich'. The three regions are known as the **emitter**, the **collector** and the **base** (see **Figure 19.15**).

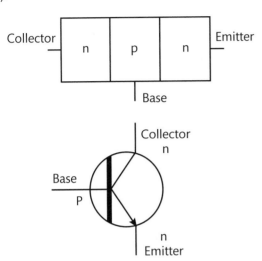

▶ **Figure 19.15** Diagram and circuit symbol for an NPN bipolar transistor

The arrow on the emitter in the circuit symbol indicates the direction of conventional current flow. You will find that NPN bipolar transistors are more widely used than PNP transistors. The PNP symbol differs only in the direction of the arrow, and circuits using PNP transistors will have opposite polarities. The diagram shows a symmetrical sandwich with identical doping of both emitter and collector regions, but this is very rarely the case in practice. Usually the emitter region is physically smaller than the collector region and has a higher doping concentration.

If you think of a system as having two connections for the input signal and two for the output signal, it follows that since a transistor has three terminals, one of them must be made common to both input and output signals. The three possible modes of connecting a transistor in a circuit are known as **common emitter**, **common base** and **common collector**. Their symbols are shown in **Figure 19.16**. You are more likely to use the common emitter configuration. The common collector circuit is also known as the emitter (or voltage) follower.

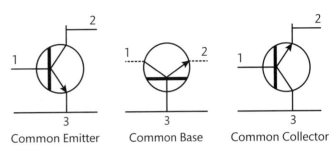

Common Emitter   Common Base   Common Collector

▶ **Figure 19.16** Three modes of bipolar junction transistor connection

The output characteristic of a transistor describes how the collector current (I_C) varies with the potential difference across the collector and emitter (V_{CE}) at a fixed value of the base current (I_B) or base–emitter voltage (V_{BE}). **Figure 19.17** shows the output characteristics for different values of the base current.

The characteristics show that the collector current is proportional to the base current for a good proportion of the working area. The ratio is known as the 'forward current transfer ratio in common emitter mode' or, more simply, the **current gain**. It can be written in a number of ways, such as

$$\frac{I_C}{I_B} = h_{fe} = \beta$$

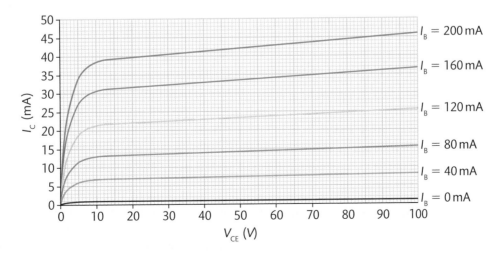

▶ **Figure 19.17** Typical bipolar junction transistor characteristics

Consider the circuit shown in **Figure 19.18**:

▸ Try simulating it and see what happens when the switch is in position A. Measure the voltage between pairs of terminals (base–emitter, base–collector, collector–emitter).

▸ What happens when the switch is in position B?

▸ Does the same happen if you build a prototype circuit?

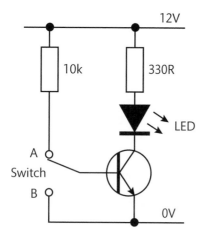

▸ **Figure 19.18** Bipolar junction transistor as a switch

Biasing a BJT

The transistor needs to be correctly biased to conduct current. From the practical activity with **Figure 19.18** you should have seen that when the switch is in position A, the collector is at a higher potential than the base, which is at a higher potential than the emitter. If you think about the junctions, then the base–collector junction is reverse biased and the base–emitter junction is forward biased. The base–emitter voltage is approximately 0.7 V. The collector–emitter voltage is very small. In this case the transistor is 'fully on'. It is said to be **saturated** because no more collector current can pass.

When the switch is in position B, the base and emitter are at the same potential and the transistor is fully off. It is said to be **cut-off**.

Transistors can be used as switches, but there is a limit to the current they can carry. Care must be taken when switching inductive loads, such as a relay, and a reverse-biased diode is usually used with a relay to prevent damage to the transistor caused by a **back e.m.f.** (electromotive force or voltage) (see **Figure 19.19**).

> **Key term**
>
> **Back e.m.f.** – a voltage in the reverse direction which results from a current through an inductor being switched.

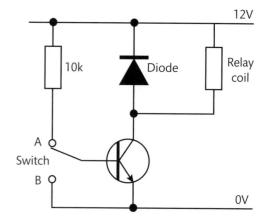

▸ **Figure 19.19** Bipolar junction transistor switching a relay

Common emitter amplifier

A common emitter amplifier, such as that shown in **Figure 19.20**, is an example of a **Class A amplifier** because the transistor amplifies both positive and negative parts of the input signal. The output is an accurate amplified copy of the input signal waveform. You need to select the component values so that the transistor is perfectly biased within its active region (**Figure 19.21**). This ensures that the output signal fits between the saturated and cut-off regions with no distortion.

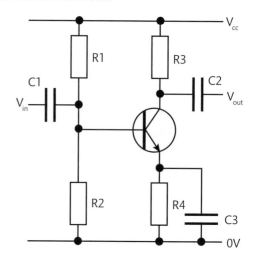

▸ **Figure 19.20** BJT common emitter amplifier

In the circuit shown in **Figure 19.20**, R1 and R2 form a potential divider circuit, which sets the bias voltage at the base. This ensures that the transistor remains conducting over the whole of the signal cycle (Class A operation). If the bias point is too low, then the output signal will enter the cut-off region and the bottom of the output signal will be clipped. Similarly, if you set the bias point too high, the transistor will saturate and the top of the output signal will be clipped. When designing a common emitter amplifier, a good starting level for the bias point is about half the supply voltage.

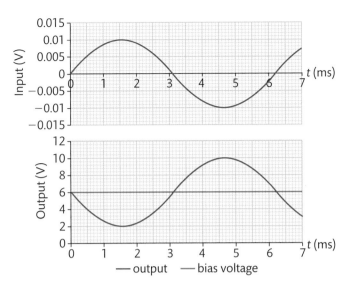

▶ **Figure 19.21** Common emitter amplifier input and output signals showing d.c. bias

The bipolar transistor is a current-controlled device. The collector load resistor R3 converts the varying collector current into a voltage. R4 is included to prevent thermal runaway. If R4 were not included, an increase in I_C would lead to a temperature rise in the transistor, which would reduce its input resistance and cause I_B to increase. This then increases I_C, and would rapidly lead to thermal breakdown of the device. With R4 included, any increase in I_C causes the potential at the emitter to rise, which decreases V_{BE}, which in turn reduces I_C, and hence V_{BE} is restored. Unfortunately, R4 prevents the very changes required in an amplifier. C3 is included to allow the a.c. signal to bypass R4 without affecting the d.c. bias conditions (see **Figure 19.22**).

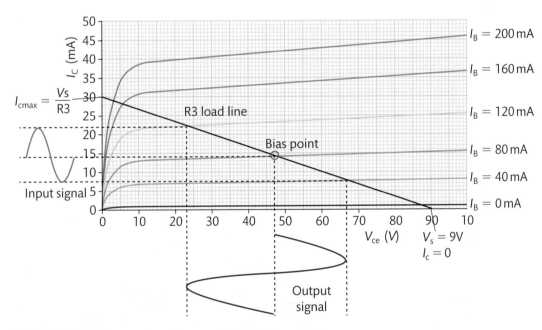

▶ **Figure 19.22** BJT common emitter amplifier load line analysis

One of the important measures of an amplifier is its **voltage gain** (A_v). You calculate voltage gain using the formula

$$\text{voltage gain} = A_v = \frac{\text{voltage output}}{\text{voltage input}}$$

Key term

Voltage gain – the ratio of voltage output to voltage input.

Voltage gain can also be expressed in decibels (dB):

$$\text{voltage gain (dB)} = 20 \log (A_v) = 20 \log \left(\frac{\text{voltage output}}{\text{voltage input}} \right)$$

Another important measure is the **bandwidth**. This is the range of frequencies for which the voltage output is above (approximately) 70% of its maximum value. These 70% thresholds are called **3 dB points**, because

$$20 \log (0.7) = -3 \text{ dB}$$

Consider the circuit shown in **Figure 19.23**.

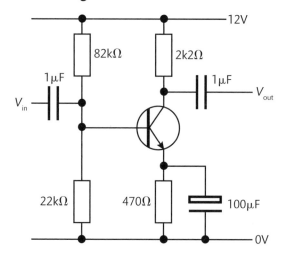

▶ **Figure 19.23** BJT common emitter amplifier with component values

▶ Try simulating the circuit and see what happens when you change the values of the components one at a time (see **Figure 19.24**).

▶ Does the amplifier work in the way you expected?

▶ Plan a layout for the circuit that works best on a prototyping system such as a breadboard.

▶ Does the real circuit behave in the same way as the simulation?

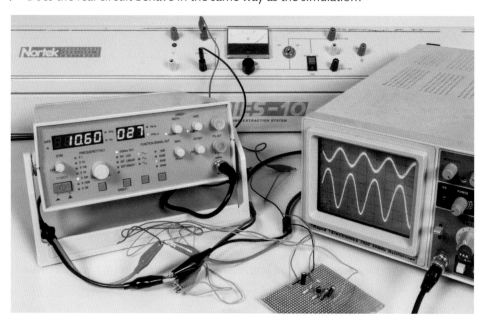

▶ **Figure 19.24** Testing a BJT common emitter amplifier

Field-effect transistors

There are two basic types of field-effect transistor (FET): the junction FET (JFET – see **Figure 19.25**) and the metal oxide semiconductor FET (MOSFET). Both types depend on the use of an applied electric field to control the current flowing through a doped channel (which could be n-type or p-type). A FET is usually simpler to fabricate, and takes up less space than a BJT. As a result, FETs are suitable for use in very large-scale integration (VLSI) devices such as memory and processors.

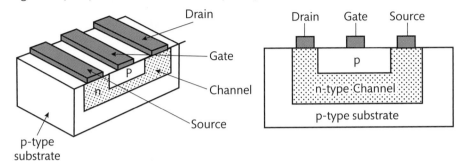

▶ **Figure 19.25** JFET construction

There are two types of MOSFET – a) depletion or b) enhancement (see **Figure 19.26**).

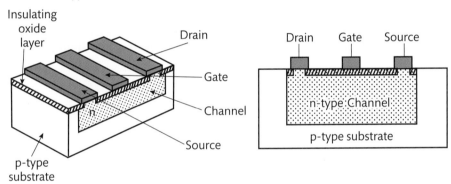

a) Depletion MOSFET

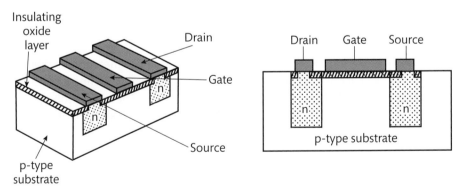

b) Enhancement MOSFET

▶ **Figure 19.26** MOSFET construction

Note the direction of the arrow on the JFET symbol in **Figure 19.27**:

▶ into the device indicates an **N-channel** device

▶ away from the device indicates a **P-channel** device.

You can analyse a FET amplifier in a similar way to the BJT by using its characteristics, its load line or small-signal models.

Figure 19.27 shows the circuit for a JFET common source amplifier.

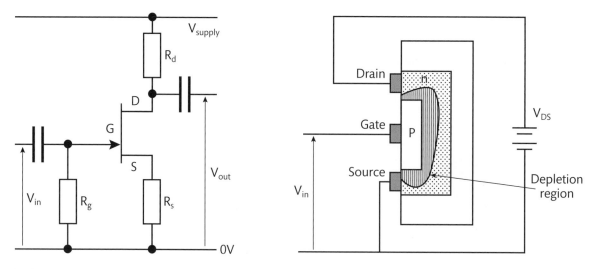

▶ **Figure 19.27** JFET common source amplifier and the effect of biasing on the channel width

A4 Operational amplifier circuits

Operational amplifiers (op amps) are the most widely used analogue integrated circuits. An op amp is a multi-stage integrated circuit containing approximately 20 transistors. The output voltage is proportional to the difference between the inputs. The two inputs are referred to as **inverting** and **non-inverting** (see **Figure 19.28**).

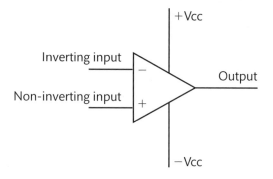

▶ **Figure 19.28** Basic operational amplifier schematic symbol

Real op amps do not have infinite gain at all frequencies. If you measure the open-loop gain of a typical operational amplifier such as a uA741, you will find it to be approximately 100 dB between 0 Hz (d.c.) and about 100 Hz. The output voltage gain then decreases linearly with frequency down to unity gain (1) at about 1 MHz, as shown in **Figure 19.29**.

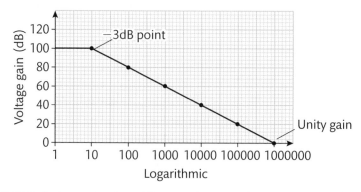

▶ **Figure 19.29** Typical operational amplifier open-loop gain–frequency characteristic

From the characteristic shown in the figure, at 10 kHz the gain is 40 dB, so

$$40 \text{ dB} = 20 \log(A_v)$$

$$\frac{40}{20} = 2 = \log(A_v)$$

$$A_v = 10^2 = 100$$

Therefore the gain–bandwidth product is $A_v B = 100 \times 10\,000 = 1\,000\,000$.

Applications of operational amplifiers

Comparator (or difference) amplifier

If the operational amplifier is used in an open-loop configuration (that is, with no feedback), the gain is very high (theoretically infinite). The linear portion of the characteristic shown in **Figure 19.30** is exaggerated to show that it is there, but it is extremely steep and you would find it very difficult to measure. (Imagine trying to balance a very long pole at its centre – a very slight movement will make it topple one way or the other.)

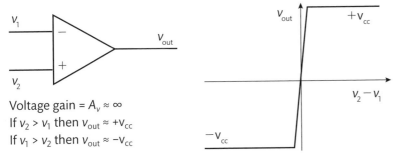

Voltage gain = $A_v \approx \infty$
If $v_2 > v_1$ then $v_{out} \approx +V_{cc}$
If $v_1 > v_2$ then $v_{out} \approx -V_{cc}$

▸ **Figure 19.30** Operational amplifier open-loop transfer function

▸ If you connect the inverting input more positive than the non-inverting input, then the output is approximately $-V_{supply}$.

▸ If you connect the non-inverting input more positive than the inverting input, then the output is approximately $+V_{supply}$.

You would expect the output to be zero when the input terminal voltages are equal. However, it is impossible to make absolutely identical input transistors, and this results in unequal bias currents. In practice you will find that op amps have two terminals to apply an input offset voltage to balance the amplifier. The power supply and input offset connections are often not shown on circuit diagrams for simplicity. You may need to refer to the manufacturer's data to find all the detail you need to use an op amp.

Inverting amplifier

The inverting amplifier (**Figure 19.31**) uses negative feedback to control the gain of the amplifier. For an ideal op amp the input resistance is infinite, which means there is no current flow between the two input terminals. If there is no current flow, then there is no potential difference between point J and ground, so J is said to be a 'virtual earth'.

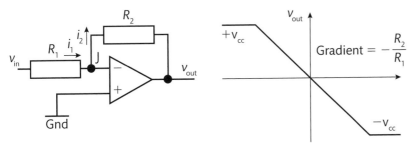

▸ **Figure 19.31** Inverting amplifier

The voltage gain for an inverting amplifier is

$$A_v = \frac{v_{out}}{v_{in}} = -\frac{R_2}{R_1}$$

The negative sign is present because this is an inverting amplifier.

Non-inverting amplifier

The non-inverting amplifier (**Figure 19.32**) also uses negative feedback to control the voltage gain. The input signal connects via the non-inverting input.

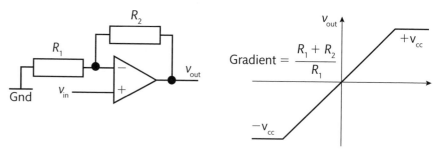

▶ **Figure 19.32** Non-inverting amplifier

In this case the voltage gain is given by

$$A_v = \frac{v_{out}}{v_{in}} = \frac{R_1 + R_2}{R_1} = 1 + \frac{R_2}{R_1}$$

Worked Example

Calculate the voltage gain of the amplifier circuits shown in **Figure 19.33**. What will be the amplitude of the output voltage of each circuit if the amplitude of the input signal is 50 mV?

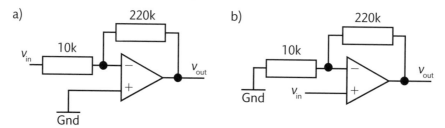

▶ **Figure 19.33** Two amplifier circuits

Solution

Circuit (a) is an inverting amplifier:

$$\text{voltage gain} = A_v = -\frac{R_2}{R_1} = -\frac{220 \times 10^3}{10 \times 10^3} = -22$$

$$v_{out} = A_v v_{in} = -22 \times (50 \times 10^{-3}) = -1.1 \text{ V}$$

Circuit (b) is a non-inverting amplifier:

$$\text{voltage gain} = A_v = \frac{R_1 + R_2}{R_1} = \frac{10 \times 10^3 + 220 \times 10^3}{10 \times 10^3} = 23$$

$$v_{out} = A_v v_{in} = 23 \times (50 \times 10^{-3}) = 1.15 \text{ V}$$

As a practical investigation:

▶ Try simulating inverting and non-inverting operational amplifier circuits. Change the values of the resistors and see if the equations for voltage gain still work.

▶ Build a prototype of each circuit and check that it behaves in the same way as the simulation.

Ⅱ   PAUSE POINT   Use the open-loop characteristic to estimate the open-loop bandwidth of the operational amplifier.

Hint   Bandwidth for the open-loop amplifier is from 0 Hz to the −3 dB point. Follow the example to calculate the voltage gain converted from the decibel value.

Extend   Estimate the bandwidth if the same operational amplifier is used to make a non-inverting amplifier with a gain of 20 dB.

Passive filters

A filter circuit allows frequencies in a certain range to pass while blocking other frequencies. There are several filter types, the most common being the **low-pass** filter, the **high-pass** filter and the **band-pass** filter. **Figure 19.34** shows the ideal behaviours of these filters.

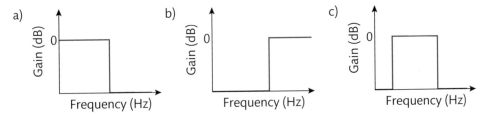

▶ **Figure 19.34** The ideal frequency response of a) a low-pass filter, b) a high-pass filter and c) a band-pass filter

A simple passive low-pass or high-pass filter can be built using a resistor and a capacitor in a potential divider arrangement. A band-pass filter is a combination of a high-pass and a low-pass filter. **Figure 19.35** shows circuit diagrams for these three types of filter.

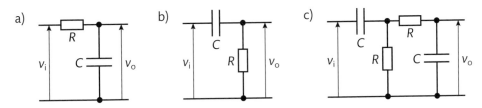

▶ **Figure 19.35** RC networks of a) low-pass, b) high-pass and c) band-pass filters

The output voltage of the filter is lower than the input voltage, so the gain factor is less than 1. This means that the circuits attenuate the signal:

$$\text{gain} = 20 \log\left(\frac{v_o}{v_i}\right) < 1$$

The frequency responses of real RC filters do not have the sharp cut-offs shown in **Figure 19.34**. This is because the reactance of the capacitor varies with frequency, causing the output voltage also to vary. Therefore the gain (attenuation) is dependent on the frequency of the input voltage. The reactive element (the capacitor) also causes a phase change between the input and output signals.

In a low-pass filter, because it is a potential divider network, the output voltage is

$$v_o = v_i \frac{X_C}{\sqrt{R^2 + X_C^2}}$$

where X_C is the capacitive reactance: $X_C = \dfrac{1}{2\pi f C}$.

▶ At low frequencies, the impedance of the capacitor is much greater than that of the resistor, so $v_o \approx v_i$

▶ At high frequencies, the impedance of the capacitor is much less than that of the resistor, so the output voltage is attenuated.

The cut-off frequency of the filter is defined as the frequency at which the output power has dropped to half the level of the input power. This happens when the impedances of the resistor and capacitor are equal, that is,

$$R = \frac{1}{2\pi f C}$$

Rearranging gives

$$f_C = \frac{1}{2\pi R C}$$

The phase angle at the cut-off frequency is given by $\Phi = -\tan^{-1}\left(\dfrac{X_C}{R}\right)$.

Because $X_C = R$, we have $\Phi = -45°$.

The combination of the plots of gain and phase angle against frequency is known as a **Bode plot**.

Figure 19.36 shows the Bode plot for a low-pass filter.

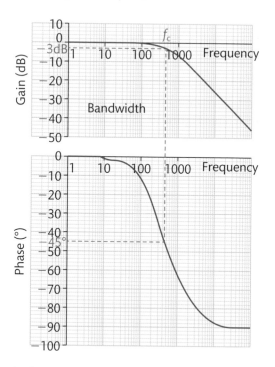

▶ **Figure 19.36** Bode plot for a low-pass filter showing the gain and the phase shift for different frequencies

Explain how a high-pass filter works.

Investigate how the voltage gain and phase angle of a high-pass RC filter vary with frequency.

Hint   Look back at the relationship between the capacitive and resistive impedances in a low-pass filter.

Active filters

An active filter uses an operational amplifier to amplify the attenuated signal from the RC filter.

Figure 19.37 shows the circuit of an active low-pass filter that uses an operational amplifier in its non-inverting configuration. The overall gain of the circuit is the product of the gain determined by the resistors R_1 and R_2 (in the way explained in the section on op-amp configurations) and the low-pass filter attenuation.

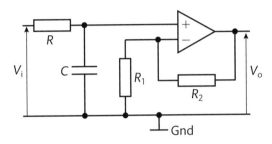

▶ **Figure 19.37** An active low-pass filter

Worked Example

Calculate component values for a low-pass filter with a cut-off frequency of 500 Hz and a low-frequency gain of 20.

Solution

For the non-inverting amplifier:

$$\text{gain} = 1 + \frac{R_2}{R_1} = 20$$

$$\frac{R_2}{R_1} = 19$$

Select a suitable value for R_1:

if $R_1 = 10\,k\Omega$ then $R_2 = 190\,k\Omega$.

This calculated value of R_2 is not one that is normally available, so use two resistors in series (such as 180 kΩ and 10 kΩ).

For the low-pass filter:

$$RC = \frac{1}{2\pi f_C} = \frac{1}{2\pi(500)}$$

Select a suitable value for R:

if $R = 10\,k\Omega$ then $C = \dfrac{1}{2\pi(500) \times 10 \times 10^3} = 31.8 \times 10^{-9}\,F = 31.8\,nF.$

This calculated value of C is not one that is normally available. The closest preferred value is 33 nF.

Explore a range of analogue devices and circuits by simulation and by building and testing circuits, and compare their theoretical and physical operation.

The circuits you investigate should include:

- a diode-based regulated full-wave rectification circuit
- a single-stage common emitter transistor amplifier circuit
- an inverting or non-inverting operational amplifier circuit.

In each investigation you should:

- simulate at least one diode, transistor and operational amplifier analogue circuit using a CAD software package
- build each type of analogue electronic circuit and take measurements to demonstrate the operational characteristics of the circuit
- evaluate the differences in operation between simulation and testing of the physical circuit for at least one diode, transistor and operational amplifier analogue circuit.

Plan

- What resources do I need for the task? How can I get access to them?
- How much time do I have to complete the task? How am I going to successfully plan my time and keep track of my progress?

Do

- I know what strategies I could employ for the task. I can determine whether these are right for the task.
- I can assess whether my approach is working and, if not, what changes I need to make.
- I can set milestones and evaluate my progress at these intervals.
- I can identify when I've gone wrong and adjust my thinking/approach to get myself back on course.

Review

- I can evaluate whether I met the task's criteria (i.e. succeeded).
- I can draw links between this learning and prior learning.
- I can explain what skills I employed and which new ones I've developed.

B Explore the safe operation and applications of digital logic devices and circuits that form the building blocks of commercial circuits

B1 Logic gates and Boolean algebra

> **Key terms**
>
> **Logic gate** – a circuit that carries out a Boolean logic function.
>
> **Boolean algebra** – a two-state logic algebra (named after George Boole).

Logic gates are the building blocks of digital circuits. The most common digital logic gates available are the TTL 74xxx series (made using BJTs) and the CMOS 40xxx series (made using MOSFETs). Both types have a large number of standard gates and functions available. Most of these are available in a number of configurations, including dual in-line package (DIP – see **Figure 19.38**) and surface-mount technology (SMT). Packages containing simple gates normally have 14 or 16 pins. More complex integrated circuit packages are likely to have more pins.

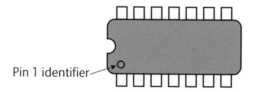
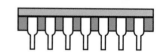

Pin 1 identifier

▶ **Figure 19.38** Logic gates in DIP configuration

The basic logic functions available as logic devices are **AND**, **OR** and **NOT** gates. The inputs and output of each gate use only two voltage levels or states: logic 1 and logic 0.

The behaviour of each type of gate is described by a Boolean statement and can be summarised in a **truth table**, which lists the output for each possible combination of inputs. The basic gates can be extended to include **NAND** (NOT AND), **NOR** (NOT OR), **XOR** (EXCLUSIVE OR) and **XNOR** (EXCLUSIVE NOR) to provide more functionality. **Table 19.1** shows the standard circuit symbols and truth tables for the different types of logic gate.

▶ **Table 19.1** Logic gates and their corresponding symbols and truth tables

Gate	Symbol	Boolean expression	Input A	Input B	Output Q
AND	ANSI & DIN	$A \cdot B$	0	0	0
			0	1	0
			1	0	0
			1	1	1
OR	ANSI >=1 DIN	$A + B$	0	0	0
			0	1	1
			1	0	1
			1	1	1
NOT	ANSI DIN	$\overline{A}$	0		1
			1		0
NAND	ANSI & DIN	$\overline{A \cdot B}$	0	0	1
			0	1	1
			1	0	1
			1	1	0
NOR	ANSI >=1 DIN	$\overline{A + B}$	0	0	1
			0	1	0
			1	0	0
			1	1	0
XOR	ANSI =1 DIN	$A \oplus B$	0	0	0
			0	1	1
			1	0	1
			1	1	0
XNOR	ANSI =1 DIN	$\overline{A \oplus B}$	0	0	1
			0	1	0
			1	0	0
			1	1	1

Logic gates can have more than one input, but they usually have only one output. For example, the 7400 has four 2-input NAND gates in a 14-pin DIP; each gate has one output. The 7410 has three 3-input NAND gates in the same-size DIP.

Comparing TTL and CMOS characteristics

The CMOS and TTL families have different electrical characteristics, summarised in **Table 19.2**, which means that the output of one type cannot usually drive (correctly operate) the input of the other. Although only two logic levels are allowed, in practice there is a range of acceptable voltages for logic 0 and logic 1. The logic levels for input and output are also slightly different. See **Figure 19.39**. The output specification is tighter than the input, to allow for signal degradation in transmission.

▶ **Table 19.2** Comparison of logic families

Property	74LS series	74HC series	74LHCT series	4000 series
Type	TTL low-power Schottky	High-speed CMOS	TTL-compatible high-speed CMOS	CMOS
Supply (+V_S)	5 V	2 V to 6 V	5 V	3 V to 15 V
Inputs	Pull up to 1 if unused	Very high impedance Connect unused inputs to 0 or 1 as appropriate	As 74HC but compatible with 74LS outputs	Very high impedance Connect unused inputs to 0 or 1 as appropriate
Outputs	Low current Use a transistor to switch higher currents	As 74LS but higher current, can source and sink 20 mA	As 74LS but higher current, can source and sink 20 mA	Very low current, can source and sink about 5 mA
Fan-out (per one output)	Can drive up to 10 74LS or 50 74HCT inputs	Can drive up to 50 CMOS, 74HC and 74HCT or 10 74LS inputs	Can drive up to 50 CMOS, 74HC and 74HCT or 10 74LS inputs	Can drive up to 50 CMOS, 74HC and 74HCT inputs or one 74LS input
Maximum frequency	35 MHz	25 MHz	25 MHz	1 MHz
Power usage	mW	μW	μW	μW

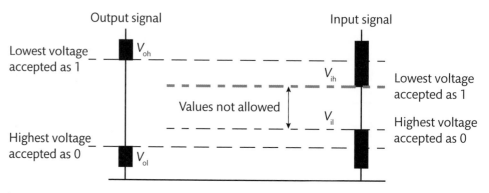

▶ **Figure 19.39** Logic levels

Key terms

TTL – transistor–transistor logic, based on BJTs.

CMOS – complementary metal oxide semiconductor logic, based on MOSFETs. **Combinational logic circuit** – a circuit made from several logic gates. For clarity, logic circuits do not always show the supply connections.

TTL was designed to be used with a regulated 5 V d.c. supply. Early CMOS could use a supply ranging from 3 V to 15 V. Developments in CMOS technology have resulted in an increase of switching speeds while maintaining lower power consumption. **Fan-out** is the number of similar inputs that can be driven by an output. The **propagation delay** is the time taken for a signal to pass from the input to the output.

B2 Combinational logic

The output of a **combinational logic circuit** is always the same for the same combination of inputs – it is a logical combination of the inputs. In a similar way to individual gates, the behaviour of the circuit can be summarised in a truth table.

Worked Example

A burglar alarm has a sensor on the door (A), a movement sensor in the lounge (B) and an alarm set switch (C). The alarm (Q) should sound if the alarm set switch is set and either the door sensor or the movement sensor (or both) is activated.

a) Write the truth table for the system.

b) Write a logic expression for when Q is logic 1.

c) Convert the Boolean equation into logic gates.

Solution

a) The truth table is shown in **Table 19.3**.

▶ **Table 19.3** Truth table for burglar alarm system

A	B	C	Q
0	0	0	0
0	0	1	0
0	1	0	0
0	1	1	1
1	0	0	0
1	0	1	1
1	1	0	0
1	1	1	1

b) The output is logic 1 if any of the following three conditions is true:

(NOT A AND B AND C), (A AND NOT B AND C), (A AND B AND C).

Therefore Q = (NOT A AND B AND C) OR (A AND NOT B AND C) OR (A AND B AND C).

c) Writing the expression from b) in Boolean algebra:

$$Q = (\overline{A} \cdot B \cdot C) + (A \cdot \overline{B} \cdot C) + (A \cdot B \cdot C)$$

Each of the bracketed terms is a 3-input AND gate. You also need:

- a 3-input OR gate to combine the three bracketed terms
- two NOT gates (inverters) to make the NOT A and NOT B terms.

Figure 19.40 shows the logic circuit diagram.

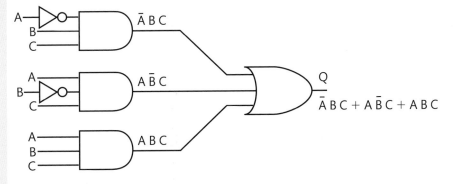

▶ **Figure 19.40** Logic circuit diagram for burglar alarm

You may have to find an alternative way to make the gate. For example, use a 3-input NOR followed by a NOT.

What would you do if you could not find the exact gates you want?

a) Investigate using two 2-input gates to make a 3-input gate.

b) It is difficult to find a TTL 3-input OR gate. Investigate making it another way by using two 2-input OR gates or a NOR gate with a NOT gate.

Hint

Use truth tables to compare the logic of the equivalent circuits.

As a practical activity, consider the burglar alarm in the worked example:

▸ Try drawing the schematic diagram and simulating it. You will have to find suitable gates in the libraries.

▸ Build a prototype and check that it works as expected.

Research

- Investigate the use of Karnaugh maps for four inputs.
- Practise finding groups and determining their Boolean expressions. Remember that if an input can be either logic 1 or logic 0, you can eliminate it from the AND (product) term.

Minimising a logic circuit

You can often reduce the number of gates used to construct a circuit by using a minimisation technique. A **Karnaugh map** is a way of writing the truth table in a slightly different way. Each box represents one line from the truth table. The grid is arranged so that each box has only one input that differs from the boxes immediately above, below, left and right. See **Figure 19.41**.

INPUTS			OUTPUT
A	B	C	Q
0	0	0	0
0	0	1	0
0	1	0	0
0	1	1	0
1	0	0	1
1	0	1	1
1	1	0	1
1	1	1	0

		AB			
		00	01	11	10
C	0			1	1
	1				1

▸ **Figure 19.41** Truth table and corresponding Karnaugh map for a 3-input logic circuit

To perform the minimisation, you circle groups that are next to each other vertically and horizontally. The size of a group can be any power of 2 (i.e. 1, 2, 4, 8 etc.). The aim is to make the groups as large as possible. In the example of **Figure 19.41** there are two overlapping groups, as shown in **Figure 19.42**.

		AB			
		00	01	11	10
C	0			1	1
	1				1

▸ **Figure 19.42** Grouping values in a Karnaugh map

▸ In the horizontal group: A is always logic 1; B can be logic 1 or logic 0, so eliminate it; C is logic 0. Therefore the Boolean expression for the horizontal group is $A \cdot \overline{C}$.

▸ In the vertical group: A is always logic 1; B is always logic 0; C can be logic 1 or logic 0, so eliminate it. Therefore the Boolean expression for the horizontal group is $A \cdot \overline{B}$.

Combining the groups, the overall Boolean expression minimises to

$$Q = A \cdot \overline{C} + A \cdot \overline{B}$$

De Morgan's laws

Sometimes you need to change the type of gate so that you use only one type. NAND or NOR are called universal gates because you can make all the other gates from them. De Morgan's laws are most simply stated as

$$\overline{A} \cdot \overline{B} = \overline{A + B}$$
$$\overline{A} + \overline{B} = \overline{A \cdot B}$$

In words, the steps are:

▸ invert the variables
▸ change the operator (AND to OR, OR to AND)
▸ invert the whole expression.

Research

Investigate the use of De Morgan's laws to make AND, OR, NOT, OR and XNOR gates with NAND gates only and with NOR gates only.

Why would you want to build a circuit using only one type of gate?

B3 Sequential logic circuits

The output of a **sequential** logic circuit depends on the combination of inputs and the current state of the outputs.

Bi-stable devices

There are several kinds of **bi-stable** circuits, also known as **flip-flops**.

The basic building block of sequential circuits is the **R-S** bi-stable circuit. This uses two NAND gates, with the output of one feeding back to the input of the other and vice versa (**Figure 19.43**). When R = logic 1, Q = 0; this is called RESET. When S = logic 1, Q = 1; this is called SET. When R = S = logic 0, Q stays at the value it was originally; this is called LATCH. The combination R = S = logic 1 is not allowed because it gives an undesired result. In addition, switching from R = S = logic 0 directly to R = S = logic 1 gives an indeterminate result depending on which gate switches faster.

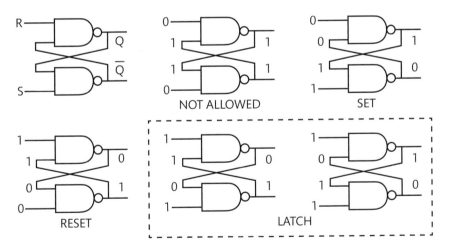

▸ **Figure 19.43** R-S flip-flop using NAND gates

The limitations of the R-S flip-flop can be overcome by using modified versions. The most commonly used modifications are the clocked **D-type** flip-flop (**Figure 19.44**) and the **J-K** flip-flop (**Figure 19.45**). D-type and J-K flip-flops usually trigger on either the rising or the trailing clock edge. You need to check the manufacturer's data to know which.

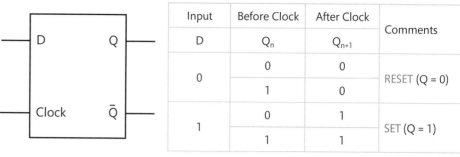

Transition Table

Input	Before Clock	After Clock	Comments
D	Q_n	Q_{n+1}	
0	0	0	RESET (Q = 0)
	1	0	
1	0	1	SET (Q = 1)
	1	1	

[Output latched at previous value if Clock = 0]

▶ **Figure 19.44** Clocked D-type flip-flop

Transition Table

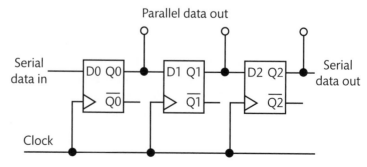

Input		Before Clock	After Clock	Comments
J	K	Q_n	Q_{n+1}	
0	0	0	0	LATCH
		1	1	
0	1	0	0	RESET (Q = 0)
		1	0	
1	0	0	1	SET (Q = 1)
		1	1	
1	1	0	1	TOGGLE (swaps over)
		1	0	

Positive edge

Inputs		Output
J	K	Q_{n+1}
0	0	Q_n
0	1	0
1	0	1
1	1	$\overline{Q_n}$

▶ **Figure 19.45** J-K flip-flop triggered on the rising clock edge

Research

Investigate the development of flip-flops from the basic R-S bi-stable circuit to practical applications of D-type and J-K flip-flops. Why do practical J-K flip-flops use a 'master–slave' arrangement?

Discover what the implications are of flip-flops being triggered on either the rising or the falling edge.

The D-type flip-flop is sometimes called a 'data-type' flip-flop because the data is clocked into the output. It may also be referred to as 'delay-type', because the movement of data from input to output is delayed until the next clock pulse.

Shift register

A **register** is an array of flip-flops used to store binary data. The most suitable flip-flop to use in a register is the D-type (see **Figure 19.46**). A J-K flip-flop can be configured to operate as a D-type by placing an inverter between the J and K inputs. A register in which the data bits can be moved to the left or right (or vice versa) is known as a **shift register**.

Parallel data out

Serial data in — D0 Q0 / Q̄0 — D1 Q1 / Q̄1 — D2 Q2 / Q̄2 — Serial data out

Clock

▶ **Figure 19.46** Three-stage shift register using D-type flip-flops

A logic 0 or logic 1 at the data input will be clocked through to Q0 on the first rising clock edge. On the next rising edge this signal is passed to Q1, and so on. At the same time, the new logic level at D0 is passed to Q0. Three bits of data can be passed in series by three clock pulses. These can be read in parallel. The shift register is converting serial data to parallel data – 'serial in parallel out' (SIPO). The data can be stored in the register by disabling the clock signal. It can also be clocked out in series by three further clock pulses – 'serial in serial out' (SISO).

Asynchronous (ripple) counter

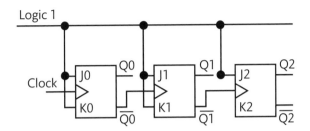

▶ **Figure 19.47** Three-stage asynchronous (ripple) counter using J-K flip-flops

The counter in **Figure 19.47** is termed **asynchronous** because the outputs Q0 to Q2 do not change simultaneously, but rather the value 'ripples' through – Q0 causes changes in Q1, which in turn causes changes in Q2. This effect causes a delay between the initial clock pulse and a settled result, known as the propagation delay. Connecting the J and K inputs together means that the output of a flip-flop toggles if held at logic 1, but does not change if held at logic 0. This is called a **T-type** (toggle) flip-flop.

Synchronous counter

The synchronous counter shown in **Figure 19.48** aims to overcome the propagation delay of a long asynchronous counter by clocking all the outputs simultaneously. You can use J-K flip-flops connected as T-types. Note how the stages connect together using the input and output of the previous stage.

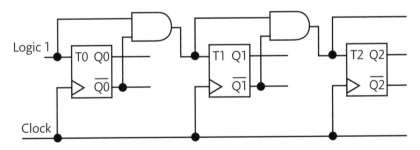

▶ **Figure 19.48** Three-stage synchronous counter (using T-type flip-flops)

⏸ PAUSE POINT   What methods do you need to use to test digital circuits?

Are there any differences in simulating, building and testing combinational and sequential circuits?

Hint   Look for a range of resources that explain the topics simply and clearly. Learn the circuit symbols for the basic gates and be able to use their truth or function tables.

Extend   Research how logic gates and circuits are used to carry out arithmetic and logic functions and how they can be used to control the flow of data.

Assessment practice 19.2

B.P4 | B.P5 | B.P6 | B.M4 | B.M5 | B.M6 | B.D2

Explore a range of digital devices and circuits by simulation and by building and testing circuits, and compare their theoretical and physical operation.

The circuits you investigate should include:

- a combinational logic circuit containing at least three inputs and five gates
- two sequential logic circuits, such as a counter and a shift register with at least three stages.

The tasks for each investigation are:

- to simulate at least one combinational logic circuit and two sequential logic circuits using a CAD software package
- to build at least one combinational logic circuit and two sequential logic circuits and take measurements to demonstrate the operational characteristics of the circuits
- to evaluate the differences in operation between simulation and testing of the physical circuit for at least one combinational logic circuit and two sequential logic circuits.

Plan

- What resources do I need for the task? How can I get access to them?
- How much time do I have to complete the task? How am I going to successfully plan my time and keep track of my progress?

Do

- I know what strategies I could employ for the task. I can determine whether these are right for the task.
- I can assess whether my approach is working and, if not, what changes I need to make.
- I can set milestones and evaluate my progress at these intervals.
- I can identify when I've gone wrong and adjust my thinking/approach to get myself back on course.

Review

- I can evaluate whether I met the task's criteria (i.e. succeeded).
- I can draw links between this learning and prior learning.
- I can explain what skills I employed and which new ones I've developed.

 Review the development of analogue and digital electronic circuits and reflect on own performance

Improving your own performance

Just doing something does not necessarily lead to learning. You need to be actively engaged and deliberately reflect on the activity to gain fully from the experience. While working through this unit you have encountered a lot of technical knowledge, developed practical and thinking skills, collected data and critically analysed the results. As a professional engineer in the future, you will need to draw upon the things you have learned to progress in your career. You need to develop the ability to continue learning as technology advances at a rapid pace. All professional engineers have to carry out continuous professional development (CPD) activities as part of maintaining their status. Reflective practice is a key tool in bringing together theory and practice and helps to place what you have learned into context. Reflection does not just mean looking back at the past; it also includes developing personal qualities and skills and assessing how they can help you reach higher levels in the future. It is a way for you to become an increasingly independent learner who knows what you want to achieve and can work purposefully towards it.

You will have become used to the concept of 'Plan–Do–Review' while carrying out assignments.

The **Plan** stage is where you identify such things as:

▶ what you need to do

▶ why you are doing it

▶ how you are going to do it

▶ what resources you need

▶ what evidence you will need to collect

▶ how you can keep records so that you can use them later.

In the **Do** stage you need to:

▶ follow your plan and meet deadlines

▶ prioritise tasks and not become distracted

▶ review progress and change your plan if necessary

▶ know when to ask for help or advice

▶ collect evidence as you generate it and keep records in an organised way, for example, by using a logbook

▶ review your work continually as you go along, and check any results that look as though they might be out of place.

In the **Review** stage you need to consider:

▶ what you did

▶ what you learned

▶ how you learned it

▶ how you felt about it

▶ whether or not it was a good way for you to learn

▶ what learning methods did not work for you

▶ what skills you developed

▶ what went well, what went less well and what you would do differently the next time

▶ whether you needed to change your plan, and why you made any changes

▶ whether your changes were effective

▶ what you would do differently in the future

▶ whether or not you met your targets.

Although it is important that you achieve your goals, the focus of reflection is more on the learning process, what you did and how you overcame any difficulties.

Assessment practice 19.3

Review and reflect on the practical activity to explore electronic devices and their use in analogue and digital logic circuits.

Prepare a 'lessons learned' report of around 500 words to explain how you applied health and safety, electronic and general engineering skills. Your report should also include an explanation of the importance and use of particular behaviours (such as communication skills).

The tasks are:

- to review and reflect on the activities that you have completed and make notes about what went well, what improvements you could make and what you would do differently next time
- to review and reflect on what you have learned from carrying out the activities in terms of knowledge, skills and techniques
- to produce a professional report.

Plan

- What am I being asked to do?
- What are the success criteria for this task?
- What am I learning? Why is this important?

Do

- I can make connections between what I'm reading/researching and the task I need to do, and I can identify the important information.
- I am recording my own observations and thoughts.
- I can identify when I've gone wrong and adjust my thinking/approach to get myself back on course.

Review

- I can draw links between this learning and prior learning.
- I can explain what skills I employed and which new ones I've developed.
- I can make informed choices based on reflection.
- I can use this experience in future tasks and learning activities to improve my planning and to monitor my progress.

Further reading and resources

Websites

Electronics Tutorials: **www.electronics-tutorials.ws**

An excellent set of short tutorials covering a wide range of devices and applications. Very clearly written, with excellent illustrations.

Everyday Practical Electronics: **www.epemag3.com**

An online magazine with useful articles and guides for the keen amateur electronics engineer.

Elektor: **www.elektor.com**

A magazine containing lots of project ideas and kits to develop skills and knowledge.

THINK ▶▶FUTURE

Simon Bentley Senior electrical engineer

I have been working as an electrical design engineer for 15 years. I have worked on a wide variety of projects, from multimillion-pound university buildings to the renovation of a Grade I listed building. Being an engineer is exciting because each day can be very different. My role involves the design of electrical services for buildings. These include power, lighting, data, fire alarm and lightning protection systems and many more. A great many devices these days can be controlled remotely, and it is important to keep up to date with new developments.

My work involves me working closely with professional engineers from other disciplines, so it is really important for me to have a working knowledge of what they do, even if I do not have their expert knowledge. Working in teams is an important skill that you need to develop as an engineer. Time management and personal organisation are also really important because contracts have to be completed on time, not only to avoid financial penalties but also to maintain the company's good reputation.

As an engineer I am always working to guidelines and regulations, so it is really important for me to link theory to practice. For example, all cables within a building have to be sized, light levels calculated and spacing of fire alarm equipment set out to guidance given within regulations. Completing my qualifications at college has really helped me understand what information I need to know and where to find it. Although I am well qualified, it is still very important to continue with professional development and keep up to date with changing technology. I regularly attend continuous professional development (CPD) events. It is important to keep my work profile as up to date as possible so that my company remains competitive.

Focusing your skills

Constructing and testing electronic circuits

It is important that you develop your knowledge and skills so that you can confidently:
- identify components from their circuit symbols
- identify real components and their values using colour codes etc.
- draw schematic diagrams using a suitable software package
- simulate analogue and digital circuits
- prepare components to connect them in prototype and permanent circuits – for example, strip and trim wires and leads, form component legs to fit

- position components in prototype circuits using breadboards and matrix boards such as Veroboard
- solder components neatly and accurately
- follow risk assessments and other guidance to work safely
- select and use appropriate measuring instruments accurately
- tabulate data and present it in suitable graphical forms such as waveform sketches, graphs, photographs
- compare measurements with expected values and investigate any anomalies
- keep accurate, organised records

Getting ready for assessment

Emily has recently completed a BTEC National in Engineering and is now studying for a Higher National Certificate (HNC). She attended college part-time as part of her apprenticeship. Emily shares her experience below.

How I got started

I started college as part of my apprenticeship. I tried to make sure that I was always on top of the work because I was worried that if I let it slip I would find it difficult to catch up.

How I brought it all together

▸ I took careful notes during the day at college.

▸ As soon as I got home I organised the notes into folders.

▸ I set aside time at home to organise my college work and to sort out anything I did not understand so I could ask about it next time. I did some college work at home and let my family and friends know that I was working at those times. Doing college work on a regular basis meant I did not have to spend a lot of time in one go. I got into the habit so it was not a problem.

▸ Practical assignments were all carried out in college. I made sure that my results looked OK and I knew what to do with them before leaving.

What I learned from the experience

I was worried when I first started the course that I might have difficulty keeping up. I thought that if I let the coursework slide I would find it difficult to catch up, because I needed to balance college, work and family life. As it turned out I did a lot better than I thought. I think being organised really helped. The college staff were really helpful.

Think about it

▸ Do you know what you have to do in each assignment?

▸ Is there anything that you need to find out before you start?

▸ Have you saved all your notes and results and stored them safely so that you can use them to do the assignment?

▸ Have you planned your time to make sure that you can meet the deadline?

▸ Have you checked that what you've written meets the assessment criteria? Aim for the highest grade.

Mechanical Behaviour of Metallic Materials 25

Getting to know your unit

Assessment
This unit will be assessed by a series of assignments set by your tutor.

Metals are a group of materials that underpin every aspect of engineering. This unit considers the mechanical characteristics of metals. It will help explain why metals are so versatile and how they can be tailored to suit particular applications by combining them in alloys and applying heat treatment processes. Of course, even the strongest materials have their limits. Understanding the mechanisms by which metals fail is important for all engineers if they are to use and apply metallic materials safely. In this unit you will measure properties such as hardness and tensile strength using testing equipment and investigate in-service failures.

How you will be assessed

This unit will be assessed by a series of internally assessed tasks set by your tutor. Throughout this unit you will find assessment practice activities to help you work towards your assessment. Completing these activities will mean that you have carried out useful research or preparation that will be relevant when it comes to your final assignment.

In order for you to complete the tasks in your assignments, it is important to check that you have met all of the Pass assessment criteria. You can do this as you work your way through each assignment.

If you are hoping to gain a Merit or Distinction grade, you should also make sure that you present the information in your assignment in the style that is required by the relevant assessment criterion. For example, Merit criteria require you to analyse and discuss, and Distinction criteria require you to assess and evaluate.

The assignments set by your tutor will consist of a number of tasks designed to meet the criteria in the table. They are likely to take the form of written reports, but may also include activities such as the following:

▶ Carrying out destructive test procedures on a range of metallic samples.
▶ Carrying out non-destructive test procedures on a range of metallic samples.
▶ Analysing case studies and assessing physical evidence to investigate in-service component failure.

Assessment criteria

This table shows what you must do in order to achieve a **Pass**, **Merit** or **Distinction** grade, and where you can find activities to help you.

Pass	Merit	Distinction

Learning aim Investigate the microstructure of metallic materials and the effects of processing on them and how these effects influence their mechanical properties

Pass	Merit	Distinction
A.P1 Explain how the microstructure of non-processed metallic materials affects the mechanical properties of the material. **Assessment practice 25.1**	**A.M1** Analyse, using an accredited data source, the microstructure of non-processed and processed metallic materials to correctly identify the material, including how the processing history affects the mechanical properties of the materials. **Assessment practice 25.1**	**A.D1** Evaluate, using an accredited data source, the microstructure of non-processed and processed metallic materials to identify the material, including how the processing history, impurities and grain boundaries affect the mechanical properties of the materials. **Assessment practice 25.1**
A.P2 Explain how the microstructure of processed metallic materials affects the mechanical properties of the material. **Assessment practice 25.1**		

Learning aim **B** Explore safely the mechanical properties of metallic materials and the impact on their in-service requirements

Pass	Merit	Distinction
B.P3 Conduct destructive tests safely on different non-processed and processed metallic samples. **Assessment practice 25.2**	**B.M2** Conduct destructive and non-destructive tests accurately on different non-processed and processed metallic samples. **Assessment practice 25.2**	**B.D2** Evaluate, using the results from safely conducted tests and an accredited data source, how the mechanical properties of processed and non-processed metallic materials affect their behaviour and suitability for different realistic applications, justifying the validity of the test methods used. **Assessment practice 25.2**
B.P4 Conduct non-destructive tests safely on at least two non-processed and processed metallic samples. **Assessment practice 25.2**		
B.P5 Explain, using the test results, how the mechanical properties of metallic materials affect their behaviour and suggest an application. **Assessment practice 25.2**	**B.M3** Analyse, using the test results and an accredited data source, how the mechanical properties of metallic materials affect their behaviour and suggest a realistic application. **Assessment practice 25.2**	

Learning aim Explore the in-service failure of metallic components and consider improvements to their design

Pass	Merit	Distinction
C.P6 Conduct a visual inspection check and at least one test safely on components that have failed in service. **Assessment practice 25.3**	**C.M4** Conduct a visual inspection check and at least one test safely and accurately on components that have failed in service. **Assessment practice 25.3**	**C.D3** Evaluate, using language that is technically correct and of a high standard, the results from safely conducted and accurate checks and tests to establish how components failed in service, recommending a design solution from a range of alternatives. **Assessment practice 25.3**
C.P7 Explain, using the results, how each component failed and how each component's design could be improved. **Assessment practice 25.3**	**C.M5** Analyse, using the results, how each component failed and justify how each component's design could be improved. **Assessment practice 25.3**	

Getting started

Work in a small group to describe all the properties that you associate with metallic materials. When you have come up with at least eight properties between you, go on to identify an engineered product or an engineering process where each of the properties you have identified is important.

 A Investigate the microstructures of metallic materials and the effects of processing on them and how these effects influence their mechanical properties

The physical properties of a metallic material are a consequence of the metallic bonds that hold the atoms together in a regular crystal structure. It is important for engineers to understand these structures and how they can be manipulated using alloying or heat treatment techniques.

A1 Types of ferrous metals and alloys

By definition, ferrous metals contain the metallic element iron (Fe). Pure iron is relatively soft and malleable and as such is unsuitable for many engineering applications. However, when alloyed with other elements, notably carbon, the characteristics of the material can be transformed.

Plain carbon steel

Steel is a general term used for a range of metal **alloys** containing iron and 0.08–1.2% carbon by weight. As the carbon content of the alloy increases, so does the strength and hardness of the steel. However, there is a corresponding reduction in toughness and ductility. **Table 25.1** lists the carbon content of plain carbon steels, as well as wrought iron and cast iron.

> **Key term**
>
> **Alloy** – a mixture of two or more metals which has enhanced mechanical properties or corrosion resistance compared with the pure metal used alone.

▶ **Table 25.1** Plain carbon steels, wrought iron and cast iron

Material	Carbon content
Wrought iron	Less than 0.08%
Low-carbon steel	Up to 0.15%
Mild steel	Between 0.15% and 0.35%
Medium-carbon steel	Between 0.35% and 0.6%
High-carbon steel	Between 0.6% and 1.2%
Cast iron	Between 1.4% and 4%

Alloy steels

An enormous range of additional alloying elements are utilised in small amounts to further enhance the mechanical properties of steel.

Structural steel

Grade S275JR is a typical hot-rolled steel used in welded, bolted and riveted structural applications.
It contains 0.21% carbon, 1.5% manganese, 0.55% copper, 0.04% sulphur and 0.04% phosphorous.

Tool steel

Grade O1 is a typical carbon–manganese hardenable tool steel used in a range of applications, including press tools, marking punches, taps and knife blades.
It contains 0.95% carbon, 1.25% manganese, 0.50% chromium, 0.50% tungsten and 0.20% vanadium.

Stainless steel

Grade 304 is a typical stainless steel in the UK.
It contains 0.08% carbon, 2.00% manganese, 18.00% chromium, 8.00% nickel, 0.045% phosphorous, 0.03% sulphur and 0.75% silicon.

Heat-resistant steel

Certain alloying metals such as molybdenum confer heat-resistant properties so that the steel maintains its key mechanical properties such as strength and creep resistance, even when operating at elevated temperatures.

> **Link**
>
> For more details about ductility, tensile strength and hardness, see Section A3. For more details about heat treatment, see Section A5. For more details about creep failure, see Section C2.

Cast iron

Grey

The most common form of cast iron contains between 2.5% and 4% carbon, the majority of which is present as graphite. This gives the characteristic dark grey appearance of a fracture surface. The presence of silicon and a gradual cooling process assist the formation of graphite.

Link

For more details about fracture surface, see Section B2.

White

In this form of cast iron, the carbon does not form areas of graphite but instead forms iron carbides such as Fe_3C (cementite). These increase the hardness of the cast iron at the expense of ductility. White cast iron is extremely brittle and therefore has limited practical applications.

Malleable cast iron

This is a form of white cast iron that is heat-treated after it has been cast in order to increase its ductility and fracture toughness.

Wrought iron

Wrought iron is a very pure form of iron which does contain carbon but in quantities less than 0.08%. This makes it relatively soft and malleable, and therefore suitable for use by blacksmiths, who were able to work it using a forge and basic hand tools. Wrought iron is the forerunner of modern mild steel. It was used in a wide range of applications such as railways, bridges and warships prior to the availability of steels. Wrought iron is no longer commercially produced in the UK, and any currently available form is more likely to be low-carbon mild steel.

Identification methods

Thousands of steel alloys have been developed over time to provide the exact characteristics required for a very broad range of applications. In order to define the composition and characteristics of these alloys, a number of national and international standards have been developed and are still in common usage.

International equivalents for the example grades of material mentioned earlier in this section are shown in **Table 25.2**.

▶ **Table 25.2** International equivalents of some example grades of steel

Case study

Material for automotive brake pipes

Few automotive systems are as important for ensuring the safety of the vehicle and its passengers as the braking system. Modern vehicles rely on a supply of pressurised hydraulic fluid reaching the braking mechanism at each wheel for the brakes to operate effectively. A system of pipes carries the fluid from the brake master cylinder under the bonnet to all four wheels.

During its production from 1959 to 2000, the classic Mini was fitted with steel brake pipes as standard. Invariably when these corroded and needed to be replaced, either copper pipes (typically almost pure copper with just 0.01% phosphorous) or cupronickel pipes (an alloy containing 10% nickel) were used by automotive technicians in local service centres.

▶ Replacement cupronickel brake pipes connecting into the master cylinder in the engine bay

Check your knowledge

1 What advantages do copper and cupronickel have over alternatives such as steel for replacement brake pipes?

2 How do manufacturers increase the longevity of mild steel pipes in a tough environment such as that underneath a car?

3 If you were an automotive manufacturer, which material would you choose for brake pipes fitted in the factory and why?

	Europe	UK	Germany	USA	USA
Standard	EN	BS	DIN	AISI/ASTM/ASME	UNS
Structural steel S275	S275 JR	43B	St44-2	1020	G10200
Tool steel O1	–	BO1	100MnCrW4	O1	T1501
Stainless steel 304	X6CrNi1810	304S15	X5CrNi18-10	304	S30400

Internet search engines provide a wealth of information on metallic materials, but don't forget that hard copies of engineering handbooks, catalogues and textbooks are also extremely useful.

What makes the materials you have investigated suitable for the applications that you have identified?

A2 Types of non-ferrous metals and alloys

Common non-ferrous pure metals

Table 25.3 lists the most commonly used non-ferrous metals. Information included here on equilibrium crystal structures will be covered in more detail in Section A4.

▶ **Table 25.3** Common non-ferrous pure metals

Non-ferrous metal	Chemical symbol	Equilibrium crystal structure at 20°C	Density (kg m^{-3})	Melting point (°C)
Aluminium	Al	FCC	2700	660
Copper	Cu	FCC	8930	1085
Gold	Au	FCC	19300	1065
Lead	Pb	FCC	11360	327
Magnesium	Mg	HCP	1740	650
Silver	Ag	FCC	10500	962
Tin	Sn	Tetrahedral	5765	232
Titanium	Ti	HCP	4500	1668
Zinc	Zn	HCP	7133	420

Non-ferrous alloys

As with pure iron, the applications of most pure non-ferrous metals are limited. Alloying is used to tailor the properties of non-ferrous metals to engineering applications.

Aluminium alloys

Aluminium in its pure form is soft, ductile and malleable with relatively low tensile strength. When alloyed with small quantities of copper, silicon, magnesium or manganese, its properties are greatly enhanced.

▶ Wrought aluminium alloys – These are commonly used to produce extrusions or rolled plates and sections. A common **wrought alloy** is:

  ▶ AW 3004 – contains 1.3% manganese and 1.1% magnesium; used in the production of drinks cans.

▶ Casting aluminium alloys – Due to its relatively low melting point, aluminium also lends itself to the production of castings. **Casting alloys** generally contain 4–13% silicon, which improves the flow when molten. A common casting alloy is:

  ▶ 319 – contains 5.5–6.5% silicon, 3–4% copper, 0.35% nickel, 0.25% titanium, 0.5% manganese, 1% iron, 0.1% magnesium and 1% zinc; used in the production of engine blocks.

Both wrought and casting alloys are available in forms which can be heat-treated to further enhance their mechanical strength and hardness.

Key terms

Wrought alloy – an alloy generally supplied in some processed form, such as billets, rods or bars, and which is suitable for machining or secondary forming processes.

Casting alloy – an alloy suitable for casting applications which generally exhibits good flow characteristics when molten.

Copper alloys

- Brass – Brasses are a group of copper alloys that use zinc as the principal alloying element, often with small amounts of other metals to enhance particular properties. Common brass alloys are:
  - general-purpose common or rivet brass – contains 32% zinc; suited to cold working applications.
  - nickel brass – contains 24.5% zinc and 5.5% nickel; used in the production of the £1 coin.
- Bronze – Bronzes are a group of copper alloys that use tin as the principal alloying element. Again, their properties can be further enhanced by the addition of small amounts of other metals. Common bronze alloys are:
  - phosphor bronze alloy C519 – contains 6% tin and 0.2% phosphorous; a wrought alloy suitable for the manufacture of springs, electrical contacts and fasteners.
  - bronze alloy C905 – contains 10% tin and 2% zinc; a common casting alloy used for pipe fittings, gears and pump components.

Magnesium

Magnesium is around 40% less dense than aluminium but exhibits similar strength and stiffness.

In magnesium alloys, aluminium, zinc, manganese and silicon are the principal alloying elements and yield improved strength, corrosion resistance and casting characteristics. Common magnesium alloys are:

- AZ91 – contains 9% aluminium, 0.7% zinc and 0.1% manganese; used as a general-purpose casting alloy.
- AZ31 – contains 3% aluminium, 1% zinc and 0.2% manganese; a wrought alloy particularly suited to extrusion.

Link

For more details about stiffness, see Section A3.

Titanium

Aluminium and vanadium are the principal alloying elements in titanium alloys.

A common titanium alloy is:

- Ti-6Al-4V – contains 6% aluminium, 4% vanadium, 0.25% iron and 0.2% oxygen; the most widely used of the titanium alloys, with applications in airframes, fasteners, biomedical implants and forged components.

PAUSE POINT Investigate the common uses of a range of non-ferrous alloys.

Hint Internet search engines provide a wealth of information on metallic materials, but don't forget that hard copies of engineering handbooks, catalogues and textbooks are also extremely useful.

Extend What makes the materials you have investigated suitable for the applications that you have identified?

Shape memory alloys (SMAs)

SMAs are a highly specialised group of alloys that exhibit unusual and surprising properties.

Super-elastic properties

These properties allow the material to be strained by up to 10% and still spontaneously return to its original shape when the stress is removed. At these strain rates conventional materials would undergo plastic deformation, and the majority of the deformation would be permanent. Super-elasticity is made possible by fully reversible movement in the material's crystal lattice (see Section A4).

Shape memory effects

Shape memory describes the ability of an apparently plastically deformed material to fully regain its original shape when heated above a certain temperature. This again relies on changes within the special crystal structure of the material that is formed in the course of complex thermomechanical treatment during manufacture.

Below a certain critical temperature the material exists in a structural form that is soft and can be easily deformed. When the material is heated above the critical temperature, a spontaneous crystalline phase change occurs. The structure of the material changes into a high-strength form, and the original shape of the material is re-formed. The material retains this shape as it is cooled to below the critical temperature. At this point the internal crystal structure spontaneously changes back to its soft and easily deformed condition. The cycle can then be repeated.

> **Link**
>
> Phases and phase changes are covered later in this Learning aim, in Section A4 under 'Metallurgical phase'.

▶ The most widely known alloy that exhibits both super-elastic and shape memory effects is nitinol, which contains 55% nickel and 45% titanium.

Applications

Typical applications of SMAs include spectacle frames, dental braces and medical implants such as bone fixings and arterial stents. They are biocompatible and so can be used in applications inside the body, and they have good mechanical properties. However, SMAs are very expensive to manufacture and have poor fatigue properties compared with conventional alloys. For example, a steel component can withstand around 100 times more fatigue loading cycles than one made from an SMA.

Identification methods

International equivalents for two of the grades of non-ferrous alloy material mentioned earlier are shown in **Table 25.4**.

▶ **Table 25.4** International equivalents of two example grades of non-ferrous alloys

	Europe	UK	Germany	USA	USA
Standard	EN	BS	DIN	AISI/ASTM/ASME	UNS
Aluminium alloy AW 3004	AW 3004	–	AlMn1Mg1	3004	A93004
Titanium alloy Ti-6Al-4V	Ti-6Al-4V	–	Ti6Al4V	4911	R56200

A3 Mechanical properties of metallic materials

The mechanical properties of materials describe their response to the application of force; see **Table 25.5** for a brief summary.

▶ **Table 25.5** Types of externally applied force

Type	Description	Illustration
Tensile	A pulling force – pulling apart, stretching	
Compressive	A squeezing force – pushing together, crushing	
Torsional	A twisting force	
Shear	A cutting or shearing force	

As the effects of a force depend on the area over which it is applied, we use **stress** when considering the mechanical properties of materials.

> **Link**
>
> An introduction to stress, strain and elastic constants was given in Unit 1, Section B2 ('Loaded components').

Elastic and plastic behaviour

When a stress is applied to a material sample, the material has a tendency to deform. The deformation is measured as a proportion of the original size of the material, known as **strain**.

▶ Elastic behaviour – If, when the stress is removed, the strain returns to zero and the material reverts to its original shape, then the material is behaving elastically.

▶ Plastic behaviour – If, when the stress is removed, the material does not return to its original shape and some degree of permanent strain remains, then the material is exhibiting plastic behaviour.

> **Key terms**
>
> **Stress** – the load distribution within a material expressed as force per unit area of material, stress $= \dfrac{\text{force}}{\text{area}}$, measured in pascals (Pa).
>
> **Strain** – the extension of a body expressed as a proportion of its original length.

Strength

Tensile strength

Ultimate tensile strength (UTS) is defined as the maximum stress that a material is able to withstand before failure. This is shown as point D on the stress–strain curve in **Figure 25.1**. For a ductile metal this does not usually represent the fracture point of the material; the fracture point occurs at point E after further rapid plastic deformation.

Ductile metals will undergo significant plastic deformation before fracture. Generally in engineering applications design loads are kept well within the region of elasticity.

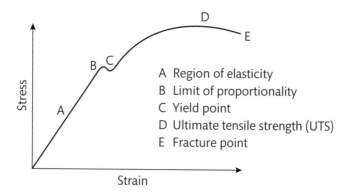

A Region of elasticity
B Limit of proportionality
C Yield point
D Ultimate tensile strength (UTS)
E Fracture point

▶ **Figure 25.1** Typical stress–strain curve of a ductile ferrous metal, showing a distinct yield point

Yield point

A useful indication of usable material strength is the yield point of the material. This point marks the transition from elastic to plastic behaviour, after which further deformation will be permanent up to the point of fracture.

In steels, the yield point tends to be a sudden, small but distinct relief in stress as the material reaches its elastic limit and permanent **slippage** occurs. This is shown as point C in **Figure 25.1**. **Strain hardening** soon restores the strength of the material and the stress level once again begins to rise.

Proof stress

For materials other than steel, a clear indication of the yield point is not obvious on the stress-strain curve. Instead, we can calculate an equivalent indicator, called the proof stress. This is the stress at which a permanent plastic strain of 0.2% occurs. It is found by locating the strain corresponding to 0.2% extension on the strain axis (point G in **Figure 25.2**), drawing a line from this point parallel to the straight (elastic) portion of the stress–strain curve, and then reading off the stress value at which this line intersects the curved portion of the stress–strain curve (point F in **Figure 25.2**).

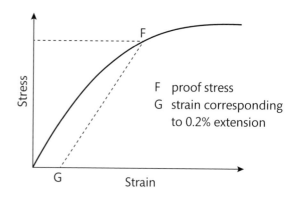

F proof stress
G strain corresponding to 0.2% extension

▶ **Figure 25.2** Stress–strain curve without a distinct yield point, for which 0.2% proof stress should be determined

Compressive strength

In metals, compressive strength is usually similar in magnitude to the stress in tension at the yield point.

Shear strength

The shear strength in metals, however, is generally different from the tensile strength. It must be considered in its own right when designing components that are loaded in multiple directions and which will therefore experience both shear and tensile stresses.

Hardness

Surface hardness is a measure of a material's ability to resist surface plastic deformation in the form of scratches, indentations or cutting. High hardness also provides good wear resistance.

Fracture toughness

Fracture toughness is a measure of a material's ability to resist fracture under **shock loading** at the sites of pre-existing defects, such as cracks or internal material defects (**voids** or **inclusions**).

> **Key terms**
>
> **Slippage** – permanent movement along slip planes within the crystal structure of the material (explained in more detail in Section A4).
>
> **Strain hardening** (work hardening) – a phenomenon where deformation of the crystal structure at high strain rates helps to block further movement or slippage, which makes the material harder.
>
> **Shock loading** – rapidly increasing dynamic loading such as that in aircraft landing gear.
>
> **Voids** – areas of empty space within a material, for example where grains are not fully in contact.
>
> **Inclusion** – an impurity in the material's microstructure.

Plasticity

Plasticity refers to the tendency of a material to undergo permanent deformation under load. This can be described in two ways:

▸ Ductility – the ability of a material to be stretched and permanently plastically deformed by tensile forces.

▸ Malleability – the ability of a material to be squeezed and permanently plastically deformed by compressive forces.

Elastic modulus

The elastic modulus (or modulus of elasticity) describes the stiffness of a material undergoing elastic deformation. Different values of elastic modulus are used in cases of direct loading and shear loading:

▸ In direct loading, the elastic modulus (also known as Young's modulus) is the gradient of the straight portion of the stress–strain curve, shown as the region of elasticity in **Figure 25.1**. The stiffness or rigidity of the material is defined as the strain produced per unit of applied stress.

▸ In shear loading, the modulus of rigidity or shear modulus is the ratio of shear stress to shear strain in the region of elasticity.

Specific stiffness

Specific stiffness indicates the resistance of a material to bending in relation to its density. It is calculated as the elastic modulus per unit density.

Fatigue limit

Fatigue is a process of material degradation caused by the effects of **cyclic loading**. Eventual cracking and fatigue failure can occur at stresses well below the UTS of the material that would apply in **static loading**.

The fatigue limit for a material is a measure of the maximum level of stress that can safely be applied to a material as a cyclic load.

> **Key terms**
>
> **Cyclic loading** – constantly varying dynamic loading such as that in a shaft running out of balance.
>
> **Static loading** – constant loading such as that in the structure of an electricity pylon.
>
> **Lattice structure** – the regular structure of atoms that make up a crystalline material.

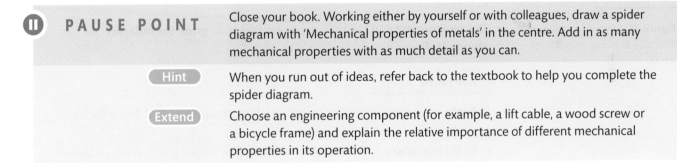

⏸ PAUSE POINT Close your book. Working either by yourself or with colleagues, draw a spider diagram with 'Mechanical properties of metals' in the centre. Add in as many mechanical properties with as much detail as you can.

> **Hint** When you run out of ideas, refer back to the textbook to help you complete the spider diagram.
>
> **Extend** Choose an engineering component (for example, a lift cable, a wood screw or a bicycle frame) and explain the relative importance of different mechanical properties in its operation.

A4 Grain structure of metallic materials

Crystal lattice structures

As the atoms of a pure molten metal cool, they start to form crystalline **lattice structures**. By far the most common of these are body-centred cubic (BCC), face-centred cubic (FCC) and hexagonal close-packed (HCP), shown in **Table 25.6** and **Figure 25.3**. Some metals only ever exist in one of these forms, but other metals can exist in more than one form depending on the prevailing conditions of temperature and pressure. Notably, iron typically exists as BCC at room temperature but spontaneously transforms to FCC when the temperature reaches a certain level. This type of behaviour will be considered in more detail later in the unit.

▸ **Table 25.6** :The three most common crystal lattice structures

Lattice structure	Abbreviation	Occurs in
Body-centred cubic	BCC	Fe, Ti
Face-centred cubic	FCC	Al, Cu, Au, Pb, Ag, Fe
Hexagonal close-packed	HCP	Mg, Zn, Ti

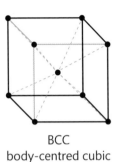 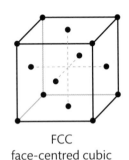 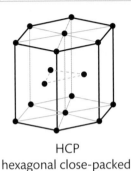

| BCC | FCC | HCP |
| body-centred cubic | face-centred cubic | hexagonal close-packed |

▸ **Figure 25.3** Body-centred cubic (BCC), face-centred cubic (FCC) and hexagonal close-packed (HCP) crystal lattice structures

Grain structure

As a molten metal cools and begins to solidify, individual atoms begin to arrange themselves into the crystal lattice structure that characterises their solid form. They form small groups of atoms that act as seed crystals onto which other atoms are able to attach themselves and thus form larger structures. In this way a crystal lattice structure grows.

Numerous seed crystals tend to form throughout the material, with each one growing until it starts to encounter its neighbours and is prevented from growing further. This leads to an overall structure that does not exist as a continuous and uniform crystal lattice but rather as collection of randomly oriented, closely packed **grains** of varying sizes.

The areas around and between grains are known as **grain boundaries** (see **Figure 25.4**).

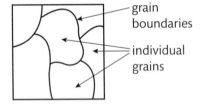

grain boundaries

individual grains

▸ **Figure 25.4** Typical metallic grain structure at a magnification of ×1000

> **Key terms**
>
> **Grain** – a defined area of uniform crystal lattice structure that forms during cooling.
>
> **Grain boundary** – the area which separates individual grains in a crystalline material.

Crystal lattice defects

Within grains the regular lattice structure rarely forms perfectly and tends to contain numerous defects.

Point defects

These are imperfections in the regular crystal lattice caused by an irregularity at one point in the crystal that has an effect on the structure immediately around it. Some common point defects are shown in **Figure 25.5**.

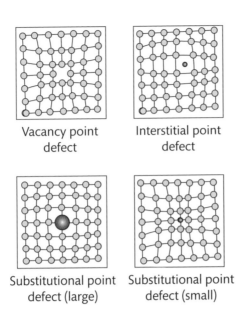

Vacancy point
defect

Interstitial point
defect

Substitutional point
defect (large)

Substitutional point
defect (small)

▶ **Figure 25.5** Point defects in the regular crystal lattice structure

> **Key terms**
>
> **Vacancy point defect** – where an atom is missing from the regular structure.
>
> **Interstitial point defect** – an atom not part of the regular lattice, often an impurity, is present in the usually unoccupied small spaces between atoms in the lattice structure.
>
> **Substitutional point defect** – an atom within the regular lattice structure is replaced by another type of atom, often an impurity, that is significantly larger or smaller than those surrounding it.

Dislocations

Line defects are imperfections that affect a whole row of atoms. A dislocation is a type of line defect where the regular crystal lattice structure has not formed correctly during cooling or has been altered by plastic deformation of the material. Dislocations take two general forms, as illustrated in **Figure 25.6**: screw dislocations and edge dislocations.

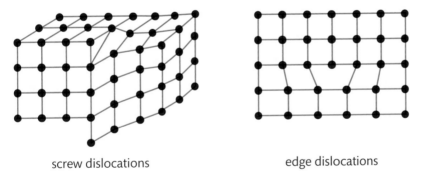

screw dislocations   edge dislocations

▶ **Figure 25.6** Two types of dislocation in the regular crystal lattice structure

When shear stresses are applied to a material, it is the movement of dislocations along planes out to the edges of the grain that is responsible for the majority of plastic deformation. This is a relatively low-energy process of breaking and re-forming just the bonds around the dislocations as they move through the material. If the movement of dislocations can be blocked or disrupted, then the material exhibits increased strength and hardness.

Planar defects

Planar defects are imperfections that involve a 2D surface of atoms in the crystal structure. The most common planar defects are the grain boundaries themselves, which are regions of imperfect crystalline arrangement. They occur where regular crystal grain structures that are formed in different orientations meet during cooling.

Slip planes

Elastic deformation

A material undergoing elastic deformation caused by an applied stress will return to its original shape once the stress is removed. This means that the bonds between the atoms in the crystal lattice are strained to accommodate the deformation but remain intact (see **Figures 25.7** and **25.8**). When the stress is removed, this relieves the strain and the lattice returns to its original shape.

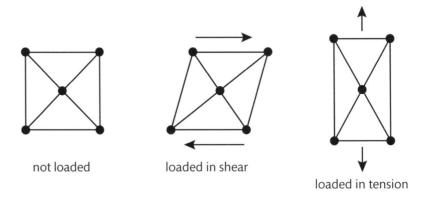

<table>
<tr><td>not loaded</td><td>loaded in shear</td></tr>
</table>

loaded in tension

▶ **Figure 25.7** Schematic examples of elastic deformation in a simple crystal lattice

Plastic deformation

Plastic deformation occurs when the applied stress not only stretches the bonds between atoms in the lattice but actually causes atoms to slide over each other and form new bonds in a different and permanently changed arrangement (see **Figure 25.8**). It is a different process from the movement of dislocations, which can cause plastic deformation with relatively low loads. This sliding along entire structural planes requires considerably greater energy as more bonds are broken and re-formed at the same time.

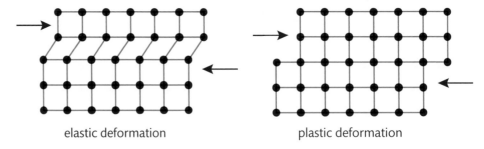

elastic deformation   plastic deformation

▶ **Figure 25.8** Elastic versus plastic deformation

<table>
<tr><td>Key term</td></tr>
<tr><td>Slip plane – a plane in a lattice structure where it is possible for slippage to occur.</td></tr>
</table>

In reality the sliding that occurs in plastic deformation, correctly referred to as slippage, acts in three dimensions along specific **slip planes** and in particular directions where the lattice structure is weakest. Preferred slip planes and slip directions differ between crystal structures, as they depend on the arrangement of atoms. Plastic deformation tends to occur by the sliding of closely packed planes in the directions of closest packing.

The three common lattice structures all have different slippage characteristics, shown in **Figure 25.9**.

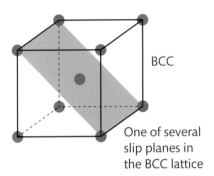
One of several slip planes in the BCC lattice

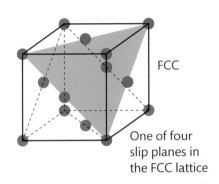
One of four slip planes in the FCC lattice

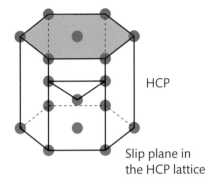
Slip plane in the HCP lattice

▶ **Figure 25.9** Slip planes in BCC, FCC and HCP lattice structures

▶ The BCC **unit cell** has no close-packed planes. The slip planes are those with the greatest spacing between them, but it is difficult for these to slide over one another. Metals with the BCC structure (e.g. iron, tungsten, chromium) therefore tend to exhibit high strength with moderate ductility.

▶ The FCC unit cell contains four close-packed planes. Any one of these can form a slip plane in three possible directions. Metals with the FCC structure (e.g. copper, lead) therefore tend to exhibit relatively low strength and high ductility.

▶ The HCP unit cell contains just one close-packed plane. This can form a slip plane in three possible directions. Metals with the HCP structure (e.g. zinc) therefore tend to have low ductility.

So far we have considered deformation only within a single regular crystal lattice, but most metals encountered in engineering are made up of multiple grains (known as polycrystalline structures). Within each grain the deformation processes of dislocation movement and slippage will take place. However, as each grain forms in a random orientation during solidification, a directional force applied to the material will affect each of the grains to a differing degree.

Any slippage or movement of dislocations will not easily cross grain boundaries. This is the main reason why materials consisting of small grains exhibit greater strength, hardness and impact resistance than materials consisting of fewer larger grains.

Surface slip bands

Cyclic loading tends to cause the formation of **slip bands** within grains at the surface of a material. Over time, **intrusions** and **extrusions** form on the surface, as illustrated in **Figure 25.10**. The intrusions can become sufficiently deep to form significant defects or micro-cracks in the material surface. These can act as **localised stress raisers** and initiate fatigue cracking. This can then propagate (spread) through the material, causing component failure at loads significantly below the UTS of the material. Fatigue is discussed in greater detail later, in Section C3 of this unit.

> **Key terms**
>
> **Unit cell** – the smallest arrangement of atoms that we can use to define the general structure of a crystalline material.
>
> **Slip band** – a section of displaced lattice structure between active slip planes. Slip bands can be observed as 'steps' in the crystal structure.
>
> **Intrusion** – a trough formed in the surface of a material.
>
> **Extrusion** – a peak pushed out from the surface of a material.
>
> **Localised** – limited to a small area.
>
> **Stress raiser** – any feature, such as a sharp corner or void, that will cause a localised increase in stress.

PAUSE POINT

Use marbles, ping-pong balls or any spherical objects that you have available to create models of the FCC, BCC and HCP crystal structures.

Hint   Use a hot glue gun to hold together your models and build them as large as you like.

Extend   Identify the close-packed planes that define the slip planes and slip directions within each structure.

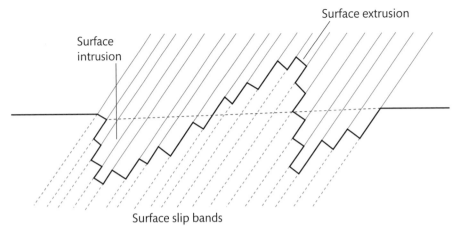

Surface intrusion

Surface extrusion

Surface slip bands

▶ **Figure 25.10** Cyclic loading causes movement in surface slip planes, leading to the formation of surface intrusions and extrusions.

Metallurgical phase

You are probably familiar with using the term 'phase change' when referring to a change of state of a material, for example, from solid to liquid or from liquid to gas.

In **metallurgy**, phases refer to a wide range of physical states related to the structure adopted by the crystal lattice at different temperatures.

A liquid metal or alloy is a mix of atoms without any defined structure. In a pure metal there will be a single type of atom, and in an alloy there could be several types of metal atoms (and other substances such as carbon in the case of steel).

When the liquid cools, solidification takes place as atoms begin to form regular crystal lattice structures. Depending on the structures formed and the **solubility** of different atoms within them, a number of different phases, or structural types, might be present.

> **Key terms**
>
> **Metallurgy** – the study of metals, including their structure, properties and uses.
>
> **Solubility** – the ability of one substance to remain dissolved in another.
>
> **Solid solution** – a metal alloy in which the main metal has alloying elements dissolved within its structure.

Although now solid, the mix of metal atoms, known as a **solid solution**, can still change. As the crystal lattice structure cools, it has a natural tendency to adopt a stable equilibrium structure or phase. At different temperatures the shape of this equilibrium structure can change. This means that a phase which is stable at high temperatures might transform to a different structural phase, with different characteristics, at a lower temperature. In alloys, a single phase which is stable at the higher temperature can transform into a mix of two or more phases as it cools.

This behaviour seems complex at first and is best understood by studying equilibrium phase diagrams, which will be considered for iron–carbon (steel) alloys and aluminium–copper alloys a little later.

Alloys

As introduced in Sections A1 and A2, an alloy is a mixture of metals (and sometimes other elements, notably carbon in steels). Often, the mechanical properties and other important characteristics, such as corrosion resistance, of a pure metal can be substantially enhanced with the addition of relatively small quantities of one or more other metals.

Eutectic alloy

Eutectics are alloys which are made up from the specific ratio of pure metal and alloying elements that gives the lowest possible melting point for the alloy. Moreover, the melting point of the eutectic alloy is uniform (the same) throughout the material. In alloys with compositions either side of the eutectic, complete melting occurs over a range of temperatures.

Interstitial solid solution

In a similar way to an interstitial point defect, the alloying element in an interstitial solid solution occupies a space between atoms in an otherwise regular lattice structure (see **Figure 25.5**).

Substitutional solid solution

In a similar way to a substitutional point defect, the alloying element in a substitutional solid solution takes the place of an atom within the regular lattice structure. These substituted atoms may be of a similar size or significantly larger or smaller than the atoms surrounding them.

Intermetallic compounds

Intermetallic **compounds** exist in metal alloys where the bonding attraction between the atoms of different metals leads to the formation of chemical compounds with a different structure from that of the pure metal crystal lattice. The strong interatomic bonding present in these compounds means that they are generally hard and brittle. If concentrated at grain boundaries or present in large quantities, the intermetallic compounds will make the alloy as a whole brittle and less ductile. However, if they are distributed evenly throughout the material as small particles, they can increase the alloy's strength and hardness.

Iron-carbon thermal equilibrium diagram

The iron-carbon phase diagram is probably the most important **equilibrium phase** diagram for engineers, as it describes the phases that form at different temperatures of this widely used and versatile alloy of iron and carbon. The part of the diagram of greatest relevance to engineers is the region describing steel alloys (up to approximately 2% carbon) – see **Figure 25.11**.

> **Key terms**
>
> Chemical **compound** – a material containing a fixed ratio of atoms strongly bonded together.
>
> **Equilibrium phases** – in metallurgy, these describe the most energy-efficient and stable crystalline structures in which atoms tend to arrange themselves at a given temperature.

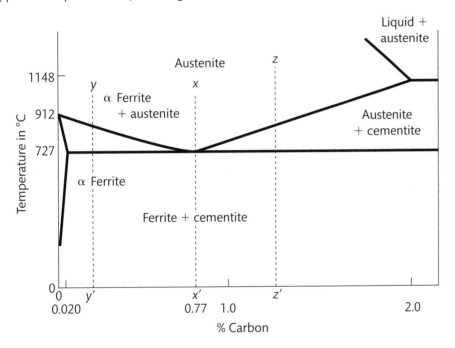

▶ **Figure 25.11** Simplified iron–carbon equilibrium phase diagram for steel alloys

The mechanical properties of steel are greatly influenced by the carbon content of the alloy. In the liquid phase, carbon is completely soluble, but it is only partially soluble in the solid crystal lattice structures that form as the alloy cools.

In the area of the iron–carbon phase diagram that refers to steel (**Figure 25.11**), the following phases can be observed.

Austenite

As pure iron cools from 1394°C to 912°C, it exists in a single phase called austenite, which has an FCC crystal lattice structure. Austenite exhibits the high ductility typical of the FCC structure and is the phase in which hot forming is carried out. Most heat treatment processes begin with the steel being heated sufficiently to transform the iron into the austenite phase.

Austenite is able to dissolve up to 2% carbon by weight.

Ferrite

Below 912°C iron transforms into a BCC crystal structure called ferrite. Only 0.02% carbon can be kept in solution with the ferrite, so if more than this was dissolved in the austenite phase, it will be forced out of solution as the ferrite forms.

Cementite

Carbon forced out of solution in the transition from austenite to ferrite forms an intermetallic compound called cementite (Fe_3C). Typical of intermetallic compounds, cementite is both hard and brittle. It can confer useful properties on the alloy if dispersed evenly but can prove problematic if present in large or localised concentrations.

Pearlite

Ferrite and cementite frequently occur together and can be observed under the microscope as alternating layers or plates. The combination of these two phases resembles mother-of-pearl when magnified, and so became known as pearlite.

Important phase transformations of steel alloys

The key transformation occurs when single-phase austenite cools and transforms into a mixture of the two phases ferrite and cementite. The nature of this transition and the properties of the materials formed will vary according to the composition of the alloy:

▸ Eutectoid (0.77% carbon) – line x to x′ in **Figure 25.11**. At temperatures above 727°C, the alloy consists of a single phase, austenite, in which all of the 0.77% carbon is dissolved in solid solution. As the material is cooled to 727°C, the transformation of austenite into ferrite and into cementite occurs simultaneously. This gives rise to a structure consisting of fine grains of ferrite and cementite (pearlite) mixed evenly throughout the material.

▸ Hypo-eutectoid (less than 0.77% carbon) – line y to y′ in **Figure 25.11**. At high temperatures, the single phase present is austenite, in which all of the carbon is dissolved in solid solution. However, as the material cools, it enters a region where austenite and ferrite are both present. As the ferrite grains begin to grow and the remaining austenite becomes increasingly rich in carbon, the eutectoid composition of 0.77% carbon will eventually be reached, and this will transform into pearlite at 727°C. The resulting grain structure consists of the ferrite that formed first and regions of pearlite.

▸ Hyper-eutectoid (greater than 0.77% carbon) – line z to z′ in **Figure 25.11**. At high temperatures, the single phase present is austenite in which all of the carbon is dissolved in solid solution. This time, as the alloy cools, it passes through a region where austenite and cementite are both present. As the cementite forms, the levels of carbon

dissolved in the austenite decrease until the eutectoid composition is reached, which will transform into pearlite at 727°C. The resulting grain structure consists of primary cementite, which formed first, and regions of pearlite.

Everything discussed so far is true only during gradual cooling, where atoms are given time to arrange themselves into their most energy-efficient or equilibrium forms at a given temperature.

Heat treatment of steels (to be discussed later) is based on disrupting the natural phase transitions that would occur in gradual cooling.

Aluminium–copper thermal equilibrium diagram

The aluminium–copper thermal equilibrium diagram (**Figure 25.12**) refers to solutions of copper in aluminium.

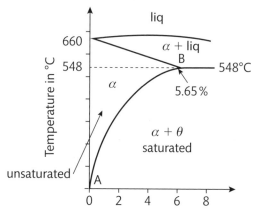

▸ **Figure 25.12** Simplified aluminium–copper equilibrium phase diagram

Solid aluminium exists in a single phase as an FCC crystal lattice. This is referred to as the α (alpha) phase. In an aluminium alloy containing between 0.2% and 5.7% copper by weight, the solubility of copper in solid solution with aluminium increases with temperature, as can be seen from the sloping phase boundary A to B in **Figure 25.12**. Anywhere to the left of this phase boundary is unsaturated, which means that all of the copper present is dissolved in the aluminium α phase.

Anywhere to the right of the phase boundary is saturated, which means that the aluminium α phase is unable to dissolve all of the copper present. The copper must therefore adopt a different form. Here that form is the intermetallic compound $CuAl_2$, which is referred to as the θ (theta) phase.

Consider an aluminium alloy containing 4% copper. At 500°C all the copper will be dissolved in the aluminium α phase. When the alloy is cooled, the saturation point is reduced by the decrease in temperature. If the alloy is cooled rapidly by quenching in water, there is no opportunity for excess copper to precipitate from (i.e. be forced out of) the solid solution, so the alloy becomes a non-equilibrium super-saturated solid solution with more copper dissolved in the aluminium phase than it would have in its natural equilibrium state.

This is the first stage in the heat treatment of aluminium–copper alloys and is known as solution treatment, which will be discussed in greater detail in Section A5.

Effects of grain structure, crystal lattice structure and alloying elements on mechanical properties

Grain structure, crystal lattice structure and alloying elements will have different effects on the mechanical properties of metal alloys.

Grain structure

▶ Small grains with regular spheroid shapes will increase toughness and tensile strength by restricting the movement of dislocations and slip planes.

▶ Concentrations of intermetallic compounds in large quantities or at grain boundaries will cause brittleness.

▶ Even distribution of intermetallic compounds throughout the grain structure will increase hardness without increasing brittleness.

Crystal lattice structure

FCC lattice structures contain the most possible slip planes active in the numerous slip directions, and so tend to have low strength and high ductility.

Alloying elements

▶ Alloying elements that form intermetallic compounds tend to increase hardness and tensile strength (as long as these are distributed evenly through the structure).

▶ Alloying elements that occupy interstitial or substitutional positions in the crystal lattice cause deformation, which disrupts the movement of dislocations and slip planes. This leads to an increase in hardness and tensile strength.

Ⅱ PAUSE POINT   Close your book. Work alone or with a group of colleagues to create a spider diagram which summarises the ways in which alloying elements, lattice structures and grain size can affect the mechanical properties of materials.

      Hint   When you run out of ideas, refer back to your notes or the textbook to fill in the details.

     Extend   Try drawing sketches or diagrams to illustrate each of the effects.

A5 Effects of processing on the mechanical properties of metallic materials

When **raw metal stock** is received from a supplier, it will already have gone through some form of processing to shape it into **plates**, **bars**, **sheets** or **billets**. Changing the external shape of the material in this way provides a convenient starting point for subsequent manufacturing operations. In addition, processing the metal has a significant impact on the internal structure, and therefore the mechanical properties, of the material.

> **Key terms**
>
> **Raw metal stock** – metals as they arrive from the manufacturer or foundry.
> **Plates** – metal sheets usually above 3 mm thickness.
> **Sheets** – metal sheets usually below 3 mm thickness.
> **Bars** – fully finished round or square bar that has been rolled to an exact size and is ready for use.
> **Billets** – part finished round or square bar that will require further shaping or processing prior to use.

Recrystallisation

When initially cast, ingots of the raw stock metal tend to form large crystal grains of uneven size and shape. They also commonly contain voids and **cavities** formed by gas bubbles or due to rapid shrinkage during cooling.

Recrystallisation is the process of re-forming and refining the grain structure of the metal. Smaller consistently shaped and sized grains tend to provide enhanced strength and toughness.

There are several forms of heat treatment that achieve recrystallisation, including sub-critical annealing, full annealing and normalising. These are all discussed later.

Key terms

Cavities – areas of empty space within a material caused by the removal of a particle, for example where an impurity has burned away during heating.

Spring back – the tendancy of a metal to recover some of its original shape once forming stresses are removed.

Hot working

Hot working processes are carried out at elevated temperatures in a phase where the metal becomes softer and more ductile. In steel, hot working is carried out when it is in the single-phase austenite form. Continuous recrystallisation within the material during formation at these temperatures allows large plastic deformations to occur without the risk of fracture. **Table 25.7** describes some common hot working processes.

Cold working

Plastic deformation below the recrystallisation temperature is known as cold working. This deforms and strains the material without any subsequent re-forming or refinement of the grain structure. Cold working can impart strain hardening to the finished component, which can yield the required strength and hardness that might otherwise have been achieved through an additional heat treatment after forming.

Cold working is generally used only for materials with a relatively low yield point (the point at which plastic flow occurs) and high ductility (the ability to undergo plastic deformation without rupturing or fracture). Materials are usually annealed prior to cold working to make them as ductile as possible.

Annealing is also sometimes performed part way through a cold forming process where work hardening needs to be relieved in order to restore ductility to the material. Annealing is discussed in detail later in this section, under 'Heat treatment of steels'.

'**Spring back**' is a phenomenon commonly encountered in cold working and must be accounted for in tooling design. Even in materials with a relatively low yield point, some elastic deformation will occur prior to reaching the plastic region. Once the forming stresses are removed, there will be a tendency for the material to spring back towards its original shape.

▶ **Table 25.7** Hot working processes

Drop forging	A large mechanical hammer is used to shape a heated billet. The process used is one of the following: • open die – the material is not fully contained and has to be manipulated by an operator between blows. • closed die – a closed die consists of two parts. The heated billet is placed in the lower die and struck with the upper die, forcing material into the die cavity and thus shaping the component. Drop forging imparts greater strength and toughness than casting or machining counterparts because of: • the formation of a fine recrystallised grain structure • the elimination of voids within the material. Impurities that do not undergo recrystallisation are distributed along grain boundaries following the lines of material flow. Where these form parallel to the surface of the component, enhanced strength and fracture resistance are obtained.
Press forging	This involves a slow squeezing action at extremely high pressure instead of the high-energy blows used in drop forging. It is able to penetrate all the way into the interior of the component, giving more uniform deformation and flow.
Rolling	Often, the first stage in the production of wrought metals, rolling involves reshaping large starting stock (such as billets or slabs) into plate, strip or specialised profiles, such as those used in the railway and construction industries. Heated billets or slabs are passed several times between two driven rolls that squeeze and elongate the material. Temperature control is important and reheating may be required between rolls. The metal must be heated and maintained at uniform high temperatures. This ensures that the finished product: • has a uniform fine grain structure • is free of any residual internal stresses caused by plastically deformed regions that were not recrystallised.
Extrusion	A heated billet of metal is forced through a shaped die to form a product with a uniform cross-section along its entire length. This process is suitable for metals with low yield strength and low working temperatures, such as aluminium.

Cold forming

The squeezing processes used in hot working processes often have a direct cold forming equivalent. Cold forming processes are used in order to obtain greater accuracy and improved surface finish. **Table 25.8** gives some common examples.

▶ **Table 25.8** Cold forming processes

Cold rolling	• Very similar to hot rolling. • Used to make a range of sheets, strips and bars with scale-free surfaces that have an excellent finish. • Can reduce material thickness by 1–50%. The greater the reduction, the greater the degree of strain (or work) hardening in the final product. This can limit the use of further cold working processes.
Cold (or impact) extrusion	• Similar to hot extrusion. • Primarily used with low-strength materials, such as lead, zinc and aluminium. • Common products include thin-walled tubes such as toothpaste tubes and those used in the production of batteries.
Coining	• Similar to closed die press forging; involves squeezing a metal blank that is totally enclosed in a set of dies. • Commonly used to manufacture coinage, as excellent detail, finish and dimensional accuracy can be achieved.

Drawing processes

Drawing processes rely on the ability of ductile metals such as copper to be plastically deformed when put under tension. Some common drawing processes are described in **Table 25.9**.

> **Key term**
>
> **Mandrel** – a mould or former over which sheet metal can be shaped.

▶ **Table 25.9** Drawing processes

Wire drawing	The cross-sectional area of a wire or rod is reduced by pulling it through a die. It is similar in some ways to extrusion, but pulling the wire through the die involves tensile rather than compressive forces. The reduction in area is between 20% and 50% in a single operation – any higher and the forces required to pull the wire through the die could exceed its tensile strength and cause the wire to break. Multiple draws may be needed to achieve the required diameter. Annealing may be required between draws to relieve strain hardening.
Spinning	A rotating sheet of ductile material is progressively formed over a shaped **mandrel**. Mounted on a lathe, the sheet and mandrel spin together and the material is manipulated by an automated roller or by hand using a rounded tool. Several passes may be required for the spun sheet to adopt the shape of the mandrel. It can then be removed and trimmed.
Shallow/ Deep drawing	These processes are used to form sheet material into cylindrical containers with closed bottoms. • Shallow drawing is where the depth of the final product is less than its diameter. • Deep drawing is where the depth of the final product is greater than its diameter. Drawing is carried out on a press where the sheet is formed by a punch and die with sufficient clearance between them to accommodate the material's thickness. During drawing, stretching and shrinkage will occur in different parts of the material. Careful component design and process control are required to prevent the occurrence of splitting, excessive thinning or buckling.

Ⅱ PAUSE POINT   Investigate the manufacture of a metal product of your choice (for example, a spanner, a car body panel or a drinks can), and explain the effects of the processes used on the microstructure of the material.

Hint   Choose a product that is familiar to you and will be straightforward to investigate.

Extend   Use sketches and diagrams to help support your explanation of what is happening to the microstructure at each stage of manufacture.

Heat treatment of steels

Heat treatment is used to manipulate the crystal lattice formed within grains and the size and shape of the grains themselves in order to influence the mechanical properties of the steel.

Sub-critical (or process) annealing

Sub-critical annealing is a process of recrystallisation carried out below the eutectoid temperature at which phase changes would start to occur in the metal. It is commonly used to relieve work hardening by causing new grains to form in areas where the crystal lattice structure has been strained during cold working.

Full annealing

This process is carried out at higher temperatures. It is used to fully re-form the crystal structure of the solid solution of ferrite and cementite and to eliminate any non-equilibrium structures present, such as **martensite**. As the crystal structure is totally re-formed, annealing also eliminates any work hardening that was present in the original crystal structure. The overall effect is to cause softening and increased ductility by eliminating hard non-equilibrium phases and re-forming a coarse structure with large (sometimes excessively large) grains.

▸ Hypo-eutectoid steels – Full annealing involves heating to around 950–1000°C and soaking for a period to ensure that the single-phase austenite structure is fully established and is of uniform temperature and composition. When cooled slowly, the resulting formation will consist of coarse pearlite surrounded by excess ferrite equilibrium structures.

▸ Hyper-eutectoid steels – The temperature required for full annealing is slightly lower, 750–800°C. If the material was fully converted to the austenite phase, then during slow cooling, cementite forming around grain boundaries would make the material excessively brittle. When cooled slowly, the resulting formation will consist of coarse pearlite with excess cementite in the form of spheroids dispersed throughout the material.

In both cases, a slow, controlled cooling process lasting several hours must be used to bring the material back below the eutectic temperature. After this, cooling to room temperature can be achieved in air. The initially slow cooling rate ensures the formation of equilibrium structures corresponding to those in the phase diagram.

> **Key term**
>
> **Martensite** – an extremely hard and brittle phase of iron and carbon formed in carbon steel during quench hardening.

Normalising

Normalising is closely related to full annealing. In materials of both hypo- and hyper-eutectoid composition, the steel must be heated to around 950–1000°C and soaked at that temperature for a period to ensure the formation of fully uniform austenite. The steel is then removed from the furnace and allowed to cool in air. This relatively rapid cooling in comparison to that used in full annealing tends to result in the formation of smaller and more uniform grains of pearlite and either excess ferrite or excess cementite.

Components formed using hot forming techniques such as forging are often normalised prior to any subsequent machining.

Through hardening

The iron–carbon equilibrium phase diagram for steel (**Figure 25.11**) describes the formation of the equilibrium structures of austenite, ferrite, cementite and pearlite, which form when sufficient time is given during cooling for the material to adopt its preferred state.

In steels containing sufficient carbon (generally above 0.3%), rapid cooling from high-temperature single-phase austenite (the same starting point required for normalising) will result in the formation of a non-equilibrium structure called martensite.

Martensite forms when the FCC austenite is unable to force out sufficient carbon to adopt the BCC ferrite form, which in equilibrium can dissolve much less carbon. This happens as a consequence of rapid cooling by quenching in water or oil. The FCC structure changes instantly to BCC and traps excess carbon interstitially within its crystal lattice, forming the highly distorted BCC structure seen in martensite. The degree of distortion is proportional to the amount of trapped carbon, which in turn is related to the rate of cooling and the minimum temperature reached. This structure is extremely hard, strong and brittle, as the severe distortion of the lattice prevents the movement of dislocations through the structure and disrupts slip planes.

Through hardening affects the full thickness of the material. However, during quenching the cooling at the interior of a large component is unlikely to be as rapid as at the surface, so some variation in hardness through the material can be expected.

Tempering

High-carbon martensite has very low toughness and ductility. As a consequence, it is far too brittle to be useful in engineering. However, an additional process called tempering can be used to increase toughness and ductility while retaining much of the hardness and strength of the material.

Martensite is a super-saturated solid solution of ferrite and carbon that is stable at low temperatures, but heating it to between 200°C and 700°C encourages it to release some of the trapped carbon and form stable ferrite and cementite structures. The rate and quantity of martensite decomposition is controlled by the temperature and the duration of the tempering process.

When the desired composition of ductile ferrite, hard but brittle cementite and very hard but very brittle martensite is achieved during tempering, any further changes can be stopped by returning the steel to room temperature.

Case hardening

Significant hardening can only be achieved in steel with greater than 0.3% carbon content. Mild steel is not hardenable, but it is tough, ductile and easier to machine than its high-carbon counterparts (even when these are fully annealed) as it contains more ductile ferrite and less hard but brittle cementite.

Often, in a particular application, the tough, ductile properties of mild steel are preferable, but there is also the requirement for the surface to be hard, perhaps to increase wear resistance. This combination of properties can be achieved through case hardening.

Case hardening relies on the **diffusion** of additional carbon into the outer surface of the mild steel component. Traditionally this was achieved by packing the component in charcoal or cast iron shavings and heating it to around 850–900°C, into the austenite phase. This carbon-rich atmosphere encourages the diffusion of the carbon atoms into solution with the austenite at the surface of the material to a depth of 0.5–1.0 mm.

Once quenched and tempered, you are left with a tempered martensite outer layer which is extremely hard and a mild steel core that enables the component to remain tough.

> **Key term**
>
> **Diffusion** – mixing that results from the natural tendency of atoms or molecules to move from areas of high concentration to areas of low concentration.

Ⅱ PAUSE POINT   Investigate the heat treatment processes carried out on a HSS slot drill (used in milling).

Hint   Your first port of call should be the workshop, to find an example of the product you are investigating. The wealth of information online provided by the manufacturers of tooling will also be helpful.

Extend   Explain the heat treatment processes in terms of what is happening at a microstructural level within the material.

Heat treatment of aluminium alloys

Precipitation hardening

Precipitation hardening is the principal method used in the heat treatment of aluminium alloys and is generally conducted in three phases:

▶ Solution treatment – This was discussed earlier when looking at the phase diagram for aluminium–copper alloys (**Figure 25.12**). It involves heating the alloy so that any θ phase intermetallic compound $CuAl_2$ present is dissolved back into the aluminium α phase. The alloy is soaked at a temperature not above the eutectic melting point for a period of time to ensure even temperatures throughout the alloy and a uniform structure.

▶ Quenching – Rapid cooling by quenching in water traps the copper in the aluminium α phase, forming an unstable non-equilibrium super-saturated solid solution. In this state the material is soft and can be readily formed or machined.

▶ Ageing – Over time, even at room temperature the unstable super-saturated solid solution will convert into the stable two-phase structure by precipitating $CuAl_2$. Often artificial ageing at elevated temperatures is used to accelerate this process.

Over-ageing

Small clusters of precipitated atoms occupying sites within the lattice structure cause considerable straining within the lattice. This blocks the movement of dislocations through the material and so increases its strength considerably. However, if these clusters grow large enough to break free from the parent lattice and form discrete well-defined grains, the useful effects of straining the crystal lattice are lost, and strength and hardness begin to decrease. This is known as over-ageing.

Heat treatment of titanium alloys

Precipitation hardening in titanium alloys is achieved in a very similar way to the precipitation hardening of aluminium alloys, using the same processes.

Alloying elements in steel

These are listed in **Table 25.10**, together with the properties they impart to the steel.

▶ **Table 25.10** Alloying elements used in steel

Alloying element	Properties conferred
Chromium	• Increases hardness penetration in heat treatment. • Increases toughness, wear resistance and corrosion resistance.
Manganese	• Present in most commercial steels. • Increases the strength of ferrite. • Increases hardness penetration in heat treatment.
Molybdenum	• Increases hardness penetration. • Increases tensile strength at elevated temperatures.
Nickel	• Increases strength of ferrite. • Increases toughness and hardenability.
Tungsten	• When combined with chromium retains high hardness at elevated temperatures. • Increases wear resistance and hardness.
Vanadium	• Inhibits grain growth during heat treatment, giving a more refined grain structure. • Increases strength and toughness.

Common alloying elements in aluminium and titanium

These are listed in **Table 25.11**.

▶ **Table 25.11** Alloying elements in aluminium and titanium

Metal	Alloying elements
Aluminium	copper, silicon, magnesium, manganese, titanium, chromium, lithium
Titanium	aluminium, vanadium

A6 Microstructure investigation of metallic materials

Using appropriate equipment, such as a hand lens, optical microscope or digital imaging system, it is possible to view grain structures, grain boundaries and even the phases present within grains.

The macrostructure of prepared samples can be viewed at low magnifications of ×10 or less through a hand lens. This is sufficient to make out larger grains and identify structures such as pearlite in carbon steel.

At higher magnifications of between ×100 and ×1000, a clear picture of the microstructure can be observed. This might include much smaller features such as individual grain boundaries or the presence of intermetallic compounds within grains.

Comparing information gained through investigation with reference data, including micrographs of typical structures, will allow you to identify the composition and processing history of a given material sample.

Figure 25.13 shows a **micrograph** of a sample of a hyper-eutectoid steel. During cooling from the austenite phase, cementite forms first at the austenite grain boundaries, seen as white bands around the grains. When sufficient carbon has been rejected from the austenite so that the remaining composition is eutectic, pearlite forms from the remaining carbon-enriched austenite. The pearlite is readily identified from its lamellar (plate-like) structure, seen here within the grains.

▶ **Figure 25.13** Micrograph of a hyper-eutectoid (1.3% carbon) steel sample, annealed at 1100°C

> **Key term**
>
> **Micrograph** – a high-resolution, highly magnified (×100 to ×1000) photograph of a polished cross-section of a crystalline solid, taken using a microscope.

Figure 25.14 shows a micrograph of the same hyper-eutectoid steel that has been quench-hardened. The presence of martensite is identifiable from the distinctive needle-like structures.

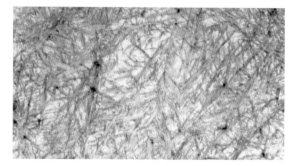

▶ **Figure 25.14** Micrograph of hyper-eutectoid (1.3% carbon) steel sample, quenched

Figure 25.15 shows a micrograph of an aluminium alloy with 4% copper content that has been over-aged during prolonged heat treatment. Grains of the intermetallic compound $CuAl_2$ can be seen outside the aluminium grains. These no longer contribute to the strain hardening effect of small clusters of $CuAl_2$ that form within grains. Over-aged aluminium has generally poor mechanical properties.

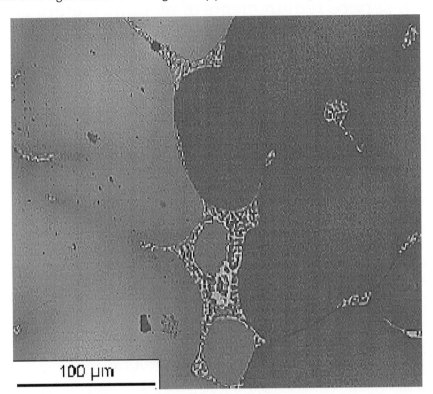

▶ **Figure 25.15** Aluminium alloy with 4% carbon showing signs of being over-aged

---

Assessment practice 25.1

A.P1   A.P2   A.M1   A.D1

You work as a technician in a quality assurance laboratory for a metals wholesaler. Your employer has asked you to evaluate six numbered metal alloy samples from a new supplier.

The samples have been suitably prepared for you so that a polished cross-section of the grain structure is visible.

For each sample you will write a short report based on your observations of the grain structure under magnification by a microscope.

Drawing on your knowledge of the microstructures of metallic materials and using a library of standard micrographs for reference, do the following:

- Identify the materials and evaluate their likely processing histories.
- Evaluate the mechanical properties that are likely to be exhibited by each sample. Do this by referring to the microstructural features you have identified.

Plan

- What will I be looking for when I carry out the observations under the microscope?
- Do I have all the equipment and reference materials that I need?

Do

- I can work in a logical and methodical way through all aspects of the assignment.
- I can identify when I've gone wrong and adjust my approach to get myself back on track.

Review

- I am clear about the effectiveness of my approach to the task and what I would do differently next time.
- I recognise the gaps in my knowledge, how they affected my performance and where I need to concentrate my efforts.

Explore safely the mechanical properties of metallic materials and the impact of their in-service requirements

B1 In-service requirements of metallic materials

Across all engineering sectors there are countless applications for metallic materials, each requiring a different combination of mechanical properties for optimal functionality, cost and performance. Some examples are given in **Table 25.12**.

▶ **Table 25.12** Examples of common in-service requirements in a variety of applications

Characteristic	Applications
High tensile strength	Pressure vessels, vehicle suspension components and any application where large loads must be supported safely and reliably.
High specific strength (strength-to-weight ratio)	Weight-critical applications, such as in aerospace and motor sport, where high strength is essential but unnecessary weight will increase fuel consumption and power requirements.
High resistance to impact loading	Impact tool bits, hammer heads and masonry drills all have to withstand high impact loads without deformation or cracking.
High hardness	Cutting tools, vice jaws, taps and dies and lathe tools all require high hardness in order to function. For any cutting tool, the tooling material must be considerably harder than the material being cut.
Toughness	Motor vehicle crumple zones, where deformation without shattering helps to absorb the high energy of an impact and so protect vehicle passengers.
Ductility	Structural components such as beams. Ductile materials give an indication by showing visible signs of deformation. Sudden brittle failure in this situation could cause a disastrous structural collapse.

PAUSE POINT Choose an engineered product and describe the in-service working conditions it will experience, including details of the environment in which it will operate.

> Hint — Think about the worst-case scenarios that the product must withstand. Don't forget about environmental factors such as temperature and weather.

> Extend — Rate the importance of a range of mechanical properties that will be relevant when selecting a material to perform in those service conditions. What other factors must you consider?

B2 Destructive test procedures

It is important to ensure that the results of mechanical testing completed by different test facilities are directly comparable and repeatable. To facilitate this, standard methodologies for carrying out testing have been developed. In the UK these are in the form of British Standards. The tests in this section use methods that follow the recommendations made in these standards. The appropriate standard for each test is identified so that you can find out the detailed guidance for carrying out destructive test procedures.

Tensile strength testing

Applicable standard:

▶ BS EN ISO 6892-1:2009 Metallic material – Tensile testing Part 1: Method of test at ambient temperature.

Selection and preparation of test specimens

Tensile testing can be performed on a wide range of material sample types and cross-sections (circular, rectangular, square, etc.). The sample must have a constant cross-sectional area over its entire length.

Sometimes material samples have machined shoulders at either end to allow secure gripping in holders. Other types of samples are not machined and come directly from manufacturing runs of products, such as wire or strip. These must be secured in different ways, such as by using screwed grips or parallel-faced jaws.

You must determine the cross-sectional area of each test piece prior to testing. This must be accurate, as any subsequent stress calculations will be heavily skewed by an error at this stage. In practice this means you should measure the area in at least three places over the sample length and find the average.

The start and end of the gauge length over which test results will be determined should also be marked on the sample. This will correspond to the length of the **extensometer**.

Tensile testing machine

There are various sorts of tensile testing machine, but they are generally aligned vertically and use a hydraulic ram to move an upper and a lower crosshead. A prepared tensile test sample is held between the upper and lower crossheads, using a gripping mechanism suitable for the type of sample. As the hydraulic ram begins to move, the force exerted on the test sample is measured by a load cell. Any associated extension is measured by an extensometer at regular intervals. Data from these two instruments is recorded throughout the test, which continues until the test sample fractures.

On completion of the test, data is analysed by plotting a force–extension graph, from which the results of the test can be determined. (A typical force–extension graph for ductile mild steel is shown in **Figure 25.18**).

Examination of fracture surfaces

The fracture surface in tensile testing can reveal the general characteristics of the material before you analyse the test data.

▶ Ductile materials exhibit extensive **necking** around the point of fracture and a characteristic cup-and-ball fracture surface (see **Figure 25.16**).

▶ Brittle materials exhibit minimal or no necking, and there is little or no plastic deformation prior to fracture (see **Figure 25.17**).

> **Key terms**
>
> **Extensometer** – a device attached to the test piece which measures extension during testing.
>
> **Necking** – a reduction in cross-section.

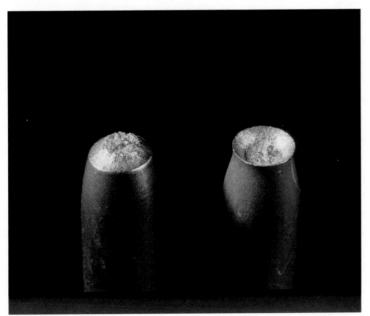

▶ **Figure 25.16** Necking with cup-and-ball fracture surface is indicative of ductile fracture

▶ **Figure 25.17** Little or no necking or plastic deformation prior to failure is indicative of brittle failure

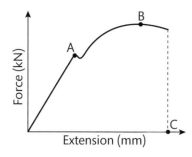

▶ **Figure 25.18** Example force-extension graph for ductile mild steel

Data analysis

Several parameters can be determined from the test results and force–extension graph, which you can include in the test report.

▶ Upper yield strength (unit: Pa) – determined from the force–extension graph. This is the maximum value of stress prior to the first decrease in force, which corresponds to point A in **Figure 25.18**, at the end of the straight (elastic) portion of the graph. The upper yield strength is obtained by dividing the force at point A by the cross-sectional area of the test piece:

$$\text{upper yield strength} = \frac{\text{force at point A}}{\text{test piece cross-sectional area}}$$

▶ Ultimate tensile strength (unit: Pa) – determined from the force–extension graph. This is the stress at point B in **Figure 25.18**, corresponding to the maximum value of force indicated during the test:

$$\text{ultimate tensile strength} = \frac{\text{force at point B}}{\text{test piece cross-sectional area}}$$

▶ Percentage elongation at fracture (unit: %) – determined from the force–extension graph. The total extension at fracture is at point C in **Figure 25.18**. Dividing this by the extensometer gauge length (the initial length of the section of the test piece examined during testing) gives:

$$\text{percentage elongation at fracture} = \frac{\text{extension at fracture}}{\text{extensometer gauge length}} \times 100$$

▶ Percentage reduction in cross-sectional area (unit: %) – describes the amount of necking observed in the test sample after fracture. If necessary, the fractured test piece should be fitted back together so that you can measure the minimum cross-sectional area after fracture:

percentage reduction in cross-sectional area

$$= \frac{\text{original cross-sectional area} - \text{minimum cross-sectional area after fracture}}{\text{original cross-sectional area}} \times 100$$

Test reporting

Your test report should include:

▶ a reference to the standard under which the test was conducted
▶ identification of the test piece
▶ type of material tested (where known)
▶ type of test piece
▶ examination of the fracture surface
▶ test results.

Hardness testing

Surface preparation

The surfaces to be tested should be smooth and even so that the indent made during testing is clearly visible and distinct. You should remove any oxide scale such as that commonly present after hot working processes. Degrease the sample surface and remove any lubricants. (Some metals such as titanium, which might otherwise stick to the indenter, will require lubrication with kerosene. Make sure that you note this in your test results.)

Brinell hardness test

Applicable standard:

▶ ISO 6506-1:2014 Metallic materials – Brinell hardness test – Part 1: Test method.

A tungsten carbide ball is pressed into the surface of the material under a static load.

The size of the ball and the test load are determined by the material being tested, with the aim of providing a clear indentation that is easy to measure and with a size between 24% and 60% of the ball diameter.

Apply the load for 10–15 seconds to ensure that any plastic deformation has time to occur fully.

When the load and ball are removed, measure the diameter of the circular indentation under magnification in each of two perpendicular directions. Then calculate the mean of these diameters.

Find the Brinell hardness number (HBW) by looking up the average indentation diameter (in mm) from Table C.1 in Annex C of the ISO 6506-1:2014 standard. **Figure 25.19** shows the format for expressing the Brinell hardness number in your test results.

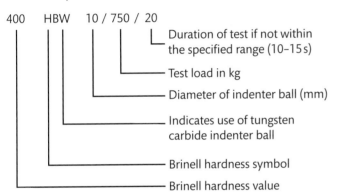

▶ **Figure 25.19** Example of how the Brinell hardness should be stated in a test report

Vickers hardness test

Applicable standard:

▶ ISO 6507-1:2005 Metallic materials – Vickers hardness test – Part 1: Test method.

The Vickers test is very similar to the Brinell test, but instead of selecting from a range of spherical indenters, a single square-based pyramid is used.

It is often easier and more accurate to measure across the diagonals of the clearly defined square indentation

obtained in the Vickers test. **Table 25.13** shows the test loads corresponding to different values on the Vickers hardness scale.

▶ **Table 25.13** Loads used in different Vickers hardness scales

Vickers hardness scale	Test load (kg)
HV 10	10
HV 20	20
HV 50	50
HV 100	100

Find the Vickers hardness number (HV) by looking up the measured diagonal of the indentation (in mm) from a data table contained in ISO 6507-4:2005 Metallic materials – Vickers hardness test – Part 4: Tables of hardness values (see Table 3 in the standard for the range HV 5 to HV 100). **Figure 25.20** shows the format for expressing the Vickers hardness number in your test results.

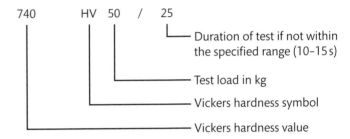

▶ **Figure 25.20** Example of how the Vickers hardness should be stated in a test report

Rockwell hardness test

Applicable standard:

▶ ISO 6508-1:2015 Metallic materials – Rockwell hardness test – Part 1: Test method.

The most common and widespread method of gauging hardness is the Rockwell hardness test. As with the Brinell and Vickers tests, it depends on the indentation of the test sample under static load, but the test method is somewhat different.

The Rockwell hardness test can be performed using a range of loads and indenters depending on the type of material being tested. The most common versions of the test are Rockwell A, B and C, summarised in **Table 25.14**.

▶ **Table 25.14** Common variants of the Rockwell hardness test

Rockwell hardness scale	Hardness unit	Indenter	Preliminary load (kg)	Total load (kg)	Typical range	Typical materials
A	HRA	Diamond cone	10	60	20–95 HRA	Thin steel, case-hardened steel
B	HRBW	Ball 1.5875 mm	10	100	10–100 HRBW	Copper alloys, aluminium alloys
C	HRC	Diamond cone	10	150	20–70 HRC	Steel, titanium

The test is conducted as follows:

1 Preload the indenter in contact with the material surface with a 10 kg preliminary load. Any indentation caused at this stage is primarily due to elastic deformation.

2 On the dial gauge, set the measuring penetration into the material to zero.

3 Apply the additional test load to the indenter according to the Rockwell scale being applied.

4 Remove this additional load after 5 seconds.

5 With the preliminary load still in place, read the Rockwell hardness number from the dial gauge.

Figure 25.21 shows the format for expressing the Rockwell hardness number in your test results.

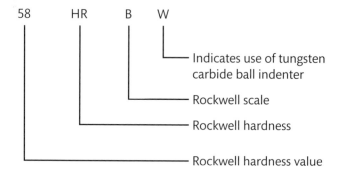

▶ **Figure 25.21** Example of how the Rockwell hardness should be stated in a test report

Test reporting

When reporting the results of hardness testing, as well as the appropriate hardness number you must include the applicable test type and scale, which indicates the conditions under which the test was carried out.

Your test report should also include:

▶ a reference to the standard under which the test was conducted

▶ identification of the sample tested

▶ temperature at which the test was carried out

▶ the hardness result in the appropriate format

▶ any deviation from the applicable standard or optional elements (such as lubrication where required)

▶ the date the test was performed.

Impact testing

Tensile strength is an important consideration for designers of systems where the load will be primarily static or where any variation in load is applied gradually, as with tow cables or bridge suspension wires. However, in most applications, engineers also need to consider behaviour under shock loading, where high rates of strain can cause brittle fractures to form in materials that are otherwise considered to be ductile.

Izod and Charpy impact testing

The Izod and Charpy tests use similar equipment and notched test specimens (depicted in **Figure 25.22**). Both tests measure the energy absorbed by a sample during an impact fracture.

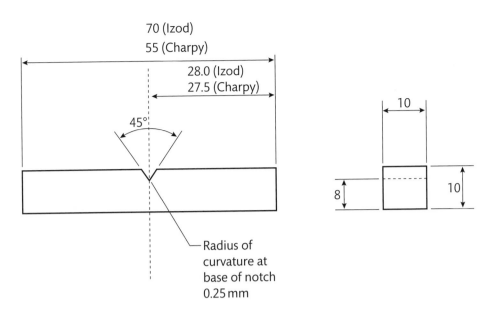

▶ **Figure 25.22** Variations of the V-notch test specimen suitable for Izod and Charpy impact tests

Applicable standards:

▶ BS 131-1:1961 Methods for notched bar tests – Part 1: The Izod impact test on metals.

▶ ISO 148-1:2009 Metallic materials – Charpy pendulum impact test – Part 1: Test method.

The main difference between the Izod and Charpy tests lies in the test piece arrangement adopted during testing, as shown in **Figures 25.23** and **25.24**.

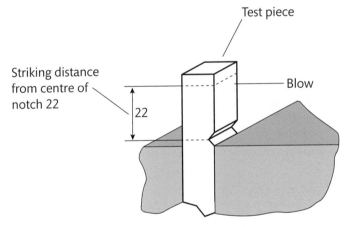

▶ **Figure 25.23** Izod impact testing uses a cantilever test piece arrangement with one end fixed

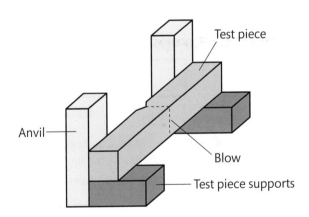

▶ **Figure 25.24** Charpy impact testing uses a simply supported test piece arrangement with both ends of the test piece fixed

The same type of impact testing machine can be used for both tests. These machines commonly have a heavily weighted pendulum that is raised to a set height and so has a known level of potential energy. When the pendulum is released, all of its potential energy is transformed into kinetic energy as it reaches the bottom of the swing and the point of impact with the test piece. The height the pendulum achieves after its impact with the test piece is recorded on the test apparatus. This is directly related to the energy remaining in the pendulum after impact, and from this the energy absorbed during the impact can be calculated. The equipment is often designed so that the tester can read the energy absorbed directly from the scale after each test.

Inspection of the fracture surface

After testing, the fracture surface can help determine the nature of the fracture and provide an indication of the notch toughness of the material. A large proportion of shear fracture generally indicates a high fracture toughness.

The fracture surface usually consists of clear regions of fibrous shear fracture and shiny crystalline fracture, shown schematically in **Figure 25.25** and in a photograph in **Figure 25.26**.

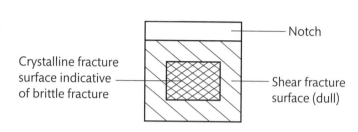

▶ **Figure 25.25** Schematic representation of the fracture surface after an impact test

▶ **Figure 25.26** Photograph of a fracture surface showing the notch, a peripheral region of dull shear fracture and a central area of shiny crystalline fracture

You can determine the percentage areas of shear and crystalline fracture by several methods, including by comparing the fracture with fracture appearance charts, such as those available in ISO 148-1:2009 Metallic materials – Charpy pendulum impact test – Part 1: Test method, an example of which is shown in **Figure 25.27**.

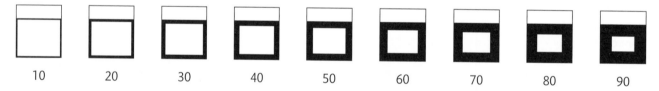

▶ **Figure 25.27** Impact test fracture appearance chart for estimating the percentage of shear present (and from which the percentage of crystalline area can also be determined)

Source: ISO 148-1:2009 Metallic materials – Charpy pendulum impact test – Part 1: Test method

Test reporting

The test report should include:

▶ a reference to the standard under which the test was conducted
▶ identification of the sample tested
▶ the temperature at which the test was carried out
▶ the type of notch used (usually the V-notch, but Charpy testing can also be performed with a U-notch)
▶ any deviation from the applicable standard, such as the size of the test piece if it is not as specified
▶ the energy absorbed by the test piece
▶ the fracture appearance, together with percentage shear and crystalline areas.

🅘🅘 **PAUSE POINT**

Investigate the types of destructive test equipment that are available for use in your centre.

Hint Ask a tutor or technician to explain how the test equipment works, read the test manual that accompanies the equipment, and familiarise yourself with any necessary safety considerations.

Extend Can you now explain to your tutor how to safely conduct a destructive test using the equipment available? If so, you might be ready to give it a try.

B3 Non-destructive test procedures

It is not always convenient to carry out destructive testing. For instance, if you manufacture a single large pressure vessel for use in the nuclear industry by welding together formed sections, it is impractical to test each weld to destruction to ensure that there are no voids, inclusions or **areas of porosity** which could cause points of weakness. Where engineered products are expensive to produce, safety-critical or both, non-destructive testing can detect or confirm the absence of a range of material and processing faults that might be cause for concern.

Surface defect detection

These methods make surface defects visible so that you can count, photograph, measure and classify them. This allows you to assess whether the component is fit for use, needs to be reworked or must be scrapped.

Dye penetrant testing

Dye penetrant testing enables you to detect surface defects that would otherwise be extremely difficult to identify using the naked eye. Both colour contrast and fluorescent dyes are commonly used. When a liquid of low **viscosity** (such as a dye) is applied to the surface of a material, it is drawn by **capillary action** into surface defects, such as micro-cracks and areas of porosity, therefore indicating their presence (see **Figure 25.28**).

▶ **Figure 25.28** Surface defects detected by dye penetrant testing

Dye penetrant testing proceeds in the following steps:

1 Thoroughly clean, degrease and dry the surface to be inspected to remove any material that might block the penetration of the dye.

2 Apply the dye to the prepared surface by painting, spraying or immersion, depending on the size of the component or area being tested.

3 Leave the dye to penetrate any surface defects for the time specified in the manufacturer's guidelines. This is referred to as the dwell time.

4 Carefully remove any excess dye following the dye manufacturer's instructions, usually by gentle cleaning with a solvent.

5 Apply a fine powder or liquid developer to the surface being tested. This will draw out any dye trapped in surface defects and begin to reveal the surface defects as areas of visible dye after around 10 minutes.

Magnetic particle testing

Magnetic particle inspection is used on **ferromagnetic** materials such as ferrous alloys, including steel. It is quick, easy to perform and requires less thorough surface preparation than dye penetrant tests.

Key terms

Areas of porosity – clusters of small voids often caused by the presence of gas bubbles during solidification.

Viscosity – the thickness of a liquid.

Capillary action – the tendency of a liquid to be drawn into narrow gaps.

Ferromagnetic – describes a material that becomes magnetised when exposed to a magnetic field.

Magnetic flux – lines along which the magnetic field acts.

Flux leakage – the forcing out of lines of flux from the surface of a material.

The component first has to be magnetised. The lines of **magnetic flux** passing through the sample are distorted by surface defects in the sample, and **flux leakage** occurs above the material surface at such points, as illustrated in **Figure 25.29**.

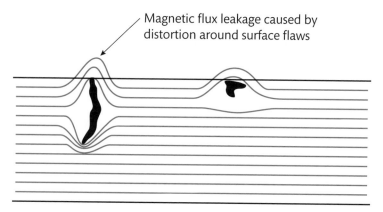

Magnetic flux leakage caused by distortion around surface flaws

▶ **Figure 25.29** Detection of surface defects in ferromagnetic materials using magnetic particle inspection

The next stage is to apply iron filings, either as a fine powder or suspended in a liquid spray, to the component surface. The iron particles will be attracted to the areas where flux leakage is occurring and form clusters around them, thus indicating the presence of material defects.

Internal defect detection using ultrasonic testing

The pulse reflection method of ultrasonic testing uses high-frequency sound energy to detect internal material defects, such as cracks, porosity or inclusions. It is often used to inspect welds in safety-critical applications.

An ultrasonic transducer/receiver in contact with the surface of the material being tested transmits ultrasonic waves into the material. When the sound waves encounter any **discontinuity** in the material structure, such as a crack or void or even the back wall of the sample, a proportion of the sound wave is reflected back towards the surface and is detected by the transducer/receiver (see **Figure 25.30**).

Key term

Discontinuity – any change in the structure of the material such as a crack or void.

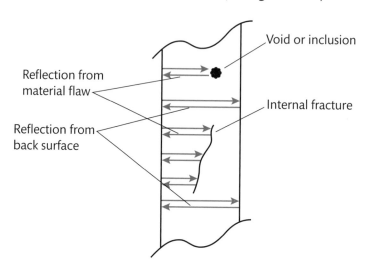

Void or inclusion

Reflection from material flaw

Internal fracture

Reflection from back surface

▶ **Figure 25.30** High-frequency sound waves being reflected by internal discontinuities

The time between the transmission of the original pulse and receipt of the reflected echo is directly related to the distance travelled through the material, and so the depth of the discontinuity can be determined. The test equipment processes this data and displays it in one of a number of ways, referred to as types A, B and C (shown in **Figure 25.31**).

Mechanical Behaviour of Metallic Materials

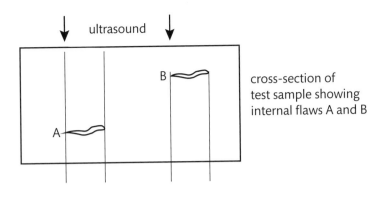

cross-section of
test sample showing
internal flaws A and B

Type A: Single pulse echo

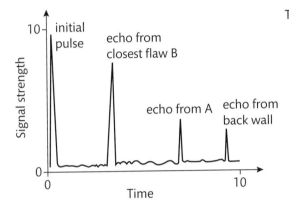

Type B: Multiple pulses
scanning across
the sample

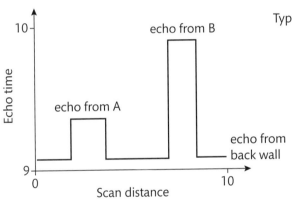

Type C: Composite image of
a complete scan of
the sample surface

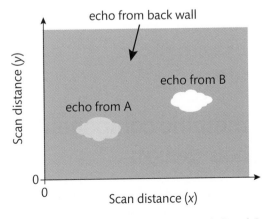

▶ **Figure 25.31** Ultrasonic testing display types A, B and C

Investigate the types of non-destructive test equipment that are available for use in your centre.

> **Hint**
>
> Ask a tutor or technician to explain how the test equipment works, read the test manual that accompanies the equipment, and familiarise yourself with any necessary safety considerations.

> **Extend**
>
> Can you now explain to your tutor how to safely conduct a non-destructive test using the equipment available? If so, you might be ready to give it a try.

Assessment practice 25.2

B.P3 B.P4 B.P5 B.M2 B.M3 B.D2

You are working as a technician in the mechanical testing laboratory of a company that manufactures industrial pipework. Your employer has asked you to perform a series of destructive tests on six material samples taken from the walls of sections of pipe manufactured using different materials.

Your employer expects you to measure a full range of mechanical properties for each material, using the necessary test equipment appropriately.

The tests you conduct should include:

Tensile test

- yield strength
- ultimate tensile strength
- modulus of elasticity
- percentage elongation at fracture
- percentage reduction in cross-sectional area at fracture
- an examination of the fracture surface.

Rockwell hardness test

- Rockwell hardness number.

Charpy impact test

- energy absorbed at impact
- percentage shear and crystalline area at the fracture surface.

You have also been asked to perform a non-destructive dye penetrant test on a section of pipe wall to ensure that no surface defects have formed during manufacture.

Record the results of all the testing you carry out and present them appropriately in a testing laboratory log book.

For each material, evaluate, based on the combined test results, the characteristics that the material will exhibit in service and suggest alternative applications where the material might be of use.

Justify the validity of the test methods used on each material.

Plan

- Am I suitably familiar with the safe use of the equipment that I need for these tests?
- Do I have all the equipment and reference materials I need?

Do

- I can work in a logical and methodical way through each test procedure, keeping careful notes as I go.
- I can identify when I've gone wrong and adjust my approach to get myself back on track.

Review

- I am clear about the overall aims of the tasks and what I was expected to produce.
- I recognise the gaps in my knowledge, how they affected my performance and where I need to concentrate my efforts in the future.

Explore the in-service failure of metallic components and consider improvements to their design

When components fail in service, the results can be disastrous. It is important for engineers to have an understanding of the in-service limitations of metallic materials so that costly mistakes can be avoided. Many lessons in how metals tend to behave and how they fail in service have been hard-learned in terms of financial cost and loss of life.

The world's first commercial passenger jet – the de Havilland Comet

When the UK company de Havilland launched the world's first production jet airliner on its maiden test flight in 1949, it was a moment that heralded a new era in long-distance travel.

The first passenger flights in 1952 were anticipated with great excitement. The Comet was on the verge of great commercial success, when a series of serious accidents began to happen. By 1954, three aircraft had been lost with all on board, after breaking up in flight. The race was on to find the cause of these tragedies.

Extensive investigations and tests were carried out in the search for answers. One test involved building facilities on the ground to simulate the pressurisation and depressurisation that a plane undergoes on take-off and landing. When this test reached somewhere in the region of 3500 cycles, the fuselage ruptured.

The cause was discovered to be the aircraft's square windows.

The Comet was designed and built before our modern understanding of fatigue failure, and the designers of the Comet had decided to fit square windows in the fuselage. At that time no-one appreciated that the sharp corners of the frames would act as stress raisers and perfect initiation sites for fatigue cracks. When the aircraft was placed in service, these tiny cracks quickly formed and then grew with every flight until the structure of the fuselage was weakened sufficiently to cause catastrophic failure.

The answer was to fit rounded windows, which is why all aircraft today have these.

▶ The de Havilland Comet

Check your knowledge

1 Investigate a major incident or disaster that was rooted in a lack of understanding of the behaviour of metallic materials. (You could use the Tay Bridge disaster of 1879 as a starting point or just type 'engineering disasters' into an internet search engine.)

2 Summarise the lessons learned as a consequence of the incident and present these to your class.

C1 Ductile and brittle fracture

Brittle fracture

Brittle fracture occurs in high-strength metals that have low toughness and poor ductility. There is little or no plastic deformation or necking before a brittle fracture.

Transgranular cleaving

A brittle fracture usually initiates at a micro-fracture, void or inclusion within the material. The fracture then propagates rapidly across the material in a plane perpendicular to the applied stress. The easiest pathway for the fracture to follow is generally along crystalline planes within the lattice structure, travelling through individual grains and splitting them apart (see **Figure 25.32**). The resulting fracture surface is flat and made up of differently oriented crystalline cleavage faces that give its characteristic crystalline appearance. Any variation in the flat surface is caused when the fracture moves from one grain into the next – the different orientations of the grains causes the fracture to change direction in order to stay on the path of least resistance through the material.

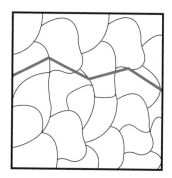

▶ **Figure 25.32** Transgranular fracture cleaving through randomly oriented crystalline grains

Intergranular fracture

Another form of brittle fracture arises when the path of least resistance is along the grain boundaries and not through the grains themselves (see **Figure 25.33**). This happens where brittle phases or impurities are concentrated at grain boundaries. For example, it can happen when cementite is present in large quantities at grain boundaries in hyper-eutectoid steels.

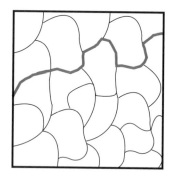

intergranular fracture

▶ **Figure 25.33** Intergranular fracture along grain boundaries

Other useful indicators may also be visible in the fracture surface. If the fracture initiated at a particular point, perhaps at a small defect in the surface of the component, it is often possible to observe ridge lines emanating from that point. These indicate the direction along which the fracture propagated (see **Figure 25.34**), and they not only signify brittle failure but also help to pinpoint the origin of the failure.

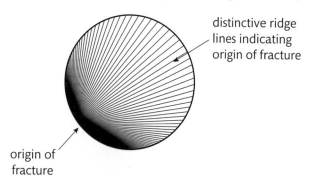

distinctive ridge lines indicating origin of fracture

origin of fracture

▶ **Figure 25.34** Ridge lines indicating the origin of the fracture

Load rate

Brittle fractures can also occur in ductile materials that are subject to the extremely high strain rates encountered during impact loading. A brittle fracture observed in what would otherwise be considered a ductile material might indicate a failure caused by an impact and not through simple overloading.

Grain size

Plastic deformation in ductile materials is mostly due to the movement of dislocations through the crystal lattice structure. In small grains, dislocations are unable to move far before they are stopped by the grain boundary. This decreases the amount of plastic deformation that can occur before fracture and so makes the material more brittle. In general, then, as grain size decreases, ductility also decreases, and so materials with small grains are likely to undergo brittle failure.

Ductile fracture

Ductile fracture is usually observed in materials with high ductility and toughness. It generally happens by transgranular cleaving through crystalline grains (as shown in **Figure 25.32**). Ductile failure is characterised by considerable plastic deformation, elongation and necking prior to fracture. The primary cause of ductile failure is overloading of the material.

The process of ductile fracture begins when small voids form at grain boundaries or around inclusions in the grain structure. As the applied stress increases, these voids become larger and begin to coalesce (join together) into larger areas of separation. Eventually the strength of the grains left supporting the load is exceeded and they fail by transgranular cleavage. During ductile failure, movement along slip planes also occurs around the edges of the fracture site, forming a shear lip. This gives rise to the fracture surface characteristics indicative of ductile failure: significant plastic deformation, necking and a cup-and-ball fracture surface with a central flat face surrounded by a shear lip (See **Figure 25.16** in Section B2).

C2 Creep failure

Creep is a phenomenon usually encountered at elevated temperatures, where the application of a constant tensile stress, well below the yield point of the metal, produces plastic deformation and elongation over time. This process can be illustrated graphically on a strain–time graph, as shown in **Figure 25.35**.

The initial strain shown in **Figure 25.35** is due to the elastic deformation caused by the applied load.

Work in a small group to summarise the differences between the characteristics of brittle and ductile fracture surfaces.

> Hint   Use annotated sketches and diagrams.

> Extend   Explain why, if a structural component is going to fail, it is always preferable for it to fail by ductile fracture.

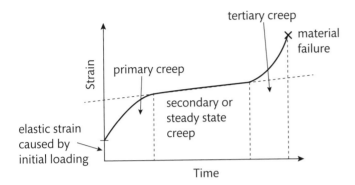

▶ **Figure 25.35** Strain–time graph illustrating the stages of creep (under constant temperature and stress)

Creep occurs in three distinct zones: primary, secondary and tertiary.

Primary creep

The primary creep rate is initially high as the movement of dislocations causes straining. This soon slows and then stops as work hardening in the material takes effect. Primary creep usually occurs in a short time period after initial loading.

Secondary creep

Secondary (or steady-state) creep is a slower process and occurs at a steady rate over a considerable period of time. There are numerous mechanisms that contribute to secondary creep. These fall into two categories: dislocation effects and diffusion effects.

Dislocation effects

These are dominant at low temperatures. As in the primary stages of creep, dislocations usually move along the same slip plane all the way to the edges of a crystal lattice, causing plastic deformation even at relatively low stresses. However, straining in itself causes more dislocations to form, which begin to get in each other's way as they move and soon block movement almost completely. This effect is called work hardening and severely limits dislocation movement along blocked slip planes. However, at high stresses, 'dislocation climb' can occur, where the dislocations are able to move between slip planes and so work their way around obstacles and continue to cause deformation.

Grain boundary sliding occurs as a result of these dislocation flow processes.

Diffusion effects

These are dominant at high temperatures approaching the melting point of the material. Diffusion creep is a high-temperature process in which atoms move to positions along the line of applied stress. This can happen inside the grain or along the grain boundaries and has the eventual effect of elongating the grains in the direction of stress, causing a corresponding permanent strain.

Grain boundary sliding occurs as a result of these diffusion flow processes.

Tertiary creep

In tertiary creep the strain rate increases rapidly. At this stage the internal structure of the material will have sustained considerable damage. Voids and cracks will be present at grain boundaries and around inclusions, and necking occurs. When the remaining intact grain structures become overloaded, they fail and the material fractures.

Temperature effects

Increasing temperature has an accelerating effect on all the mechanisms involved in creep. At or near the melting point of a material, creep rates can be significant. Blades in gas turbines or jet engines are able to perform in these environments only through careful temperature management and the development and use of high-temperature creep-resistant materials.

Below a certain temperature threshold it is unlikely that creep will be observed in a given metal. As a rule of thumb for metals with high melting points, the temperature threshold is approximately 40% of the melting point. For metals with low melting points, such as lead, creep can be observed at room temperature.

Applied stress

Increasing stress also has an accelerating effect on all the mechanisms involved in creep. However, for a material to exhibit creep, it must be loaded above its limiting creep stress at any given temperature. Below the limiting creep stress, little or no creep will be observed.

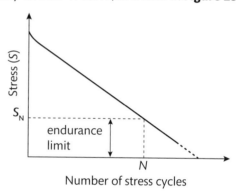

PAUSE POINT

Investigate the conditions in which turbine compressor blades are required to operate. Which of these operating conditions might accelerate the mechanisms of creep? How are the blades manufactured to limit the effects of creep?

Hint   Generally speaking, creep is reduced as grains get larger.

Extend   Why is the effective control of creep so important in this particular application?

Grain size

In general, large grain sizes increase resistance to creep. Indeed, at the limits of materials technology, jet engine turbine blades, which operate in very hostile environments with high stress and temperatures at or near the melting point of the material, are manufactured as a single crystal to eliminate any effects related to movement at grain boundaries.

C3 Fatigue failure

Fatigue failure is a phenomenon encountered in components that are subject to cyclic loading. This might be in the form of full stress reversal, like that encountered by vehicle suspension springs; random loading, like that experienced by a chain holding a vessel at anchor; or vibration, like that experienced by an engine mounting.

In all these instances, fatigue can cause premature failure at loads considerably below the normal tensile strength of the material. Although eventual failure can come with little warning, fatigue cracking takes some time to develop and propagate through the material until the remaining intact cross-section can no longer support peak loading and the material fractures.

In ferrous metals such as steels, it is found that there are stress levels below which cyclic loading will not produce fatigue fracturing. This is known as the fatigue limit and can be found from a graph of stress versus number of cycles, or S–N curve, as shown in **Figure 25.36**.

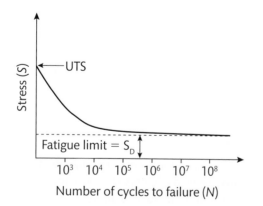

▶ **Figure 25.36** S–N curve for a typical ferrous metal, showing the fatigue limit (S_D)

In non-ferrous metals, often no such fatigue limit can be determined, and any level of cyclic loading is likely to eventually lead to fatigue failure. In such cases a different approach is taken that requires more careful monitoring, which in effect means counting the number of loading cycles the material undergoes while in service. The endurance limit is the stress which can be endured for a given number of loading cycles. This too can be displayed graphically on an S–N curve, as shown in **Figure 25.37**.

▶ **Figure 25.37** S–N curve for a material that does not exhibit a fatigue limit and instead is characterised using an endurance limit (S_N)

The initiation sites where fatigue fracturing begins are microscopic stress fractures which form at points of high stress concentration. Common stress raisers include:

▶ internal voids, inclusions or areas of porosity formed during manufacture or processing (**Figure 25.38a**)

▶ external surface defects such as intrusions (**Figure 25.38b**), sharp corners, abrupt changes in cross-section (**Figure 25.38c**) or tooling marks.

The formation of intrusions was discussed in Section A4, under 'Surface slip bands'.

Once a micro-fracture is formed, it acts as its own stress raiser. With each peak in the loading cycle, a little more damage is caused at the tip of the fracture as it finds a path through the microstructure of the first few crystalline grains it encounters. Once it is several grains long, the fracture is well established and will continue to propagate. The time until final fracture depends on the material and loading conditions. Once the fracture has reduced the surface area of intact material by so much that it becomes over-stressed at peak loads, the material will fail.

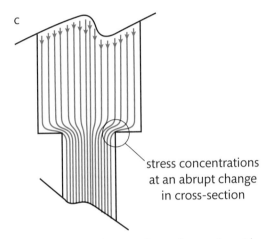

a

stress concentrations
around a void formed
at a grain boundary

b

stress concentrations
around a surface intrusion

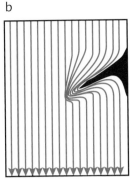

c

stress concentrations
at an abrupt change
in cross-section

▶ **Figure 25.38** Stress concentrations can form: a) around a void at a grain boundary; b) around a surface intrusion; c) at an abrupt change in cross-section

The fracture surface characteristics indicative of fatigue cracking are quite distinctive – burnished **beach marks** spreading from the point of initiation and a crystalline fracture surface where the final tensile failure took place (see **Figure 25.39**).

C4 Corrosion mechanisms

Corrosion is a process of chemical degradation in metals that gradually affects their physical appearance and removes their mechanical properties. The corrosion mechanism that we are most familiar with is the formation of rust on iron and steel.

Chemical fundamentals of corrosion

The most common corrosion mechanism encountered by engineers is called electrochemical corrosion and is caused when metal atoms lose electrons and become **ions** which are then transported away from the material in a liquid **electrolyte**. By far the most common electrolyte encountered by metals in service is water. It is no coincidence that cars tend not to rust in hot, dry countries.

The electrochemical cell

A simple electrochemical cell consists of two dissimilar metals that are in electrical contact with each other and also in contact with a liquid electrolyte.

At the **anode**, the metal atoms will lose electrons to become metal ions that then enter the liquid electrolyte. The liberated electrons flow directly from the anode to the **cathode** if the two are in direct electrical contact.

If the anode is iron for example, an atom of iron will lose two electrons and move into the solution as an ion, while the liberated electrons pass directly to the cathode. This is described by the reaction

$$Fe \rightarrow Fe^{2+} + 2\,e^-$$

When the liberated electrons from the anode reach the cathode, they combine with elements of the electrolyte present to form negative ions. These negative ions then combine with the positive ions formed at the anode that are now available in the electrolyte to form the corrosion by-product.

> **Key terms**
>
> **Beach marks** – fracture surface markings indicative of fatigue failure.
>
> **Ions** – electrically charged particles formed when atoms either lose or gain electrons.
>
> **Electrolyte** – a solution through which ions are able to move.
>
> **Anode** – a positive electrode.
>
> **Cathode** – a negative electrode.

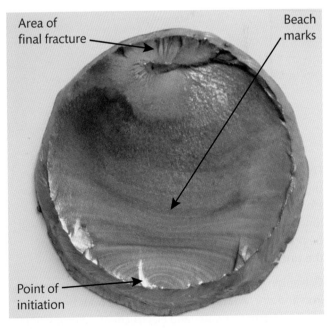

Area of
final fracture

Beach
marks

Point of
initiation

▶ **Figure 25.39** A fatigue fracture surface

If the anode is iron and the electrolyte is water, the electrons at the cathode combine with water and oxygen to form hydroxide ions in the following reaction:

$$H_2O + \tfrac{1}{2}O_2 + 2\,e^- \rightarrow 2(OH)^-$$

The hydroxide ions formed at the cathode can now combine with the iron ions that were formed at the anode to form the precursor to rust, called iron(II) hydroxide:

$$2(OH)^- + Fe^{2+} \rightarrow Fe(OH)_2$$

At this stage in the formation of rust, the iron(II) hydroxide reacts further with the oxygen and water to form iron(III) hydroxide:

$$2\,Fe(OH)_2 + H_2O + \tfrac{1}{2}O_2 \rightarrow 2\,Fe(OH)_3$$

Over time, iron(III) hydroxide dries and crystallises into the red flaky hydrated iron(III) oxide that we all recognise as rust:

$$2\,Fe(OH)_3 \rightarrow Fe_2O_3 \cdot H_2O + 2\,H_2O$$

Electrochemical series for metals

When an electrochemical cell is set up using electrodes of dissimilar metals, which metal that will form the anode and which will form the cathode is determined by their relative positions in the electrochemical series, which is summarised in **Table 25.15**.

▶ **Table 25.15** The electrochemical series for commonly encountered metals in engineering

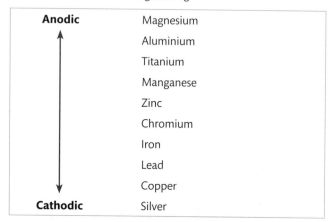

Anodic	
↑	Magnesium
	Aluminium
	Titanium
	Manganese
	Zinc
	Chromium
	Iron
	Lead
↓	Copper
Cathodic	Silver

For instance, if a cell is established with iron and zinc, then zinc will form the anode as it is above iron in the electrochemical series. Zinc will be the metal that loses ions into the electrolyte and becomes corroded. Iron will be protected from being corroded by the presence of the zinc.

Galvanising takes advantage of this relationship between iron and zinc. A layer of zinc on steel provides both a physical barrier to water and electrochemical protection from corrosion.

Dry corrosion

Dry corrosion takes place in the absence of water. For example, the reaction of steel with atmospheric oxygen does not occur readily, at least not at room temperature. However, at elevated temperatures such as those encountered during hot forming processes or heat treatment, steel will react directly with atmospheric oxygen to form a bluish-black layer of hard scale composed mainly of Fe_3O_4.

Other metals, notably chromium, aluminium and titanium, form **passivating** oxide layers very readily upon contact with atmospheric oxygen, even at room temperature. However, these stable layers of oxide protect the remaining metal beneath from further oxidation and are actually advantageous in most circumstances.

> **Key term**
>
> **Passivating** – making a material passive by forming an inert protective layer over its surface.

Galvanic action

In practical applications, the effects of electrochemical corrosion in localised areas of a component or assembly are collectively referred to as galvanic action.

The electrochemical behaviour of different alloys in specific working environments (when exposed to different electrolytes and varying concentrations) can be complex and may vary from the standard electrochemical series. Galvanic series are used to predict the electrochemical behaviour of widely used engineering alloys in common electrolytes, such as fresh water and seawater, or other electrolytes that might be present in industrial environments.

There are three significant mechanisms of localised galvanic corrosion.

▶ Composition cells form between dissimilar metals that are in contact with each other. The metal highest in the galvanic series for that operating environment becomes the anode and undergoes corrosion.

▶ Concentration cells form between areas with different electrolyte concentrations. The same metal becomes anodic when in contact with a high-concentration electrolyte and becomes cathodic when in contact with a low-concentration electrolyte.

▶ Stress cells develop between regions of the same metal that are stressed by different amounts. Areas of high stress become anodic and areas of lower stress become cathodic.

 **PAUSE POINT** If a copper water pipe is joined with lead solder, which of the two metals will tend to corrode?

> **Hint** Establish the anode by referring to the electrochemical series.
>
> **Extend** Which of the three types of galvanic corrosion cell is being established here?

Types of corrosion

Table 25.16 lists the types of corrosion you are likely to encounter.

▶ **Table 25.16** Types of corrosion encountered in engineering applications

Type of corrosion	Cause/Description	Effects/Where encountered
Hydrogen embrittlement	Caused by hydrogen atoms diffusing into the surface of a metal from a hydrogen-rich external environment (which may exist during processing or while in service). The H atoms become dissolved in the metal's solid solution crystalline structures. Once present in sufficient quantities, H atoms may combine into their molecular form H_2 or combine with the carbon present in steel to form methane (CH_4). These forms of hydrogen are too large to diffuse further and become trapped, causing increased stresses inside or between the grains.	Can lead to the formation of micro-cracks within the structure and can make the component more prone to brittle failure. High-strength steel alloys, titanium and aluminium can all be affected by hydrogen embrittlement.
Surface corrosion	Probably the most common but least problematic form of general corrosion encountered in engineering. Light surface rust on a steel structure is almost inevitable but also predictable and controllable.	Will not cause significant weakening of the structure if kept in check by regular painting or other measures to slow its progress.
Crevice corrosion	Galvanic corrosion caused by the formation of a concentration cell between areas of high and low oxygen concentration in a water electrolyte. The poorly oxygenated region acts as the anode and the normally oxygenated region acts as the cathode. Elsewhere in the water electrolyte, oxygen levels are maintained through diffusion of atmospheric oxygen into solution.	Often occurs under tape, wood or dirt in contact with the metal surface or between lap joints, under washers and under rivet heads, where trapped water will become depleted in oxygen.
Intergranular corrosion	A type of galvanic corrosion caused by the formation of a composition cell between alloy phases or between precipitates present at grain boundaries and the grains themselves. The grain boundaries become anodic and corrode.	Intergranular corrosion between the layers causes them to delaminate, leading to the appearance of flaky blisters as shown in **Figure 25.40**. Exfoliation is a severe form of intergranular corrosion encountered in metals which have undergone cold rolling and so consist of layers of slightly flattened grains.
Bimetallic corrosion	A type of galvanic corrosion caused by the formation of a composition cell when dissimilar metals are in contact with each other. The metal highest in the galvanic series will form the anode and be corroded.	Commonly encountered when steel fixings are used to secure aluminium components.
Pitting corrosion	Severe localised corrosion in otherwise corrosion-resistant materials – those protected from corrosion by a passive oxide layer at their surface. A corrosion pit can be initiated when damage occurs to the passivation layer and a galvanic combination cell is established with the exposed metal. The pit's continued growth is encouraged by the presence of chloride ions (plentiful in seawater), which cause the acidity within the pit to increase and the corrosion process to accelerate.	Leads to the formation of deep pits and eventually holes in otherwise corrosion-resistant materials such as stainless steel and aluminium alloys. Can be very damaging and cause the failure of pipes and pressure vessels with little warning.
Stress corrosion	A type of galvanic corrosion caused by the formation of a stress cell between areas of differential stressing in a component. The high-stress region becomes the anode and corrodes.	Stress cells can form between areas of residual stress formed by cold working operations and adjacent unworked areas that have lower residual stress.

▶ **Figure 25.40** Surface signs of severe intergranular corrosion causing exfoliation in aluminium

⏸ **PAUSE POINT**

Keep an eye out for examples of corrosion that you encounter in your local environment. Once you start looking, you will see the signs everywhere. Take photos of any signs of corrosion you encounter and bring them in to discuss. With your colleagues, decide on the likely mechanism by which the corrosion is happening.

Hint

Signs of corrosion are commonly seen on older cars and vans, street furniture, gate hinges and so on.

Extend

How might some of the mechanisms of corrosion that you have encountered be prevented?

C5 Design considerations to help prevent component failure

Knowledge of the component's operating environment

Table 25.17 describes the effects of a component's operating environment on its performance.

▶ **Table 25.17** Effects of a component's operating environment

Temperature	Has a significant effect on the behaviour of failure mechanisms such as creep. Materials operating at high temperatures that undergo phase changes have significantly different mechanical properties from those operating at room temperature.
Wet or dry	The amount of water (or other electrolyte) in the operating environment has a significant impact on the number of possible corrosion mechanisms that will need to be considered.
Loading conditions	• Static loading requires materials with an appropriate UTS. • With dynamic loading, impact resistance becomes an important consideration. • Cyclic loading makes materials prone to fatigue failure, which must be accounted for.

Correct choice of material

Only when engineers have a full understanding of the component's operating environment can they begin the material selection process. In commercial engineering, cost is always a consideration, and engineers choose materials that will provide the most cost-effective combination of the required mechanical properties and resistance to creep, fatigue and corrosion.

Design features

We have seen that certain design features, such as sharp corners or abrupt changes in cross-section, introduce stress raisers which, in combination with cyclic loading, can form fatigue crack initiation sites. Any problematic features should be designed out of the component to avoid such problems. Even leaving a component with a poor surface finish is not just a matter of aesthetics. Machining marks can act as crack initiation sites, and so engineers need to avoid these by specifying an appropriate roughness for the surface finish at the design stage.

Higher-quality materials

The selection of materials is done on the basis of achieving the required properties at minimum cost. Materials processed to eliminate all inclusions or areas of porosity within their structure may require the use of alternative manufacturing techniques such as forging, which will add to the overall cost of the component. However, it may sometimes be more cost-effective to use a carefully processed and heat-treated inexpensive material than to use a more expensive material that does not require the same degree of processing.

Surface treatment and finishes

These are usually used for aesthetic reasons and to form a physical barrier to prevent the ingress of water, oxygen or other chemical agents that will enable corrosion of the material. Surface treatments and finishes might be in the form of paint, powder coating or plating with a corrosion-resistant metal such as chromium.

Sealants or gaskets are often used between contact surfaces of dissimilar metals to prevent corrosion through galvanic action.

Other beneficial effects can be obtained through the use of surface treatments. In the manufacture of automotive suspension springs, the surfaces are **shot peened** after heat treatment, which involves covering the surface of the material with small **dints**. This obliterates any marks made during manufacturing, strain hardens the surface of the spring and introduces residual compressive stresses. All of these help to prevent the formation of surface micro-cracks during service, which can act as initiation sites for fatigue fractures.

> **Key terms**
>
> **Shot peened** – a process of causing multiple dints in the surface of a metal by bombarding it with a stream of high velocity metal shot.
>
> **Dints** – slight hollows or indentations made in a hard, even surface by a blow or the application of high pressure.

Assessment practice 25.3

C.P6   C.P7   C.M4   C.M5   C.D3

You are working as a part of a technical team investigating the failure of a series of mechanical components in plant machinery that your company manufactures.

You have been supplied with three failed components that appear to have fractured in different ways and so perhaps through different mechanisms.

Your supervisor has asked you to apply your knowledge of the different characteristics of failure mechanisms in metals to determine the cause of each failure.

Carry out a detailed visual inspection and perform other testing procedures (e.g. hardness tests) to support your findings.

Present all of your findings in a detailed technical report, fully evaluating the conclusions you have reached on the most likely failure mechanism in each case. In addition, make detailed suggestions on measures to limit the likelihood of similar failures happening in future. This may mean changes in component design, material or processing techniques.

Plan

- Am I clear on what the outcome of the assignment should look like?
- What key elements do I need to include in the written report?

Do

- I can work methodically and logically through each stage of the assignment.
- I can effectively plan the key areas I need to include in a written report so that I don't miss anything important.

Review

- I am able to recognise weaknesses in my approach to the task and find ways to address these next time.
- I can prioritise which of the gaps in my knowledge are most important and need to be addressed first.

THINK ▶FUTURE

Simon Kirkpatrick
Mechanical testing technician

I have worked for the three years since leaving school in the quality control laboratories of a manufacturer of steel wire. The products we make are very specialised and are exported all over the world. Our product range includes lift cables, tow lines and the wire that is used in pre-stressed concrete structures.

Obviously if any of these fail in service then the consequences could be horrific. That's why we take the job of quality control so seriously. Samples are taken at regular intervals for every batch of wire strand that is manufactured. It is part of my job to carry out tensile tests on these samples and check that the results are above the minimum requirements for that particular type of wire. Only then is that batch allowed to be spun together with other wire strands into thicker cables.

We also have rigs that test the creep resistance of our concrete reinforcing wire over several weeks to make sure that it won't start to elongate during use. This is important, but nowhere near as much fun as tensile testing wire samples to destruction!

I very much enjoy my job, the kinds of work I get to do and the sense of teamwork that we have across the whole site. I know that if I don't do my job quickly I will be holding up the wire-spinning production unit. I also know that I have quite a responsible job, as it's up to me and my supervisor to spot any problems and prevent substandard wire from being made into finished products that lives might later depend on.

Focusing your skills

Working safely in a new environment

Whenever you are asked to perform practical tasks using new equipment or in an unfamiliar working environment, it is important that you follow some simple rules:

- Obey all the general safety guidance that you are given when first entering the workshop.
- Wear the appropriate personal protective equipment (PPE).
- Never use a piece of machinery or equipment that you are not trained to use. Even if you have used a tensile testing machine before, make sure you are familiar with the particular type in use.

- If you need a refresher because it's been a while since you last used a piece of equipment, don't be afraid to ask.
- Always obey workshop safety posters. These are often there to remind you to wear ear or eye protection.
- Know where the emergency exits are and know the route to the closest exit from where you will be working.
- Know where the emergency stop buttons are located.
- Always use common sense.
- Never take unnecessary risks.

Getting ready for assessment

Eve is working towards a BTEC National in Engineering. She was given an assignment for Learning aim A which asked her to use her knowledge of the microstructures of metals to identify a series of material samples. The assignment went on to ask her to relate the microstructure in each case to the overall mechanical properties of the material.

Her findings had to be written up in a technical report, in which she needed to:

▶ identify the materials and evaluate their likely processing histories

▶ evaluate the mechanical properties likely to be exhibited by each sample, based on the microstructural features that were identified.

Eve shares her experience below.

How I got started

First, I had a good look at what I would need in order to carry out the assignment. I had already used a hand lens and microscope during our lessons looking at the structures visible in polished cross-sections of various materials, so I was comfortable with that. I also knew that we would have access to some example micrographs of different types of materials processed in different ways. I knew that to get the highest grade I had to identify the samples correctly, and I was confident I could do that. At our centre we have a digital camera fitted to one of the microscopes, so I was able to use that to take nice big photographs of the samples to include in my report.

The second part of the assignment was about the way different microstructures affect mechanical properties. This relied more on having good class notes and carrying out some additional research on a few websites. To start with I made a mind map to help me structure my thoughts on the general effects of different microstructural features. This gave me something easy to refer back to when I was considering each sample in turn.

How I brought it all together

I wanted my report to look professional, as if it had been produced in a real testing lab, so I used a simple Arial font and included a footer on each sheet containing my name and the page number. I added a title page which included one of the micrographs that I had taken and wrote a short introduction. I included all the micrographs I had taken of the samples and annotated them to support what I was saying about the identification of the types of material. I included a similar reference micrograph in each case and pointed out the similarities to my samples.

While I was writing the parts of the report on mechanical properties, I ticked off the points I had covered on my mind map to make sure I remembered to mention everything.

What I learned from the experience

I think my assignment went really well, and that was mostly down to good preparation and having a clear idea of what I was going to do. Keeping organised notes of the things we covered in class really helped too. I am going to buy a notebook to keep all my class notes together, as I did lose the notes from a couple of lessons that were on loose bits of paper.

It's important to keep a copy of the assignment and the unit specification at hand when doing your work. The specification 'Essential information for assessment decisions' page is really useful as it explains what the assessment criteria actually mean in practice, and tutors use this when marking the work.

Think about it

▶ Have you planned what you need to do in the time available for completing the assignment to make sure that you meet the submission date?

▶ Do you have all your class notes on the microstructures of metals and their effects on mechanical properties?

▶ Is your report written in your own words and referenced clearly where you have used quotations or reference information such as example micrographs?

Glossary

3-8 decoder – an integrated circuit that can provide eight controllable outputs from three inputs.

3D rapid prototyping (additive manufacturing) – a method of synthesising a 3D prototype, where material is added layer by layer to build an object.

Acceleration – the rate at which the velocity of an object changes over time. Unit: metre per second squared (ms^{-2}).

Active low (or **active high**) – identifies whether an input pin requires a logic 0 (low) or a logic 1 (high) to function.

Activity cost pool – a category of costs.

Activity driver – a factor within an activity cost pool that causes costs to increase.

ADC (analogue-to-digital converter) – hardware that converts continuous physical quantities, such as a voltage, into a digital number representing the value's amplitude. A common example in a microcontroller system is the conversion of temperature (analogue) into an 8-bit value that can be digitally displayed.

Alloy – a mixture of two or more metals that has enhanced mechanical properties or corrosion resistance compared with the pure metal used alone.

Anode – a positive electrode.

Arbitration – the process that determines which master device controls the I^2C data bus.

Areas of porosity – clusters of small voids often caused by the presence of gas bubbles during solidification.

Arithmetic progression – a sequence of numbers where the difference between any two successive terms is a constant. This constant is called the **common difference** (d).

Arithmetic series – the sum of the terms in an arithmetic progression.

ASCII (American Standard Code for Information Interchange) – a set of universally accepted codes that represent characters used in computers.

Back e.m.f. – a voltage in the reverse direction that results from a current through an inductor being switched.

Barrier potential – the minimum forward bias voltage needed for conduction to start.

Base – the term that is raised to an index, power or exponent. For example, in the expression 4^3 the base is 4.

Base number – the number of digits used in a numbering system. You have been using base 10 (the decimal system) since you learned to count. It has 10 digits: 0, 1, 2, 3, 4, 5, 6, 7, 8, 9. Base 2 (the binary system) has 2 digits: 0 and 1.

Batch – three or more products manufactured or services delivered together.

BCD to seven-segment decoder – an integrated circuit that has four binary inputs representing coded decimal values (0000 = 1, 0001 = 2, ..., 1000 = 8, 1001 = 9) and a controlled output that matches the pattern necessary to illuminate the corresponding segments. 'BCD' stands for 'binary coded decimal'. This decoder can even be built into the seven-segment displays.

Beach marks – fracture surface markings indicative of fatigue failure.

Bill of materials (BOM) – a table containing information about the parts in a drawing. CAD software can automatically generate this for you.

Binomial – a polynomial with two terms.

Boolean algebra – a two-state logic algebra (named after George Boole).

Bus width – the size of an information channel for accessing or transferring data. The larger the bus the more data can be transferred at any time. An 8-bit data bus can transmit 8 bits of data at any one time and a 16-bit data bus can transmit 16 bits of data at any one time.

Capillary action – the tendency of a liquid to be drawn into narrow gaps.

Carbon footprint – a measure of the environmental impact of a particular individual or organisation, measured in units of carbon dioxide.

Casting alloy – an alloy suitable for casting applications that generally exhibits good flow characteristics when molten.

Cathode – a negative electrode.

Chemical compound – a material containing a fixed ratio of atoms strongly bonded together.

Class width – the difference between ucb and lcb.

Classes – the groups into which the data is organised.

Client brief – outlines the client's expectations and requirements for the product.

CMOS – complementary metal oxide semiconductor logic, based on MOSFETs.

Coefficient – a number or symbol that multiplies a variable. For example, in the expression $3x$, 3 is the coefficient of the variable x.

Coefficient of linear expansion (α) – a material property that describes the amount by which the material expands upon heating with each degree rise in temperature. Unit: K^{-1}.

Combinational logic circuit – a circuit made from several logic gates. For clarity, logic circuits do not always show the supply connections.

Competitive advantage – what an organisation does that allows it to outperform competitors.

Compiler – a program that converts the code created into machine code (1s and 0s) that your microcontroller can interpret. Different compilers are required for different IDEs and microcontrollers.

Complex number – a number that has the format $a \pm jb$. The 'complex operator' or 'imaginary unit' j satisfies the equation $j^2 = -1$

Concurrent forces – forces that all pass through a common point.

Conductor – a material that has very low resistance and easily passes an electric current. Most good conductors are metals, such as copper, aluminium and silver.

Contract out – to make a contract with an external supplier to carry out a specific activity.

Convergent series – a series for which S_n heads towards a specific value as n (the number of terms in the sum) increases.

Coplanar forces – forces acting in the same two-dimensional plane.

Critical path analysis (CPA) – a method of task management that allows activities to be organised so that a project can be completed in the shortest possible time.

Cyclic loading – constantly varying dynamic loading such as that in a shaft running out of balance.

Decommission – to remove or withdraw from service.

Degree (symbol: °) – one degree is $\frac{1}{360}$ th of a complete circle. A complete circle contains 360°.

Degree – the highest power of the independent variable in a polynomial function. A polynomial of degree 2 is called quadratic, of degree 3 is called cubic, of degree 4 is called quartic, and so on.

Density – the mass per unit volume of a material, calculated by dividing the mass by the volume.

Depreciation – the reduction of an object's value over time; this could be related to wear and tear on machines or the decreasing market value of a product.

Derivative – the result of applying differentiation to a function.

Design – a drawing and/or specification to communicate the form, function and/or operational workings of a product prior to it being made or maintained.

Detail view – an enlarged, more detailed view of part of a drawing.

Determinant – a useful value calculated from the elements of a **square matrix**.

Differentiation – the process of calculating the gradient of a function at a given point, or calculating the rate of change of the dependent variable at a given value of the independent variable. The area of mathematics dealing with differentiation is called differential calculus.

Diffusion – mixing that results from the natural tendency of atoms or molecules to move from areas of high concentration to areas of low concentration.

Direction – the orientation of the line in which the force is acting (the line of action).

Discontinuity – any change in the structure of the material such as a crack or void.

Displacement – the straight-line distance between the start and finish positions of a moving object. Unit: metre (m).

Divergent series – a series for which S_n does not settle down to a definite value as n increases.

Electrolyte – a solution through which ions are able to move.

Engineering drawing – a drawing that conveys technical information using standard layouts, projections and symbols.

Entity – a discrete element of a drawing, such as a line or circle.

Equation – used to equate two expressions that have equal value, such as $a^2 + 3 = 19$ or $t - 1 = 3a + 12$. Equations can easily be recognised because they always contain an equals sign.

Equilibrant – the force that when applied to a system of forces will produce equilibrium. This force will be equal in direction and magnitude to the resultant but have the opposite sense.

Equilibrium phases – in metallurgy, these describe the most energy-efficient and stable crystalline structures in which atoms tend to arrange themselves at a given temperature.

Exponential function – a function of the form $f(x) = a^x$ with $a > 0$.

Expression – a mathematical statement such as $a^2 + 3$ or $3a - t$. Expressions can easily be recognized because they do not contain an equals sign.

Extensometer – a device attached to the test piece that measures extension during testing.

Extrusion – a peak pushed out from the surface of a material.

Factor of safety – describes how much stronger a system is than it needs to be for an expected load.

Faraday's laws of induction – combined, these laws state: 'When a magnetic flux through a coil is made to vary, an e.m.f. is induced. The magnitude of this e.m.f. is proportional to the rate of change of flux.'

Feasibility study – an assessment of a proposed project to see if it can be achieved in practice.

Ferromagnetic – describes a material that becomes magnetised when exposed to a magnetic field.

Fit for purpose – good enough for the job it was intended to do; fulfills the specifications or requirements.

Fitness watch – a watch with advanced features to support sport and fitness activities, such as a heart monitor, workout programmes, a sleep monitor, an accelerometer and GPS; most give immediate on-screen displays of live data.

Flow chart – a diagram consisting of boxes and arrows that shows a workflow or process.

Flux leakage – the forcing out of lines of flux from the surface of a material.

Force majeure – a clause that releases a company from its obligations under a contract if unpredictable, extraordinary circumstances prevent it from fulfilling the contract.

Forward bias – describes a diode where the anode is more positive than the cathode.

Frame (or **packet**) – a group of a specific number of data bits that includes a **header**, which indicates the start and the end of the data, the number of bits being transmitted and, in some instances, a form of error checking.

Function – a portion of code that has a specific function, also called a subroutine.

Function – the relationship between the input to a system and the corresponding output. In general, if the input is the variable x and the output is the variable y, then we can write the function as $y = f(x)$.

Functions – areas of activity into which a business can be organised.

Gantt chart – a chart that shows the periods of time planned for the different elements of a project, providing a graphical view of the project schedule.

Geometric progression – a sequence of numbers where each element is multiplied by a constant value to form the next element in the sequence. This constant multiplier is called the **common ratio** (r).

Geometric series – the sum of the terms in a geometric progression.

Gradient – also called 'slope', measures how steep a line is. It is calculated by picking two points on the line and dividing the change in height by the change in horizontal distance, or change in y value ÷ change in x value.

Grain – a defined area of uniform crystal lattice structure that forms during cooling.

Grain boundary – the area that separates individual grains in a crystalline material.

Header – a collection of bits within a frame or packet that provides information about the data within it.

Header file – a file that has a .h extension and includes declarations of pre-written functions, macros and global variables. In the example code given here, the header file allows the compiler to understand what TRISB and PORTB mean.

Hybrid car – a car with a petrol engine and an electric motor.

IC footprint – identifies the locations, pattern and distance between the pins on a device.

Inclusion – an impurity in the material's microstructure.

Indemnity – an obligation of one party to pay compensation for a particular loss suffered by another party.

Index – the term to which the base is raised. For example, in the expression 4^3 the index is 3. The index may also be called the power or exponent. The plural of 'index' is 'indices'.

Instantaneous gradient or **rate of change** – the gradient at a specific point on a curve, which represents the rate of change of the dependent variable at a specific value (instant) of the independent variable.

Insulator – a material that has very high resistance. It is generally a non-metal. Examples of good insulators include plastics and rubber.

Insurance – a payment made to guard against uncertain losses.

Integral – the result of applying integration to a function.

Integrated development environment (**IDE**) – a software application that provides a set of tools to aid programmers in the development of software. A typical IDE consists of a source code editor, a compiler, automation and simulation tools, and a debugger.

Integration – the inverse process of differentiation, or the process of summing small increments to find the whole. The area of mathematics dealing with integration is called integral calculus.

Interlock – a physical or electronic lock that prevents something from operating until guards are in place. For example, a CNC lathe will usually have an interlock to prevent the spindle from operating until the access door is closed.

International Organization for Standardization (ISO) – an independent organisation that develops international standards for quality, safety and efficiency.

Interrupt – a signal sent to a specific microcontroller input pin to mark the occurrence of an input event that requires specific immediate attention. An interrupt requests the internal processor to stop the current function and to execute a code associated with input. Interrupts can be given a priority order.

Interstitial defect – found in the small spaces between atoms in the regular lattice structure.

Intrusion – a trough formed in the surface of a material.

Inverse of a function or operation – carrying out the opposite actions in the reverse order.

Inverse of a matrix – a matrix which when multiplied by the original matrix gives the identity (**unity**) **matrix**.

Ions – electrically charged particles formed when atoms either lose or gain electrons.

ISO 9000 – a quality assurance system for manufacturing and service industries.

ISO 14000 – an environmental management system.

Issue – an event that is currently having an effect on project processes or outcome.

Iteration – repeating a step, process, version or adaptation of something, usually with the aim of improving it.

Latent heat – heat energy causing a change of state of a substance without a change in temperature.

Lattice structure – the regular structure of atoms that make up a crystalline material.

Layer – a means of overlaying different information on a common drawing outline.

Lenz's Law – an induced current always acts in such a direction so as to oppose the change in flux producing the current.

Liability – being legally or financially responsible for something.

Localised – limited to a small area.

Logarithm – the power (or exponent) to which a base number is raised to give a particular value: $number = base^{logarithm}$.

Logic gate – a circuit that carries out a Boolean logic function.

Logic probe – a pen-like device that indicates by means of audio and/or visual output whether a logic 1 or 0 is output on a device pin.

Lower class boundary (lcb) – the lowest value of the class.

Maclaurin series – a Taylor series with $a = 0$.

Magnetic flux – lines along which the magnetic field acts.

Magnitude – the size of a force.

Majority carriers – the charge carriers due to doping, which are holes in p-type material and electrons in n-type material.

Manufacture – to make a product for commercial gain.

Master – a device in the I^2C bus system that drives the serial clock line.

Matrix – a set of data arranged in rows and columns.

Metallurgy – the study of metals, including their structure, properties and uses.

Micrograph – a high-resolution, highly magnified (x100 to x1000) photograph of a polished cross-section of a crystalline solid, taken using a microscope.

Midpoint – median of the class.

Milestone – an important point, stage or event in the progress or development of something.

Mind map – a web of thoughts written out on paper or drawn on a computer, with a key word or idea in the centre, linked by lines to related ideas.

Minority carriers – the holes in n-type material and electrons in p-type material due to thermally generated hole–electron pairs in the silicon lattice.

Modification – a change or alteration, usually to make something better.

Moment – the tendency of a force to rotate the object on which it acts.

Moment of inertia – an object's resistance to a change in its rotational direction; this is dependent on the distribution of mass within the object with respect to the axis of rotation.

n factorial (n!) – n multiplied by $(n - 1)$ multiplied by $(n - 2)$ and so on, until multiplied by 1.

Necking – a reduction in cross-section.

Non-concurrent forces – forces that do not all pass through the same point.

Non-value-added activities – activities that are essential to a product and its production but do not directly add value to it; for example, product inspection, review and reporting.

Normal reaction – the force that acts perpendicular to a surface upon an object that is in contact with the surface.

Observable emotions – emotions that are displayed so that others can see, rather than internal feelings.

Operation – a single step in a manufacturing process. For example, this could be marking out the position of a hole when manufacturing a product or removing an access panel when delivering an engineering service.

Orthogonal drawing – a drawing that represents a 3D object in several 2D views, typically plan, front and side views in a specific layout. For instance, in third-angle orthographic projection, the plan view is positioned above the front view, with side views on either side of the front view.

Overheads – costs that relate to extra services involved in the manufacture of a product, such as equipment costs and heating or lighting; they are not directly related to the cost of labour or materials.

Parameter passing – a technique used to transfer a value from one function to another so that it can be used by the receiving function. For example, in the code for task 1b 'Delay_ms(1000);' takes the value 1000 and passed it to the Delay_ms function.

Parity bit – a bit of data added to the end of a set of binary data to make the number of 1s in the data an even or odd number. This is used in the simplest form of error checking.

Passivating – making a material passive by forming an inert protective layer over its surface.

PDCA – a four-step quality-management process where you plan, do, check and act in a continuous cycle to continually improve a product.

Performance parameters – the capabilities that a system must have in order for it to be successful or to achieve its goals.

Phase – the physically distinctive form of a substance: solid, liquid or vapour.

PIC (peripheral interface controller) – a family of specialised microcontrollers.

Polynomial – a function that is a sum of non-negative integer powers of the independent variable x. A polynomial of degree n is usually written as $f(x) = a_n x^n + a_{n-1} x^{n-1} + a_{n-2} x^{n-2} + \ldots + a_2 x^2 + a_1 x + a^0$ where all the a values are real numbers, called the coefficients of the polynomial.

Principal moment – the tendency of a force to rotate an object about a turning point, also known as torque; it is calculated by multiplying the force by the perpendicular distance of the force from the turning point.

Prismatic – having the form of a prism, that is, a 3D shape with identical ends and the same cross-sections all along its length.

Processor speed – usually refers to the maximum number of calculations per second that the microcontroller processor can perform. Microcontroller processor speeds can be variable and set by the user using a crystal depending on the application. This will be covered later as you build your own microcontroller system.

Product – the final tangible outcome of a manufacturing process, often referred to in economics as 'goods' (e.g. a car, television or chair).

Product – the result of multiplying two or more factors together.

Product design specification (PDS) – a document detailing the requirements in various areas that a product, service, system or process must meet.

Progression – a sequence in which each term is related to the previous term or terms by a uniform law. All progressions are sequences, but not all sequences are progressions.

Project life cycle – a series of activities that are necessary to fulfill a project's objectives.

Project log – a document to record the progress made, key activities and decisions taken during the development of a product.

Project management – carefully planning and guiding a project's processes from start to finish of the project.

Project manager – the person who manages a project, solves problems and issues, organises and guides the project team, ensures that resources are used effectively, and carries responsibility for successful project completion.

Protocol – a specific set of rules that are applied when two electronic devices communicate with each other to ensure effective, efficient communication.

Prototype – a first or preliminary version of something, from which other versions will be developed.

PWM (pulse-width modulation) – a modulation procedure that creates a representation of an analogue signal from a digital output. PWM adjusts the mark-to space ratio of a square wave, which when connected via an R/C network will generate an analogue signal. A common application of PWM is to control the speed of a winch motor.

Quality assurance – the set of planned activities for ensuring that the quality requirements of a product or service are met.

Quality control – the testing and monitoring of activities that are used to check the quality of a product or service outcome.

Radian (symbol: rad or ^{c}) – one radian is the angle **subtended** at the centre of a circle by two radii of length r that describe an arc of the same length r on the circumference. A complete circle contains 2π rad.

Resultant – the force that represents the combined effect of all the forces in a system.

Rate of change – how fast the dependent variable changes as the independent variable increases. It should always be stated in the form 'the rate of change of ... with respect to ...'. For example, velocity is the rate of change of displacement with respect to time. The shorthand 'w.r.t.' is often used for 'with respect to'.

Read – where a microcontroller receives an input and follows a specific set of instructions depending on the input.

Reductionism – a way of thinking that breaks down a problem into simpler parts.

Register – a reserved section of memory that can be written to and read from. Registers typically control specific operations of the microcontroller.

Reverse bias – describes a diode where the anode is more negative than the cathode.

'Right first time stimulus' – a quality management concept which asserts that preventing defects and errors is better and more cost effective than correcting them late.

Risk – a future event that could adversely affect project processes or outcome.

Semiconductor – a material such as silicon (Si), germanium (Ge) or gallium arsenide (GaAs) that has electrical properties somewhere between those of a conductor and an insulator. Pure semiconductor materials are called **intrinsic** semiconductors. **Extrinsic** semiconductors contain very small amounts of impurity chosen to increase the conductivity. The introduction of measured amounts of impurity is called **doping**.

Sense – the direction along the line of action in which the force acts.

Sensible heat – heat energy that causes a change in the temperature of a substance.

Sequence – a set of numbers written in order according to some rule. Each value in the sequence is called an element, a term or a member.

Series – the sum of the terms in a sequence.

Service – activities that provide some intangible benefit to a customer (e.g.

processing a credit card payment or performing an MOT inspection on a car).

Shock loading – rapidly increasing dynamic loading such as that in aircraft landing gear.

Signed variable – a variable that can take both negative and positive values. For example, a signed 8-bit number has 256 possible values, from –128 to +127.

Slave – a device that only responds to instructions from a master device and cannot initiate any data transfer.

Slip band – a section of displaced lattice structure between active slip planes. Slip bands can be observed as 'steps' in the crystal structure.

Slip plane – a plane in a lattice structure where it is possible for slippage to occur.

Slippage – permanent movement along slip planes within the crystal structure of the material.

Snap – a function that is used to make the cross-hairs/cursor on the screen jump to grid points. A grid can be set to various sizes and used with the snap function to enable accurate and speedy drawing production. Snap can also be set to jump to the ends of lines, midpoints of lines or centres of circles.

Solid solution – a metal alloy in which the main metal has alloying elements dissolved within its structure.

Solubility – the ability of one substance to remain dissolved in another.

Spatial awareness – the ability to visualise a flat 2D drawing as a 3D object.

Static loading – constant loading such as that in the structure of an electricity pylon.

Storyboard – a graphic sequence of events used for planning operations, e.g. for 3D modelling on CAD systems or for additive manufacturing.

Strain – the extension of a body expressed as a proportion of its original length.

Strain hardening (work hardening) – a phenomenon where deformation of the crystal structure at high strain rates helps to block further movement or slippage, which makes the material harder.

Stress – the load distribution within a material expressed as force per unit area of material, stress = , measured in pascals (Pa).

Stress raiser – any feature, such as a sharp corner or void, that will cause a localised increase in stress.

Subject of an equation – a single term becomes the subject of an equation when it is isolated on one side of the equation with all the other terms on the other side. For example, in the equation $y = 4x + 3$, the y term is the subject.

Substitutional defect – a different atom taking the place of an atom that would normally be there.

Subtend – to form an angle between two lines at the point where they meet.

Support reactions – the forces that are maintaining the equilibrium of a beam or structure.

Sustainability – the endurance of systems and processes in the environment; the maintenance of a balance in natural systems by avoiding the depletion of resources and reducing the generation of waste.

Synectics – a way of applying creativity in problem solving.

Syntax – a set of rules that govern the structure, symbols and language used in specific type of computer code.

Tangent – a straight line with a slope equal to that of a curve at the point where they touch.

Team – a group containing three or more individuals who have a common objective or shared goal.

Technical specification – detailed description of the technical requirements for a product, service, system or process. This will become the reference point for the project design.

Test plan – a document detailing the objectives, target market, personnel and processes for a specific test of a software or hardware product.

Thermistor – a thermally sensitive resistor that is manufactured to give predictable and accurate variations in resistance based on the external temperature of the device.

Tracker band – a thinner version of a fitness watch, usually limited to keeping track of steps, distance and calories; most trackers can synchronise with your phone, tablet or computer so that you can store and access the data on other devices.

Troubleshoot – to work out in a logical way what and where the problem is and suggest how to resolve it.

TTL – transistor–transistor logic, based on BJTs.

Unit cell – the smallest arrangement of atoms that we can use to define the general structure of a crystalline material.

Unity matrix – a matrix with a 1 in each position in the leading diagonal and 0 in every other position.

Unsigned variable – a variable that can take only positive values (or be zero). For example, an unsigned 8 bit number has 256 possible values, from 0 to 255.

Upper class boundary (ucb) – the highest value of a class.

Vacancy defect – a gap in the regular lattice structure caused by a missing atom.

Value – how well a product meets the needs of customers in relation to its price.

Value-added activities – activities that directly affect a product and add value to the outcome; these might include activities related to the product's design, manufacturing and innovative features.

Variable – a memory location that will be used in a particular function (local variable) or throughout the program (global variable).

Velocity – the rate at which the displacement of an object changes over time. Unit: metre per second (ms^{-1}).

Viscosity – the thickness of a liquid.

Void – an empty space or gap in the material's microstructure.

Voltage gain – the ratio of voltage output to voltage input.

Warranty or **guarantee** – a written promise to repair or replace a product if it does not work or breaks down, usually within a specified time frame.

WORD – a fixed number of data bits, typically 8 bits, or a byte of data.

Work in progress (WIP) – an incomplete product or project that is still being worked on, and which has a total cost allocated to it, largely for labour and materials.

Wrought alloy – an alloy generally supplied in some processed form, such as billets, rods or bars, and which is suitable for machining or secondary forming processes.

Index